Small-Scale Gas to Liquid Fuel Synthesis

Small-Scale Gas to Liquid Fuel Synthesis

Edited by
Nick Kanellopoulos

CRC Press
Taylor & Francis Group
Boca Raton London New York

CRC Press is an imprint of the
Taylor & Francis Group, an **informa** business

CRC Press
Taylor & Francis Group
6000 Broken Sound Parkway NW, Suite 300
Boca Raton, FL 33487-2742

First issued in paperback 2020

ISBN-13: 978-1-4665-9938-3 (hbk)
ISBN-13: 978-0-367-73849-5 (pbk)

Visit the Taylor & Francis Web site at
http://www.taylorandfrancis.com

and the CRC Press Web site at
http://www.crcpress.com

Contents

SECTION I   Integration of Innovative Membranes and Sorbents with the GTL Process

SECTION II Integration of Innovative Catalysts with the GTL Process

SECTION III   Innovative Oxidative and Nonoxidative GTL Processes

Preface

It is estimated that a large fraction of the proven natural gas (NG) reserves are found in locations from where it is not cost-efficient to transport them. It should be noted that if these stranded NG reserves could be converted to synthetic fuels, they would generate around 250 billion barrels of synthetic oil, a quantity equal to one-third of the Middle East's proven oil reserves. In addition, the U.S. Energy Information Administration report on technically recoverable shale gas resources indicates a potential of about 23,000 trillion cubic feet of world resources, of which a large part is situated in areas that are difficult to be connected to NG pipelines. A more economical exploitation of these resources requires developing efficient methodologies to achieve long-distance transport of NG. Gas to liquid (GTL) technologies are among the most effective for achieving this objective.

The major hurdle for the extensive application of GTL conversion technologies is the high cost of the state-of-the-art GTL processes, the largest part of which is associated to air fractioning/syngas production. Therefore, there is the need to improve these technologies, on the one hand, and to explore alternative routes to conventional GTL process, on the other hand, producing higher-value products such as raw materials for the chemical industry.

This book overviews some of the exciting new developments in this area, based on (but not limited to) research activities in the framework of two large European projects, namely, "Innovative Catalytic Technologies & Materials for Next Gas to Liquid Processes" (NEXT-GTL) and "Oxidative Coupling of Methane followed by Oligomerization to Liquids" (OCMOL). These two projects investigated different routes, from the development of a novel energy-efficient syngas process scheme to routes for direct conversion of methane to methanol or aromatics and for the oxidative coupling of methane followed by its subsequent oligomerization to liquids. These frontier researches share not only the common general objectives but also the concept of how the progress in this highly challenging area derives from combining advanced process engineering tools to a tailored design of nanoporous membranes, catalysts, and sorbents.

Today, nanoporous materials realize a compromise in performance between demanding operational conditions and limited capabilities of materials. Manipulating porous solids so as to exhibit desired characteristics (engineered pore size, geometry, specific active sites, etc.) is currently at the cusp between tailoring and design. It is shown that *engineering in* the required nanopore structure and properties can lead to the establishment of *next-generation* GTL processes.

This book describes well the concept of integrated process engineering at the nano- and macroscale as the key approach for innovation. The first two sections of the book are devoted to the recent advances in the development of membranes, catalysts, and sorbents with tailored properties for GTL processes.

Next-generation GTL technologies are presented in the third section of the book, where innovative oxidative and nonoxidative GTL processes are discussed together with results on a novel membrane-assisted process scheme for syngas production.

The impressive list of internationally known contributors is what makes this book stand out. We thank every one of them for their invaluable contribution. Special thanks to Jill Jurgensen, Allison Shatkin, and their colleagues from Taylor & Francis Group for their help and patience.

Last but not the least, we thank the European Commission for funding the cited EU projects (NEXT-GTL and OCMOL), the three networks of excellence (IDECAT, INSIDE-PORES, and NANOMEMPRO), and the derived virtual institutes (ERIC aisbl, ENMIX aisbl, and EMH), which actively promoted and collaborated to the successful realization of the aforementioned projects and this book. Special thanks go to Dr. Helge Wessel, who was in charge of the integrated projects, for his valuable assistance and patience.

Gabriele Centi
Nick K. Kanellopoulos
Guy Marin
Joris Thybaut

Editor

Dr. Nick K. Kanellopoulos is the research director of the Membranes for Environmental Separations Laboratory. His research interests are the development, characterization, and evaluation of the performance of novel nonporous membranes. He is the author or coauthor of more than 170 publications in peer-reviewed journals, and he is the editor of 5 books in the field of nanoporous membrane applications. He has received a total funding of €12 million from over 50 European and national programs, and he participates in two high-tech companies in the field of nanoporous materials. He participated in the National Representation Committee of Greece for the FP6-NMP and FP7-NMP European programs in nanotechnology from 2001 to 2009. He was the coordinator of the European Network of Excellence in nanotechnology in-situ study and development of processes involving nano-porous solids (inside-pores.gr) and of the committee for the preparation and submission of the proposal for a Greek National Nanotechnology program. He has been a member of the National Committee for Nanotechnology in Brussels over the past 10 years, a Fulbright scholar, and president of the Greek Fulbright Scholars Association. Since 2011, he has been the president of the National Research Center Demokritos, and since 2013, he has been the chairman of the Committee of the Presidents of the Greek Research Centers.

Contributors

Salvatore Abate
Dipartimento di Chimica Industriale
ed Ingegneria dei Materiali
and
INSTM Laboratory of Catalysis for
Sustainable Production and Energy
Università di Messina
Messina, Italy

Jens Aßmann
Bayer Technology Services GmbH
Leverkusen, Germany

Christos Agrafiotis
Institute of Solar Research
German Aerospace Center
Cologne, Germany

Sonia Aguado
Faculty of Chemistry
Department of Chemical Engineering
University of Alcalá
Madrid, Spain

Duncan Akporiaye
SINTEF Materials and Chemistry
Oslo, Norway

M.A. Arribas
Consejo Superior de Investigaciones
Científicas
Instituto de Tecnología Química
Universitat Politècnica de València
Valencia, Spain

Pablo Beato
Haldor Topsøe
Lyngby, Denmark

Rainer Bellinghausen
Bayer Technology Services GmbH
Leverkusen, Germany

A.S. Bobin
Institute of Research on Catalysis
and Environment
Claude Bernard Lyon University
Villeurbanne, France

and

Department of Heterogeneous Catalysis
Boreskov Institute of Catalysis
Novosibirsk State University
Novosibirsk, Russia

L.N. Bobrova
Boreskov Institute of Catalysis
Novosibirsk, Russia

Pablo del Campo Huertas
Department of Chemistry
Innovative Natural Gas Processes
and Products
Centre of Research Based Innovation
University of Oslo
Oslo, Norway

Jürgen Caro
Institute for Physical Chemistry and
Electrochemistry
Leibniz University Hannover
Hannover, Germany

Gabriele Centi
Dipartimento di Chimica Industriale
ed Ingegneria dei Materiali
and
INSTM Laboratory of Catalysis for
Sustainable Production and Energy
Università di Messina
Messina, Italy

C. Daniel
Institute of Research on Catalysis
and Environment
Claude Bernard Lyon University
Villeurbanne, France

David Farrusseng
Institute of Research on Catalysis
and Environment
Claude Bernard Lyon University
Villeurbanne, France

Yu. E. Fedorova
Boreskov Institute of Catalysis
Novosibirsk, Russia

V.V. Galvita
Laboratory for Chemical Technology
Ghent University
Ghent, Belgium

Jorge Gascon
Department of Chemical Engineering
Delft University of Technology
Delft, the Netherlands

T.S. Glazneva
Boreskov Institute of Catalysis
and
Novosibirsk State University
Novosibirsk, Russia

Carlos A. Grande
SINTEF Materials and Chemistry
Oslo, Norway

Canan Gücüyener
Department of Chemical Engineering
Delft University of Technology
Delft, the Netherlands

Emiel J.M. Hensen
Department of Chemical Engineering
and Chemistry
Schuit Institute of Catalysis
Eindhoven University of Technology
Eindhoven, the Netherlands

G. Iaquaniello
KT—Kinetics Technology SpA
Rome, Italy

A. Ishchenko
Boreskov Institute of Catalysis
and
Novosibirsk State University
Novosibirsk, Russia

Finn Joensen
Haldor Topsøe
Lyngby, Denmark

Nick K. Kanellopoulos
Institute of Nanoscience and
Nanotechnology
Demokritos National Research Center
Athens, Greece

Freek Kapteijn
Department of Chemical Engineering
Delft University of Technology
Delft, the Netherlands

Georgios N. Karanikolos
Department of Chemical Engineering
The Petroleum Institute
Abu Dhabi, United Arab Emirates

P.N. Kechagiopoulos
Laboratory for Chemical Technology
Ghent University
Ghent, Belgium
Abu Dhabi, United Arab Emirates

Leonid M. Kustov
N.D. Zelinsky Institute of Organic
    Chemistry
Russian Academy of Sciences
and
Department of Chemistry
Moscow State University
Moscow, Russia

Anastasios Labropoulos
Institute of Nanoscience and
    Nanotechnology
Demokritos National Research Center
Athens, Greece

Fangyi Liang
Institute for Physical Chemistry and
    Electrochemistry
Leibniz University Hannover
Hannover, Germany

Karl Petter Lillerud
Department of Chemistry
Innovative Natural Gas Processes
    and Products
Centre of Research Based Innovation
University of Oslo
Oslo, Norway

Anna Lind
SINTEF Materials and Chemistry
Oslo, Norway

Guy B. Marin
Laboratory for Chemical Technology
Ghent University
Ghent, Belgium

A. Martínez
Consejo Superior de Investigaciones
    Científicas
Instituto de Tecnología Química
Universitat Politècnica de València
Valencia, Spain

Cristina Martínez
Consejo Superior de Investigaciones
    Científicas
Instituto de Tecnología Química
Universidad Politécnica de Valencia
Valencia, Spain

Juan Salvador Martinez-Espin
Haldor Topsøe
Lyngby, Denmark

N.V. Mezentseva
Boreskov Institute of Catalysis
and
Novosibirsk State University
Novosibirsk, Russia

C. Mirodatos
Institute of Research on Catalysis
    and Environment
Claude Bernard Lyon University
Villeurbanne, France

Leslaw Mleczko
Bayer Technology Services GmbH
Leverkusen, Germany

Giorgia Mondino
SINTEF Materials and Chemistry
Oslo, Norway

S. Moussa
Consejo Superior de Investigaciones
    Científicas
Instituto de Tecnología Química
Universitat Politècnica de València
Valencia, Spain

L. Olivier
Institute of Research on Catalysis
    and Environment
Claude Bernard Lyon University
Villeurbanne, France

Unni Olsbye
Department of Chemistry
Innovative Natural Gas Processes
    and Products
Centre of Research Based Innovation
University of Oslo
Oslo, Norway

E. Palo
KT—Kinetics Technology SpA
Rome, Italy

S.N. Pavlova
Boreskov Institute of Catalysis
Novosibirsk, Russia

Siglinda Perathoner
Dipartimento di Chimica Industriale
    ed Ingegneria dei Materiali
and
INSTM Laboratory of Catalysis for
    Sustainable Production and Energy
Università di Messina
Messina, Italy

M. Teresa Portilla
Consejo Superior de Investigaciones
    Científicas
Instituto de Tecnología Química
Universidad Politécnica de Valencia
Valencia, Spain

Martin Roeb
Institute of Solar Research
German Aerospace Center
Cologne, Germany

V.A. Rogov
Boreskov Institute of Catalysis
and
Novosibirsk State University
Novosibirsk, Russia

Dorota Rutkowska-Zbik
Jerzy Haber Institute of Catalysis
    and Surface Chemistry
Polish Academy of Sciences
Krakow, Poland

V.A. Sadykov
Boreskov Institute of Catalysis
and
Novosibirsk State University
Novosibirsk, Russia

A. Salladini
Processi Innovativi
Rome, Italy

Christian Sattler
Institute of Solar Research
German Aerospace Center
Cologne, Germany

N.N. Sazonova
Boreskov Institute of Catalysis
Novosibirsk, Russia

Y. Schuurman
Institute of Research on Catalysis
    and Environment
Claude Bernard Lyon University
Villeurbanne, France

M. Yu. Smirnova
Boreskov Institute of Catalysis
Novosibirsk, Russia

Stian Svelle
Department of Chemistry
Innovative Natural Gas Processes
    and Products
Centre of Research Based Innovation
University of Oslo
Oslo, Norway

Shewangizaw Teketel
Department of Chemistry
Innovative Natural Gas Processes
    and Products
Centre of Research Based Innovation
University of Oslo
Oslo, Norway

and

Department of Chemical and
    Biomolecular Engineering
Center for Catalytic Science and
    Technology
University of Delaware
Newark, Delaware

Christiaan H.L. Tempelman
Department of Chemical Engineering
    and Chemistry
Schuit Institute of Catalysis
Eindhoven University of Technology
Eindhoven, the Netherlands

Joris W. Thybaut
Laboratory for Chemical Technology
Ghent University
Ghent, Belgium

Renata Tokarz-Sobieraj
Jerzy Haber Institute of Catalysis and
    Surface Chemistry
Polish Academy of Sciences
Krakow, Poland

Yvonne Traa
Institute of Chemical Technology
University of Stuttgart
Stuttgart, Germany

A.C. van Veen
Laboratory of Industrial Chemistry
Ruhr Universität Bochum
Bochum, Germany

Charitomeni Veziri
Institute of Nanoscience and
    Nanotechnology
Demokritos National Research Center
Athens, Greece

Ørnulv Vistad
SINTEF Materials and Chemistry
Oslo, Norway

Z. Yu. Vostrikov
Boreskov Institute of Catalysis
Novosibirsk, Russia

Dennis Wan Hussin
Institute of Chemical Technology
University of Stuttgart
Stuttgart, Germany

Marius Westgård Erichsen
Department of Chemistry
Innovative Natural Gas Processes
    and Products
Centre of Research Based Innovation
University of Oslo
Oslo, Norway

Małgorzata Witko
Jerzy Haber Institute of Catalysis
    and Surface Chemistry
Polish Academy of Sciences
Krakow, Poland

Section I

Integration of Innovative Membranes and Sorbents with the GTL Process

1 Status of Research and Challenges in Converting Natural Gas

Salvatore Abate, Gabriele Centi, and Siglinda Perathoner

CONTENTS

This chapter introduces methane catalytic chemistry, evidencing how there is a changing landscape for research in this area due to new scientific developments and industrial opportunities deriving from the availability of natural gas (related to shale gas basins and stranded NG wells) as well as socioeconomic pulling forces. This chapter introduces the different possibilities of direct conversion of natural gas (e.g., not passing through syngas formation) and analyzes in more detail the recent developments on gas-phase processes: oxidative coupling, direct methane conversion to methanol or formaldehyde, methane aromatization, and methane conversion via halomethane.

1.1   INTRODUCTION

Methane is one of the main actual raw materials for chemical industry but essentially through its conversion to syngas (mainly by steam reforming). Syngas are then converted to methanol and further to a range of products. The direct conversion (e.g., without passing through syngas) of methane to chemicals such as methanol, olefins, and aromatics has been for a long time a challenge for the chemical industry. Although these direct routes have been extensively investigated in the past, some main drivers have recently stimulated a renewal of interest in this area:

- The large natural gas (NG) reservoirs in the world, mainly in the Middle East and Russia, but with about one-third of them (stranded NG resources) not directly exploitable (via pipeline or liquefaction/regasification), which thus requires to develop efficient systems for the gas to liquid (GTL) conversion [1].
- The discovery and rapid proliferation of shale gas basins in North America (and other geographical areas as well) that is driving up NG supplies, with the effect of depressing gas prices and altering oil–gas price parity [2], but creating new opportunities to use them as raw material. Unconventional gas (gas shales, tight gas sands, and coalbed methane) represents a potential of about 330 Tcm (trillion cubic meters) [3].
- The scientific advances in both homogeneous and heterogeneous catalysts, and biocatalysts as well, which opened the doors to the development of new innovative solutions at scientific level, in some cases already tried to be exploited by companies.

The expanding production of shale gas, notwithstanding the large environmental concerns that have limited the production in areas such as Europe for now, resulted in a decoupling of the NG price market costs from that of oil in some areas such as the United States. In the United States (but not in Europe, where the impact of shale gas on NG costs is less relevant), chemical companies are actively looking to new opportunities created from the large availability of shale gas (methane, ethane) to expand production capacity for ethylene, ethylene derivatives (i.e., polyethylene, polyvinyl chloride), ammonia, methanol, propylene, and chlorine [4].

It must be commented, however, that according to various analysts, for example, Hughes [5], the so-called shale revolution is nothing more than a bubble, driven by record levels of drilling, speculative lease and flip practices on the part of shale energy companies, fee-driven promotion by the same investment banks that fomented the housing bubble, and unsustainably low NG prices. Shale gas production has grown explosively to account for nearly 40% of U.S. NG production, but production has reached a plateau since December 2011. About 80% of shale gas production comes from five plays, several of which are in decline. The very high decline rates of shale gas wells require continuous inputs of capital—estimated at $42 billion per year to drill more than 7000 wells—in order to maintain production. In comparison, the value of shale gas produced in 2012 was just $32.5 billion. High collateral environmental impacts have led to severe protests from citizens, resulting in moratoriums in some U.S. states. Shale gas production growth has been offset by declines in conventional gas production, resulting in only modest gas production growth overall. Moreover, the basic economic viability of many shale gas plays is questionable in the current gas price environment.

Therefore, great care should be put in considering actual low prices for NG to continue in the future and thus, based on these low prices, the estimation of the profitability of producing chemicals using NG as raw material. Nevertheless, a number of U.S. companies are actively investigating the possibility to develop a commercially viable gas-to-ethylene (GTE) process [6].

1.1.1 Revamped Industrial Interest on Valorization of NG to Chemical Products

Dozens of patents were issued recently on this topic (Figure 1.1), and also some attempts to pass on a commercial scale. Siluria Technologies, a venture capital U.S. company claiming to be the first company to provide the industry with an economically viable and commercially scalable alternative to petroleum, is building a

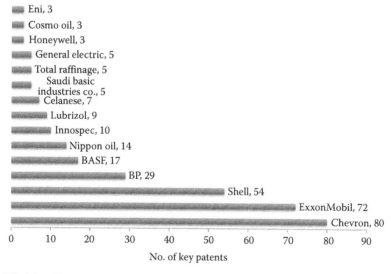

FIGURE 1.1 Global oil majors and the number of patents in converting methane to ethylene. (Based on the data reported in Dallas, K., The story of ethylene... now starring natural gas, Cleantech Blog, 2011, http://www.cleantechblog.com/blog/2011/11/the-story-of-ethylene-now-starring-natural-gas.html.)

$15 million demonstration plant at Braskem's site in La Porte, Texas, to show how oxidative coupling of methane (OCM) can produce ethylene by an economic route. Siluria's OCM process costs $1 billion a year less than the average costs at naphtha crackers that produce ethylene and $250 million cheaper than ethane cracking, based on feedstock costs since 2010, according to Siluria CEO. The catalyst is based on the use of virus-based templates that guide the growth of nanowire catalysts made of inorganic crystals, according to Siluria limited information diffused.

Honeywell International and Dow Chemical are also developing catalytic conversion processes for NG. Honeywell (UOP) announced the proof of concept for a new technology for one-step methane conversion into ethylene. In 2008, Dow Chemical put up over $6.4 million for methane activation research led by teams at Northwestern University and the U.K.'s Cardiff University. Catalysis design firm Catalytica spent 5 years and over $10 million to develop a sophisticated catalyst and process to turn methane into methanol, but its process proved too costly.

Nineteen blue chip companies and six universities and government labs have also announced new developments in this area [6]. Among them there are the following:

- Fertilizer Research Institute (Poland) is currently operating a pilot methane to ethylene facility based on OCM.
- Quantiam Technologies (Alberta, Canada) is also working on the development of novel catalyst for methane conversion (in alliance with BASF, IRAP).
- Carbon Sciences (Santa Barbara, CA) is instead developing an indirect route, where the first stage is the production of syngas by CO_2 reforming of methane on new catalysts.
- Synfuels has developed a process in which methane is transformed to acetylene in a pyrolysis reactor, and then in a liquid-phase hydrogenation, acetylene is converted to ethylene [7]. Figure 1.2 shows the simplified flowsheet of the Synfuels GTE process. In a variation of the process, with a stage of oligomerization, the process can produce gasoline (Synfuels GTL process).
- Broken Hill Proprietary (BHP) in collaboration with Commonwelth Scientific and Industrial Research Organization (CSIRO) has developed methane to ethylene process based on a fluidized bed oxidative coupling zone and a pyrolysis zone in the same reactor.

These indications evidence how there is a worldwide interest in exploiting the chemical use of NG, particularly methane, although light alkanes present in NG (ethane and propane, in particular) also show interesting opportunities not only as feed for crackers (ethane) but also for direct routes to olefins, such as by catalytic dehydrogenation (propane, especially), interesting in particular for on-site olefin production.

1.1.2 A GLIMPSE ON SOME RECENT SCIENTIFIC ADVANCES ON DIRECT METHANE CONVERSION

Parallel to the industrial interest on the valorization of NG, a large increase in scientific papers on this topic was observed, leading to also some interesting breakthroughs. We highlight here some of them as an introduction to more specific discussion reported later.

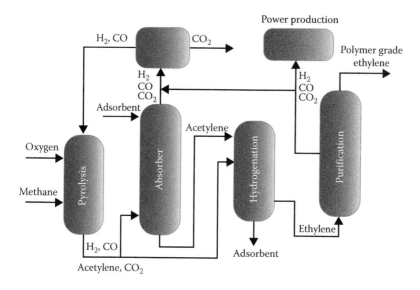

FIGURE 1.2 Simplified flowsheet of the Synfuels GTE process. (Based on the process scheme reported in Synfuel web site, Synfuels GTE Presentation, 2012, http://www.synfuels. com/WebsiteGasRichConferPaper.pdf.)

Xu et al. [8] showed the possibility of room temperature activation of methane over Zn-modified H-ZSM-5 zeolites (Figure 1.3a). Zhou et al. [9] have reported a theoretical study showing the possibility in Mo/H-ZSM-5 zeolite catalysts of a reaction mechanism of methane dehydrogenation and coupling to ethylene (Figure 1.3b). Solomon and coworkers [10,11] have shown the analogies between methane monooxygenase (MMO) enzymes (able to convert methane to methanol) and copper–zeolites, and how this analogy can be useful to understand the ability of some specific sites in Cu-ZSM-5 to convert methane directly to methanol, although the closing of the catalytic cycle requires to operate at different temperatures the generation of the active species and its reactivity with methane to give methanol (Figure 1.3c).

He et al. [12] have proposed a novel catalytic route for the conversion of methane to propylene via monohalogenomethane. CeO_2 is the most efficient catalyst for the oxidative chlorination and bromination of methane, while for the second step, an F-modified H-ZSM-5 is highly selective and stable for the conversion of CH_3Cl or CH_3Br into propylene.

$$CH_4 + HCl\ (HBr) + 1/2O_2 - cat.\ A \rightarrow CH_3Cl\ (CH_3Br) + H_2O \qquad (1.1a)$$

$$CH_3Cl\ (CH_3Br) - cat.\ B \rightarrow 1/3C_3H_6 + HCl\ (HBr) \qquad (1.1b)$$

$$Net: CH_4 + 1/2O_2 - cat. \rightarrow 1/3C_3H_6 + H_2O \qquad (1.2)$$

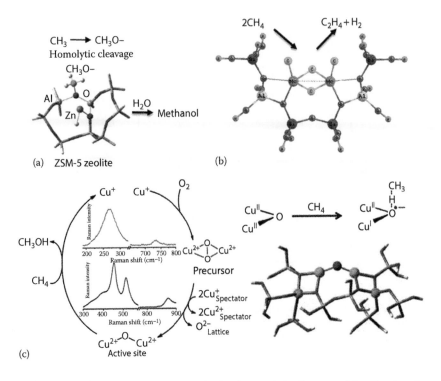

FIGURE 1.3 Examples of recent scientific developments in the area of the direct methane conversion. (a) Room temperature activation of methane over Zn modified H-ZSM-5 zeolites. (Adapted from Xu, J. et al., *Chem. Sci.*, 3, 2932, 2012.) (b) Methane dehydrogenation and coupling to ethylene in Mo/HZSM-5 zeolite catalysts. (Adapted from Zhou, D. et al., *J. Phys. Chem. C*, 116, 4060, 2012.) (c) Proposed reaction mechanism and active sites in methane conversion to methanol on Cu-ZSM-5 catalysts. (Adapted from Vanelderen, P. et al., *J. Catal.*, 284, 157, 2011.)

Wei et al. [13] reviewed recently the conversion of methane to higher hydrocarbons via Me halide as the intermediate. After the production of halomethane, they could be transformed to gasoline and light olefins over modified zeolites and silicoalumi-nophosphate (SAPO) molecular sieves. High conversion efficiency and selectivity indicated the feasibility of industrial application.

These examples show the renewed interest in methane conversion and the wide range of new scientific breakthrough, opening new scientific perspectives in this field. This chapter discusses selected recent developments in the last 3–4 years in this area, with the focus on catalytic aspects of direct methane valorization. Indirect methane conversion via syngas (and further conversion of the products of reaction) will not be discussed here, neither aspects related to H_2 production.

Hammond and coworkers [14] have recently made a minireview on oxidative methane upgrading. Table 1.1 summarizes the potential routes for methane valorization to chemicals, which will be discussed in part in this chapter together with major related issues/challenges.

TABLE 1.1

Potential Routes for Methane Valorization to Chemicals

Type of Reaction	Products (Main)	Main Issues/Challenges
Selective oxidation	Methanol	Low productivity, close the cycle at the same reaction temperature
	Formaldehyde	Selectivity, productivity
	Formic acid	Selectivity, limited market
	Methyl bisulfate	Use of oleoum
Oxidative carbonylation	Acetic acid	Competing with current industrial processes
Oxidative coupling	Light olefins (+light alkanes)	Overcoming the conversion vs. selectivity ceiling
Aromatization or alkylation	BTX (benzene, toluene, xylenes)	Thermodynamic limitations, deactivation

Source: Adapted from Hammond, C. et al., *Angew. Chem. Int. Ed.*, 51, 5129, 2012.

1.1.3 DIRECT AND INDIRECT ROUTES IN METHANE CONVERSION

There are two main routes for converting methane to transportable liquid fuels or chemicals, namely, indirect and direct routes. Commercial catalytic technologies are based currently on the indirect route that involves a two-step process in which methane is first converted to synthesis gas by steam reforming (Equation 1.3), CO_2 reforming (Equation 1.4; this is still not commercial but used in some pilot scale processes), partial oxidation (Equation 1.5; catalytic partial oxidation is a new area of development), followed by either Fischer–Tropsch (FT) synthesis of hydrocarbons (Equation 1.6) or methanol synthesis (Equations 1.7), and subsequent conversion to hydrocarbons

$$CH_4 + H_2O \rightarrow CO + 3H_2 \quad (\Delta H^\circ_{298\,K} = +206 \text{ kJ mol}^{-1}) \tag{1.3}$$

$$CH_4 + CO_2 \rightarrow 2CO + 2H_2 \quad (\Delta H^\circ_{298\,K} = +247 \text{ kJ mol}^{-1}) \tag{1.4}$$

$$CH_4 + \tfrac{1}{2}O_2 \rightarrow CO + 2H_2 \quad (\Delta H^\circ_{298\,K} = -36 \text{ kJ mol}^{-1}) \tag{1.5}$$

$$nCO + 2nH_2 \rightarrow (-CH_2)_n + nH_2O \quad (\Delta H^\circ_{298\,K} = -165 \text{ kJ mol}^{-1}) \tag{1.6}$$

$$CO + 2H_2 \rightarrow CH_3OH(g) \quad (\Delta H^\circ_{298\,K} = -90.8 \text{ kJ mol}^{-1}) \tag{1.7a}$$

$$CO_2 + 3H_2 \rightarrow CH_3OH(g) + H_2O \quad (\Delta H^\circ_{298\,K} = -49.5 \text{ kJ mol}^{-1}) \tag{1.7b}$$

The direct route is a one-step process in which methane reacts with O_2 or another oxidizing species to give the desired product, for example, methanol (Equation 1.8), formaldehyde (Equation 1.9; this is a route which has loss of interest), or ethylene Equation 1.10 (as discussed earlier, this is a route under active research). The direct route is more energy efficient than the indirect route since it bypasses the energy-intensive endothermic steam-reforming step of syngas formation.

$$CH_4 + (1/2)O_2 \rightarrow CH_3OH(g) \quad (\Delta H°_{298\,K} = -126.4 \text{ kJ mol}^{-1}) \quad (1.8)$$

$$CH_4 + O_2 \rightarrow HCHO(g) + H_2O(g) \quad (\Delta H°_{298\,K} = -276 \text{ kJ mol}^{-1}) \quad (1.9)$$

$$2CH_4 + O_2 \rightarrow H_2C = CH_2(g) + 2H_2O(g) \quad (\Delta H°_{298\,K} = -67 \text{ kJ mol}^{-1}) \quad (1.10)$$

These oxidation reactions compete in all the reactions in which oxygen is present in the feed and are the more thermodynamically favored reactions (Equation 1.11a and b):

$$CH_4 + (3/2)O_2 \rightarrow CO + 2H_2O(g) \quad (\Delta H°_{298\,K} = -519.6 \text{ kJ mol}^{-1}) \quad (1.11a)$$

$$CH_4 + 2O_2 \rightarrow CO_2 + 2H_2O(g) \quad (\Delta H°_{298\,K} = -802.6 \text{ kJ mol}^{-1}) \quad (1.11b)$$

1.2   GAS-PHASE CONVERSION PROCESSES

1.2.1   Oxidative Coupling

OCM involves the reaction of CH_4 and O_2 over a catalyst at high temperatures to form C_2H_6 as a primary product and C_2H_4 as a secondary product. Some higher hydrocarbons may also form, because OCM is essentially a radical (homogeneous) chemistry although the catalyst is important for the activation, as well as some of the quenching reactions of the radicals. Mixed homogeneous–heterogeneous mechanisms are thus present. This is a main issue for scaling-up the results.

1.2.1.1   State of the Art of Research

A detailed review on recent advances in direct methane conversion routes to chemicals and fuels, including OCM, has been made by Fierro and coworkers [15]. Takanabe [16] reviewed the recent results in the catalytic reforming of methane with carbon dioxide and in the oxidative coupling, focusing on the mechanistic aspects and the development of efficient and durable catalysts for the two reactions. Arndt et al. [17] have focused their attention in reviewing the behavior and characteristics of a single catalyst: $Mn–Na_2WO_4/SiO_2$, one of the very few catalysts for the OCM, whose stability over extended times on stream has been reported by different research groups.

Baerns and coworkers [18] have made a statistical analysis (1870 data sets on catalyst compositions and their performances in the OCM), with over 1000 full-text references from the last 30 years, to select 18 catalytic key elements. All oxides of the selected elements, which positively affect the selectivity to C2 products, show strong basicity. Analysis of binary and ternary interactions between the selected key elements shows that high-performance catalysts are mainly based on Mg and La oxides. Alkali (Cs, Na) and alkaline-earth (Sr, Ba) metals used as dopants increase the selectivity of the host oxides, whereas dopants such as Mn, W, and the Cl anion have positive effects on the catalyst activity. The maximal C2 selectivities for the proposed catalyst compounds range from 72% to 82%, and the respective C2 yields range from 16% to 26%.

About one-third of the references available in literature (mostly made from 1985 to 1995 in laboratory-scale catalytic fixed-bed reactors) focuses on the intrinsic reactions between reactants and catalyst surface, over hundreds of different catalytic materials and numerous metal or metal-oxide loadings. Although an upper bound of Y_{C2} (yield of C2 products, e.g., ethane and ethylene) of about 25% has been predicted on the kinetics, at least 24 different catalysts providing $Y_{C2} \geq 25\%$ have been reported (Figure 1.4a). Some of these catalysts are close to the target for the industrial application of the OCM process: single-pass conversions of methane of at least 30% and S_{C2} (selectivity of C2 products) of around 80%.

Although extensive research has been done on the OCM reaction in the last 30 years, many fundamental aspects, which determine the choice of catalytic components, for example, distribution between surface-to-gas phase reactions, the participation of nonequilibrium sites in the OCM process, as well as the essential features for an optimal catalyst composition, remain unknown.

From the catalyst research and development perspective, the question about key components of a catalyst that leads to high C2 selectivity has to be first answered. For identifying the key components that contribute to S_{C2} and Y_{C2}, first the catalytic performances of single unsupported oxides without any promoters have to be analyzed. The results of the statistical analysis of literature data sets by Zavyalova et al. [18] are summarized in Figure 1.4b in the form of a periodic table, in which their S_{C2} ranges are indicated. To analyze the performance data of multicomponent catalysts and identify the highly significant elements of the catalysts, Zavyalova et al. [18] made a multiway analysis of variance (ANOVA) for main effects. The results can be summarized as follows:

- Elements that correlate positively with both Y_{C2} and S_{C2} are alkali metals (Li, Na, Cs), alkaline-earth metals (Sr, Ba, Mg, Ca), as well as La, some other metals (Bi, Mo, Ga, Nd, Re, Yb), and fluorine
- Elements that correlate positively only with Y_{C2} are Cl, W, Ti, Mn, Y, and Sm
- Elements that correlate negatively with both Y_{C2} and S_{C2} are Ag, V, K, and C (derived from carbonate species)

A further refinement of the analysis and the identification of elements that contribute to the best catalytic performance, the quantitative values of Y_{C2} and S_{C2} were approximated with a piecewise-constant function as applied in regression tree analyses. This analysis showed that the main groups of catalysts that comprise the well-performing materials are based on alkali and alkaline-earth metals (Li, Na, Mg, Sr, Ba, Ca) as well as on La, W, and F and Cl anions.

The group of well-performing catalysts has an overall mean of $S_{C2} = 65\%$ and $Y_{C2} = 19.5\%$. The mean value of S_{C2} is higher than the mean value of 65% for the whole group of the catalysts in the presence of the following elements:

- Mean $S_{C2} = 69\%$ in the presence of Mg or promoters Cl, B, and S
- Mean $S_{C2} = 68\%$ in the presence of the combination La*Sr
- Mean $S_{C2} = 66\%$ in the presence of W

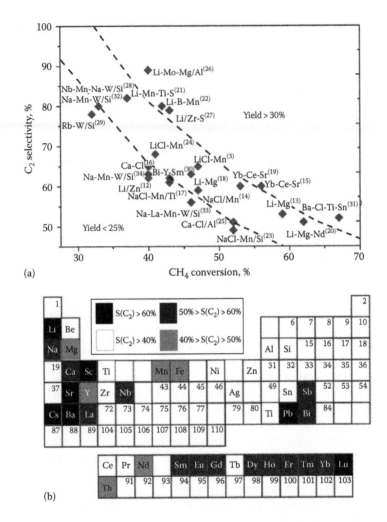

FIGURE 1.4 (a) Elemental compositions of OCM catalysts with $Y_{C2} \geq 25\%$ reported in the literature. All the catalysts were tested in a fixed-bed reactor in the co-feed mode under atmospheric pressure at temperatures from 943 to 1223 K, $P_{CH_4}/P_{O_2} = 1.7$–9.0, and contact times from 0.2 to 5.5 s. (b) S_{C2} of various unsupported single oxides tested in the OCM reaction in the co-feed mode. (Y, yield; S, selectivity). (Adapted from Zavyalova, U. et al., *ChemCatChem*, 3, 1935, 2011.)

The mean value of Y_{C2} is higher than the mean value of 19.5% for the group of well-performing catalysts in the presence of the following elements:

- Mean $Y_{C2} = 21.7\%$ in the presence of promoters such as Cl, B, and S
- Mean $Y_{C2} = 20.9\%$ in the presence of Al or Si oxide supports
- Mean $Y_{C2} = 20.3\%$ in the presence of Li

In addition, the nature of promoters and supports apparently plays an important role. Therefore, the catalytic key elements selected on the basis of the statistical results by Zavyalova et al. [18] can be divided into three major groups with respect to their function in a catalyst:

- Main components: Sr, Ba, Mg, Ca; La, Nd, Sm; Ga, Bi, Mo, W, Mn, Re
- Dopants: Li, Na, Cs
- Promoters: F, Cl

Mn was selected as the catalytic key element potentially suitable for the design of high-performance OCM catalysts because it was identified as a highly significant element with respect to the variance of Y_{C2} and it has a positive correlation with Y_{C2}. In addition, 11 from the 24 reported catalysts with $Y_{C2} \geq 25\%$ include Mn (underlined in Figure 1.4a). Re and Ga were also selected as significant elements, although few data are present in literature.

A further aspect analyzed by Zavyalova et al. [18] regarded the analysis of inter-actions between the catalytic key components. The results indicate the following aspects in terms of molar fractions (x) of the main components with positive effect on the mean value of S_{C2} within the well-performing catalysts:

- $78\% < x(Mg) < 90\% \rightarrow$ mean S_{C2} of 71%–77%
- $x(La) > 67\% \rightarrow$ mean $S_{C2} = 75\%$
- $x(Mn) > 21\% \rightarrow$ mean $S_{C2} = 67\%$

Molar fractions of the main components with positive effect on the mean value of Y_{C2}:

- $x(Mn) > 22\% \rightarrow$ mean $Y_{C2} = 24.5\%$
- $48\% < x(La) < 89\% \rightarrow$ mean Y_{C2} of 19.2%–19.9%
- $76\% < x(Mg) < 89\% \rightarrow$ mean Y_{C2} of 19.6%–21.3%
- $x(W) < 16\% \rightarrow$ mean $Y_{C2} = 19.8\%$
- $x(Ba) > 30\% \rightarrow$ mean $Y_{C2} = 20.1\%$
- $x(Sr) < 2.5\% \rightarrow$ mean $Y_{C2} = 20.4\%$

Thus, it may be possible to conclude that the majority of high-performance catalysts can be divided into three main groups:

- Mg oxide doped by alkali metals (Cs, Na), by Mn, and/or promoted by the Cl anion
- La oxide doped by Na or alkaline-earth metals (Sr, Mg, Ba)
- Mn oxide doped by Na, W, and/or promoted by the Cl anion

The catalysts based on Mg and La may possess the highest values of $S_{C2} = 72\%$–82% and $Y_{C2} = 16\%$–26%, whereas Mn-based catalysts from the third group result in high $Y_{C2} = 20\%$–26% and are less selective with $S_{C2} = 53\%$–67% but more active as compared with the catalysts from the first and second groups. The following recommendations for the design of high-performance OCM catalysts can be made. A high-performance OCM catalyst is a multicomponent material that may consist of the following:

1. Host oxides
   a. Mg oxide with molar fraction 76% < x <89%, or
   b. La oxide with molar fraction 67% < x <89%, or
   c. Mixtures of La oxide with Mg, Sr, or Ba oxides
2. Dopants that positively affect S_{C2}: Na and/or Cs
3. Dopants that positively affect Y_{C2}: Mn and/or W
4. Cl anion as a promoter that positively affects Y_{C2}

Some particular catalyst compositions with the highest mean values of Y_{C2} and S_{C2} (Table 1.2) can be identified as follows (oxides of): (1) Na/Mg, (2) Na-CsCl/Mg, (3) Na/Mn-Mg, (4) La-Ba-Mg, (5) Na/La, (6) Sr-La, (7) NaCl/Mn, and (8) Na/W-Mn.

We have analyzed in detail the work of Zavyalova et al. [18], because it is the most comprehensive analysis of the very intensive past OCM experiments (a unique situation in catalysts, where nearly 2000 data sets are available in literature) and thus provides an excellent state of the art. The conclusions are that Mn- and Mo-containing catalysts, on which recent open and patent literature focused attention, were identified as promising catalysts. It was also confirmed that alkaline-earth oxides and rare-earth oxides both doped with alkali metal oxides have a high potential for application. The Li–MgO system, the original catalysts identified for OCM, was, however, not identified as a promising high-performance material; this is in agreement with the conclusion of an extended review on Li–MgO catalysts by Arndt et al. [19].

A new strategy proposed from these results is based on the use of synergetic effects in multicomponent materials based on strongly basic oxides (Mg, La) with dopants having positive effects on both C2 selectivity (Cs, Na, Sr, Ba) and catalyst activity (Mn, W, Cl anion). Catalysts with the compositions identified perform close to the target required for an industrially applied OCM process. Further improvement

---

TABLE 1.2

Selected Catalyst Compositions Based on the Significant Combinations of the Key Elements with the Highest Mean Values of Y_{C2} (Yield to C2) and S_{C2} (Selectivity to C2)

Catalyst and Elemental Composition (mol%)	Mean S_{C2} (%)	Mean Y_{C2} (%)
Na (9–15) – Mg (85–91)	79	25
Na (3–10) – Cs (3–10) – Mg (60–90) – Cl (5–20)	82	20
Na (4–25) – Mn (2–20) – Mg (57–92)	72	18
Na (7–75) – La (25–93)	71	22
Sr (1–59) – La (61–99)	73	18
Na (17–33) – Mn (34–67) – Cl (17–33)	53	26
Na (50–53) – W (24–27) – Mn (26–27)	67	20

Source: Adapted from Zavyalova, U. et al., *ChemCatChem*, 3, 1935, 2011.

in the performance of the catalysts based on the selected key components is expected by applying an evolutionary catalyst development approach.

However, the obvious question is whether this possible incremental increase of the performances (after some extensive literature review and patent data) can achieve the necessary industrial targets, also taking into account that scaling-up to industrial levels the performances of catalysts strongly depending on homogeneous gas-phase radical reactions (and related fluidodynamic and heat/mass transfer aspects) could be a major issue. Thus, there are two alternatives: one is the further refinement of the identified multicomponent catalysts, and the other is to explore conceptually new catalysts, based on different reaction mechanisms and possibly operating at lower temperatures (as shown in Figure 1.3b, for example). It should also be commented that selectivity to C2 indicates the sum of ethane and ethylene, but ethylene is the target product. Thus, the overall process is the separation of C2 from unreacted methane and other C3+ products, followed by the selective dehydrogenation of an ethane–ethylene mixture, which is not an easy task.

Even considering the increasing cost gap between crude oil and NG, the yield of 30% in C2 products for an industrial process appears to be optimistic and should be revised with updated technoeconomic estimations and sensitivity analyses. More attention has to be given also to the development of innovative separation processes, based on membranes or other solutions, for the selective separation of C2 fraction. The issue of selective dehydrogenation of ethane in the C2 fraction has to be analyzed in more detail also.

In addition to those already cited, other recent reviews on OCM confirm the renewed interest in the topics. Kondratenko and Baerns [20,21] have reviewed the catalytic conversions of methane to C2 hydrocarbons, that is, ethane and its consecutive product ethylene, as well as to methanol and formaldehyde, with a focus on the type of catalysts and the mechanisms of these reactions, including a comparison with the mechanistic aspects of enzymatic conversion of methane to methanol. More details on the enzymes that activate O_2 at carboxylate-bridged nonheme diiron clusters residing within ferritin-like, 4-helix-bundle protein architectures (e.g., soluble MMO [sMMO]) were discussed by Krebs et al. [22].

Havran et al. [23] have analyzed the literature data on direct conversion of methane and carbon dioxide into higher value products, including new alternative routes such as photocatalytic conversion and dielectric barrier discharges. Martinez et al. [24] have reviewed the role of zeolites (and zeotypes) as well as ordered silica mesostructures in providing well-defined, active, and selective sites in the catalytic routes to convert methane (direct routes) or synthesis gas (indirect routes) into clean fuels and platform chemicals. Sinev et al. [25] have analyzed the mechanisms of OCM over oxide catalysts and the corresponding progress in its kinetic description. The latter becomes essential at the stage of scaling-up and optimization of the process in pilot and industrial reactors. The role of homogeneous–heterogeneous mixed mechanisms and kinetics description was emphasized. In particular, some important features of the OCM process can be described if several elementary reactions of free-radical species (formation and transformation) with surface active sites are included into the detailed scheme of methane oxidation in gas. However, some important features, such as a nonadditive character of the reciprocal influence of methane and ethane in the case of their simultaneous presence in the reaction mixture, cannot yet be described and comprehended in the framework of schemes developed so far. It was thus indicated, and outlined, the need to develop advanced kinetic models, accounting the main principles of catalyst functioning (redox nature of active sites) and pathways of product formation (via free radicals). Usachev et al. [26] instead put attention on process aspects of the partial oxidation of light alkanes into syngas and OCM into C2 hydrocarbons. The problems of these processes (high cost of pure oxygen; safety; activity, selectivity, and stability of catalysts; temperature regime; coke formation and other by-products; insufficient level of methane transformation into ethane and ethylene) are highlighted, with some indications about the possible solutions of these problems to enable the practical use of light alkanes processes.

A good review of the prospects in the direct conversion of methane to fuels and chemicals was also made by Holmen [27], while Kondratenko [28] has examined the strategies for direct conversion of methane to higher-valued petrochemicals and olefinic feedstocks, including OCM to ethane and ethylene, methane oxidation to C1 oxygenates (methanol and formaldehyde) and their derivatives, as well as methane conversion to acetic acid.

1.2.1.2 Comparison with Alternative Processes

Ren et al. [29] have made a very valuable comparison of energy use, CO_2 emissions, and production costs in light olefins production through steam cracking with respect to methane conversion via methanol and oxidative coupling. They also present a state of the art in the industrial development. The block flowsheet considered in OCM is reported in Figure 1.5 and shows the various steps that have to be considered in this apparently

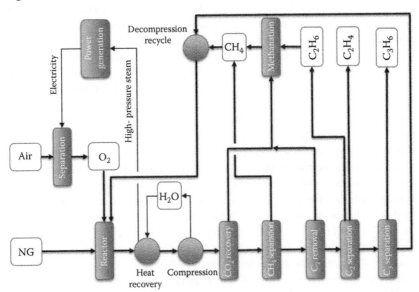

FIGURE 1.5 Block flowsheet for a generic process of OCM which includes electricity cogeneration and CO_2 methanization steps. (Elaborated from the scheme presented by Ren, T. et al., *Energy*, 33, 817, 2008.)

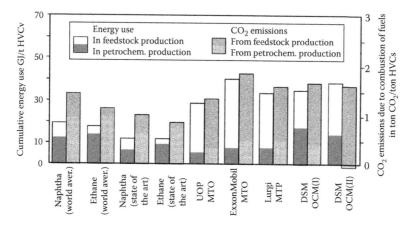

FIGURE 1.6 Comparison of cumulative process energy use (per ton of high value chemicals—HVCs, excluding the carbon content of HVCs) and estimated CO_2 emissions by steam cracking and C1 routes (emissions due to combustion in ton CO_2 per ton of HVCs, excluding the carbon content of HVCs). (Elaborated from the scheme presented by Ren, T. et al., *Energy*, 33, 817, 2008.)

simple process, although often not fully recognized. Figure 1.6 reports the summary of the comparison, in terms of cumulative process energy and estimated CO_2 emissions.

Production cost analyses by Ren et al. [29] are summarized in Figure 1.7. The results of this comparative analysis of the energy use, CO_2 emissions, and production costs of C1 technologies with respect to steam cracking indicate that methane-based routes use more than twice as much process energy than state-of-the-art steam cracking routes do (the energy content of products is excluded). The methane-based routes can be economically attractive in remote, gas-rich regions where NG is available at low prices. This estimation, however, has to be revised based

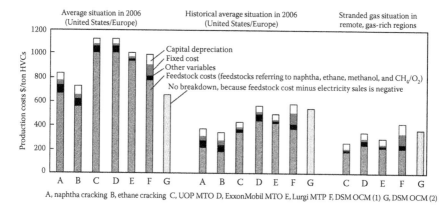

FIGURE 1.7 Comparison of production costs by steam cracking and C1 routes (all in US$ of year 2000). A, naphtha cracking; B, ethane cracking; C, UOP MTO; D, ExxonMobil MTO; E, Lurgi MTP; F, DSM OCM(1); G, DSM OCM(2). (Adapted from Ren, T. et al., *Energy*, 33, 817, 2008.)

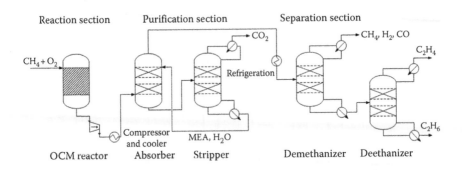

FIGURE 1.8 Flowsheet for the OCM process. (Adapted from Salerno, D. et al., *Comp. Aid. Chem. Eng.*, 31, 410, 2012.)

on actual price differential between crude oil and NG, which has largely increased in the last 2 years. While several possibilities for energy efficiency improvement do exist, Ren et al. [29] concluded that none of the NG-based routes are likely to become more energy efficient or lead to less CO_2 emissions than steam cracking routes do. Oxidative coupling routes were considered immature due to low ethylene yields and other problems: (1) high energy use in separation and recycling (see block diagram in Figure 1.5) and (2) high temperature needs (>750°C) to have good selectivities and kinetics, which make the process non–thermally efficient and cause severe catalyst stress.

An updated technoeconomic analysis for OCM and downstream processes was made recently by Salerno et al. [30]. In a previous paper (Salerno et al. [31]), they also made a technoeconomic analysis for ethylene and methanol production by OCM. The process flowsheet considered for OCM is reported in Figure 1.8.

Three main sections are considered: reaction, purification, and separation. The reactor is continuously fed with NG and O_2. The feed is preheated at 700°C, catalytic partial oxidation at a pressure of 115 kPa, and the reaction is carried out at 850°C. The exothermic reaction heat has to be immediately removed using the transfer line exchanger to take away by vaporization high-pressure boiler feedwater, which is separated in the steam drum and subsequently superheated in the conversion section to high-pressure superheated steam. The reaction products are compressed in a multicompaction section to 1090 kPa and cooled down to 40°C later on. In the purification section, the reactor effluent gases are cooled and fed into the bottom stages of a series of absorber columns that uses monoethanolamine (MEA) as absorber solution. The MEA is then regenerated in stripper columns releasing CO_2 captured. Finally, the ethylene separation section consists of two cryogenic distillation columns.

The estimated OCM process economics (Table 1.3), considering or not the recycle of methane, indicate a payout period of >10 years, which is relatively high (simulated processes consider a production of 396 tons day^{-1} of 99% purity ethylene from 2593 tons day^{-1} methane as feedstock). However, payout period can be reduced to about 8 years (which is more promising) by using unreacted methane to produce methanol and formaldehyde (direct conversion under 30 bar pressure).

TABLE 1.3

OCM Process Economics

Investment (M€)	OCM Process	
	Without CH₄ Recycle	With CH₄ Recycle
Total project capital cost	170.7	183.9
Total operating cost	266.1	255.3
Total raw material cost	83.8	81.3
Total utilities cost	154.9	147.7
Total products sales (C_2H_4)	272.4	272.4
Payout period (years)	12.2	10.6

Source: Adapted from Salerno, D. et al., *Comp. Aid. Chem. Eng.*, 31, 410, 2012.

1.2.1.3 Reaction Network, Mechanism, and Recent Advances in Catalysts

The reaction network and mechanism are determined from the presence of a mixed homogeneous–heterogeneous reaction. Methyl radicals formed at the surface of the catalyst enter the gas phase where they combine to form ethane. At atmospheric pressure, this coupling occurs mainly in the void space between catalyst particles. In addition to coupling, the gas-phase radicals may enter the chain reactions that result in the formation of CO and subsequently CO_2. A simplified reaction scheme is shown below.

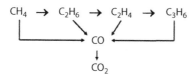

Isotopic labeling experiments have shown that at small conversion levels, most of the CO_2 is derived from CH_4, but at commercially significant conversion levels, C_2H_4 would be the dominant source of CO_2. Additional experiments have shown that this occurs mainly via a heterogeneous reaction. One of the challenges in catalyst development (and reaction design as well) is to modify a material so that the secondary reaction of C_2H_4 will be inhibited while the activation of CH_4 will still occur. This would require different sites, although this is not demonstrated. Both porous ceramic and dense ionic membranes and mixed conducting ones have been applied for the OCM reaction [32]. Porous membranes such as alumina, zirconia, or Vycor glass have high stability for the reactions, but low oxygen selectivity. By contrast, the oxygen selectivity for dense ionic or mixed-conducting oxide membranes is theoretically infinite. These membranes can be prepared in the form of ceramic hollow-fibers showing high oxygen permeation rate and suited for commercialization [33], but the high operative costs make still not economic the process.

Godini et al. [34] have analyzed the performance of the OCM process based on different alternative reactor structures including fixed-bed reactor, two different feeding-structures of porous packed bed membrane reactor, and different conceptual

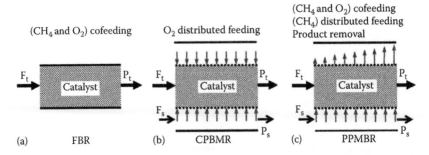

FIGURE 1.9 Schematic representation of the three reactor concepts for the OCM process. (a) FBR: provides a fast conversion; tube side feed (F_t) = N_2, CH_4, O_2. (b) CPBMR: high methane to oxygen ratio by maintaining a low oxygen concentration; tube side feed (F_t) = N_2, CH_4; shell side feed (F_s) = N_2, O_2; shell side product (P_s) = mainly O_2 and N_2; low permeable tubular membrane. (c) PPBMR: high methane to oxygen ratio by keeping the methane concentration high; product removal; tube side feed (F_t) = N_2, CH_4, O_2; shell side feed (F_s) = N_2, CH_4; shell side product (P_s) = all components (mainly CH_4 and removed products); relatively high permeable tubular membrane; tube side product for all reactors (P_t) = all components (reactants, products, and inert gas). (Adapted from Godini, H.R. et al., *Ind. Eng. Chem. Res.*, 51, 7747, 2012.)

network combinations of them (Figure 1.9). In addition to different configurations, the effect of various operating parameters was evaluated on three types of catalysts for which literature data on the kinetics were available: La_2O_3/CaO (Cat I), Mn/ Na_2WO_4/SiO_2 (Cat II), and PbO/Al_2O3 (Cat III). The investigated parameters are temperature, membrane thicknesses, types of catalysts, amount of inert packing in the catalyst bed, the flow rate of oxygen-rich stream entering the reactor system (total methane to oxygen ratio), distribution of the oxygen-rich stream through the reactor blocks (local methane to oxygen ratio), distribution of the total methane-rich feed stream into the reactor blocks, and the contact time represented by the reactor length. A high C2 yield of 31% and C2 selectivity of 88% were observed at the same time on optimized reactor/reaction conditions. The optimization of composition of reactants along the catalytic bed has thus a large influence on performances and to reach target objectives, although scaling-up the presented reactor concepts is challenging.

An extended microkinetic model for methane oxidative coupling (OCM) including the so-called catalyst descriptors has been used by Thybaut et al. [35] for the simulation of experimental data on various catalysts in different setups. The model allows the selection of the optimal operating conditions for catalyst evaluation. Figure 1.10 shows an example of simulation showing the maximum C2 yield (about 27%), which can be estimated as a function of catalyst reaction enthalpy of hydrogen abstraction of methane and the chemisorption enthalpy of oxygen. Effects of operating conditions and catalyst texture properties, such as feed flow rate, temperature, pressure and porosity, Brunauer–Emmett–Teller (BET) surface area, and tortuosity, may be also investigated using the model.

The morphology and microstructure aspects of Li/MgO catalysts for OCM were analyzed by Schlögl group [36]. Samples with low Li content exhibit a hierarchical pore structure of tubular MgO particles. Upon calcination at 1073 K, these particles undergo a change in shape, from cubic via truncated octahedral to platelet,

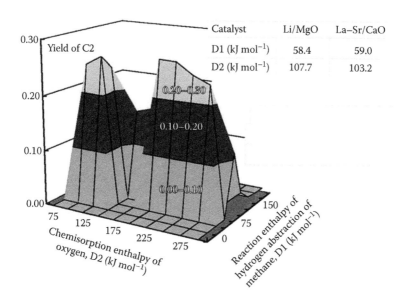

Catalyst	Li/MgO	La–Sr/CaO
D1 (kJ mol^{-1})	58.4	59.0
D2 (kJ mol^{-1})	107.7	103.2

FIGURE 1.10 Yield of C2 products vs. catalyst descriptors D1 (reaction enthalpy of hydrogen abstraction of CH$_4$) and D2 (chemisorption enthalpies of O$_2$); under 1069 K. Operating conditions: T = 1069 K, p = 108 kPa, CH$_4$/O$_2$ = 3.0, W/F$_{t,0}$ = 11.0 kg s mol^{-1}. (Adapted from Thybaut, J.W. et al., *Catal. Today*, 159, 29, 2011.)

depending on Li content. The role of Li is as flux in this transformation. The modification of the primary particle morphology leads to a drastic change in secondary structure from open sponges to compact sintered plates upon addition of Li at loadings above 10%. The microstructure of the primary particles reveals two families of high-energy structures, namely, edge-and-step structures and protrusions on flat terraces. A relation was found between catalytic activity in OCM and the transformation from cubic to complex-terminated particles. Based on these findings, sites active for the coupling reaction of methane were related to the protrusions arising from segregation of oxygen vacancies to the surface of MgO. Figure 1.11a reports the correlation observed between surface morphology of the primary MgO particles (as a function of Li loading) and the catalytic behavior (yield C2 for m^2). In both extreme samples, shallow protrusions are observed, whereas the MgO typical step-and-terrace structures are not essential for the C2 productivity. This trend seems to support the localization of electrophilic oxygen at step edges [37] and of nucleophilic oxygen [38] at locations of segregated oxygen vacancies.

Figure 1.11b shows the hierarchical structure of the Li/MgO catalyst with a model of the Li/MgO$_x$ islands active sites [39]. At low concentration and temperature, the Li incorporates into the MgO matrix most likely in combination with oxygen vacancies. Those Li-induced defects give rise to a distinct change in the optical response of the system. Annealing above 500 K triggers the segregation of substitutional Li toward the surface, where it first accumulates in mixed Li–Mg oxide islands and later forms phase-separated Li$_x$O clusters. The segregation is accompanied by a considerable surface roughening, as observed by the different microscopic techniques.

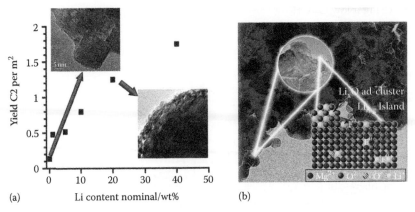

FIGURE 1.11 (a) Correlation observed between surface morphology of the primary MgO particles (as a function of Li loading) and the catalytic behavior (yield C2 for m²). (b) The hierarchical structure of the Li/MgO catalyst with a model of the Li/MgO$_x$ islands active sites. (Adapted from Zavyalova, U. et al., *ChemCatChem*, 3, 949, 2011.)

At temperatures above 1050 K, the Li desorbs from the surface, leaving behind a characteristic pattern of rectangular holes. However, even prolonged heating is insufficient to completely remove the Li from the MgO matrix. The formation of Li$^+$O$^-$ centers, which are commonly connected to the catalytic activity of the doped oxide, could not be approved by electron paramagnetic resonance (EPR) spectroscopy, either in thin-film or in powder samples. This finding is consistent with the results of density functional theory (DFT) calculations that indicated the thermodynamic instability of such centers in a large pressure and temperature window. However, the development of Li$^+$O$^-$ centers might still be possible at nonequilibrium conditions

A high C2 yield (30.7%) was reported by Caro group [40] using chlorinated perovskite $Ba_{0.5}Sr_{0.5}Fe_{0.2}Co_{0.8}O_{3-\delta}Cl_{0.04}$ (BSFCCl0.04) as catalyst in OCM but using N_2O as the oxidant. The introduction of Cl into BSFC enhanced the selectivity for Me radical formation. Besides, using N_2O instead of O_2 also, to some extent, prevents deep oxidant of C2 in gas phase and enhances CH_4 conversion. Finally, Schmidt-Szalowski et al. [41] examined three kinds of plasma-catalytic hybrid systems for OCM. Acetylene and soot were the main products formed in the homogeneous plasma system (without a catalyst) and with alumina-ceramic particles. In the presence of Pt and Pd supported on alumina ceramic particles, two effects were noticed: ethylene and ethane became the main products instead of acetylene, and the soot formation was strongly reduced.

In conclusion, even with the extensive experimentation in the past which still continue today also inspired from new advances in the understanding and simulation, the best results are in the 25%–30% C2 yield. There are some attempts to commercialize OCM technology, but important hurdles to be overcome in this process are still present:

- Since C2 hydrocarbons are much more reactive than methane, high selectivity in the process can be obtained only at low methane conversion.
- As the reaction is conducted at high temperatures (ca. 800°C), a catalyst with high thermal and hydrothermal stability is required.

- Using a low O_2/CH_4 molar ratio in the feed, the selectivity to C2 hydrocarbons is high, but CH_4 conversion is rather low.
- Due to the low concentration of ethylene in the exit stream, the cost of its separation is high, thereby rendering its separation uneconomical.

1.2.2 METHANE DIRECT CONVERSION TO METHANOL OR FORMALDEHYDE

The direct conversion of methane into C1 oxygenates (CH_3OH and $HCHO$) involves partial oxidation under specific reaction conditions. This reaction operates at 350°C–500°C under fuel-rich mixtures with the oxidant to minimize the extent of combustion reactions. Under these conditions, the gas-phase oxidation reactions of methane operate at high temperatures, which are detrimental to the control of selectivity to C1 oxygenates.

1.2.2.1 Noncatalytic Conversion

At high pressure, in the absence of a catalyst, it is known that it is possible to obtain good selectivities to methanol, but together with a number of other products (formaldehyde, in particular) that have never allowed to commercialize the process, due to separation costs [42]. However, the method was patented even recently, for example, by Olah and Surya Prakash [43]. The rate-limiting step is the first H abstraction to form methyl radicals. Thus, initiators and sensitizers can be incorporated into the reaction mixture with a view to decrease the energy barrier of this H abstraction. Methanol yields as high as 7%–8% are obtained in the absence of catalysts operating at 350°C–500°C and 50 bar. As reactor inertness is essential for obtaining good selectivity to methanol, the feed gas should be isolated from the metal wall by using quartz and Pyrex glass-lined reactors.

1.2.2.2 Catalytic Conversion: Metal Oxides

Considerable efforts have been made in the past devoted to developing active and selective catalysts, but neither the product yield of C1-oxygenates nor the complete mechanism of the reactions has been clarified. The selective O insertion into CH_3, or other fragments resulting from the first H abstraction of the CH_4 molecule, was usually conducted on redox oxides of the MoO_3 and V_2O_5 type [44,45].

In these systems, the key to maximizing catalytic performances is to maintain isolated metal oxide structures on a silica substrate in a slightly reduced state [46]. The presence of these partially reduced oxides allows the reduction–oxidation cycles of catalytic surfaces to proceed more rapidly and smoothly. Additionally, gas-phase radical initiators—that is, nitrogen oxides—strongly enhance catalyst performance. The performance of V_2O_5/SiO_2 catalyst is significantly altered by adding small amounts of NO in the gas feed [47]. The yield to C1 oxygenates reaches 16% at atmospheric pressure, which is between the highest values reported. The strong effect of NO appears to be due to the chain propagation of the radical reactions in the close vicinity of the catalyst bed. The reactions induced by radicals are strongly affected by the reaction conditions. In particular, the methyl-methylperoxo ($CH_3 \cdot / CH_3O_2 \cdot$) radical equilibrium depends on the concentration of oxygen and on temperature and is shifted to O-containing radicals as temperature decreases.

A limited number of studies on the use of metal-oxide catalysts for the direct methane to methanol conversion have been published recently, confirming that most researchers do not consider this route feasible. A recent development has been reported by Carlsson et al. [48]. They have studied the catalytic partial oxidation of methane under periodic control conditions, that is, deliberate changes of the gas composition, over iron molybdate catalyst, a typical formaldehyde synthesis catalyst. The study was made feeding low concentrations of CH_4 and thus quite far from conditions relevant for application. In addition, the results were not quantified but reported only as arbitrary mass spectrometer intensity. The experiments were made by pulsing of oxygen (0.25 vol%) to a constant concentration of CH_4 (1.0 vol%) on a $Fe_2(MoO_3)_4$ catalyst. Formaldehyde forms in these experiments, indicating that transient operation of the gas mixture can be a potential method to control product selectivity.

A recent DFT theoretical study [49] clarifies some aspects related to the question whether metal-oxo species are the key active species. To answer this questions, they have performed DFT calculations to study systematically the methane-to-methanol reaction catalyzed by $MO(H_2O)_p^{2+}$ complexes (M = V, Cr, Mn, Fe, Co, p = 5 and M = Ni, Cu, p = 4) in gas phase. Methane-to-methanol reaction follows a rebound mechanism in two steps (Figure 1.12a): (1) H abstraction leading to a $(MOH)^{2+}$ species and (2) carbon radical collapse onto this species.

Figure 1.12b reports the *lowest lying acceptor* (LLA) orbital for Mn and Fe complex, which determines the electrophilic properties of the MO^{2+} moiety. These LLA can be either a $\sigma*$ orbital or a $\pi*$ orbital. These orbitals are illustrated with plots of the $3\sigma*(\alpha)$ and the $2\pi*(\alpha)$ for Mn and Fe complexes in Figure 1.12b.

The first step in the methane-to-methanol reaction (Figure 1.12a) is the kinetic controlling step for all the complexes studied by Michel and Baerends [49]. The activation barrier and the transition state geometry of this H abstraction step are directly

(a) R   RC   I   PC   P

(b)

FIGURE 1.12 (a) H abstraction/O rebound mechanism scheme. R is the reactants. RC is the reactants complex. I is the intermediate. PC is the products complex. P is the products. (b) $2\pi x*(\alpha)$ (left side) and $3\sigma(\alpha)*$ (right side) molecular orbitals of the $MnO(H_2O)_5^{2+}$ and $FeO(H_2O)_5^{2+}$ in the ground state. (Adapted from Michel, C. and Baerends, E.J., *Inorg. Chem.*, 48, 3628, 2009.)

correlated to the nature of lowest acceptor orbital. The calculated energy profiles for the V, Cr, Mn, Fe, Co, Ni, and Cu complexes indicate the following main aspects:

- The σ^* controlled H abstraction reactions present linear transition states and π^* controlled H abstraction reactions present bent transition states.
- The σ^* controlled reactions present lower activation barriers than the π^* controlled reactions for a given lowest acceptor orbital energy.
- The activation barrier is directly correlated to the lowest vacant orbital energy provided the Pauli repulsion remains the same (same number of ligands, similar number of electrons) and provided the lowest acceptor orbital remains of the same type (σ^* or π^*).

Iron is so special because of the nature of its lowest acceptor orbital: a low-lying $3\sigma^*(\alpha)$, which is particularly efficient in promoting the H abstraction step. However, the Fe complex does not emerge as the only one from the study by Michel and Baerends [49]. The cobalt complex should be as efficient as iron for the Fenton chemistry under the proper experimental conditions. Indeed, cobalt-based compounds have already been used successfully in oxidation of cyclohexane and Co-zeolites are known to be active in methane conversion.

Another question concerns the charge effect. Let us take two isoelectronic complexes: $FeO(H_2O)_5^{2+}$ and $MnO(H_2O)^{5+}$. The electronic structures of those two complexes are very similar: same spin state (quintet), same lowest acceptor orbital ($3\sigma^*(\alpha)$). However, the charge decrease from the Fe to the Mn complex induces orbitals lying higher in energy: $\varepsilon 3\sigma^*(\alpha)$ $(FeO(H_2O)_5^{2+})$ = -13.9 eV and $\varepsilon 3\sigma^*(\alpha)$ $(MnO(H_2O)^{5+})$ = -6.7 eV. As expected, the activation barrier of the H abstraction step catalyzed by $MnO(H_2O)^{5+}$ is much higher than the one catalyzed by $FeO(H_2O)_5^{2+}$ (ΔE_1 (Mn) = 105.3 kJ mol^{-1} vs. ΔE_1 (Fe) = 9.2 kJ mol^{-1}). The late transition metal complexes should be less sensitive to this charge effect. Indeed, with a lowest acceptor orbital lying at -9.7 eV, the $CuO(H_2O)^{3+}$ complex seems to be a promising active species. CuO^+ has already been postulated in some enzymes and biologically relevant systems as a possible intermediate [50]. As commented later, iron and copper zeolites are particularly interesting for methane oxidation to methanol. Thus, this study provides new insights on their activity.

The results also indicate that the MnO^{2+} moiety is a candidate to perform C–H hydroxylation. As commented and remarked earlier, for example, by Takanabe and Iglesia [51], Mn is a critical element for methane activation in OCM reaction. In addition, Chen et al. [52] reported the use of MnO_2 as catalyst investigated for the direct conversion of methane in liquid phase in the frame of an UOP LLC (Honeywell) project. However, it is unclear whether the recent Honeywell announcement of a breakthrough in methane liquid-phase oxidation to methanol is related to this type of catalysts.

1.2.2.3   Catalytic Conversion: Zeolites and Related Materials

Transition metal ion–containing zeolites possess unique properties in methane selective oxidation. Panov and coworkers [53] first reported the gas-phase selective oxidation of methane to methanol on iron-containing ZSM-5 catalysts. The authors showed that Fe-containing zeolites (with MFI topology) were able to activate methane at ambient temperatures using nitrous oxide (N_2O) as an oxidant. It was found that following thermal autoreduction at 900°C, Fe-containing ZSM-5 could be activated with N_2O at 250°C, leading to the formation of a highly oxidative α-oxygen species [54]. Along with oxidizing benzene to phenol at high yield and selectivity, it was also found that this catalyst was also able to form a surface-bound methoxy species through interaction with methane at only 25°C.

While high conversions are observed (80%) and greater than 90% methanol selectivity achieved, industrial application of this process is not possible given the nature of the oxidation product, a chemisorbed methoxy species ($H_3C-O-_{(ads)}$). While phenol is sufficiently thermally stable to desorb from the catalyst surface under steady-state/continuous reaction conditions (250°C), the methoxy species undergoes deeper oxidation at such conditions, and high selectivities to CO_x are observed [55]. To obtain any significant methanol selectivity, the *N_2O-activated* catalyst must be first cooled prior to the introduction of methane. However, under these mild conditions, the reaction product does not desorb from the solid surface, and no methanol is detected at the reactor outlet. To liberate methanol from the solid matrix, a third hydrolysis/extraction step with a suitable solvent is required; as such, the zeolite is simply used as an oxygen-storage material. The solvent extraction also forces an additional regeneration cycle to be added between catalytic cycles. In addition, the use of N_2O as oxidation makes the process costly.

This approach to methane oxidation was later investigated by other groups [56] and extracted methanol yields upward of 208 μmol g^{-1} were later obtained (notably for 895 μmol Fe, CH_3OH/Fe ~ 0.25). However, a steady-state catalytic process that yields partial oxygenated species at any significant selectivity has still not been reported.

Significant progresses have been made recently using copper-containing zeolites. In 2005, Groothaert et al. [57] reported the ability of Cu-exchanged ZSM-5 to activate methane at relatively low reaction temperatures (125°C) (Figure 1.13).

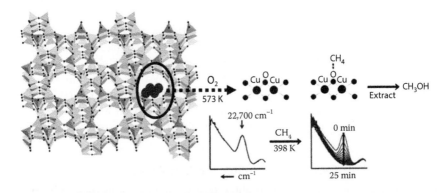

FIGURE 1.13 Simplified scheme of the formation of methanol from methane on Cu-ZSM5 and of the characterizing bands observed by Groothaert et al. [57].

There are some common aspects to the Fe-MFI zeolite chemistry. Treating Cu-ZSM-5 in O_2 at high temperatures (>500°C) leads to the formation of an oxygenated Cu center. The latter can later interact with methane at lower temperatures (<200°C). Again, no methanol was detected at the reactor outlet, most likely due to strong chemisorption of the reaction intermediate on the solid surface. In further analogy to the Fe-ZSM-5/N_2O system, it was again found that thermal desorption leads exclusively to CO/CO_2, and that additional solvent extraction or hydrolysis is a prerequisite for selective product formation to be observed, albeit at the destruction of the active site. However, this breakthrough paper [57] shows that (1) it is possible to operate with O_2 rather than N_2O and (2) the temperatures between the steps are more compatible to develop a process based on cyclic movement of the catalyst between the different temperature zones, especially if the further development in catalyst can reduce the temperature gap between the different steps and increase the productivity (in the 10 µmol g⁻¹ range, thus lower than for Fe-zeolites) catalyst.

Various papers have been published recently on this topic. Alayon et al. [58] have shown that also copper-exchanged mordenite (Cu-MOR) is active at 200°C in the methane conversion to methanol. The amount of methanol that could be extracted from Cu-MOR (calcined in O_2 at 450°C and then reacted with methane at 200°C), determined by taking out the material after methane interaction and stirring in water at r.t., was 13 µmol MeOH per g Cu-MOR, thus well comparable with the value of 11 µmol g⁻¹ using a 1:1 water/acetonitrile extraction medium reported earlier [57]. Vanelderen et al. [59] analyzed the dependence of the amount of methanol formed as a function of the Cu/Al ratio in Cu-ZSM-5 (Figure 1.14a) and put in relation with the formation a new absorption band at 22,700 cm⁻¹.

Much of the recent studies were focused on identifying this spectroscopic feature and the type of copper active sites responsible for the mechanism of methane activation. A multitude of spectroscopic and theoretical techniques have been used to probe the Cu species present within the catalyst. From these investigations, a number of mono- and binuclear oxo- or peroxo-bridged Cu

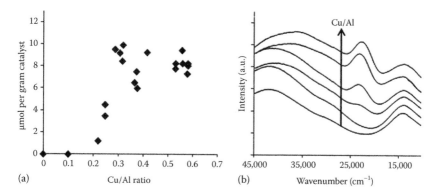

FIGURE 1.14 (a) Amount of methanol extracted per gram of Cu-ZSM-5 sample, as a function of the Cu/Al ratio. (b) UV-vis-NIR absorption spectra of the of Cu-ZSM-5 samples (Si/Al = 12) activated in O_2 at 450°C. (Adapted from Vanelderen, P. et al., *J. Catal.*, 284, 157, 2011.)

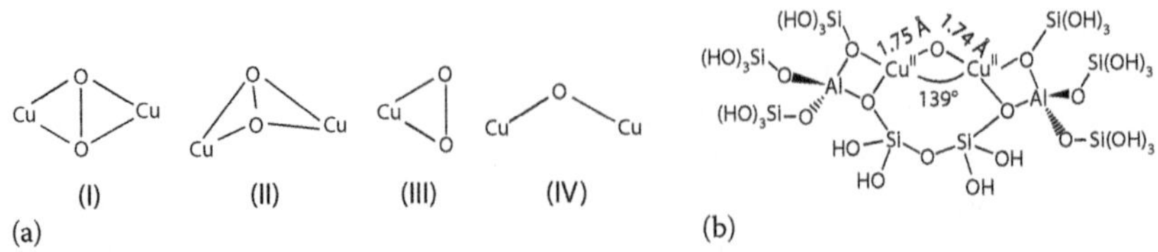

FIGURE 1.15 (a) Mononuclear and binuclear copper-oxo species proposed to be the active site within Cu-ZSM-5 for the stoichiometric aerobic oxidation of methane. (b) Large structural model constructed from part of the 10 MR (member ring) of Cu-ZSM-5 used for DFT calculations of the Cu_2O intermediate. (Adapted from Vanelderen, P. et al., *J. Catal.*, 284, 157, 2011; Hammond, C. et al., *ChemSusChem*, 5, 1668, 2012.)

species have so far been proposed and/or identified (Figure 1.15a, I–IV) and assigned as the active species of the material.

Recent research has proposed that a bent mono(μ-oxo) dicopper center (IV) (Figure 1.15b) is the key core present within Cu-exchanged ZSM-5 [61], although this is contentious and still the subject of much debate. It has been proposed [62] that the high O–H bond strength in the so-formed Cu–O(H)–Cu species renders the initial C–H abstraction step endothermic by only 13 kcal mol^{-1}. Himes and Karlin [62] also remarked the biomimetic characteristics of these systems (and those earlier reported as the Shilov Pt catalysts) in comparison to monooxygenase enzymes (sMMO employs a high-valent nonheme dinuclear iron (IV) site to activate dioxygen to oxidize methane (Figure 1.16b.B); for pMMO, there is still debate, but trinuclear Cu clusters are probably the active centers).

A theoretical picture of methane oxidation at copper-oxo sites emerges in which unpaired electron density roughly localized on a Cu-bound oxo ligand may be critical for reactivity (Figure 1.16a). Mononuclear cupryl or one-electron-reduced Cu^{II}–(O)(O(H))–Cu^{III} dinuclear species (calculated to be significantly more reactive toward hydrocarbon bonds than its Cu^{III}–Cu^{III} congener) fit the bill. The presence of other species appears to be critical for the O_2 reduction mechanism. Probably, the four-electron reduction of O_2 by the two active-site copper ions is followed by e– transfer from two spectator Cu^I ions.

Additionally, it has been proposed that asymmetric electron density surrounding the monobridged dicopper center facilitates a lower-energy reaction pathway via the formation of a Cu-oxyl (cupryl) radical center. Although not yet experimentally observed, such cupryl centers are believed to possess extreme reactivity toward C–H bonds [62]. This is confirmed from the earlier comments on the results of the study by Michel and Baerends [49]. A recent DFT study by Fellah and Onal [63] has investigated the C–H bond activation of methane on $[(SiH_3)_4AlO_4(M, MO)]$ (where M = Ag, Au, Cu, Rh, and Ru) cluster models representing ZSM-5 surfaces. The following activity order of clusters with respect to their activation barriers could be indicated: Au ≫ Rh > Cu = Ru > Ag for metal-ZSM-5 clusters and Ag > Cu > Au ≫ Rh > Ru for metal–O–ZSM-5 clusters. Therefore, activation barriers based on transition-state calculations showed that Ag–O–, Cu–O–, and Au–O–ZSM-5 clusters (4, 5, and 9 kcal mol^{-1}, respectively) are more active than all the other clusters for C–H bond activation of methane. The same group [64] has made DFT study of C–H bond activation of methane on a bridge site of M–O–M–ZSM-5 (M = Au, Ag, Fe, and Cu) clusters. Activation barriers for

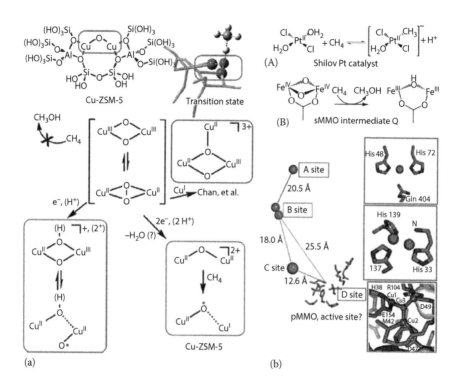

(a)

(b)

FIGURE 1.16 (a) Hypothesized methane-oxidizing copperoxy species. (*Upper*) The Cu-ZSM-5 active site model and C–H abstraction transition state. (*Lower*) Proposed evolution of methane-oxidizing Cu-O cores by reduction of species inert toward alkane C–H. (b) Synthetic and biological examples of methane-oxidizing metal species (A) and possible pMMO active sites, as determined by x-ray crystallography or modeling (B). sMMO, soluble methane monooxygenase; pMMO, particulate methane monooxygenase. (Adapted from Himes, R.A. and Karlin, K.D., *PNAS*, 106, 18877, 2009.)

C–H bond activation of methane on Au–O–Au–ZSM-5 and Ag–O–Ag–ZSM-5 clusters were calculated as 4.83 and 4.79 kcal mol^{-1}, respectively. These values are lower than the activation barrier values for C–H bond activation on Cu–O–Cu–ZSM-5 and Fe–O–Fe–ZSM-5, which are 9.69 and 26.30 kcal mol^{-1}, respectively. Activation process is exothermic on Au–O–Au–, Cu–O–Cu–, and Fe–O–Fe–ZSM-5 clusters, whereas it is endothermic on Ag–O–Ag–ZSM-5 cluster.

The problem of active and spectator Cu species in the oxidation of methane to methanol was analyzed by Weckhuysen and coworkers [65]. They commented that Cu–O species on the outer surface are not involved in the oxidation reaction, while copper inside the channels are responsible for the selective conversion of methane to methanol. However, they have not provided more insights about the role of the different copper species inside the zeolite cavities. When copper ions are only inside the channels, quantitative estimations indicate that only around 5%–10% of the copper sites (which are already low; the copper loading is 1–2 wt.%) are active. Thus, the key for the development is the identification of how to maximize these sites and the identification of the possible cocatalytic role of the other copper species, for example, in O_2 activation, as commented earlier.

DFT calculations to examine the energetics of the dioxygen intermediates inside a 10-ring of Cu–ZSM-5 showed that the properties of the $O_2 \cdots Cu_2$ complexes [66], such as the O_2 bridging modes and O_2 activation, are strongly affected by the locations of the 2 Al atoms within the 10-ring. In particular, the $O_2 \cdots Cu_2$ complexes have either end-on or side-on bridging modes depending on the substituted Al positions. On the other hand, the steric hindrances of a ZSM-5 cavity determines the properties of the *bis*(μ-oxo)dicopper complexes containing a diamond Cu_2O_2 core. By restricting its Cu_2O_2 core to a 10-ring of ZSM-5 in which the 2 Al atoms are second-nearest neighbors, each Cu cation is tetrahedral. On the other hand, the Cu cations have almost square planar coordination inside ZSM-5, where the Al atoms are the fourth-nearest neighbors.

The different Cu coordination environments are responsible for the different levels of stability. The planar diamond Cu_2O_2 core is about 31 kcal mol^{-1} more stable with respect to the tetrahedral case. Since the ZSM-5 nanospaces directly influence the stability of the *bis*(μ-oxo)dicopper complexes by changing the Cu coordination environments, zeolite confinement effects on the *bis*(μ-oxo) dicopper complexes are more noticeable than those in the $O_2 \cdots Cu_2$ cases. This also indicates that the spatial constraint from the ZSM-5 should significantly contribute to the stability of the reaction intermediates formed during the dioxygen activation. Kuroda and coworkers [67], studying the C–H activation of CH_4 in CuMFI zeolite, showed that there is a specific interaction between Cu^+ in MFI and CH_4, even at temperatures around 300 K. However, single copper ions have been only considered. As shown in Figure 1.16a, the dicopper sites may further transform to copper-oxyl (radical), that is, *cupryl* character, the species described by Kuroda and coworkers [67] and other authors [68] to have high reactivity toward C–H bonds.

It is still not conclusively clarified why the active sites in Cu–ZSM-5 catalysts selectively oxidize methane to methanol, neither what is the nature of the reaction cycle. Remarkable, however, is the analogy with enzymes (MMO), which are able to oxidize CH_4 to CH_3OH at room temperature and under atmospheric pressure [69–71]. Shilov and Shteinman [72] have made an extensive review recently on the biomimetic approach to methane hydroxylation.

From the application point of view, the issues of Cu–ZSM-5 methane oxidation to methanol are the need to operate at different temperatures to complete the cycle and especially the low productivity. There are new methods to control the location of Al pairs by synthesis, at least partially, but this method has still not been applied to control the nature and structure of transition metal ions in zeolites. There is the possibility to tune these materials and optimize their performances to develop novel industrial catalysts for this challenging reaction of direct methane transformation to methanol, which can open new frontiers in sustainable energy and chemistry. Martinez et al. [24] and Cejka et al. [73] have recently discussed the new opportunities offered from zeolite and micro-/mesoporous materials in methane conversion.

1.2.3 Nonoxidative Conversion

In the absence of an oxidant, the activation of the C–H bonds of the highly stable CH_4 molecule requires high temperatures. Under such conditions, radical reactions in the gas phase prevail. However, as the strength of the C–H bond in the resulting C-containing products is weaker than methane, the products will be more reactive than methane. This means that the challenge in methane conversion is related to selectivity rather than to reactivity. In order to circumvent theses hurdles, several different approaches based on catalysis and reaction engineering have been proposed and tested:

- Methane cracking, to produce CO-free hydrogen and elemental carbon [74]
- Combination of methane conversion processes (pyrolysis and hydrogenation, aromatization and steam reforming)
- Methane aromatization (also indicated as catalytic pyrolysis of methane [27] or methane dehydroaromatization [DHA]) [75]

The latter will be the focus of the discussion in the following sections.

1.2.3.1 General Aspects on Methane Aromatization

Various reviews have been published on this topic. In addition to those that have included this topic in discussing the conversion of methane [15,27,76,77], more specific reviews are also present. Ismagilov et al. [75] have analyzed recent development of methane DHA over Mo/ZSM-5 catalysts. This review focuses on the range of issues dealing with the effect of catalyst composition, preparation, pretreatment, and operation conditions on the physicochemical properties and activities of Mo/ZSM-5 catalysts in DHA reaction. The concepts of the reaction mechanism and the nature of the active molybdenum forms are reviewed. Various aspects of the Mo/ZSM-5 deactivation under reaction conditions and methods of their regeneration are also discussed, as well as some approaches for improvement of the Mo/ZSM-5 performance. Xu et al. [78] have also reviewed the use of Mo/zeolites in methane aromatization.

Guo et al. [79] have analyzed the advances in recent research on nonoxidative aromatization of methane in the presence of propane over different modified H-ZSM-5 catalysts. Aspects included are the thermodynamics of the reaction (an important aspect being methane aromatization, a reversible reaction) and the mechanism of reaction, with the focus on the influence of propane on methane activation. To be mentioned also is the recent review of Bao and coworkers [80], which is focused on the in situ solid-state NMR for heterogeneous catalysis and which presents the progresses in understanding of methane aromatization catalysts by this technique, in addition to catalysts for olefin selective oxidation and metathesis.

Methane can be dehydrogenated on catalysts in the absence of gas-phase oxygen to produce highly value-added chemicals such as benzene and other aromatics and hydrogen. Although the reaction takes place at temperatures lower than those applied in thermal pyrolysis, the yield is limited by thermodynamics, being a reversible

TABLE 1.4

Catalytic Performance of H-ZSM-5 Supported Re and Mo Catalysts

		Selectivity (%)			
Catalyst	Conv. CH$_4$ (%)	C2	Benzene	Naphthalene	Coke
5% Re	7	4	48	11	33
2% MoO$_3$	6	4	50	—	43
2% Mo	9	3	57	15	15
4% Mo	10	2	65	18	3

Source: Adapted from Holmen, A., *Catal. Today*, 142, 2, 2009.

Temperature: 973 K (950 K for 4% Mo); pressure 0.1 MPa (0.3 MPa for 5% Re). Time of reaction: 1–2 h.

reaction. The CH$_4$ conversion at equilibrium under 1 bar and 700°C is about 12.2% and almost double at 800°C.

$$6CH_4 \rightarrow 9H_2 + C_6H_6 \ (\Delta H°_{298K} = +88.7 \text{ kJ mol}^{-1} \text{ CH}_4) \tag{1.12}$$

Early studies on this reaction were about 20 years ago, but still the original bifunctional Mo/H-ZSM-5 catalyst remains between the most active and selective in this reaction, although catalyst lifetime has been improved by a better design. Mo/ZSM-5 remains the most widely used catalysts, but various other bifunctional catalysts consisting of narrow pore zeolites and a metal oxide MO$_x$ (M = Mo, W, V, Cr) phase have been also developed, with a trend in the activity Mo > W > Fe > V > Cr.

Some typical experimental results are shown in Table 1.4.

1.2.3.2 Reaction Mechanism and Nature of the Active Sites

The mechanism of the reaction seems to involve the conversion of CH$_4$ to C$_2$H$_4$ (C$_2$H$_2$) on molybdenum carbide or oxycarbide and further conversion of C$_2$H$_4$ (C$_2$H$_2$) to aromatic products over the acidic sites within the channels of the zeolite. A recent DFT study [81] has proven that Mo$_2$(CH$_2$)$_5$/ZSM-5 is the effective active center for methane activation. The C–H bond dissociation occur on the π orbital of Mo = CH$_2$ with an activation energy of 106 kJ mol^{-1} (Figure 1.17).

Recent studies using ^{95}Mo NMR have demonstrated that the carburized molybdenum species originating from the exchanged Mo species are the active centers [82]. They showed that (1) a new species centered at about −250 ppm in ^{95}Mo NMR can also be observed, attributed to an exchanged molybdenum species (molybdenum atom that migrates into zeolitic channels and anchor on Brönsted acid sites), (2) all of the carburized molybdenum species originated from this species, and (3) a linear relationship is present between this species aromatics formation rate, while not for MoO$_3$ crystallites and total amount of Mo (Figure 1.18). This demonstrates that the carburized molybdenum species originating from the exchanged Mo species are the active centers for the methane DHA reaction.

According to the high-resolution transmission electron microscopy (HRTEM) data [83], molybdenum oxide is located on the external surface of the impregnated 2% Mo/ZSM-5 catalyst, in the form of clusters 1–5 nm in size. When molybdenum

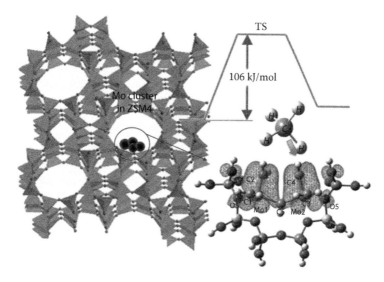

FIGURE 1.17 Proposed active site of Mo/ZSM-5 catalysts for methane aromatization. (Adapted from Xing, S. et al., *Chin. J. Catal.*, 31, 415, 2010.)

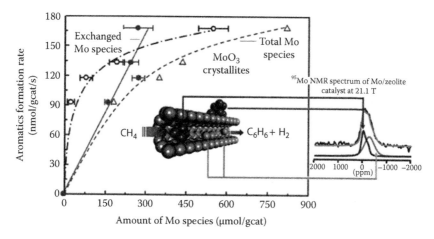

FIGURE 1.18 Correlating the aromatics formation rate with different molybdenum species. On the top, view of the model of migration of Mo species inside the channels of zeolite and ^{95}Mo NMR-related spectra. (Adapted from Zheng, H. et al., *J. Am. Chem. Soc.*, 130, 3722, 2008.)

concentration increases over about 10%, molybdenum oxide particles with dimensions as high as 100 nm appear on the external zeolite surface. During treatment in methane or hydrogen–methane mixture at about 700°C for the activation, MoO_3 first reduces to MoO_2, followed by the formation of a stable molybdenum carbide phase, *hcp* β-Mo_2C, supported on the zeolite. A treatment with a hydrogen–butane mixture resulted in the MoO_3 reduction to a metastable molybdenum carbide phase *fcc* α-MoC_{1-x}/ZSM-5 with preferential formation of molybdenum oxycarbide ($MoO_xH_yC_z$) at the intermediate stages [75].

The Brønsted acid sites of the zeolite play a different role [84]. In addition to providing anchoring points, the molybdenum inside the channels of the ZSM-5 performs ethylene oligomerization and catalyzes the dimerization of benzene to naphthalene and coke; their density determines the anchoring mode of the molybdenum inside the zeolite channels: monomer at low Si/Al ratio and dimer at high Si/Al ratio, with the former being more active.

1.2.3.3 Deactivation

The catalysts are easily deactivated during on-stream operation. The activated methane molecules could readsorb and become anchored to the active Mo species to form coke. Further dehydrogenation and oligomerization of monocyclic aromatic products could also lead to the deposition of aromatic-type carbon species on the Brønsted acid sites. The carbon deposits associated with the Mo species are reactive and reversible; nevertheless, the coke formed on the Brønsted acid sites is inert and irreversible. The latter species is responsible for the deactivation of the Mo/H-ZSM-5 catalyst [85]. In the course of the reaction, carbon is accumulated on the catalyst surface and activity progressively drops. Graphitic carbon and additionally encapsulated carbon, originated mainly from the successive dehydrogenation and polymerization of CH_x radicals or naphthalene graphitization on the surface, were observed with the HRTEM.

Coke and large hydrocarbons are formed on the external surface inhibiting the formation of complexes inside the pores. Several methods have been proposed for increasing the catalyst stability such as post–steam treatment, adding CO/CO_2 to the feed stream (although different results have been obtained by CO/CO_2 treatment), and selective silanation of the external acid sites on H-ZSM-5 by means of large organosilane molecules [27]. It has also been claimed that, by careful design of the catalyst precursor, the activity and stability can be increased without adding extra CO_2 [86]. A pilot plant study has also been carried out in order to demonstrate the technology [87].

1.2.3.4 Coupling with Membrane

Several groups have used hydrogen selective transport membranes for coupling of methane. H_2 can thereby be removed by oxidation eliminating the thermodynamic constraints. Theoretical studies of the system have shown promising results [88]. A marked enhancement of the conversion of methane as well as of the formation of benzene, toluene, naphthalene, and hydrogen was observed when a membrane was used together with Mo/H-ZSM-5 [89]. However, the synthesis and operation of membranes are not straightforward. A problem also lies in the fact that eliminating H_2, the rate of deactivation becomes faster.

1.2.3.5 Ways of Improving the Methane Dehydroaromatization Process

Different ways to improve the performances were explored. The following are from the analysis of Ismagilov et al. [75]:

- *Addition of a second metal in the zeolite matrix in addition to molybdenum.* Addition of copper (Mo/Cu/ZSM-5) improves activity and stability. The presence of copper increased the concentration of Mo^{5+} ions in the catalyst and decreased the zeolite dealumination rate and its carbonization, thus increasing the catalyst lifetime. The character of the carbonaceous deposits also changed: their oxidation temperature decreased, and there were more carbon radicals in the sample. The addition of Co, W, Zr, or Ru to Mo/ZSM-5 catalysts was shown to increase their activity and selectivity. The use of platinum increases the stability of Mo/ZSM-5 catalyst due to the lower concentration of carbonaceous deposits. The addition of Zn or La was also found to decrease the carbonization rate of Mo/zeolite catalysts.

- *Dealumination* of the parent zeolite is an alternative way to reduce the formation of the carbonaceous residues during reaction. The zeolite dealumination decreases the concentration of the carbonaceous deposits. The selectivity to the carbonaceous deposits was observed to decrease from 37.9% to 18.2% over dealuminated Mo/ZSM-5 catalyst at the same methane conversion (9.5%), whereas the yield of aromatic hydrocarbons increased by 32%.

- *Creation of an hierarchic pore structure*, by controlled desilylation, in order to reduce mass transfer limitations. Cui et al. [90] showed recently that depending on the space velocity, three rate-controlling regions are present in Mo/ZSM-5 catalysts for methane aromatization: external mass transfer, intracrystalline diffusion, and kinetic desorption controlling steps of the reaction.

- The *effect of silanation* (with 3-aminopropyl-trietoxysilane) of the external Si–OH groups on H-ZSM-5 zeolite crystals, to avoid the building of carbon-type species on the external surface.

- *Adding small amounts of CO, CO_2, O_2, or H_2O to the feed*, to reduce the rate of carbon formation. The positive effect was observed only in a narrow concentration range of the added reagent.

- *Improvement of process technology and engineering.* In addition to the cited use of membranes, it is possible to combine the aromatization of methane with that of propane, as indicated in the cited review of Guo et al. [79]. Multiple recycling of CH_4 with collection of the aromatic product was also investigated to increase the overall methane conversion [91]. If the reaction was carried out under flow recycling conditions, the benzene formation rate improved up to four times. The benzene yield increased when the circulating factor was increased.

1.2.3.6 Advances in Catalysts

In addition to ZSM-5, interesting performances have been shown by MCM-22 and derived materials. Xu et al. [92] reported how ITQ-2-like (derived by MCM-22 precursor by treatment with H_2O_2) shows interesting performances in methane aromatization. Gu et al. [93] have developed structured MCM-22/silicon carbide (SiC) catalyst for methane aromatization. Chu et al. [94,95] have reported the synthesis and an enhanced stability of nestlike hollow hierarchical MCM-22 microspheres (MCM-22-HS) (Figure 1.19). The MCM-22-HSs were obtained by one-pot rotating hydrothermal synthesis using carbon black as a hard template via self-assembly combined hydrothermal

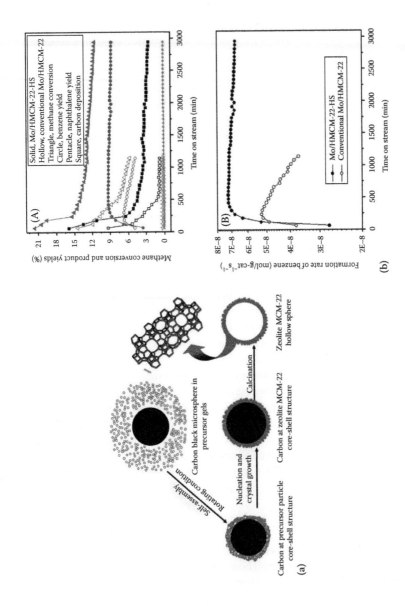

FIGURE 1.19 (a) Schematic representation of the growth process of the MCM-22-HS. (b): (A) Catalytic performances of Mo/HMCM-22-HS and conventional Mo/HMCM-22 catalysts for methane DHA reaction; (B) formation rates of benzene at 973 K on these two catalysts under space velocity of 1500 mL/(g h). (Adapted from Chu, N. et al., *Chem. Mater.*, 22, 2757, 2010.)

crystallization. The shell of MCM-22-HS was hierarchically constructed by many flaky MCM-22 crystals. The Mo/MCM-22-HS exhibited significantly improved methane conversion, yield of benzene product, and catalyst lifetime for methane DHA. The exceptional catalytic performance was attributed to the hollow and hierarchical structure. In general, working on catalyst pore structure and hierarchy is one of the research directions to improve the performances and stability [94–98].

Many recent studies focused on the use of alternative transition metals to Mo or the addition of a second transition metal to Mo/zeolite catalysts. Among these studies, the following are worth citing. Xu et al. [99] have shown that the addition of Fe to Mo/H-ZSM-5 catalyst promotes stability in methane aromatization. Over short time frames in the continuous CH_4 feed mode, the formation of Fe-induced carbon nanotubes is the origin of its promotional effect, while under the long-term periodic CH_4–H_2 switch operation, the catalytic involvement of Fe in the surface coke removal during the H_2 flow periods is the main cause of Fe-modified catalyst being much more stable than the unmodified Mo/H-ZSM-5 catalyst. The same group [100] have also investigated various 0.5 wt.% M-5 wt.% Mo/H-ZSM-5 catalysts (M = Cr, Mn, Fe, Co, Cu, Zn, Ru or Pd), prepared by coimpregnation, for the nonoxidative methane DHA under required severe conditions of 1 atm, 1073 K, and 10,000 mL g^{-1} h^{-1}, in periodic CH_4–H_2 switch operation mode.

Liu et al. [101] have investigated the catalytic performance of Zn-based/H-ZSM-5 catalysts for the methane DHA under supersonic jet expansion (SJE). Under these conditions, the Zn/H-ZSM-5 catalyst exhibited high catalytic activity. A new reaction mechanism that involves an active $ZnO–CH_3^+...-HZnO$ intermediate formed as a result of synergetic action between ZnO and H-ZSM-5 has been proposed for CH_4 dissociation and dehydrogenation. Under atmospheric pressure, however, the catalytic activity of Zn/H-ZSM-5 is low. Co- and Zn-impregnated H-ZSM-5 catalysts were also shown by Liu et al. [102] to be excellent catalysts for the aromatization of methane by using propane as coreactant.

Zn-modified H-ZSM-5 zeolites are also investigated using solid-state NMR and theoretical calculations by Xu et al. [103] that indicate the possibility of room temperature activation of methane. They found for the first time that the activation of methane resulted in the preferential formation of surface methoxy intermediates at room temperature, which mediated the formation of methanol and its further conversion to hydrocarbons, in agreement with the indications by Liu et al. [101], although suggesting that an oxygen-containing dizinc cluster center in an open shell was responsible for homolytic cleavage of the C–H bond of methane at room temperature, leading to the formation of methyl radicals. The zeolite matrix readily trapped the methyl radicals by forming the surface methoxy intermediates for further selective conversion. In parallel to the homolytic cleavage pathway, heterolytic dissociation of methane was also observed on isolated Zn^{2+} ions at room temperature, which gives rise to zinc methyl species.

1.2.4 METHANE CONVERSION VIA HALOMETHANE

An alternative route for the conversion of methane to light olefins or oxygenates is the old route (known from three decades, but revised taking into account last

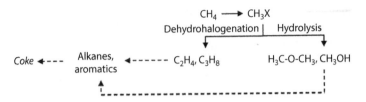

FIGURE 1.20 Potential reaction network for the conversion of methane to olefins via halagenoalkanes. (Adapted from Hammond, C. et al., *ChemSusChem*, 5, 1668, 2012.)

developments) involving the intermediate production of halomethanes (CH_3X) through the reaction between methane and a suitable halogen (typically Cl_2/Br_2, or HCl/HBr). Unlike OCM (direct OCM), whereby methyl radicals generated by the solid catalyst terminate and subsequently dehydrogenate to yield olefins, the reaction between methane and a suitable halogen can yield olefins either through the DHHC (dehydrohalogenative coupling) of two equivalents of monohalogenoalkanes or the condensation of methanol or dimethyl ether produced in situ through haloalkane hydrolysis (Figure 1.20).

Indeed, it has been shown that by substituting a methanol/dimethyl feedstock with a monohalogenated alkane, a similar final product distribution is observed over the industrial benchmark MTO (methanol-to-olefin) catalyst (H-SAPO-34) [104]. By feeding CH_3Br over a SAPO-34 catalyst, light olefins are the main reaction product, with the selectivity to propylene decreasing and to ethylene increasing with increasing temperature. However, there is increased methane and decreased higher hydrocarbons production, C_{4-6}, which is consistent with increased hydrocarbon cracking.

Brominated species, RBr, are decreased with increasing temperature as dehydrohalogenation takes place. Halogenated alkanes may be considered analogues substrates for methanol-based upgrading processes, for example, MTO. However, the use of halogens increase largely (1) fixed costs due to the need of special materials, and (2) operational costs, due to separation and halogenated products disposal. The advantage of this pathway, relative to direct OCM, is the relatively low energy barrier for the initial reaction step, that is, halogenation, which allows the reaction to proceed under milder reaction conditions.

As earlier cited, He et al. [12] have proposed a novel catalytic route for the conversion of methane to propylene via monohalogenomethane. Wei et al. [13] reviewed recently the conversion of methane to higher hydrocarbons via Me halide as the intermediate. They have indicated that high conversion efficiency and selectivity indicated the feasibility of industrial application. However, although haloalkanes provide an interesting alternative route for methane-to-olefin technology, some significant hurdles must be overcome.

Firstly, for significant light olefin selectivity to be observed, it is imperative that the first step of the reaction (halogenation) proceeds at exceptional selectivity. Indeed, it has been demonstrated that polyhalogenated alkanes are a major precursor to heavier hydrocarbons, such as long-chain alkanes and aromatics, and ultimately result in significant coking of the catalyst [105]. Attempts to isolate monohaloalkanes at high

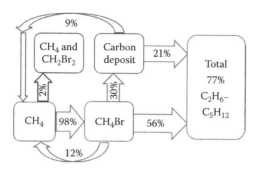

FIGURE 1.21   Flow diagram for a single-pass conversion of methane (without recycling results) into >75% C2–C5 hydrocarbons. (Adapted from Osterwalder, N. and Stark, W.J., *ChemPhysChem*, 8, 297, 2007.)

selectivity have so far involved the development of nonradical-based catalytic systems [106,107], to avoid the expected homogeneous free-radical chain reactions [108].

Lercher group [107] showed that mixtures of $LaOCl$ and $LaCl_3$ are promising catalysts for oxidative chlorination of methane to methyl chloride. Metal dopants with redox character such as Co and Ce, which form stable chlorides under reaction conditions, lead to a higher rate of methane conversion, but the formed methyl chloride is strongly adsorbed and directly oxidized to CO leading to low methyl chloride selectivity. Doping with nickel weakens, in contrast, the interaction with methyl chloride leading to high methyl chloride selectivity.

The alternative, but more costly, is the utilization of Br_2 as opposed to Cl_2 as halogen source; the weaker C–Br bond results in a more controlled halogenation pathway and typically higher monohalogenated selectivities, albeit at the expense of conversion [109]. Figure 1.21 shows how the direct bromination of methane offers a quite selective (>98%) route toward methane activation, but shifts the problem of fuel production to converting and handling corrosive methyl bromide. The utilization of Br_2 presents an additional advantage, given that the so-formed monohalogenated liquid product is significantly easier to handle than the gaseous Cl_2 analogue.

Alternatively, the use of HCl/HBr as halogen source decreases the propensity of gas-phase radical reactions based on the homolytic cleavage of the X–X (X = Cl or Br) bond [107]. To date, however, only low yields for halogenation have been achieved, although the DHHC step itself is much more facile.

An additional challenge within this field is the regeneration or recycling of the halogen source to yield a fully closed catalytic cycle. For example, in its most simple form, that is, the halogenation pathway to olefins, four equivalents of HX per mol of olefin are formed, which presents additional environmental (viz., low atom efficiency) and reactor design issues. The most interesting approach to overcome these issues involves the oxidative regeneration of the halogen, although an integrated process has not yet been exemplified.

$$2CH_4 + 2X_2 \rightarrow 2CH_3X + 2HX \tag{1.13a}$$

$$2CH_3X \rightarrow C_2H_4 + 2HX \tag{1.13b}$$

$$4HX + O_2 \rightarrow 2H_2O + 2X_2 \qquad (1.13c)$$

$$Net: 2CH_4 + O_2 \rightarrow C_2H_4 + 2H_2O \qquad (1.14)$$

An interesting novel development reported by He et al. [12] developed novel and efficient catalysts for a new two-step route for the production of propylene from methane via CH_3Cl or CH_3Br. CeO_2 is an efficient and stable catalyst for the oxidative chlorination and bromination of methane to CH_3Cl and CH_3Br. The catalytic properties of CeO_2 are dependent on its morphology or the exposed crystalline planes. The modification of CeO_2 nanocrystals by FeO_x or NiO enhances the selectivity of CH_3Cl or CH_3Br formation. For the second step, an F-modified H-ZSM-5 is highly selective and stable for the conversions of both CH_3Cl and CH_3Br into propylene. Representative results are shown in Figure 1.22 for both the oxidative chlorination

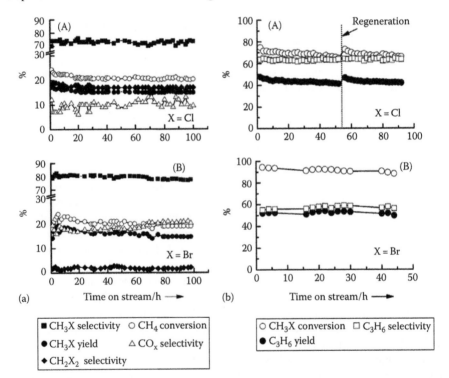

FIGURE 1.22 (a) Dependences of catalytic performances on time on stream for the oxidative chlorination of CH_4 over the 15 wt.% FeO_x–CeO_2 nanorods (a.A) and the oxidative bromination of CH_4 over the 10 wt.% NiO–CeO_2 nanocubes (a.B). Reaction conditions: (a.A) catalyst (0.50 g), T = 753 K, $CH_4/HCl/O_2/N_2/He$ = 4/2/1/1.5/1.5, total flow rate = 40 mL min^{-1}; (a.B) catalyst (1.0 g), T = 873 K, 40 wt.% HBr aqueous solution 4.0 mL h^{-1}, $CH_4/O_2/N_2$ = 4/1/1 (flow rate = 15 mL min^{-1}). (b) Catalytic performances of F-modified H-ZSM-5 versus time on stream for the conversions of CH_3Cl (b.A) and CH_3Br (b.B). Reaction conditions: (b.A) catalyst (0.30 g), T = 673 K, P(CH_3Cl) = 3.3 kPa, flow rate = 15 mL min^{-1}; (b.B) catalyst (0.10 g), T = 673 K, P(CH_3Br) = 9.2 kPa, flow rate = 11 mL min^{-1}. (Adapted from He, J. et al., *Angew. Chem. Int. Ed.*, 51, 2438, 2012.)

of CH_4 over the 15 wt.% FeO_x–CeO_2 nanorods (Figure 1.22a.A) and the oxidative bromination of CH_4 over the 10 wt.% NiO–CeO_2 nanocubes (Figure 1.22a.B), and for the catalytic conversion of CH_3Cl (Figure 1.22b.A) and CH_3Br (Figure 1.22b.B) over F-modified H-ZSM-5. The good stability has to be remarked.

The haloalkane approach is also a viable method of converting methane to methanol, as recently reported by Li and Yuan [110]. Unlike early reports, whereby haloalkanes were hydrolyzed in situ to yield methanol or dimethyl ether, in this case Br_2 acts as an oxygen shuttle, facilitating the partial oxidation of methane to methanol with molecular oxygen under the influence of dibromo(dioxo)molybdenum(VI) supported on Zn-MCM-48. The key to this process is Br_2 metathesis from the MoO_2Br_2/ MoO_3 and $ZnBr_2$/ZnO couples, which first yields the haloalkane intermediate via methane activation and subsequently rescavenges the halogen to regenerate the catalyst and yield the oxidized product (methanol or dimethyl ether for Mo and Zn, respectively).

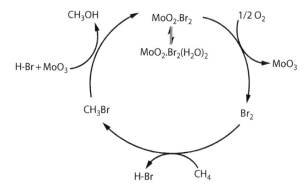

Although effective, some inherent limitations to the approach, for example, low methanol selectivity (maximum 50% at 14% conversion), the loss of Br_2 during the reaction, and poor Br_2 metathesis efficiency still require significant improvement.

Although haloalkane approaches are a promising method for the low-energy conversion of methane to partial oxygenates or olefins, the corrosive and environmental limitations associated with halogenated compounds, more facile coking, and the typically poor rates and selectivities observed render this approach rather unfavorable, particularly with respect to the optimal MTO process.

1.3 FINAL COMMENTS AND PERSPECTIVES

Although new economic opportunities and scientific developments have pushed a renewed interest on the direct methane conversion to chemicals (mainly light olefins, methanol, and aromatics) and fuels, there is still a number of technical hurdles. The indirect methane conversion routes, through the formation of syngas, are still the only commercial route. However, considering that, for example, in methanol process up to 40% of the cost is related to syngas production, it is clear that a direct route would be rather attractive.

Given that few catalytic systems of real industrial relevance have been developed, the greatest challenge facing this area of catalysis remains, therefore, the identification or development of new catalytic systems that may yet provide the required breakthroughs. For direct OCM in particular, it is a reality that an incredible number of systems and approaches have already been investigated. The best catalysts give a C2 yield of about 30%, which is above the 25% limit considered in the past as the breakthrough for industrial exploitation. However, a more realistic technoeconomic assessment would be required, in view of the large investments necessary and the moving political and economic scenarios. Being reasonably an OCM technology applicable in remote areas, where NG is available but not possible to transport via pipeline and through liquefaction/regasification (stranded NG resources), a technology to transport ethylene would be also necessary. Liquefied ethylene requires temperatures below $-104°C$. This is already a commercial technology, but further improvements are necessary.

In addition, OCM produces an ethane and ethylene mixture, with the former being often the more abundant. Selective dehydrogenation of ethane to ethylene in the OCM reactor mixture is still a challenge. Separation of the products and recycle of ethane together with methane and the insertion of specific ethane dehydrogenation sites in OCM catalyst would be preferable. However, besides that, in terms of catalyst design, this approach has two consequences. First, the use of methane + ethane feeds has to be evaluated (rather than methane only), and second the selectivity to ethylene, rather than the yield to C2 should be the technical target. These aspects are scarcely recognized in literature, but clearly a clue to progress in OCM reaction.

An effort has been made to use the unique wide range of catalytic data on OCM reaction to derive the best catalysts combinations, but based on C2 yield rather that ethylene yield. Thus, a revision of these results from this perspective would be welcome. Furthermore, it would be necessary to combine these indications with advances in understanding the reaction mechanism, to put on more solid bases the design of new catalysts.

A large effort has also been made on kinetic and microkinetic modeling of OCM reaction. This is certainly worthwhile, but the need is to use (and validate) this effort in scaling-up the reactors, because a large industrial reactor in the presence of homogeneous–heterogeneous reactions such as in the case of OCM is still a challenge. New reactor design from this perspective would be necessary, which also consider the severe safety problems present. This implies also to optimize catalyst design together with reactor design. Some of the results presented here also show that an optimal catalyst in a fixed bed reactor could not be ideal in a microreactor, for example. In the case of OCM, due to the peculiar mixed reaction mechanism, this effect would be even more critical. Optimal catalyst texture and porosity should be also optimized from this perspective. Catalyst stability is the further element of this mosaic. Therefore, even the large effort in catalysts development and testing, still effort would be necessary, but much more focalized on critical aspects with respect to the past.

Conceptually, new catalysts would also be welcome, and some of the efforts in this direction have been reviewed here, but it is difficult to expect really new breakthroughs after such a large R&D effort. However, the combination of OCM with membranes, for the possibility of distributed O_2 profile, or selective separation of one or more products could open interesting perspectives. More research in this direction is necessary. It is noteworthy that Honeywell (UOP) has recently announced to have proof of concept for a new catalyst/process to produce ethylene directly from methane, although details are unknown.

While ethylene is a valuable product, actual market would welcome propylene. Rather, minimal effort has been made to analyze the formation of C3 products rather than C2. Propane dehydrogenation to propylene would be also easier than ethane dehydrogenation.

Finally, the need of high temperature operations for OCM is a major issue because it makes the process energetically inefficient, aside from creating problems in materials. Low-temperature operations using nonthermal plasma coupled to catalysts would be interesting from this perspective, but still the developments are not satisfactory. While radical (homogeneous) plasma chemistry is known, the coupling of this chemistry with a solid catalyst, and the design of catalysts specific for this application (rather than use conventional catalysts as made up to now), is still at the early stage. An effort in this direction is necessary, before concluding whether or not this direction could be promising.

The alternative is the indirect process of methane pyrolysis to acetylene followed by the selective hydrogenation of the latter to ethylene. GTE process developed by *Synfuels* is interesting and claimed to be competitive, but a better energy and safety assessment of this route would be necessary together with more solid economic data.

A renewed interest in halomethane (via CH_3Cl or CH_3Br) route is present in literature, as well as in some companies such as Dow Chemical. Significant improvements have been made recently in catalysts for the selective conversion of halomethane to light olefins and to a less extent to produce the halomethane intermediates. However, selectivity data in both cases are still unsatisfactory, although they can be reasonably improved. However, the critical questions still remaining are how to (1) handle toxic and corrosive intermediates/products from an industrial perspective, (2) use Cl_2 or Br_2 (with related safety issues), (3) dispose halogenated by-products. An additional challenge within this field is the regeneration or recycling of the halogen source to yield a fully closed catalytic cycle. Stability of the catalysts has been demonstrated on a short term, but it is likely that longer-term deactivation is present.

Although haloalkane approaches are a promising method for the low-energy conversion of methane to partial oxygenates or olefins, the corrosive and environmental limitations associated with halogenated compounds, more facile coking, and the typically poor rates and selectivities observed render this approach still unsatisfactory, particularly with respect to the optimal MTO process.

OCM is thus a highly desirable procedure for its ability to cleanly and efficiently produce olefins in the absence of crude oil. However, given the stagnation in research, indirect routes—particularly via methanol and MTO technology—are becoming a more promising alternative. Thus, partial oxidation is envisaged to be the major area of research over the coming decade.

REFERENCES

1. Centi, G. and Perathoner, S. Recent developments in gas-to-liquid conversion and opportunities for advanced nanoporous materials. In *Nanoporous Materials: Advanced Techniques for Characterization, Modeling, and Processing*, Chapter 14, pp. 481–511, Kanellopoulos, N. (ed.). Boca Raton, FL: CRC Press, 2011.
2. EIA—U.S. Energy Information Administration. Annual Energy Outlook 2012, with Projections to 2035. DOE/EIA-0383 (June 2012). http://www.eia.gov/forecasts/aeo/pdf/0383(2012).pdf. Accessed on October 10, 2014.
3. IEA—International Energy Agency. Golden Rules for a Golden Age of Gas. World Energy Outlook—Special report on unconventional gas. IEA, 2012. http://www.worldenergyoutlook.org/media/weowebsite/2012/goldenrules/WEO2012_GoldenRulesReport.pdf. Accessed on October 10, 2014.
4. American Chemistry Council. Shale Gas, Competitiveness, and New US Chemical Industry Investment: An Analysis Based on Announced Projects, May 2013. http://chemistrytoenergy.com/sites/chemistrytoenergy.com/files/shale-gas-full-study.pdf. Accessed on October 10, 2014.
5. Hughes, J.D. *Drill, Baby, Drill: Can Unconventional Fuels Usher in a New Era of Energy Abundance?* Santa Rosa, CA: Post Carbon Institute, 2013.
6. Dallas, K. The story of ethylene... now starring natural gas. Cleantech Blog, 2011. http://www.cleantechblog.com/2011/11/the-story-of-ethylene-now-starring-natural-gas.html. Accessed on October 10, 2014.
7. Synfuel web site. Synfuels GTE Presentation, 2012. http://www.synfuels.com/WebsiteGasRichConferPaper.pdf. Accessed on October 10, 2014.
8. Xu, J., Zheng, A., Wang, X., Qi, G., Su, J., Du, J., Gan, Z., Wu, J., Wang, W., and Deng, F. Room temperature activation of methane over Zn modified H-ZSM-5 zeolites: Insight from solid-state NMR and theoretical calculations. *Chem. Sci.* 2012, 3, 2932–2940.
9. Zhou, D., Zuo, S., and Xing, S. Methane dehydrogenation and coupling to ethylene over a Mo/HZSM-5 catalyst: A density functional theory study. *J. Phys. Chem. C* 2012, 116, 4060–4070.
10. Vanelderen, P., Hadt, R.G., Smeets, P.J., Solomon, E.I., Schoonheydt, R.A., and Sels, B.F. Cu-ZSM-5: A biomimetic inorganic model for methane oxidation. *J. Catal.* 2011, 284, 157–164.
11. Solomon, E.I., Ginsbach, J.W., Heppner, D.E., Kieber-Emmons, M.T., Kjaergaard, C.H., Smeets, P.J., Tian, L. and Woertink J.S. Copper dioxygen (bio)inorganic chemistry. *Faraday Discuss.* 2011, 148, 11–108.
12. He, J., Xu, T., Wang, Z., Zhang, Q., Deng, W., and Wang, Y. Transformation of methane to propylene: A two-step reaction route catalyzed by modified CeO_2 nanocrystals and zeolites. *Angew. Chem. Int. Ed.* 2012, 51, 2438–2442.
13. Wei, Y., Zhang, D., Liu, Z., and Su, B.-L. Methyl halide to olefins and gasoline over zeolites and SAPO catalysts: A new route of MTO and MTG. *Cuihua Xuebao* (in English) 2012, 33, 11–21.
14. Hammond, C., Forde, M.M., Ab Rahim, M.H., Thetford, A., He, Q., Jenkins, R.L., Dummer, N.F. et al. Direct catalytic conversion of methane to methanol in an aqueous medium by using copper-promoted Fe-ZSM-5. *Angew. Chem. Int. Ed.* 2012, 51, 5129–5133.
15. Alvarez-Galvan, M.C., Mota, N., Ojeda, M., Rojas, S., Navarro, R.M., and Fierro, J.L.G. Direct methane conversion routes to chemicals and fuels. *Catal. Today* 2011, 171, 15–23.
16. Takanabe, K. Catalytic conversion of methane: Carbon dioxide reforming and oxidative coupling. *J. Jpn. Petrol. Inst.* 2012, 55, 1–12.
17. Arndt, S., Otremba, T., Simon, U., Yildiz, M., Schubert, H., and Schomaecker, R. $Mn-Na_2WO_4/SiO_2$ as catalyst for the oxidative coupling of methane. What is really known? *Appl. Catal. A Gen.* 2012, 425–426, 53–61.

18. Zavyalova, U., Holena, M., Schlögl, R., and Baerns, M. Statistical analysis of past catalytic data on oxidative methane coupling for new insights into the composition of high-performance catalysts. *ChemCatChem* 2011, 3, 1935–1947.
19. Arndt, S., Laugel, G., Levchenko, S., Horn, R., Baerns, M., Scheffler, M., Schlögl, R., and Schomaecker, R. A critical assessment of Li/MgO-based catalysts for the oxidative coupling of methane. *Catal. Rev. Sci. Eng.* 2011, 53, 424–514.
20. Kondratenko, E.V. and Baerns, M. Oxidative coupling of methane. In *Handbook of Heterogeneous Catalysis* (2nd edn.), Vol. 6, pp. 3010–3023, Ertl, G. (ed.). Weinheim, Germany: Wiley-VCH, 2008.
21. Kondratenko, E.V. and Baerns, M. Catalysis of oxidative methane conversions. *RSC Nanosci. Nanotechnol.* 2011, 19(*Nanostruct. Catal.*), 35–55.
22. Krebs, C., Bollinger Jr., J.M., and Booker, S.J. Cyanobacterial alkane biosynthesis further expands the catalytic repertoire of the ferritin-like 'di-iron-carboxylate' proteins. *Curr. Opin. Chem. Biol.* 2011, 15, 291–303.
23. Havran, V., Dudukovic, M.P., and Lo, C.S. Conversion of methane and carbon dioxide to higher value products. *Ind. Eng. Chem. Res.* 2011, 50, 7089–7100.
24. Martinez, A., Prieto, G.O., Garcia-Trenco, A., and Peris, E. Advanced catalysts based on micro- and mesoporous molecular sieves for the conversion of natural gas to fuels and chemicals. In *Zeolites and Catalysis*, Vol. 2, pp. 649–685, Cejka, J., Corma, A., and Zones, S. (eds). Weinheim, Germany: Wiley-VCH, 2010.
25. Sinev, M.Y., Fattakhova, Z.T., Lomonosov, V.I., and Gordienko, Y.A. Kinetics of oxidative coupling of methane: Bridging the gap between comprehension and description. *J. Nat. Gas Chem.* 2009, 18, 273–287.
26. Usachev, N.Y., Kharlamov, V.V., Belanova, E.P., Starostina, T.S., and Krukovskii, I.M. Oxidative processing of light alkanes: State-of-the-art and prospects. *Russ. J. Gen. Chem.* 2009, 79, 1252–1263.
27. Holmen, A. Direct conversion of methane to fuels and chemicals. *Catal. Today* 2009, 142, 2–8.
28. Kondratenko, E.V. Fundamentals of the oxidative conversion of methane to higher hydrocarbons and oxygenates. *DGMK Tagungsber.* 2008, 2008-3 (Preprints of the DGMK-conference "Future Feedstocks for Fuels and Chemicals"), 45–58.
29. Ren, T., Patel, M.K., and Blok, K. Steam cracking and methane to olefins: Energy use, CO_2 emissions and production costs. *Energy* 2008, 33, 817–833.
30. Salerno, D., Arellano-Garcia, H., and Wozny, G. Techno-economic analysis for the synthesis of downstream processes from the oxidative coupling of methane reaction. *Comp. Aid. Chem. Eng.* 2012, 31, 410–414.
31. Salerno, D., Arellano-Garcia, H., and Wozny, G. Techno-economic analysis for ethylene and methanol production from the oxidative coupling of methane process. *Comp. Aid. Chem. Eng.* 2011, 29, 1874–1878.
32. Bouwmeester, H.J.M. Dense ceramic membranes for methane conversion. *Catal. Today* 2003, 82, 141–150.
33. Tan, X. and Li, K. Inorganic hollow fibre membranes in catalytic processing. *Curr. Opin. Chem. Eng.* 2011, 1, 69–76.
34. Godini, H.R., Jaso, S., Xiao, S., Arellano-Garcia, H., Omidkhah, M., and Wozny, G. Methane oxidative coupling: Synthesis of membrane reactor networks. *Ind. Eng. Chem. Res.* 2012, 51, 7747–7761.
35. Thybaut, J.W., Sun, J., Olivier, L., Van Veen, A.C., Mirodatos, C., and Marin, G.B. Catalyst design based on microkinetic models: Oxidative coupling of methane. *Catal. Today* 2011, 159, 29–36.
36. Zavyalova, U., Geske, M., Horn, R., Weinberg, G., Frandsen, W., Schuster, M., and Schlögl, R. Morphology and microstructure of Li/MgO catalysts for the oxidative coupling of methane. *ChemCatChem* 2011, 3, 949–959.

37. Trionfetti, C., Babich, I.V., Seshan, K., and Lefferts, L. Presence of lithium ions in MgO lattice: Surface characterization by infrared spectroscopy and reactivity towards oxidative conversion of propane. *Langmuir* 2008, 24, 8220–8228.

38. Buyevskaya, O.V., Rothaemel, M., Zanthoff, H.W., and Baerns, M. Transient studies on reaction steps in the oxidative coupling of methane over catalytic surfaces of MgO and Sm_2O_3. *J. Catal.* 1994, 146, 346–357.

39. Myrach, P., Nilius, N., Levchenko, S., Gonchar, A., Risse, T., Dinse, K.-P., Boatner, L.A. et al. Temperature-dependent morphology, magnetic and optical properties of Li-doped MgO. *ChemCatChem* 2010, 2, 854–862.

40. Liu, H., Wei, Y., Caro, J., and Wang, H. Oxidative coupling of methane with high C2 yield by using chlorinated perovskite $Ba_{0.5}Sr_{0.5}Fe_{0.2}Co_{0.8}O_{3-\delta}$ as catalyst and N_2O as oxidant. *ChemCatChem* 2010, 2, 1539–1542.

41. Schmidt-Szalowski, K., Krawczyk, K., Sentek, J., Ulejczyk, B., Gorska, A., and Mlotek, M. Hybrid plasma-catalytic systems for converting substances of high stability, greenhouse gases and VOC. *Chem. Eng. Res. Des.* 2011, 89, 2643–2651.

42. Gesser, H.D. and Hunter, N.R. A review of C1 conversion chemistry. *Catal. Today* 1998, 42, 183–189.

43. Olah, G.A. and Surya Prakash, G.K. Selective oxidative conversion of methane to methanol, dimethyl ether and derived products. U.S. Patent 7,705,059, 2010.

44. Navarro, R.M., Pena, M.A., and Fierro, J.L.G. Methane oxidation on metal oxides. In *Metal Oxides: Chemistry and Applications*, Chapter 14, pp. 463–490, Fierro, J.L.G. (ed.). Boca Raton, FL: CRC Press, 2006.

45. Tabata, K., Teng, Y., Takemoto, T., Suzuki, E., Banares, M.A., Pena, M.A., and Fierro, J.L.G. Activation of methane by oxygen and nitrogen oxides. *Catal. Rev. Sci. Eng.* 2002, 44, 1–58.

46. Chempath, S. and Bell, A.T. A DFT study of the mechanism and kinetics of methane oxidation to formaldehyde occurring on silica-supported molybdena. *J. Catal.* 2007, 247, 119–126.

47. Barbero, J.A., Alvarez, M.C., Banares, M.A., Pena, M.A., and Fierro, J.L.G. Breakthrough in the direct conversion of methane into c1-oxygenates. *Chem. Commun.* 2002, 11, 1184–1185.

48. Carlsson, P.-A., Jing, D., and Skoglundh, M. Controlling selectivity in direct conversion of methane into formaldehyde/methanol over iron molybdate via periodic operation conditions. *Energy Fuels* 2012, 26, 1984–1987.

49. Michel, C. and Baerends, E.J. What singles out the FeO^{2+} moiety? A density-functional theory study of the methane-to-methanol reaction catalyzed by the first row transition-metal oxide dications $MO(H_2O)_p^{2+}$, M = V–Cu. *Inorg. Chem.* 2009, 48, 3628–3638.

50. Pitié, M., Boldron, C., and Pratviel, G. DNA oxidation by copper and manganese complexes. *Adv. Inorg. Chem.* 2006, 58, 77–1330.

51. Takanabe, K. and Iglesia, E. Rate and selectivity enhancements mediated by OH radicals in the oxidative coupling of methane catalyzed by $Mn/Na_2WO_4/SiO_2$. *Angew. Chem. Int. Ed.* 2008, 47, 7689–7693.

52. Chen, W., Kocal, J.A., Brandvold, T.A., Bricker, M.L., Bare, S.R., Broach, R.W., Greenlay, N., Popp, K., Walenga, J.T., and Yang, S.S. Manganese oxide catalyzed methane partial oxidation in trifluoroacetic acid: Catalysis and kinetic analysis. *Catal. Today* 2009, 140, 157–161.

53. Sobolev, V.I., Dubkov, K.A., Panna, O.V., and Panov G.I. Selective oxidation of methane to methanol on a FeZSM-5 surface. *Catal. Today* 1995, 24, 251–252.

54. Starokon, E.V., Dubkov, K.A., Pirutko, L.V., and Panov, G.I. Mechanisms of iron activation on Fe-containing zeolites and the charge of α-oxygen. *Top. Catal.* 2003, 23, 137–143.

55. Zecchina, A., Rivallan, M., Berlier, G., Lamberti, C., and Ricchiardi, G. Structure and nuclearity of active sites in Fe-zeolites: Comparison with iron sites in enzymes and homogeneous catalysts. *Phys. Chem. Chem. Phys.* 2007, 9, 3483–3499.

56. Wood, B.R., Reimer, J.A., Bell, A.T., Janicke, M.T., and Ott, K.C. Methanol formation on Fe/Al-MFI via the oxidation of methane by nitrous oxide. *J. Catal.* 2004, 225, 300–306.

57. Groothaert, M.H., Smeets, P.J., Sels, B.F., Jacobs, P.A., and Schoonheydt, R.A. Selective oxidation of methane by the bis(μ-oxo)dicopper core stabilized on ZSM-5 and mordenite zeolites. *J. Am. Chem. Soc.* 2005, 127, 1394–1395.

58. Alayon, E.M., Nachtegaal, M., Ranocchiari, M., and van Bokhoven, J.A. Catalytic conversion of methane to methanol over Cu-mordenite. *Chem. Comm.* 2012, 48, 404–406.

59. Vanelderen, P., Hadt, R.G., Smeets, P.J., Solomon, E.I., Schoonheydt, R.A., and Sels, B.F. Cu-ZSM-5: A biomimetic inorganic model for methane oxidation. *J. Catal.* 2011, 284, 157–164.

60. Hammond, C., Conrad, S., and Hermans I. Oxidative methane upgrading. *ChemSusChem* 2012, 5, 1668–1686.

61. Woertink, J.S., Smeets, P.J., Groothaert, M.H., Vance, M.A., Sels, B.F., Schoonheydt, R.A., and Solomon, E.I. A [Cu$_2$O]$^{2+}$ core in Cu-ZSM-5, the active site in the oxidation of methane to methanol. *PNAS* 2009, 106, 18908–18913.

62. Himes, R.A. and Karlin, K.D. A new copper-oxo player in methane oxidation. *PNAS* 2009, 106, 18877–18878.

63. Fellah, M.F. and Onal, I. C-H bond activation of methane on M- and MO-ZSM-5 (M = Ag, Au, Cu, Rh and Ru) clusters: A density functional theory study. *Catal. Today* 2011, 171, 52–59.

64. Kurnaz, E., Fellah, M.F., and Onal, I. A density functional theory study of C-H bond activation of methane on a bridge site of M-O-M-ZSM-5 Clusters (M = Au, Ag, Fe and Cu). *Micropor. Mesopor. Mater.* 2011, 138, 68–74.

65. Beznis, N.V., Weckhuysen, B.M., and Bitter, J.H. Cu-ZSM-5 zeolites for the formation of methanol from methane and oxygen: Probing the active sites and spectator species. *Catal. Lett.* 2010, 138, 14–22.

66. Yumura, T., Takeuchi, M., Kobayashi, H., and Kuroda Y. Effects of ZSM-5 zeolite confinement on reaction intermediates during dioxygen activation by enclosed dicopper cations. *Inorg. Chem.* 2009, 48, 508–517.

67. Itadani, A., Sugiyama, H., Tanaka, M., Ohkubo, T., Yumura, T., Kobayashi, H., and Kuroda, Y. Potential for C-H activation in CH$_4$ utilizing a CuMFI-type zeolite as a catalyst. *J. Phys. Chem. C* 2009, 113, 7213–7222.

68. Shiota, Y. and Yoshizawa, K. Comparison of the reactivity of bis(μ-oxo)Cu^{II}Cu^{III} and Cu^{III}Cu^{III} species to methane. *Inorg. Chem.* 2009, 48, 838–845.

69. Chan, S.I. and Yu, S.S.-F. Controlled oxidation of hydrocarbons by the membrane-bound methane monooxygenase: The case for a tricopper cluster. *Acc. Chem. Res.* 2008, 41, 969–979.

70. Balasubramanian, R., Smith, S.M., Rawat, S., Yatsunyk, L.A., Stemmler, T.L., and Rosenzweig, A.C. Oxidation of methane by a biological dicopper centre. *Nature* 2010, 465, 115–119.

71. Himes, R.A., Barnese, K., and Karlin, K.D. One is lonely and three is a crowd: Two coppers are for methane oxidation. *Angew. Chem. Int. Ed.* 2010, 49, 6714–6716.

72. Shilov, A.E. and Shteinman, A.A. Methane hydroxylation: A biomimetic approach. *Russ. Chem. Rev.* 2012, 81, 291–316.

73. Cejka, J., Centi, G., Perez-Pariente, J., and Roth, W.J. Zeolite-based materials for novel catalytic applications: Opportunities, perspectives and open problems. *Catal. Today* 2012, 179, 2–15.

74. Choudhary, T.V., Aksoylu, E., and Goodman, D.W. Nonoxidative activation of methane. *Catal. Rev.* 2003, 45, 151–203.

75. Ismagilov, Z.R., Matus, E.V., and Tsikoza, L.T. Direct conversion of methane on Mo/ZSM-5 catalysts to produce benzene and hydrogen: Achievements and perspectives. *Energy Env. Sci.* 2008, 1, 526–541.

76. Yu, C. and Shen, S. Progress in studies of naturals gas conversion in China. *Petrol. Sci.* 2008, 5, 67–72.
77. Martinez, A. Application of zeolites in the production of petrochemical intermediates. In *Zeolites: From Model Materials to Industrial Catalysts*, Chapter 10, pp. 227–261, Cejka, J., Perez-Pariente, J., and Roth, W.J. (eds.). Kerala, India: Transworld Research Network Publishers, 2008.
78. Xu, Y., Lu, J., Wang, J., and Zhang, Z. Mo-based zeolite catalysts and oxygen-free methane aromatization. *Progr. Chem.* 2011, 23, 90–106.
79. Guo, J., Lou, H., and Zheng, X. Energy-efficient coaromatization of methane and propane. *J. Nat. Gas Chem.* 2009, 18, 260–272.
80. Zhang, W., Xu, S., Han, X., and Bao, X. In situ solid-state NMR for heterogeneous catalysis: a joint experimental and theoretical approach. *Chem. Soc. Rev.* 2012, 41, 192–210.
81. Xing, S., Zhou, D., Cao, L., and Li, X. Density functional theory study on structure of molybdenum carbide and catalytic mechanism for methane activation over ZSM-5 zeolite. *Chin. J. Catal.* 2010, 31, 415–422.
82. Zheng, H., Ma, D., Bao, X., Hu, J.Z., Kwak, J.H., Wang, Y., and Peden, C.H.F. Direct observation of the active center for methane dehydroaromatization using an ultrahigh field 95Mo NMR spectroscopy. *J. Am. Chem. Soc.* 2008, 130, 3722–3723.
83. Matus, E.V., Ismagilov, I.Z., Sukhova, O.B., Zaikovskii, V.I., Tsikoza, L.T., Ismagilov, Z.R., and Moulijn, J.A. Study of methane dehydroaromatization on impregnated Mo/ZSM-5 catalysts and characterization of nanostructured Mo phases and carbonaceous deposits. *Ind. Eng. Chem. Res.* 2007, 46, 4063–4074.
84. Tessonnier, J.-P., Louis, B., Rigolet, S., Ledoux, M.-J., and Pham-Huu, C. Methane dehydro-aromatization on Mo/ZSM-5: About the hidden role of Brønsted acid sites. *Appl. Catal. A Gen.* 2008, 336, 79–88.
85. Navarro, R.M., Pena, M.A., and Fierro, J.L.G. Hydrogen production reactions from carbon feedstocks: Fossil fuels and biomass. *Chem. Rev.* 2007, 107, 3952–3991.
86. Tessonnier, J.-P., Louis, B., Ledoux, M.-J., and Pham-Huu, C. Green catalysis for production of chemicals and CO-free hydrogen. *Catal. Commun.* 2007, 8, 1787–1792.
87. Ichikawa, M. Recent progress on the methane-to-benzene catalytic technology and its industrial application. In Plenary Lecture, *Eighth Natural Gas Conversion Symposium*, Natal, Brazil, May 27–31, 2007.
88. Li, L., Borry, R., and Iglesia, E. Reaction-transport simulations of non-oxidative methane conversion with continuous hydrogen removal—Homogeneous-heterogeneous reaction pathways. *Chem. Eng. Sci.* 2006, 56, 1869–1881.
89. Kinage, A.K., Ohnishi, R., and Ichikawa, M. Marked enhancement of the methane dehydrocondensation toward benzene using effective Pd catalytic membrane reactor with Mo/ZSM-5. *Catal. Lett.* 2003, 88, 199–202.
90. Cui, Y., Xu, Y., Suzuki, Y., and Zhang, Z.-G. Experimental evidence for three rate-controlling regions of the non-oxidative methane dehydroaromatization over Mo/HZSM-5 catalyst at 1073 K. *Catal. Sci. Technol.* 2011, 1, 823–829.
91. Matus, E.V., Ismagilov, I.Z., Sukhova, O.B., Tsikoza, L.T., Kerzhentsev, M.A., and Ismagilov, Z.R. In *Proceedings of Third International Conference (Catalysis: Theory and Practice)*, Novosibirsk, Russia, 2007, 299–300.
92. Xu, L., Xing, H., Wu, S., Guan, J., Jia, M., and Kan, Q. Synthesis, characterization and catalytic performance of a novel zeolite ITQ-2-like by treating MCM-22 precursor with H_2O_2. *Bull. Mater. Sci.* 2011, 34, 1605–1610.
93. Gu, L., Ma, D., Hu, G., Wu, J., Wang, H., Sun, C., Yao, S., Shen, W., and Bao, X. Fabrication and catalytic tests of MCM-22/silicon carbide structured catalysts. *Dalton Trans.* 2010, 39, 9705–9710.

94. Chu, N., Wang, J., Zhang, Y., Yang, J., Lu, J., and Yin, D. Nestlike hollow hierarchical MCM-22 microspheres: Synthesis and exceptional catalytic properties. *Chem. Mater.* 2010, 22, 2757–2763.

95. Chu, N., Yang, J., Wang, J., Yu, S., Lu, J., Zhang, Y., and Yin, D. A feasible way to enhance effectively the catalytic performance of methane dehydroaromatization. *Catal. Commun.* 2010, 11, 513–517.

96. Chu, N., Yang, J., Li, C., Cui, J., Zhao, Q., Yin, X., Lu, J., and Wang, J. An unusual hierarchical ZSM-5 microsphere with good catalytic performance in methane dehydroaromatization. *Micropor. Mesopor. Mater.* 2009, 118, 169–175.

97. Xu, C., Liu, H., Jia, M., Guan, J., Wu, S., Wu, T., and Kan, Q. Methane non-oxidative aromatization on Mo/ZSM-5: Effect of adding triethoxyphenylsilanes into the synthesis system of ZSM-5. *Appl. Surf. Sci.* 2011, 257, 2448–2454.

98. Martinez, A., Peris, E., Derewinski, M., and Burkat-Dulak, A. Improvement of catalyst stability during methane dehydroaromatization (MDA) on Mo/HZSM-5 comprising intracrystalline mesopores. *Catal. Today* 2011, 169, 75–84.

99. Xu, Y., Wang, J., Suzuki, Y., and Zhang, Z.-G. Improving effect of Fe additive on the catalytic stability of Mo/HZSM-5 in the methane dehydroaromatization. *Catal. Today* 2012, 185, 41–46.

100. Xu, Y., Wang, J., Suzuki, Y., and Zhang, Z.-G. Effect of transition metal additives on the catalytic stability of Mo/HZSM-5 in the methane dehydroaromatization under periodic CH4-H2 switch operation at 1073 K. *Appl. Catal. A Gen.* 2011, 409–410, 181–193.

101. Liu, B.S., Zhang, Y., Liu, J.F., Tian, M., Zhang, F.M., Au, C.T., and Cheung, A.S.-C. Characteristic and mechanism of methane dehydroaromatization over Zn-based/HZSM-5 catalysts under conditions of atmospheric pressure and supersonic jet expansion. *J. Phys. Chem. C* 2011, 115, 16954–16962.

102. Liu, J.F., Liu, Y., and Peng, L.F. Aromatization of methane by using propane as co-reactant over cobalt and zinc-impregnated HZSM-5 catalysts. *J. Molec. Catal. A Chem.* 2008, 280, 7–15.

103. Xu, J., Zheng, A., Wang, X., Qi, G., Su, J., Du, J., Gan, Z., Wu, J., Wang, W., and Deng, F. Room temperature activation of methane over Zn modified H-ZSM-5 zeolites: Insight from solid-state NMR and theoretical calculations. *Chem. Sci.* 2012, 3, 2932–2940.

104. Zhang, A., Sun, S., Komon, Z.J.A., Osterwalder, N., Gadewar, S., Stoimenov, P., Auerbach, D.J., Stucky, G.D., and McFarland, E.W. Improved light olefin yield from methyl bromide coupling over modified SAPO-34 molecular sieves. *Phys. Chem. Chem. Phys.* 2011, 13, 2550–2555.

105. Yilmaz, A., Zhou, X.P., Lorkovic, I.M., Ylmaz, G.A., Laverman, L., Weiss, M., Sun, S.L. et al. Bromine mediated partial oxidation of ethane over nanostructured zirconia supported metal oxide/bromide. *Micropor. Mesopor. Mater.* 2005, 79, 205–214.

106. Liu, Z., Huang, L., Li, W.S., Yang, F., Au, C.T., and Zhou, X.P. Higher hydrocarbons from methane condensation mediated by HBr. *J. Mol. Catal. A Chem.* 2007, 273, 14–20.

107. Peringer, E., Salzinger, M., Hutt, M., Lemonidou, A.A., and Lercher, J.A. Modified lanthanum catalysts for oxidative chlorination of methane. *Top. Catal.* 2009, 52, 122–1231.

108. Degirmenci, V., Yilmaz, A., and Uner, D. Selective methane bromination over sulfated zirconia in SBA-15 catalysts. *Catal. Today* 2009, 142, 30–33.

109. Osterwalder, N. and Stark, W.J. Direct coupling of bromine-mediated methane activation and carbon-deposit gasification. *ChemPhysChem* 2007, 8, 297–303.

110. Li, F.B. and Yuan, G.Q. Hydrated dibromodioxomolybdenum(VI) supported on Zn-MCM-48 for facile oxidation of methane. *Angew. Chem. Int. Ed.* 2006, 45, 6541–6544.

2 Oxygen Transporting Membranes in Syngas Production

Jürgen Caro and Fangyi Liang

CONTENTS

2.1   INTRODUCTION

The traditional way to produce synthesis gas is methane steam reforming (MSR) according to $CH_4 + H_2O \rightarrow 3H_2 + CO$. In following high- and low-temperature water gas shift steps according to $CO + H_2O \rightarrow CO_2 + H_2$, the desired H_2/CO ratio of 2 can be established. In the autothermal reforming, the endothermic steam reforming (SR) is coupled with some methane partial oxidation according to $CH_4 + (1/2)O_2 \rightarrow CO + 2H_2$. However, also in the case of autothermal reforming, the H_2/CO ratio is near 3 and has to be established.

The only reaction that gives immediately a synthesis gas of the composition H_2/CO ratio of 2—as required for methanol synthesis or for diesel production after Fischer and Tropsch—is the partial oxidation of methane (POM) according to $CH_4 + (1/2)O_2 \rightarrow CO + 2H_2$. However, the cost factor here is the production of pure oxygen, which is usually produced in a Linde plant near the POM. Oxygen transporting membranes (OTMs) can be used to separate pure oxygen from air. Two concepts are possible on how OTMs can be used for the synthesis of gas production on the basis of methane due to the POM:

1. In an oxygen separator with an OTM, oxygen is separated from air, and in a second apparatus, POM takes place.
2. In a membrane reactor, the two steps of separation and catalytic partial oxidation are combined in one apparatus following the idea of process intensification.

In this chapter, concepts 1 and 2 will be discussed.

2.2   TWO-APPARATUS CONCEPTS: OXYGEN SEPARATION FROM AIR AND METHANE PARTIAL OXIDATION IN TWO DIFFERENT APPARATUS

2.2.1   BACKGROUND

The supply of pure oxygen is the largest cost factor in the POM reaction $CH_4 + (1/2)O_2 \rightarrow CO + 2H_2$. In large quantities, the price of liquid oxygen in 2001 was approximately \$0.21/kg. Since the primary cost of production is the energy cost of liquefying the air, the production cost will change as energy cost varies.

Two major methods are employed to produce 100 million tons of O_2 extracted from air for industrial uses annually. The most common method is to fractionally distill liquefied air into its various components, with N_2 distilling as a vapor while O_2 is left as a liquid. The other major method of producing O_2 gas involves passing a stream of clean, dry air through one bed of a pair of identical zeolite molecular sieves, which adsorbs the nitrogen and delivers a gas stream that is 90%–93% O_2. Simultaneously, nitrogen gas is released from the other nitrogen-saturated zeolite bed, by reducing the chamber operating pressure and diverting part of the oxygen gas from the producer bed through it, in the reverse direction of flow. After a set cycle time, the operation of the two beds is interchanged, thereby allowing for a continuous supply of gaseous oxygen to be pumped through a pipeline. This is known as pressure swing adsorption. Oxygen gas is increasingly obtained by these noncryogenic technologies (see also the related vacuum swing adsorption).

A new and so far not commercialized air separation technology is based on oxygen transporting dense ceramic membranes to produce pure O_2 gas. The production of high-purity oxygen using OTMs can be used by different techniques: (1) pressure-driven process, (2) vacuum process, and (3) a combination of both. The vacuum operation is more economical to produce oxygen from air compared to the pressure operation and a membrane process using a sweep gas. The oxygen permeability of OTMs can be well explained by the Wagner theory for oxygen ion bulk diffusion as a rate-limiting step.

So far, the most studied oxygen-permeable materials are single-phase perovskite-type materials since these materials possess high oxygen permeability. However, most of these perovskite-type materials have stability problems in a low oxygen partial pressure atmosphere, and especially in a CO_2-containing atmosphere. Dual-phase membranes that consist of an oxygen ionic conducting phase and an electronic conducting phase in a microscale phase mixture are considered to be promising substitutes for the single-phase mixed ionic–electronic conducting (MIEC) materials. In addition to good phase stability in CO_2 and reducing atmospheres, their chemical compositions can be tailored according to practical requirements. However, due to the low oxygen permeability of dual-phase membranes, the new chemical composition of dual-phase membranes should be further investigated.

In the following, the practice-relevant testing of three different OTMs of different geometry and different chemical composition is described.

2.2.2 HOLLOW FIBER MEMBRANES OF THE COMPOSITION BCFZ OBTAINED BY SPINNING

Dense $BaCo_{0.4}Fe_{0.4}Zr_{0.2}O_{3-\delta}$ (BCFZ) perovskite hollow fiber membranes were manufactured by a phase inversion spinning process. As shown in Figure 2.1, the sintered fiber has a thickness of around 0.17 mm with an outer diameter of 1.10 mm and an inner diameter of 0.76 mm [1]. The fibers used here had a length of 30 cm. The production of high-purity oxygen was studied in a flow-through permeator (Figure 2.2) and in a dead-end permeator (see Figure 2.7 later in the Chapter) with BCFZ hollow fiber membranes at high temperatures using (1) pressurized air/O_2-enriched air as feed, (2) reduced pressure on the permeate side, and (3) a combination of both.

FIGURE 2.1 SEM micrograph of the dense BCFZ hollow fiber membrane produced by a spinning technique after sintering. The surface of the fiber can be coated, thus a surface enlargement for a higher oxygen transport can be obtained. (From Liang, H. et al., *Chem. Eur. J.*, 16, 7898, 2010.)

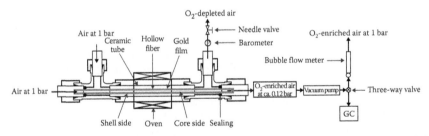

FIGURE 2.2 Permeator in flow-through geometry for the production of oxygen-enriched air.

Pure oxygen is produced in a dead-end permeator using O_2-enriched air as pressurized feed and using again a vacuum pump to establish a reduced oxygen pressure of about 0.05 bar on the permeate side. The oxygen permeation fluxes increase with increasing temperature, pressure on the feed side (Figure 2.3), and oxygen concentration in the fed O_2-enriched air (Figure 2.4). An oxygen permeation flux of 10.2 cm^3 (STP) cm^{-2} min^{-1} was reached using O_2-enriched air with 50 vol.% O_2 at 5 bar as feed and reduced pressure of 0.05 bar on the permeate side at 900°C (Figure 2.4). A high oxygen purity up to 99.9 vol.% and high permeation rates of almost 10 cm^3 (STP) cm^{-2} min^{-1} were obtained at 900°C in a 150 h operation under the conditions of 0.05 bar on the oxygen side and oxygen-enriched air at 4 bar on the feed side (Figure 2.5).

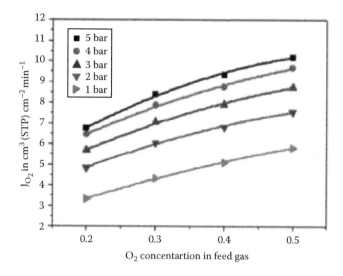

FIGURE 2.3 Influence of the oxygen partial pressure difference on the production of pure oxygen at different temperatures. Feed/shell side: air flow = 150 cm^3 min^{-1} with 20 vol.% O$_2$ at different pressures. Permeate/core side: pure O$_2$ at about 0.05 bar.

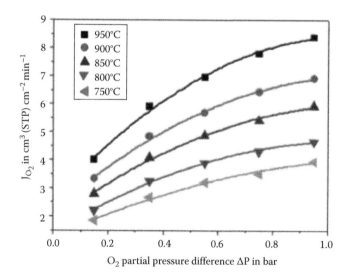

FIGURE 2.4 Influence of oxygen concentration in the fed air at different feed pressures on the production of pure oxygen for different feed pressures at 900°C. Feed/shell side: O$_2$-enriched air flow rate = 150 cm^3 min^{-1}. Permeate/core side: pure O$_2$ at about 0.05 bar.

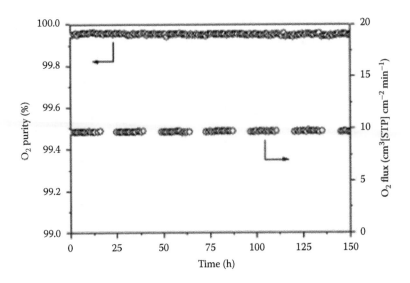

FIGURE 2.5 Long-term operation of the production of pure oxygen: O_2 purity and oxygen permeation flux of the BCFZ hollow fiber membrane at 900°C. Feed/shell side: air flow rate = 150 cm³ min⁻¹ at 4 bar. Permeate/core side: pure O_2 at about 0.05 bar.

2.2.3 TUBES OF THE COMPOSITION BSCF OBTAINED BY EXTRUSION IN DEAD-END GEOMETRY

Stability and oxygen permeation behavior of dead-end $Ba_{0.5}Sr_{0.5}Co_{0.8}Fe_{0.2}O_{3-\delta}$ (BSCF) tube membranes with an outer diameter of 10 mm, an inner diameter of 8 mm, and a length of 400 mm (Figure 2.6) were studied in a dead-end membrane permeator, as shown in principle in Figure 2.7 [3]. Good BSCF stability was found when the BSCF membrane was operated at temperatures ≥850°C and oxygen with a purity of almost 100 vol.% was obtained. Oxygen recovery increases

FIGURE 2.6 Dead-end BSCF tube membrane with an outer diameter of 1 cm and 1 mm wall thickness as produced by IKTS.

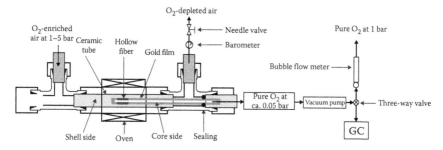

FIGURE 2.7   Permeator in dead-end geometry for the production of high-purity oxygen.

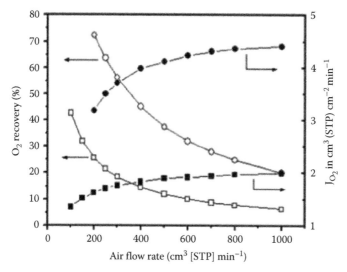

FIGURE 2.8   Influence of the flow rate of the fed air (20 vol.% O_2) on oxygen recovery and oxygen permeation flux for two feed pressures at 950°C. Feed side: air at 1 bar (■, □) and 5 bar (●, ○) with different flow rates. Permeate side: pure O_2 at 0.1 bar (fluctuating between 0.08 and 0.12 bar).

with increasing air pressure but decreases with increasing air flow rate (Figure 2.8). Oxygen permeation flux and space-time yield of the high-purity oxygen production can be raised using oxygen-enriched air as feed at relatively low feed pressure of a few bar (Figure 2.9).

2.2.4 PLATES OF THE COMPOSITION 40 wt.% NiFe₂O₄–60 wt.% Ce₀.₉Gd₀.₁O₂₋δ AS DUAL-PHASE DISK MEMBRANE

Different synthesis methods for the preparation of a carbon dioxide–resistant dual-phase membrane of the composition 40 wt.% $NiFe_2O_4$–60 wt.% $Ce_{0.9}Gd_{0.1}O_{2-\delta}$ (40NFO–60CGO) with NFO as electron conductor and CGO as oxygen ion conductor have been evaluated [4]. The membrane prepared by a direct one-pot synthesis shows the smallest grains and the highest oxygen permeation fluxes in the range 900°C–1000°C. On the opposite, the membranes prepared by mixing the NFO and CGO powders in a mortar or by ball-milling give larger grains and a lower oxygen

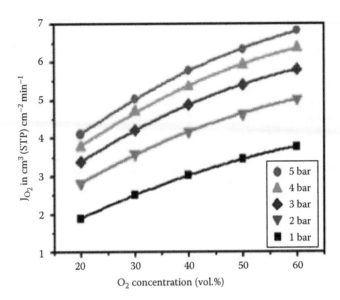

FIGURE 2.9 Influence of oxygen concentration in the fed air on the production of oxygen for different feed pressures at 950°C in the vacuum process. Feed side: O_2-enriched air flow rate = 500 cm³ (STP) min⁻¹. Permeate side: pure O_2 at 0.1 bar (fluctuating between 0.08 and 0.12 bar).

permeation flux (Figures 2.10 and 2.11). High-temperature XRD in CO_2-containing atmospheres showed that the NFO and CGO phases in 40NFO–60CGO remained unchanged and no carbonate was formed. In long-time permeation studies with CO_2 as sweep gas, we also found that the dual-phase membrane 40NFO–60CGO 03 was CO_2 stable and an average oxygen flux of 0.30 mL min⁻¹·cm² was obtained at 1000°C for a 0.5 mm thick membrane (Figure 2.12). This value is comparable with $La_2NiO_{4+\delta}$ as well as $La_2Ni_{0.9}Fe_{0.1}O_{4+\delta}$, which are promising CO_2-stable membrane materials.

2.2.5 PRACTICE-RELEVANT TESTING OF 1 cm TUBES OF THE COMPOSITION BSCF OBTAINED BY EXTRUSION IN DEAD-END GEOMETRY

For the proper use of the 1 cm tubes of the composition BSCF obtained by extrusion in dead-end geometry (purchased from the former hitk, now Fraunhofer IKTS, Germany), different operation modes have been tested under practice-relevant conditions:

- *Pressure process*: The air as feed is pressurized, and the permeate is pure oxygen at 1 bar. They have a driving force for oxygen permeation; the oxygen partial pressure on the feed side of the membrane must be >1 bar, which is the case if air with 20% O_2 is pressurized to 5 bar. Also oxygen-enriched air can be used, for example, 50 vol.% O_2-containing air at P > 2.5 bar.

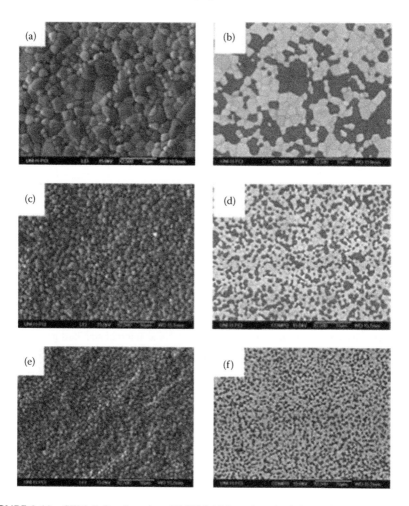

FIGURE 2.10 SEM (left column) and BSEM (right column) of the surface grain structure of the membranes 40NFO–60CGO 01, 02, and 03, prepared by powder mixing in a mortar (sample 01: a and b), ball-milling (sample 02: c and d), and one-pot method (sample 03: e and f), respectively.

- *Vacuum process*: The oxygen partial pressure on the permeate side is reduced by vacuum pumps to an oxygen partial pressure P < 0.1 bar. There is no need to pressurize the air on the feed side.

When the pressure on the permeate side is fixed at 1 bar, pure oxygen can be obtained at atmospheric pressure, if the oxygen partial pressure difference across the membrane is higher than 1 bar. The oxygen partial pressure ratio can be increased by elevating the pressure of atmospheric air with 21 vol.% O_2 or by using O_2-enriched air. O_2-enriched air of different oxygen content can be produced by using either organic polymeric hollow fiber membranes [5] or perovskite hollow fiber membranes [6,7]. Figure 2.13 shows the oxygen permeation rate as a logarithmic function

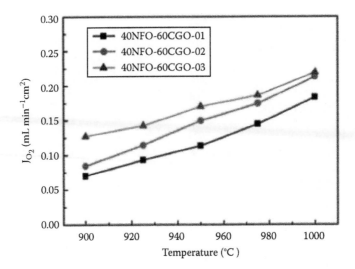

FIGURE 2.11 Temperature dependence of the oxygen permeation fluxes through 40NFO–60CGO membranes prepared by different methods without coating: 01 powder mixing in a mortar, 02 ball-milling, and 03 one-pot method. Feed: 150 mL min^{-1} 50 vol.% O_2 with 50 vol.% N_2. Sweep: 29 mL min^{-1} He + 1 mL min^{-1} Ne as internal standard gas. Thickness of membranes = 0.6 mm.

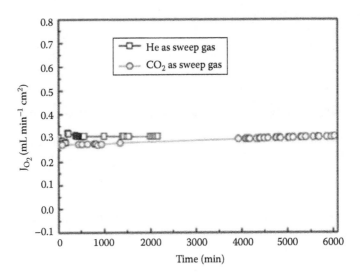

FIGURE 2.12 Oxygen permeation flux of the 40NFO–60CGO 03 composite membrane with $La_{0.8}Sr_{0.2}CoO_{3-\delta}$ (LSC) coating as a function of time at 1000°C, membrane thickness = 0.6 mm. Feed: 150 mL min^{-1} synthetic air with 20 vol.% O_2. Sweep: 29 mL min^{-1} He + 1 mL min^{-1} Ne as an internal standard gas.

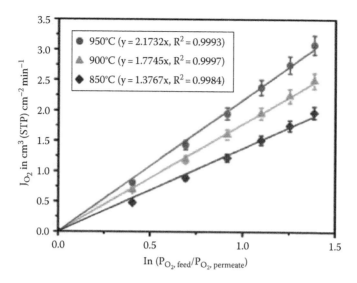

FIGURE 2.13 *Pressure process*: Oxygen permeation rate as a logarithmic function of the oxygen partial pressure ratio on the feed and permeate sides at different temperatures for the pressure-driven process. Feed side: air flow rate = 500 cm^3 (STP) min^{-1} at different oxygen partial pressure differences. Permeate side: pure O$_2$ at 1 bar.

of the oxygen partial pressure ratio representing the driving force of the process. The oxygen partial pressure difference was established by elevating the pressure of the O$_2$-enriched air with 50 vol.% oxygen as feed. For example, an oxygen partial pressure difference of 1.0 bar across the membrane can be established by using compressed oxygen-enriched air with 50 vol.% oxygen at 4 bar on the feed side while keeping the permeated oxygen at 1 bar. It was found that the oxygen permeation fluxes at all temperatures increase linearly as a logarithmic function of the oxygen partial pressure ratio, which is in agreement with the Wagner theory, assuming that the oxygen transport is limited by oxygen ion bulk diffusion. According to the Wagner theory, the oxygen ionic conductivity at various oxygen partial pressure ratios and temperatures can be calculated by the following.

$$\sigma_i = -J_{O_2} \frac{4^2 F^2 L}{RT} \left(\ln \frac{P_2}{P_1} \right)$$

Similar results have been published recently for BSCF membranes [8].

Figure 2.14 shows the calculated oxygen ionic conductivity as a function of temperatures at different oxygen partial pressures on the feed and permeate sides for the pressure-driven process (Figure 2.13). It can be found from Figure 2.14 that the oxygen ionic conductivity increases linearly with increasing reciprocal temperature, which is again in agreement with the prediction from the Wagner theory for bulk diffusion of oxygen ions as a rate-limiting step. Moreover, the oxygen ionic conductivity increases also with increasing oxygen partial pressure difference between

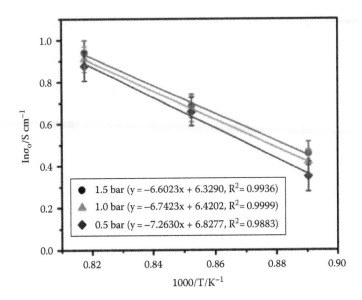

FIGURE 2.14 *Pressure process*: Calculated oxygen ionic conductivity as a function of temperature at different oxygen partial pressures on the feed and permeate sides for the pressure-driven process (data from Figure 2.13).

the feed and permeate sides. The oxygen ionic conductivity of the dead-end BSCF tube at 900°C is ca 1.9 S cm^{-1}. A similar value of an oxygen ionic conductivity of ca 1.4 S cm^{-1} was reported by Chen et al. [9] for a 1.1 mm thick BSCF disk membrane at 900°C.

Figure 2.15 shows the influence of the oxygen concentration in the fed air on the oxygen production at different feed pressures at 950°C. It was found that the oxygen permeation fluxes increase with increasing both the oxygen concentration in the fed air and the feed pressure, suggesting that any increase of the oxygen concentration in the fed air has the same effect as increasing the feed pressure. It can be expected from the Wagner theory that the oxygen permeation flux increases with the feed pressure, but the corresponding incremental growth for a stepwise increase of the feed pressure drops for a constant pressure step with increasing absolute pressure. For example, the same pressure difference of 1 bar from 1 to 2 bar and from 4 to 5 bar between feed and permeate gives the different oxygen permeation fluxes an enhancement of 0.9 and 0.3 cm^3 (STP) cm^{-2} min^{-1}. Wang et al. reported that the oxygen permeation flux can be increased by elevating the air pressure until 50 atm [10]. The compression cost in the pressure-driven process can have a negative impact on the economics. However, the same holds true for the production of oxygen-enriched air as feed. On the other hand, the space-time yield of the permeator in the production of oxygen will increase. It has to be noted that a high feed pressure can be problematic in practice due to, for example, the mechanical stability and the sealing of the membranes. Therefore, a relative low pressure on the feed side is desired, and the oxygen permeation flux can be raised by using oxygen-enriched air as feed. For example, the oxygen permeation flux increases

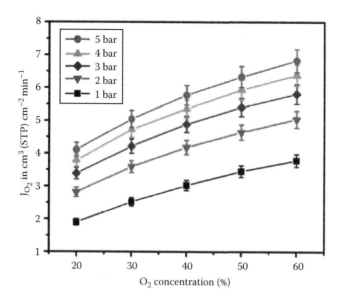

FIGURE 2.15 *Vacuum process*: Influence of oxygen concentration in the fed air on the production of oxygen for different feed pressures at 950°C in the vacuum process. Feed side: O_2-enriched air flow rate $= 500$ cm^3 (STP) min^{-1}. Permeate side: pure O_2 at 0.1 bar (fluctuating between 0.08 and 0.12 bar).

from 4.1 to 6.8 cm^3 (STP) cm^{-2} min^{-1} when increasing the oxygen concentration in the fed O_2-enriched air from 20 to 60 vol.% at 5 bar and 950°C.

Similar to Figure 2.13 for the pressure-driven process, Figure 2.16 shows the oxygen permeation fluxes in the vacuum and pressure-driven processes as a function of the logarithm of the oxygen partial pressure ratios on the feed and permeate sides from 850°C to 950°C. The straight lines of J_{O_2} vs. $\ln(P_{O_2, \text{feed}}/P_{O_2, \text{permeate}})$ for all temperatures indicate again that oxygen permeation can be described well by the Wagner theory for oxygen ion diffusion in the bulk as a rate-limiting step. In a recent paper by Kovalevsky et al. [11], the role of surface oxygen exchange on oxygen permeation through a BSCF membrane was studied and the corresponding O_2 flux limitations were expected for a membrane thickness below 1 mm.

It can be found from Figure 2.16 that a higher oxygen permeation flux was obtained in the vacuum process using synthetic air with 20 vol.% oxygen at 1 bar as feed gas compared to the pressure-driven process with an oxygen partial pressure difference of 0.5 bar (air with 50 vol.% oxygen at 3 bar on the feed side, oxygen pressure of 1 bar on the permeate side). Moreover, in the vacuum process, the oxygen permeation fluxes can be significantly enhanced when raising the oxygen concentration in the feed from 20 to 50 vol.%. In addition, in a combination of both the vacuum and the pressure-driven processes, the oxygen permeation fluxes can be further enhanced by using oxygen-enriched air with 50 vol.% oxygen.

Due to the difficulties in sealing of the membrane at high temperatures, there are still few reports on the separation of pure oxygen by using OTMs. In this work, a high oxygen purity of almost 100 vol.% (gas chromatography does not detect any N_2)

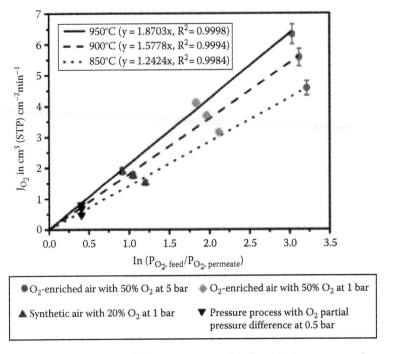

FIGURE 2.16 *Vacuum process*: The oxygen permeation fluxes in the vacuum and pressure-driven processes as function of the logarithm of the partial pressure ratios on feed and permeate sides from 850°C to 950°C. Permeate side: pure oxygen at 0.1 bar (fluctuating between 0.08 and 0.12 bar) by using a vacuum pump or 1 bar in the pressure-driven process. Feed side: air gas flow rate = 500 cm³ (STP) min⁻¹.

was found at operation temperatures ≥850°C. The high oxygen purity indicates that the BSCF tube and the BSCF disk forming the dead-end part of the membrane have been sealed perfectly. Moreover, the open side of the dead-end tube membranes was simply sealed by silicone rubber rings outside the high-temperature zone. Therefore, the geometry of dead-end tube membranes can solve the problem of the high-temperature sealing.

2.2.6 TESTING OF A LAB PERMEATOR FOR OXYGEN SEPARATION

Oxygen transport through oxygen transporting dense ceramic membranes is described by the Wagner equation. By integration of the Wagner equation, the following simple expression for the oxygen flux J_{O_2} through a ceramic membrane of thickness L with different oxygen partial pressures P_1 and P_2 on the feed and permeate side of the membrane is obtained. Assuming the ionic conductivity σ_i of the ceramic membrane as constant, the oxygen flux J_{O_2} is a simple function of $\ln P_1/P_2$ as a driving force:

$$J_{O_2} = -\frac{RT\sigma_i}{4^2F^2L}\ln\frac{P_1}{P_2}$$

From the earlier considerations, the following operation modes are possible:

1. Normal air of 1 bar as feed (P_{O_2} = 0.2 bar) and a vacuum pump producing a reduced pressure of 0.1 bar on the permeate side: ln (P_1/P_2) = 0.69.
2. Compressed normal air of 20 bar on the feed side (P_{O_2} = 4 bar) and atmospheric pressure of 1 bar on the permeated side ln (P_1/P_2) = 1.39.
3. Slightly compressed air of 10 bar as feed (P_{O_2} = 2 bar) and slightly reduced pressure of 0.4 bar on the permeate side: ln (P_1/P_2) = 1.65.

Comparing these three operation modes, different German groups and consortia identified the combined pressure–vacuum operation mode 3 as technologically and economically superior in comparison with modes 1 and 2 (Oxycoal-AC project: http://www.oxycoal.de; Helmholtz Alliance mem-Braine: http://www.mem-brain-allianz.de, Cooretec-Project: http://www.cooretec.de). It is generally accepted that the electrical energy that is necessary for the production of oxygen is 0.4 kW h N^{-1} m^3 oxygen in the cryogenic air separation after Linde and 0.36 kW h N^{-1} m^3 in the pressure swing adsorption (95 vol.% oxygen) [12].

The earlier equation indicates that a doubling of the feed pressure results in the same increase of the oxygen flux like a decrease of the oxygen pressure on the permeate side to the half by a vacuum pump. However, the pressure process consumes more energy for the compression of air than the suction of pure oxygen with vacuum pumps and the compression of air to ambient pressure. The energy demand for both processes was calculated based on the adiabatic compression energy and an addition of 45% for usual energy losses. According to Kriegel et al. [15], the results are shown in Figure 2.17. Since a surplus of free heat energy can be assumed for these industrial processes, the calculations do not contain the energy for heating the membrane and gases to the separation temperature of 900°C.

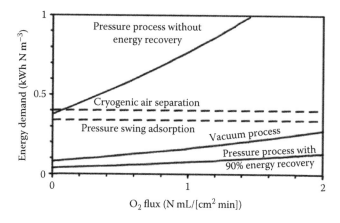

FIGURE 2.17 Comparison of the energy demand for different oxygen production processes; membrane separation is simulated for the pressure process with and without energy recovery as well as the vacuum process at 900°C. (From Kriegel, R. et al., *10th Int. Conf. Inorg. Membranes*, August 18–22, 2008, Tokyo, Japan.)

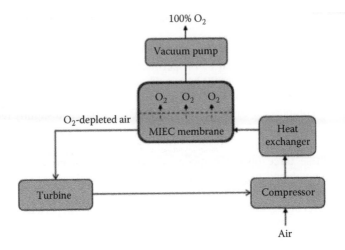

FIGURE 2.18 The three-end concept of oxygen production with OTM combined with the turbine.

According to Kriegel et al. [15], it follows from Figure 2.17 that the compression of air needs nearly five times more energy than the vacuum process because of the higher gas volume. Therefore, a saving of 80% of the compression energy is necessary for the pressure process to overcome the energetic advances of the vacuum process. This could be reached by the use of a high-efficient decompression turbine, as shown in Figure 2.18. Besides, the calculated efficiency of a prospective Advanced Adiabatic Compressed Air Energy Storage Plant promised by RWE amounts to only 70% (http://www.rwe.com/generator.aspx/konzern/fue/strom/ergiespeicherung). Air Products claims that when using oxygen selective ceramic membranes, the cost of producing oxygen from air can be lowered by 30% [13].

Different operation modes have been tested: slightly pressurized air on the feed side of the membrane at different pressures, and (1) vacuum or (2) pure oxygen at 1 bar on the permeate side. It follows from Figure 2.19 that for equal oxygen partial pressure difference between feed and permeate sides of the perovskite membrane, the vacuum process gives much higher oxygen fluxes. In both the vacuum and the pressure-driven processes, oxygen-enriched air with 50 vol.% O_2 as feed has been used.

Figure 2.19 shows the dependence of the oxygen permeation flux on the oxygen partial pressure difference at 950°C when using 50 vol.% O_2-enriched air as feed gas. If the oxygen pressure on the permeate side is fixed (at 0.1 or 1 bar), the oxygen partial pressure difference across the membrane can be adjusted by the feed pressure and oxygen content of the feed gas. For the same oxygen partial pressure difference, the oxygen permeation flux in the vacuum process is significantly higher than that in the pressure-driven process. This experimental finding can be explained again by the Wagner theory since the oxygen permeation flux J_{O_2} is proportional to the logarithmic ratio of the oxygen partial pressures P_1 and P_2 on feed

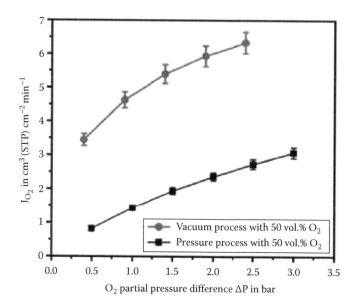

FIGURE 2.19 Oxygen fluxes as a function of oxygen partial pressure difference in the vacuum and pressure-driven processes at 950°C. Permeate side: pure oxygen at reduced pressure of 0.1 bar using a vacuum pump or 1 bar in the pressure-driven process. Feed side: feed gas flow rate = 500 cm³ (STP) min⁻¹ with 50 vol.% O_2 at different pressures.

and permeate side if all other conditions are fixed. As an example, a higher oxygen permeation flux can be obtained for the oxygen partial pressure difference of 0.5 bar in the vacuum process with the driving force (ln (0.6/0.1)) than for the oxygen partial pressure difference of 3.0 bar in the pressure-driven process with the driving force (ln (3/1)). Here, our experimental results of high-purity oxygen production is in good agreement with the straightforward process analysis by Tan and Li [14] and Kriegel et al. [15], showing that the vacuum operation is more economical to produce oxygen from air compared to the pressure operation and a membrane process using a sweep gas. However, Tan et al. [14] reported that a relative leakage of 44% existed in the oxygen production with the vacuum operation using $La_{0.6}Sr_{0.4}Co_{0.2}Fe_{0.8}O_{3-\delta}$ at 1180°C.

Let us assume an oxygen flux of 3 m³ h⁻¹ m² (such value is obtained only for BCFZ hollow fibers for the test condition pressurized air of 3 bar as feed, reduced pressure of 0.1 bar on the permeate side, see Table 2.1); a permeator of 1 m³ is expected to produce per h at 900°C the following amounts of 99.9% O_2:

- Hollow fibers with an outer diameter of 1 mm, assuming a packing density of one fiber per 4 mm²: 2300 m³ oxygen
- Tubes with an outer diameter of 1 cm, assuming a packing density of one tube per 4 cm²: 230 m³ oxygen
- Plates of 1 mm thickness and 5 mm distance between neighboring plates: 110 m³ oxygen

TABLE 2.1

Testing of Different Membranes Made from Different Materials in Different Geometries

	7 Bar Air → 1 Bar	3 Bar → Vac.[a]	1 Bar → Vac.[a]
BSCF tubes 1 mm wall thickness	0.5	2.8	1.1
BCFZ hollow fibers of 0.17 mm wall thickness	0.8	4.5	1.7
NFO–CGO disks with 0.5 mm thickness	0.013	0.069	0.027

For these three membranes, the following oxygen fluxes have been obtained: oxygen (>99.9%) flux in mL min^{-1} cm^2 (if you multiply by 0.6, you have the number in m^3 h^{-1} m^2) at 900°C. The pressures on the feed side and on the permeate side are always given.

[a] Vacuum means reduced pressure of 0.1 bar.

With respect to their geometry, OTMs are semicommercially available in plate, hollow fiber, capillary, and tube geometries (Table 2.1). However, plate-shaped geometries are difficult to seal, and hollow fibers and capillaries do not show a sufficient mechanical stability—they can easily break by pressure waves or mechanical vibrations. Therefore, 1 cm in diameter and 50 cm long BSCF perovskite tubes with 1 mm wall thickness have been selected for the erection of the lab-scale separator (Figure 2.20). One end of the tubes was sealed with a disk from the same BSCF material. To find the limits of performance and stability of the scaled-up membrane permeator with the three 50 cm long perovskite tubes, systematic temperatures,

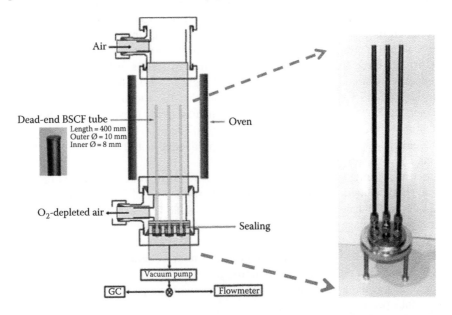

FIGURE 2.20 Three-tube dead-end oxygen permeator. The three BSCF perovskite tubes are 1 cm in diameter and 50 cm long with 1 mm wall thickness (producer: IKTS, former HITK, Germany).

oxygen partial pressure on feed and permeate sides, oxygen content on the feed side, and oxygen pressure on the permeate side have been varied.

Slightly pressurized air of a pressure <5 bar on the feed side and reduced pressure (vacuum) of 0.1 bar on the permeate side are recommended. At 900°C, 4 mL oxygen (STP) cm^{-2} min^{-1} can be produced if the air pressure on the feed side is 5 bar and the oxygen pressure of the permeate side is kept constant at 0.1 bar by using vacuum pumps.

The oxygen produced has a high purity (oxygen content higher 99.99%), which means that by gas chromatography no traces of impurities could be detected. The separator was operated 1 week at 900°C without problems.

2.3 PROCESS INTENSIFICATION: INTEGRATION OF OXYGEN IN SITU SEPARATION FROM AIR AND CPOM TO SYNGAS IN ONE APPARATUS

2.3.1 BACKGROUND

Supplying the oxygen for hydrocarbon activation through a solid oxide membrane has technical, economical, and environmental advantages over the direct use of air as oxidant. Safety benefits result from the separation of the hydrocarbon and oxygen. Since the membrane is impermeable to nitrogen, NO_x formation is avoided sui generis. Due to the success made in air separation by MIEC perovskite-type membranes, it is expected that a membrane reactor combining OTMs with selective catalytic layers for partial oxidation of hydrocarbons may be the first commercial application of a catalytic membrane reactor [16]. It has been estimated that, for industrial use, oxygen-conducting MIEC membranes should have an O_2 flux larger than 1 mL$_N$ min^{-1} cm^{-2} [17,18]. Oxidation in the reactor may take place with *molecular oxygen* if this is desorbed from the membrane into the reaction zone. It may also occur more selectively with *lattice oxygen* from the membrane if the membrane material itself is a catalyst, or if it is coated with a catalyst layer (Figure 2.21). If this layer consists of metal oxides, lattice oxygen ions can be transported from the membrane to the catalyst where a Mars–van Krevelen oxidation mechanism may take place. For this concept, partial oxidation must be faster than the surface exchange, that is, adsorption,

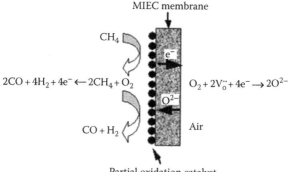

FIGURE 2.21   MIEC membrane in the POM to synthesis gas.

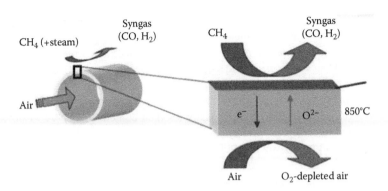

FIGURE 2.22 Principle of POM to syngas by using oxygen transporting hollow fiber membranes with air on the core side of the hollow fiber.

ionization, and desorption of molecular oxygen [19]. The principle of POM to syngas by using oxygen transporting hollow fiber membranes with air on the core side of the hollow fiber is shown in Figure 2.22.

Synthesis gas (H_2 and CO) is usually prepared by the strongly endothermic MSR, $CH_4 + H_2O = 3H_2 + CO$, $\Delta_R H° = +206$ kJ mol^{-1}, at pressures between 15 and 30 bar and temperatures between 850°C and 900°C. An alternative is the slightly exothermic POM, $CH_4 + (1/2) O_2 \rightarrow 2H_2 + CO$, $\Delta_R H° = -36$ kJ mol^{-1}, giving a lower H_2/CO ratio of 2 as required for methanol synthesis or the Fischer–Tropsch process. The POM reaction can take place in an MIEC membrane reactor with oxygen from the surrounding air (Figure 2.20). Due to the potential economic and operational benefits of the production of synthesis gas in an MIEC membrane reactor, worldwide alliances coordinated by Air Products and Praxair have launched big R&D programs (see Section 2.4). The key to success in the industrial implementation of this process will be the development of a ceramic membrane with high long-term operational stability.

In numerous previous POM studies using MIEC membranes, the membrane stability turned out as a critical problem. In a tubular $La_{0.6}Sr_{0.4}Co_{0.2}Fe_{0.8}O_{3-\delta}$ membrane reactor with a packed-bed SR catalyst, excellent conversions (>96%) and CO selectivities (>97%) were reported for diluted methane (6% methane in He) as the feed [9,20]. However, after a few hours of operation, mechanical failure of the membrane occurred. A tubular $La_{0.2}Sr_{0.8}Co_{0.8}Fe_{0.2}O_{3-\delta}$ membrane with a Rh-based catalyst was tested in the POM, and the brittle membrane broke a few minutes after the introduction of methane [21]. Both the oxygen partial pressure and the grain size were found to influence the mechanical failure behavior of $La_{0.2}Sr_{0.8}Cr_{0.2}Fe_{0.8}O_{3-\delta}$ perovskites [22,23]. In another study on $La_{0.2}Sr_{0.8}Cr_{0.2}Fe_{0.8}O_{3-\delta}$, a loss of strength under mild conditions was reported [24,25]. The operation of a tubular $SrCoFeO_{3-\delta}$ perovskite membrane lasted only a few minutes due to lattice expansions, which resulted in cracking [26]. On $SrCoFeO_{3-\delta}$ perovskites, two types of fractures were observed: one type resulted from a strong oxygen gradient across the membrane and the other was the result of chemical decomposition of the perovskite in reducing atmosphere [27]. A $La_{0.2}Sr_{0.8}Co_{0.1}Cr_{0.1}Fe_{0.8}O_{3-\delta}$ membrane cracked in the POM after 350 h at 900°C [18].

Different concepts were developed to prolong the membrane lifetime at low O_2 partial pressure. A bilayered membrane consisting of a Sm-doped CeO_2 protective layer on the reducing side of the tubular $La_{0.6}Sr_{0.4}Co_{0.8}Fe_{0.2}O_{3-\delta}$ membrane led to an improved stability in the POM [28]. Recently, bilayered membranes with a protective layer of $Ce_{0.8}Gd_{0.2}O_2$ on $La_{1-x}Sr_xCoO_3$ and $La_{0.9}Sr_{0.1}FeO_3$ were proposed [29]. Using a tubular YSZ-promoted $SrCo_{0.4}Fe_{0.6}O_{3-\delta}$ perovskite membrane, failure was observed after 4 h. By adding a small amount of oxygen to the methane feed, the lifetime of the membrane was extended from 4 to 70 h [30]. Several concepts are being followed to improve stability such as reducing the relative amount of cobalt in the perovskite and codoping the material with less reducible ions like Zr^{4+} or Ga^{3+} [31]. Moreover, new cobalt-free perovskite membrane materials like $BaCe_{0.15}Fe_{0.85}O_{3-\delta}$ [32] and $Ba_{0.5}Sr_{0.5}Zn_{0.2}Fe_{0.8}O_{3-\delta}$ [33] were developed.

During the last few years, remarkable progress was made in the development of stable MIEC materials. Using $La_{0.2}Ba_{0.8}Fe_{0.8}Co_{0.2}O_{3-\delta}$ in a disk-shaped reactor for the POM reaction, oxygen permeation fluxes of 4.4 mL$_N$ min^{-1} cm^{-2} over 850 h were obtained [34]. Impressive results in the POM were obtained recently by the Dalian group studying BSCF tubular [35] and disk [36,37] membranes. Using a CH_4 feed diluted by 20% He, the CO selectivity was 95% at 94% methane conversion, and the stability of the membranes was of the order of 500 h. For better stability under syngas conditions, the material $BaCo_{0.4}Fe_{0.4}Zr_{0.2}O_{3-\delta}$ was developed [38]; stable syngas production for more than 2200 h was obtained with a membrane reactor of this type [39]. Eltron Research developed a material with Brownmillerite structure of the general composition $A_2B_2O_5$. The corresponding membrane reactor was continuously operated over 1 year in syngas production at 900°C. A very high oxygen flux of 10–12 mL$_N$ min^{-1} cm^{-2} linked to a syngas production rate of about 60 mL$_N$ cm^{-2} min^{-1} was reported [40].

2.3.2 HOLLOW FIBER MEMBRANES OF THE COMPOSITION BCFZ OBTAINED BY SPINNING

The reaction of methane with oxygen in a membrane reactor is called *partial oxidation*. However, there is experimental evidence that methane is first oxidized to CO_2 and H_2O and then these products of the total oxidation are reduced by unreacted methane on conventional Ni-based catalysts to CO and hydrogen according to dry reforming (CO_2) and SR (H_2O), respectively. Therefore, synthesis gas formation from methane in an MIEC perovskite membrane reactor is called an *oxidation-reforming process* [41]. Figure 2.23 shows the different zones in a membrane reactor.

By tuning the oxygen supply via the membrane and the processing of this oxygen by tuning the methane flux, kinetic compatibility can be reached and the effluent product flux consists of about 1/3 CO and 2/3 H_2, as shown in Figure 2.24 [42]. Also, Figure 2.25 shows that for the reaction parameters applied, in a temperature window 850°C–900°C, high-quality synthesis gas can be produced. On the other hand, the nonreacted feed methane and the products hydrogen and carbon monoxide as reducing gas can damage the perovskite OTM by reducing the metal oxides of the perovskite to its pure metals (cf. Ellingham diagram). On the other hand, by having an inert porous coating direct on the OTM, the release of oxygen from the perovskite avoids the direct contact of reducing gases like CH_4, H_2, or CO with the perovskite surface [43].

The problem of the hot sealing of tubular OTMs is avoided in a reactor design by Air Liquid (see Figure 2.26). In this design (Figure 2.26), the OTM is used in a

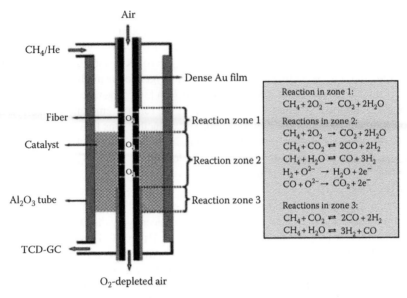

FIGURE 2.23 Hollow fiber BCFZ perovskite membrane in the POM to syngas. Ni-based SR catalyst around the fiber on the methane side. Depending on the interplay of oxygen released from the fiber and the catalyst, three reaction zones can be distinguished.

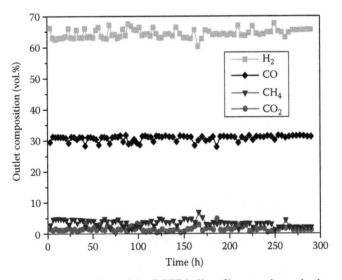

FIGURE 2.24 Long-time testing of the BCFZ hollow fiber membrane in the syngas production by POM. Experimental details: 850°C, core side: air with 150 mL min^{-1}, shell side: 55 mL CH$_4$ min^{-1} + 5 mL steam min^{-1}. Product gas composition: 65% H$_2$, 31% CO, 2.5% CH$_4$, 1.5% CO$_2$, which corresponds to X(CH$_4$) = 95%, S(CO) = 96%, S(CO$_2$) = 4%.

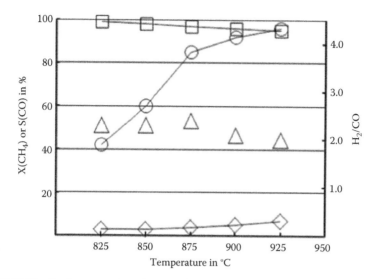

FIGURE 2.25 In situ separation of oxygen from ambient air for the CPOM: performance of a BCFZ hollow fiber membrane reactor in the CPOM as a function of temperature with CH_4 conversion (▢), CO selectivity (≤), CO_2 selectivity (↓), and H_2/CO ratio (ρ). (After Wang, H.H. et al., *Catal. Commun.*, 7, 907, 2006.) Experimental details: Flow rate on the core side of the BCFZ hollow fiber membrane = 150 mL min⁻¹ air, flow rate on the shell side = 20 mL min⁻¹ (10 mL min⁻¹ CH_4 + 10 mL min⁻¹ He). 0.88 cm² effective membrane area, 0.8 g Ni/Al_2O_3 SR catalyst as packed bed on the shell side.

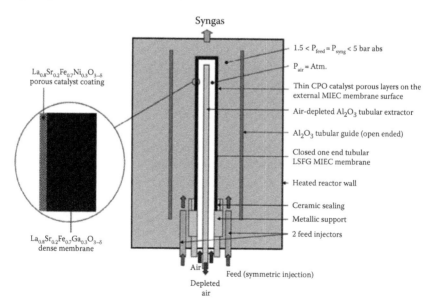

FIGURE 2.26 Schema of the CPOM to synthesis gas in an OTM reactor with dead-end tubes for elevated pressure (up to 5 bar) with air on the core side of the OTM, and methane as feed on the shell side. (From Delbos, C. et al., *Proceedings of the Ninth International Conference on Catalysis in Membrane Reactors*, Lyon, France, June 28–July 2, 2009, pp. S05–S03.)

dead-end configuration. The OTM was $La_{0.8}Sr_{0.2}Fe_{0.7}Ga_{0.3}O_{3-\delta}$ with a porous catalyst layer of $La_{0.8}Sr_{0.2}Fe_{0.7}Ni_{0.3}O_{3-\delta}$ for catalytic partial oxidation of methane (CPOM) brought about by dip coating. The membrane reactor was operated at a pressure of 3 bar with a steam to methane ratio of 1:1 for at least 142 h. During this operation, severe microstructural and chemical evolutions on both catalyst and OTM were observed. However, these degradations seem to have little impact on the reactor performance.

2.3.3 PEROVSKITE MEMBRANE STABILITY TESTS FOR CONCEPT OF OXYGEN IN SITU SEPARATION AND METHANE PARTIAL OXIDATION TO SYNTHESIS GAS USING HOLLOW FIBER MEMBRANES OF THE COMPOSITION BCFZ OBTAINED BY SPINNING

In this section, results of stability tests of BCFZ hollow fiber membranes in the POM to syngas will be reported [46].

The dense BCFZ perovskite hollow fiber membrane was prepared by phase inversion spinning as described in Section 2.2 followed by sintering. Two ends of the hollow fiber were coated by Au paste (C5754, Heraeus). After sintering at 950°C for 5 h, a dense Au film was obtained. Therefore, such Au-coated hollow fibers can be sealed outside the oven at room temperature and the uncoated part can be kept in the middle of the oven ensuring isothermal conditions for the oxygen transport. Air of 150 mL min^{-1} was fed to the core side and a mixture of CH_4 and He was fed to the shell side. Under these conditions, the fiber was subjected to synthesis gas production in different reactor configurations, as shown in Figure 2.27. In Configuration A, no catalyst was used. In Configuration B, a commercial Ni-based SR catalyst (Süd-Chemie) was coated on the fiber by a dip-coating procedure. In Configuration C, 0.4 g Ni-based SR catalyst was packed around the active oxygen transporting part of the hollow fiber membrane. In Configuration D, 0.6 g Ni-based SR catalyst was packed around and behind the uncoated part of the hollow fiber membrane.

2.3.3.1 POM Performance in the BCFZ Hollow Fiber Membrane Reactor without Catalyst (Configuration A)

Perovskites show some catalytic activity and selectivity to light hydrocarbons, for example, for the POM, the oxidative coupling of methane and the oxidative dehydrogenation of ethane. Especially in the oxidative coupling of methane, perovskites show a rather high selectivity to C_2. Therefore, it is informative to investigate the catalytic performance of the pure BCFZ hollow fiber itself without any additional catalyst.

Table 2.2 shows that under our reaction conditions, only CO_2 rather than other carbon-containing products (CO and C_{2+} hydrocarbons) were observed in the outflow when the BCFZ perovskite hollow fiber membrane was used in the POM at 875°C without any catalyst. The methane conversion is lower than 3%. Some amount of unreacted oxygen was found at the outlet of the membrane reactor together with unconverted methane.

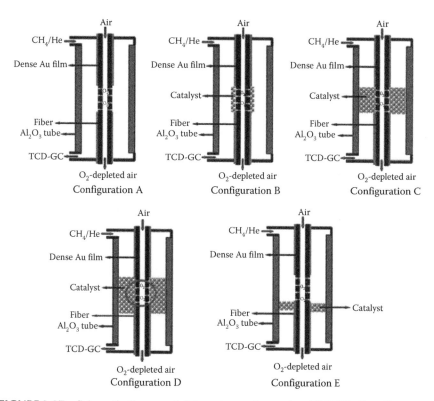

FIGURE 2.27 Schematic diagram of different operation modes of BCFZ hollow fiber membrane reactor for POM. The circle indicates the broken position of the BCZ fiber during POM.

TABLE 2.2

Catalytic Performance of the BCFZ Hollow Fiber Membrane Reactor without Reforming Catalyst in the POM

Methane Concentration in the Feed (%)	CO Selectivity (%)	CO$_2$ Selectivity (%)	CH$_4$ Conversion (%)
10	0	100	1.5
20	0	100	2.6
50	1	99	2.4

Air flow rate on the shell side: 150 mL min^{-1}, total flow rate on the core side: 30 mL min^{-1}, CH$_4$ concentration: 10–50 vol%, membrane surface area: 3.3 cm^2, temperature: 875°C.

Similar results were reported by Balachandran et al. who found that the permeated oxygen reacted with methane yielding CO$_2$ and H$_2$O in a SrFeCo$_{0.5}$O$_x$ tubular membrane reactor in the absence of a reforming catalyst [26]. The presence of CO$_2$, H$_2$O, CH$_4$, and O$_2$ was also reported by Tsai et al. in the effluent gas of a La$_{0.2}$Ba$_{0.8}$Fe$_{0.8}$Co$_{0.2}$O$_{3-\delta}$ disk-shaped membrane reactor without catalyst [35]. Therefore, it is reasonable to use a suitable catalyst, which will catalyze the POM.

2.3.3.2   CPOM Performance in the BCFZ Hollow Fiber Membrane Reactor Coated with Ni-Based SR Catalyst (Configuration B)

The surface of the hollow fiber was coated with the Ni-based SR catalyst by dip coating with a solution of the crashed catalyst. The modified hollow fiber membrane was characterized by field-emission scanning electron microscopy (SEM) (Figure 2.28). From the cross section, it can be concluded that the catalyst layer is homogeneous and a few micrometers thick. However, the catalyst layer shows some porosity, so oxygen permeation can occur.

Table 2.3 shows the CO and CO_2 selectivities as well as the methane conversion as a function of the methane flux on the shell side. Compared with Configuration A, almost no gaseous oxygen was found in the outflow of the reactor in Configuration B. The CH_4 conversion is much higher than in the absence of the catalyst, which indicates that the permeated oxygen had reacted with methane, thus decreasing the

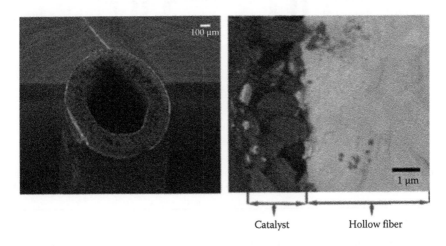

Catalyst   Hollow fiber

FIGURE 2.28   SEM images of the modified hollow fiber membrane with Ni-based catalyst layer.

TABLE 2.3

Catalytic Performance of the POM in the Modified BCFZ Hollow Fiber Membrane Reactor Coated with Ni-Based Catalyst according to Configuration B

CH_4 Flow Rate (mL min⁻¹ cm²)	CO Selectivity (%)	CO_2 Selectivity (%)	CH_4 Conversion (%)
5	0.1	99.9	76.5
10	0.1	99.9	59.2
15	0.2	99.8	63.7
20	7.2	92.8	63.2
25	9.7	90.3	56.7

Air flow rate on the core side: 150 mL min⁻¹, total flow rate on the shell side: 10–50 mL min⁻¹, methane concentration: 50%, membrane surface area: 3.3 cm², catalyst amount: 0.4 g, temperature: 875°C.

oxygen partial pressure and increasing the driving force for oxygen permeation. As a result, a higher oxygen flux through the hollow fiber membrane compared with the uncoated fiber is obtained. However, the CO selectivity is very low. This means that the deep oxidation of methane is the controlling reaction.

Dissanayake et al. [47] studied a Ni-based catalyst for the POM in the conventional packed-bed reactor. They found that an effective catalyst bed consists of three different regions. In the first one at the reactor inlet, the catalyst is in contact with the $CH_4/O_2/He$ feed mixture and the catalyst is present as $NiAl_2O_4$, which has only moderate activity for the complete oxidation of methane to CO_2 and H_2O. In the second region, NiO/Al_2O_3 is present, where the complete oxidation of methane to CO_2 occurs, resulting in a strong temperature increase in this section of the bed. As a result of the complete consumption of O_2 in the second region, the catalyst in the third region consists of a reduced Ni/Al_2O_3 phase. Formation of CO and H_2, corresponding to the thermodynamic equilibrium at the catalyst bed temperature, occurs in this final region via reforming reactions of CH_4 with the CO_2 and H_2O produced during the complete oxidation on the NiO/Al_2O_3 phase.

In our experiment, a thin layer of Ni-based SR catalyst was coated on the surface of the hollow fiber membrane (see Figure 2.28). Because of the continuous oxygen permeation through the membrane, Ni can easily be oxidized to NiO. Therefore, the Ni species of the catalyst layer are presumably NiO or $NiAl_2O_4$, which are active for the total oxidation of methane to produce CO_2 and H_2O instead of the POM to synthesis gas. So it can be easily understood that a lot of CO_2 was observed in the coated hollow fiber membrane reactor. To increase the CO selectivity, more catalyst is needed for CH_4 reforming with CO_2 and H_2O and additional catalyst will be packed in the section behind the active part of the hollow fiber membrane.

It should be pointed out that in Configuration B, the coated BCFZ hollow fiber membrane survived only 5 h when it was operated as a reactor for the POM reaction at 875°C. One possible reason for this behavior is the high CO_2 concentration near the membrane surface. Details of the effects of CO_2 on microstructure and oxygen permeation of a different perovskite are reported elsewhere [48]. The consideration of Ellingham diagrams allows estimating critical conditions for the operation of materials containing alkaline-earth cations [49,50]. For the conditions chosen in the experiments under discussion, it is reasonable to assume that the structure of the hollow fiber membrane was destroyed by $BaCO_3$ formation because of the high CO_2 partial pressure. An alternative explanation for the degradation of the fiber could be the local temperature increase near the membrane surface caused by the total oxidation of methane.

2.3.3.3 CPOM Performance in the BCFZ Hollow Fiber Membrane Reactor Packed with Ni-Based Catalyst (Configurations C and D)

The Ni-based SR catalyst was applied as a packed bed (~0.16–0.4 μm particle size) around the perovskite fiber (shell side, Configuration C). From Table 2.4, it can be seen that the CH_4 conversion is considerably higher than in the absence of catalyst (Table 2.2). This result suggests that the oxygen permeation flux in Configuration C is much higher than without the Ni-based catalyst. Furthermore, the CO selectivity is much higher than in Configuration B. However, the CO selectivity is lower than 82%

TABLE 2.4

Catalytic Performance of the Membrane Reactor with a BCFZ Hollow Fiber with a Ni-Based SR Catalyst (Configurations C and D) in the CPOM Reaction

Catalyst Position	Temperature (°C)	CO Selectivity (%)	CO_2 Selectivity (%)	CH_4 Conversion (%)	H_2/CO Ratio
Around the fiber	825	81.6	18.4	38.7	1.86
	850	79.4	20.6	49	1.85
	875	74.6	25.4	60.9	1.85
	900	60.6	39.4	65.9	1.64
	925	49.1	50.9	71.4	1.4
Around and	825	98.0	2.0	43.9	2.33
behind the fiber	850	98.4	1.6	59.6	2.38
	875	97.8	2.2	85	2.48
	900	97.1	2.9	92.2	2.22
	925	96.6	3.4	95.8	2.08

Air flow rate on the shell side: 150 mL min^{-1}, total flow rate on the shell side: 20 mL min^{-1}, methane concentration 50%, catalyst amount: 0.4 g (for cf. C) and 0.6 g (cf. D), membrane surface area: 0.56 cm^2.

and decreases with increasing temperature. A considerable amount of unreacted methane is also found. At 925°C, both CO_2 and CO selectivities reach 50% at only 70% CH_4 conversion. These experimental findings indicate that the methane reforming with the produced CO_2 and H_2O was not complete.

In the case of additional catalyst behind the active part of the membrane (Configuration D in Figure 2.27), a considerable improvement can be stated and CO and H_2 become the main reaction products (cf. Table 2.4). At 925°C, the CO selectivity is above 97% at roughly 96% CH_4 conversion and the H_2/CO ratio is around 2.08 as expected for POM. The temperature increase in the CH_4 conversion is ascribed to the temperature-accelerated oxygen permeation flux.

The changes in the catalytic performance of the membrane reactor in the presence of the Ni-based SR catalyst can be explained as follows: As the BCFZ membrane has very low intrinsic activity for methane oxidation (Table 2.2), methane conversion occurs nearly exclusively over the Ni-based SR catalyst. Previous research showed that synthesis gas formation from methane in a mixed conducting perovskite membrane reactor may be called an *oxidation-reforming process*, that is, a total oxidation followed by steam and dry reforming steps. In the second step, the concentration of H_2O and CO_2 is reduced by further SR and dry reforming of unreacted CH_4 giving CO and H_2 [31]. By applying more and more catalysts, the catalyst-based residence time is enlarged so this oxidation-reforming process can be completed.

2.3.3.4 Operational Stability and Performance of the Membrane Reactor in the CPOM (Configurations D and E)

The stability of the hollow fiber membrane was studied for two configurations (D and E, in Figure 2.27). Figure 2.29 shows the stability measurements for the

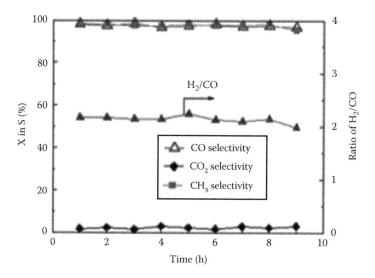

FIGURE 2.29 Performance and stability of the fiber during POM reaction in Configuration D. Air flow rate on the core side: 150 mL min^{-1}, total flow rate on the shell side: 20 mL min^{-1}, methane concentration: 50%, catalyst amount: 0.6 g, membrane surface area: 0.56 cm^2, temperature: 875°C.

POM reaction in the hollow fiber membrane reactor at 875°C in Configuration D. CO and H$_2$ are the main reaction products. CO selectivity of 97% with CH$_4$ conversion of 96% is obtained and the H$_2$/CO ratio is 2.0 as expected for the POM. However, the hollow fiber membrane was broken after 9 h time on stream. The defect occurred at a position in which the membrane was in contact with the Ni-based SR catalyst (as shown in Configuration D in Figure 2.27 with a circle).

The destroyed membrane was studied by SEM and energy-dispersive x-ray spectroscopy (EDXS). As can be seen from Figure 2.23, the fiber became amorphous and porous, especially on its outer parts in contact with the catalyst/methane side. From the initial wall thickness of 170 μm, only 40 μm remained as a dense perovskite phase showing some cracks. About 110 μm of the fiber in contact with the catalyst and about 20 μm on the air side became porous. Different positions of the spent hollow fiber membrane were examined by EDXS. Surprisingly, in the outer and to some extent in the middle parts of the cross section of the porous fiber, Al from the SR catalyst was found. The Al diffusion into the perovskite fiber may be a possible reason for its failure. No Al was observed in the spectrum of the inner intact part (40 μm) of the hollow fiber membrane (Figure 2.30).

By avoiding physical contact between fiber and catalyst, the stability of the membrane in the POM reaction should be improved. In Configuration E, the catalyst was positioned only behind the oxygen permeation zone where the fiber was coated with Au. Figure 2.31 shows the stability measurements for the POM reaction in the hollow fiber membrane reactor at 875°C. A CO selectivity of above 90% with a methane conversion of 93% and a H$_2$/CO ratio around 2 are obtained. It is obvious that the fiber in this case survives longer than in Configuration D because the Au film avoids

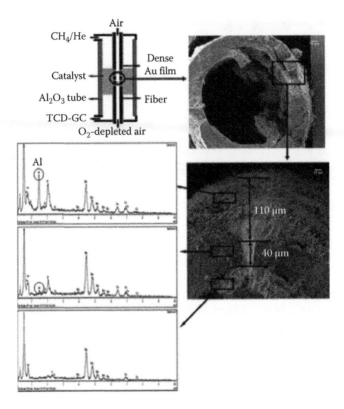

FIGURE 2.30 SEM and EDXS of spent fiber after 9 h operation in the CPOM in Configuration D (cf. Figure 2.27).

the contact of catalyst and fiber. Compared with the other configurations, the stability of the fiber in Configuration E is significantly improved. By separating catalyst and fiber, the hollow fiber membrane reactor could be steadily operated more than 300 h in the POM [43].

It follows from literature analysis that in most of the POM studies, the catalyst was in direct contact with the membrane and usually problems such as the impact of reducing atmospheres and solid-state reactions between catalyst and membrane were not mentioned. In this study, we found the best choice to pack the catalyst behind the membrane. At first sight, a coupling of the three steps (1) oxygen permeation, (2) membrane total oxidation, and (3) carbon dioxide/SR with methane looks promising for several reasons. The continuous consumption of the permeated oxygen will establish a constant high driving force for the oxygen flux and the combination of the exothermal combustion with the endothermal reforming could guarantee optimal heat management in Configurations C and D. However, there are also reasons speaking against these concepts, as high concentrations of reducing gases produced by the reforming step (CO, H_2) could reduce the BCFZ fiber and the direct contact between catalyst and fiber can initiate an exchange of mobile species between BCFZ and catalyst by solid-state ion diffusion (Fe, Co).

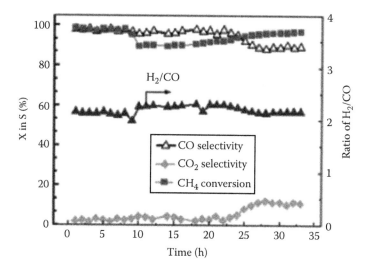

FIGURE 2.31　Stability of the fiber during CPOM reaction in Configuration E. Air flow rate on the core side: 150 mL min⁻¹, total flow rate on the shell side: 20 mL min⁻¹, methane concentration: 50%, catalyst amount: 0.4 g, membrane surface area: 0.56 cm², temperature: 875°C.

Obviously, in Configuration E—our best configuration—the oxygen permeation zone of the BCFZ hollow fiber is in contact with methane and one could assume a reduction of the perovskite by methane. However, this reduction does not take place, most probably due to a continuous flow of oxygen released from the perovskite. In other words, there is always a nonzero oxygen partial pressure at the perovskite surface, which means that no pure methane as reducing agent is in contact with the perovskite. Furthermore, the benefit of Configuration E is that no major CO_2, CO, or H_2 concentrations are in contact with the unprotected perovskite fiber because the formation of these products is kinetically suppressed in the homogeneous gas-phase reaction without catalyst. And even if these products are formed, the oxygen flow from the perovskite surface would reduce their impact.

In the conventional technology, pure oxygen is needed for the CPOM. This oxygen is conventionally produced by cryogenic fractionation technology, which requires a large-scale plant and high-operation costs. In the membrane reactor technology, the membrane can be used in situ to produce pure oxygen from air, which is competitive to the conventional technology.

The CPOM to syngas was investigated at 825°C–925°C using a membrane reactor based on BCFZ hollow fibers with and without SR catalyst in different catalyst arrangements. It was found that to get a high CO selectivity, an additional amount of catalyst was needed behind the active membrane area to ensure that the reforming reactions can take place completely. Stability studies of the hollow fiber membrane in the POM reaction showed that the stability of the fiber can be significantly improved if the contact between fiber and catalyst is avoided. A stable reactor concept is proposed with an arrangement of the reforming catalyst behind the oxygen permeation zone. In this case, the BCFZ perovskite is not attacked by strongly reducing components like CO and H_2 or by carbonate formation with CO_2 because the formation of these components at the perovskite site is kinetically suppressed without a catalyst. Furthermore, the continuous release of oxygen from the perovskite surface results in a nonzero oxygen partial pressure at the surface and prevents a reduction of the perovskite by, for example, methane.

2.4 CONCLUSIONS AND SUMMARY

Since the 30% energy savings in oxygen production in comparison with established separation technologies is a high driving force, numerous R&D projects have been launched worldwide in the last 20 years. However, most of them failed because of insufficient materials stability at the operation temperature of 800°C–900°C. A major problem is the gas tight sealing of the planar (stacks) or tubular OTMs. One of the rare successful developments—maybe the worldwide unique successful development—is the pilot plant of Air Products that started operation in 2012. Figure 2.32 shows a photograph.

Air Products carried out first experiments in 2006, and as of July 2012, the pilot plant has operated for a total on-stream time of over 1000 days with oxygen purities reaching 99.9%. The team has successfully produced up to 5 tons of oxygen per day. Continued operation of this unit has verified thermal and pressure cycle performance. The heart of the plant is the planar membrane wafers, as shown in Figure 2.33.

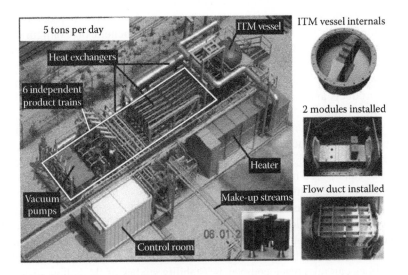

FIGURE 2.32 Aerial photo of the Air Products pilot plant (subscale engineering prototype). (From Repasky, J.M. et al., *International Pittsburgh Coal Conference 2012*, Pittsburgh, PA, October 15–18, 2012.)

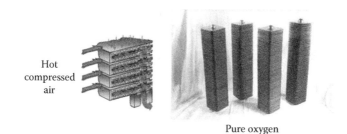

Hot compressed air

Pure oxygen

FIGURE 2.33 Planar membrane wafer design: Four cut-away wafers are shown on a common oxygen withdrawal tube. Hot compressed air flows between each adjacent pairs of wafers; oxygen is withdrawn at the bottom. Each of the four modules is capable of producing 1 ton of oxygen per day (http://www.airproducts.com/~/media/Files/PDF/industries/itm-oxygen-technology-280-12-058-glb.pdf).

REFERENCES

1. T. Schiestel, M. Kilgus, S. Peter, K.J. Caspary, H. Wang, J. Caro, *J. Membr. Sci.* 258 (2005), 1.
2. H. Liang, F. Jiang, O. Czuprat, K. Efimov, A. Feldhoff, S. Schirrmeister, T. Schiestel, H.H. Wang, J. Caro, *Chem. Eur. J.* 16 (2010), 7898.
3. F. Liang, H. Jiang, H. Luo, R. Kriegel, J. Caro, *Catal. Today* 193 (2012), 95.
4. H. Luo, K. Efimov, H. Liang, A. Feldhoff, H. Wang, J. Caro, *Angew. Chem. Int. Ed.* 50 (2011), 759.
5. R. Spillman, *in* R.D. Noble, S.A. Stern (eds.). Economics of gas separation membranes, *Membrane Science and Technology*, Chapter 13, Vol. 2. Elsevier, New York, 1996, pp. 589–668.
6. H.H. Wang, S. Werth, T. Schiestel, J. Caro, *Angew. Chem. Int. Ed.* 44 (2005), 6906.
7. F.Y. Liang, H.Q. Jiang, T. Schiestel, J. Caro, *Ind. Eng. Chem. Res.* 49 (2010), 9377.
8. R. Kriegel, *in* J. Kriegesmann (ed.), *Handbuch Technische Keramische Werkstoffe*, HvB-Verlag Ellerau, 2010, p. 1–46.
9. W. Jin, S. Li, P. Huang, N. Xu, J. Shi, Y.S. Lin, *J. Membr. Sci.* 166 (2000), 13.
10. H.H. Wang, R. Wang, D.T. Liang, W.S. Yang, *J. Membr. Sci.* 243 (2004), 405.
11. A.V. Kovalevsky, A.A. Yaremchenko, V.A. Kolotygin, A.L. Shaula, V.V. Kharton, F.M.M. Snijkers, A. Buekenhoudt, J.R. Frade, E.N. Naumovich, *J. Membr. Sci.* 380 (2011), 68.
12. Cooretec-report, Forschungs- und Entwicklungskonzept für emissionsarme fossil befeuerte Kraftwerke, BMWI-Document 527, 2003, pp. 1–46.
13. P.A. Armstrong, D.L. Bennet, E.P.T. Förster, E.E. van Stein, *The Gasification Technology Conference*, Washington, DC, October 3–6, 2004.
14. X.Y. Tan, K. Li, *AIChE J.* 53 (2007), 838.
15. R. Kriegel, W. Burckhardt, I. Voigt, M. Schulz, E. Sommer, *Proceedings of the 10th International Conference on Inorganic Membranes (ICIM10)*, Tokyo, Japan, August 18–22, 2008.
16. M.P. Dudukovic, *Catal. Today* 48 (1999), 5.
17. B.C.H. Steele, *Mater. Sci. Eng.* B13 (1992), 79.
18. H.J.M. Bouwmeester, *Catal. Today* 82 (2003), 141.
19. H.J.M. Bouwmeester, A.J. Burggraaf, *in* A.J. Burggraaf, L. Cot (eds.), *Fundamentals of Inorganic Membrane Science and Technology*, Membrane Science and Technology Series, Vol. 4. Elsevier, Amsterdam, the Netherlands, 1996, pp. 435–528.
20. W. Jin, X. Gu, S. Li, P. Huang, N. Xu, J. Shi, *Chem. Eng. Sci.* 55 (2000), 2617.

21. U. Balachandran, J.T. Dusek, R.L. Mieville, R.B. Poeppel, M.S. Kleefisch, S. Pei, T.P. Kobylinski, C.A. Udovich, A.C. Bose, *Appl. Catal.* A 133 (1995), 19.
22. G. Majkic, L.T. Wheeler, K. Salama, *Solid State Ionics* 146 (2002), 393.
23. G. Majkic, L.T. Wheeler, K. Salama, *Solid State Ionics* 164 (2003), 137.
24. N. Nagendra, R.F. Klie, N.D. Browning, S. Bandhopadhyay, *Mater. Sci. Eng.* A 341 (2003), 236.
25. N. Nagendra, S. Bandhopadhyay, *J. Eur. Ceram. Soc.* 23 (2003), 1361.
26. U. Balachandran, J.T. Dusek, P.S. Maiya, B. Ma, R.L. Mieville, R.B. Poeppel, M.S. Kleefisch, C.A. Udovich, *Catal. Today* 36 (1997), 265.
27. S. Pei, M.S. Kleefisch, T.P. Kobylinski, J. Faber, C.A. Udovich, V. Zhang-McCoy, B. Dabrowski, U. Balachandran, *Catal. Lett.* 30 (1995), 201.
28. E.D. Wachsmann, T.L. Clites, *J. Electrochem. Soc.* 149 (2002), A424.
29. M. Schroeder, *Phys. Chem. Chem. Phys.* 7 (2005), 166.
30. X. Gu, L. Yang, L. Tan, W. Jin, L. Zhang, N. Xu, *Ind. Eng. Chem. Res.* 42 (2003), 795.
31. W. Yang, H. Wang, X. Zhu, L. Lin, *Top. Catal.* 35 (2005), 155.
32. X. Zhu, H. Wang, W. Yang, *Chem. Commun.* (2004), 1130.
33. H. Wang, C. Tablet, A. Feldhoff, J. Caro, *Adv. Mater.* 17 (2005), 1785.
34. C.Y. Tsai, A.G. Dixon, W.R. Moser, Y.H. Ma, *AIChE J.* 43 (1997), 2741.
35. H. Wang, Y. Cong, W. Yang, *Catal. Today* 82 (2003), 157.
36. Z. Shao, G. Xiong, H. Dong, W. Yang, L. Lin, *Sep. Purif. Technol.* 25 (2001), 97.
37. H. Dong, Z. Shao, G. Xiong, J. Tong, S. Sheng, W. Yang, *Catal. Today* 67 (2001), 3.
38. J.H. Tong, W.S. Yang, B.C. Zhu, R. Cai, *J. Membr. Sci.* 203 (2002), 175.
39. J.H. Tong, W.S. Yang, R. Cai, B.C. Zhu, L.W. Lin, *Catal. Lett.* 78 (2002), 129.
40. A.F. Sammells, M. Schwartz, R.A. Mackay, T.F. Barton, D.R. Peterson, *Catal. Today* 56 (2000), 325.
41. C. Chen, S. Feng, S. Ran, D. Zhu, W. Liu, H.J.M. Bouwmeester, *Angew. Chem. Int. Ed.* 42 (2003), 5196.
42. J. Caro, K.J. Caspary, C. Hamel, B. Hoting, P. Kölsch, B. Langanke, K. Nassauer et al., *Ind. Eng. Chem. Res.* 46 (2007), 2286.
43. H.Q. Jiang, F. Liang, O. Czuprat, K. Efimov, A. Feldhoff, S. Schirrmeister, T. Schiestel, H.H. Wang, J. Caro, *Chem. Eur. J.* 16 (2010), 7898.
44. H.H. Wang, C. Tablet, T. Schiestel, S. Werth, J. Caro, *Catal. Commun.* 7 (2006), 907.
45. C. Delbos, G. Lebain, N. Richet, C. Bertail, *Proceedings of the Ninth International Conference on Catalysis in Membrane Reactors*, Lyon, France, June 28–July 2, 2009, pp. S05–S03.
46. H.H. Wang, A. Feldhoff, J. Caro, T. Schiestel, S. Werth, *AIChE J.* 55 (2009), 2657.
47. D. Dissanayake, M.P. Rosynek, K.C.C. Kharas, J.H. Lunsford, *J. Catal.* 132 (1991), 117–127.
48. M. Arnold, H.H. Wang, J. Martynczuk, A. Feldhoff, *J. Am. Ceram. Soc.* 90 (2007), 3651–3655.
49. A. Feldhoff, J. Martynczuk, H.H. Wang, *Progr. Solid State Chem.* 35 (2007), 339–353.
50. A. Feldhoff, J. Martynczuk, M. Arnold, H.H. Wang, *Solid State Sci.* 10 (2008), 689–701.
51. J.M. Repasky et al., *International Pittsburgh Coal Conference 2012*, Pittsburgh, PA, October 15–18, 2012.

3 Recent Developments in Membrane Technologies for CO_2 Separation

Charitomeni Veziri, Anastasios Labropoulos, Georgios N. Karanikolos, and Nick K. Kanellopoulos

CONTENTS

3.1   INTRODUCTION

Separation of carbon dioxide from other gases is critically important for many industrial applications, such as hydrogen production [1–3], natural gas purification [4], biogas purification [5], CO_2 scrubbing of power plant combustion exhausts [6], and capture of CO_2 from advanced power generation sources such as the integrated gasification combined cycle (IGCC) [7,8]. Carbon capture and storage (CCS) is an efficient way to reduce CO_2 concentration in the atmosphere. It is a three-step process including capture of waste CO_2 from other emissions before entering the atmosphere, CO_2 transportation to a storage site, and its permanent storage. Among them, the CO_2 capture is the most challenging key step in which new adsorbent materials need to be developed. Three main approaches have been proposed for the separation of CO_2/CH_4 and CO_2/N_2 mixtures: absorption with liquid solvents, membranes, and adsorption using porous solids. Conventional adsorbent materials rely on either chemisorption or physisorption to capture CO_2. Amine scrubbing has been extensively used in order to remove CO_2 from gases and utilizes alkanolamines such as monoethanolamine (MEA) in aqueous solutions as the adsorbent, relying on the chemical reaction between the amine group and CO_2 to generate carbamate or bicarbonate [9]. The biggest problem with amine scrubbing, however, is that large amounts of heat are needed to release absorbed CO_2 during adsorbent regeneration. Moreover, the amine scrubbing solutions are corrosive and chemically unstable upon heating. Additionally, because of their liquid form, their handling is considerably more difficult than that of solid adsorbents. In amine-functionalized adsorbents such as mesoporous molecular sieves in solid form although they partially overcome some of the aforementioned limitations, their parasitic energy waste is still pretty high [10]. In contrast, physisorption between solid adsorbents and CO_2 molecules is a reversible process that requires much less energy for desorption. Traditional adsorbents such as zeolites and activated carbons have been extensively studied for CO_2 capture [11,12].

High porosities endow activated carbon and charcoal with CO_2 capture capacities of 10%–15% by weight. However, their CO_2/N_2 selectivities are relatively low. For zeolitic materials, on the other hand, while they offer CO_2/N_2 selectivities 5–10 times greater than those of carbonaceous materials, their CO_2 capacities are 2–3 times lower [13,14]. Moreover, zeolite performance is compromised in the presence of water vapor. To be competitive with liquid solvents, solid sorbents must exhibit substantially greater capacities and selectivities for CO_2 than currently available physical sorbents and be less sensitive during their explosion to steams. Advanced research in the development of new adsorbent materials may overcome many of the limitations of the currently available adsorbents. Some of these approaches are discussed here.

Membrane-based separation has recently attracted significant attention for CO_2 capture applications because it has the potential to offer high energy savings, lower operating cost, and small unit footprint. Many membranes for selective CO_2 separation have been developed during the last several decades [15]. The major advantage of membrane intensification for CO_2 capture is the low energy consumption, due to the elimination of the energy-consuming regeneration step. However, as in the case of absorption and adsorption, economic considerations dictate if membrane systems can be applied to recover CO_2 from flue gas. Membrane permeation is generally pressure driven, that is, the feed gas is compressed and/or the permeate channel operates under vacuum and/or a sweep gas is employed to enhance the separated gas yield. Due to the low partial pressure of CO_2 in the flue gas, higher driving pressures are needed and this constitutes a major challenge for the membrane-based separation compared to liquid and solid absorbents that are thermally regenerated.

Over recent years, membrane technologies have been also successfully applied industrially [16], in many cases replacing traditional, energy-demanding, and environmentally polluting separation techniques. Intensification of processes utilizing membranes has been applied for water desalination [17,18], oxygen/nitrogen separation [19–22], and carbon dioxide separation from natural gas [23–25]. In order for membranes to be rendered as economically viable and exhibit the necessary level of sustainability for utilization in gas separation processes at the industrial scale, they are required to exhibit high-flux and high-selectivity factors. The goal is to obtain a high permeability of one species and a high selectivity of that species over any other in the mixture.

Separation in membranes is carried out by exploiting differences in properties such as particle size of the gaseous or liquid mixture components and chemical affinity. Membranes, which in most cases consist of thin polymeric films, owe their selectivities to the relative rates at which chemical species permeate. Differences in permeation rates are generally due (in the case of porous membranes) to the relative sizes of the permeating molecules or (in the case of dense membranes) their solubilities and/or diffusion coefficients (i.e., mobilities) in the membrane material. Because permeation rates vary inversely with membrane thickness, membranes are made to be as thin as possible without compromising mechanical strength (which is frequently provided by nonselective, porous support layers).

Gas transport can occur through several mechanisms [26–34] depending on the type of material that is being used. Solution diffusion occurs in polymers and is based on the solubility and mobility factors, which favor the transport of the most condensable component and the component having the smallest molecules, respectively. Molecular sieving occurs in microporous ceramic barriers, carbon molecular sieves, and zeolites and is based primarily on higher diffusion rates of the species having smaller kinetic diameter, while adsorption and condensation ability differences become important for mesoporous membranes.

Polymers used for CO_2/N_2 separation membranes have to meet certain criteria [35]. One is their sufficient CO_2 permeation rate through the membrane; so a reasonable gas flux is achieved during separation. The second criterion is the separation of carbon dioxide from other gases. The third one is that the polymeric membrane needs to provide good thermal and mechanical properties; hence, the separation can be conducted effectively and sometimes at elevated temperatures. The preceding criteria are commonly met by synthesis of a block copolymer system. The copolymer usually processes a hard block and a soft block. The hard block can be synthesized by polymer with well-packed and more rigid structures; therefore, it forms a glassy segment of the polymer chain. On the other hand, the soft block can be synthesized from a polymer with more flexible chains, and that can form rubbery segments on the polymer chain. When a polymeric membrane is formed by the use of these copolymers, glassy polymer segments will form a structural frame and provide mechanical support. If the hard block is formed by high-temperature polymers such as polyimides, it can also provide better thermal resistance. On the other hand, the rubbery segments usually form continuous microdomains, and the flexible chain structure allows the transportation of gas, hence providing a good permeability. Usually, the balance of the hard and soft block ratio provides the good separation without loss of permeability. The glass segments usually have lower free volume, while the rubbery segments provide higher free volumes that increase gas permeation rates. The high free volume can also be achieved by intruding bulky structures into the soft blocks. Examples of polymers used in the construction of gas separation membranes include polyacetylenes, polyaniline, poly(arylene ether)s, polyacrylates, polycarbonates, polyetherimides, poly(ethylene oxide), polyimides, poly(phenylene oxide)s, poly(pyrrolone)s, and polysulfones.

Recently, Wiley and coworkers [36] published the results of extensive calculations that explore the dependence of CO_2 capture costs on membrane selectivity, permeability, and unit price. Most significantly, for membranes to be competitive with amine-based absorption for capturing CO_2 from flue gases, their CO_2/N_2 selectivities (i.e., permeability ratios) must be in the 200 range. With rare exception, the selectivities of available polymers fall well below that. While many have selectivities of 50–60, they present low permeances [35]. Once again, cost-effectiveness may be achievable only when separation is promoted by a CO_2-selective chemical reaction. It has been demonstrated that by virtue of their reversible reactions with CO_2, amines can raise the CO_2/N_2 selectivity of polymeric membranes to 170 while also boosting CO_2 fluxes [36]. If these encouraging results are sustainable for extended periods of operation, such systems are expected to merit serious consideration for CO_2 capture.

A widely investigated type of membranes for CO_2 separation pertains to the supported liquid membranes (SLMs). In general, the SLM systems are porous membranes whose pores are saturated with a solvent mixture having high absorption capacity for a specific gas species. The gas is dissolved at the feed/SLM interface, diffuses through the liquid phase within the pores, and desorbs at the opposite membrane surface. Therefore, the solvent mixture acts as a gas carrier and the supporting membranes are also called *facilitated-transport membranes* (FTMs). In particular, SLMs containing CO_2 absorbents such as aqueous amines, which serve as CO_2 carriers, have attracted considerable attention because they facilitate CO_2 transport through the membrane material [1,37,38]. SLM systems combine the processes of extraction and stripping, and the amount of solvent in a process based on SLMs is also much less than that in the conventional solvent extraction process.

Although aqueous amine-based FTMs offer promising separation performance through high CO_2 permeability and selectivity, they have serious disadvantages, namely, the loss of volatile amines and the evaporation of water from the membrane, which lead to alteration of amine concentration. Especially for separations at elevated temperatures, the SLM systems are quite unstable since they experience significant solvent loss due to vaporization. Partial dissolution of CO_2 and other acidic gases in the aqueous solutions constitutes an additional disadvantage. Moreover, application of low-pressure gradients (<10 kPa) might cause displacement of the liquid phase from the pore network and substantially deteriorate their separation efficiency. Therefore, these systems suffer from low stability that renders them unsuitable for gas separation processes at relatively high pressures and temperatures.

On the other hand, CO_2 can be successfully separated from other gas components by utilization of ionic liquids (ILs) due to the fact that the quadrupole moment of the CO_2 molecules interacts with the electrical charges of the ion pairs [39,40]. This specific interaction provides an enhancement in the solubility over other gases such as N_2 or CH_4. The ILs are low-melting-point salts that are liquid at or slightly above room temperature. If they remain liquid at room temperature and below, they are known as room temperature ILs (RTILs). The counterions are poorly coordinated due to the large size and asymmetric structure of the cation. At least one ion has a delocalized charge and one component is organic, which prevents the formation of a stable crystal lattice. As a result, these solvents are liquids below 100°C, or even at room temperature. Properties, such as melting point, viscosity, bulk conductivity, solubility, chemical affinity, and hydrophilic/hydrophobic character of these solvents, as well as their thermal behavior and thermal stability, are determined by the nature of the organic cation (kind of charged ring or group and possible substituents) and by the nature of the counteranion. A characteristic property of RTILs is their negligible vapor pressures, and thus they present no measurable solvent loss due to volatilization (in general, they are considered as nonvolatile solvents).

3.2 SUPPORTED IONIC LIQUID MEMBRANES FOR CO_2 SEPARATION

Considering the wide variety of asymmetric cations and anions forming ILs, an extremely large number of ion pair combinations are possible, by varying the cation or the anion components. Hence, it is clear that ILs can be synthesized in a huge range of structures. This incredible flexibility provides the opportunity to *tune* or adjust the physical/chemical and thermophysical properties of these solvents and thus renders them capable of being implemented to a wide range of engineering and chemical synthesis/catalysis applications. The properties of ILs that make them of interest for these applications include nonflammability, negligible vapor pressure, good conductivity, stable liquid state, and high thermal stability over a wide temperature range. The unique properties of the ILs also allow one to obtain materials with very different morphologies. These include the mixtures of ILs, polymers, gels (ionogels), and composite structures including mixed-matrix membranes. Applications of RTILs include, but are not limited to, *green* solvents for reactions, gas absorption and capture in bulk fluid, membrane separation processes, and electrochemical systems. Fuel cell membranes take advantage of the fact that the ILs are electrically charged fluids. The IL can be impregnated into a membrane [41,42] or synthesized as the membrane [43] to improve the performance. The resulting membrane has conductivity properties, and this can be useful in a range of applications.

The use of RTILs as media for gas separations appears especially promising. The solubility and diffusion of several gases, particularly CO_2, N_2, and CH_4, has already been studied—experimentally and computationally—in numerous RTILs in the bulk phase [44–60], and the results showed very promising CO_2 absorption capacity and high CO_2/N_2 separation selectivities for energy and environmental applications such as CCS processes. The solubility of other gases of industrial importance such as hydrocarbons (e.g., for paraffin/olefin separations), H_2 (CO_2/H_2 separations), and SO_2 is also of particular interest for ILs. Process temperature and the chemical structures of the cation and anion have significant impacts on gas solubility and gas pair selectivity. It is also possible to use one of the ions in the IL as a complexing agent giving a maximum loading much greater than loadings typically observed in standard solvent systems. These so-called *task-specific* ILs (TSILs) are capable of absorbing very high quantities of a specific gas. For example, Davis and coworkers reported the synthesis of an IL containing an amine functionality that was capable of reversibly absorbing 0.5 mol of carbon dioxide per mole of IL [61]. This absorption capacity was more than 100 times greater than the physical CO_2 uptake in alkyl-imidazolium-based [Rmim][Tf_2N] solvents at similar temperature and pressure conditions. More recently, 1:1 complexation has been reported [62] illustrating the fact that functionalized RTILs are a very interesting class of nonvolatile chemical (reactive) solvents for CO_2 separations.

However, one of the major drawbacks of RTILs is their high viscosity, which leads to very slow CO_2 diffusion and makes their industrial scale application unfeasible. As a more convenient solution, immobilization of a thin layer of IL in sufficiently permeable membranes has already evidenced the possibility to overcome the problem of slow diffusivity and improve the efficiency of CO_2 removal from flue gas. Up-to-date RTILs have been immobilized in macroporous polymeric supports [8,37,63–72] and scarcely on inorganic nanoporous ones [73,74]. Previous testing of polymer-based, supported IL membranes (SILMs) showed promising results with permeabilities/selectivities that were consistently above the upper bound of a Robeson plot for CO_2/N_2 separation [75,76].

A major advantage of using RTILs in a SILM configuration is the negligible vapor pressure of RTILs. As opposed to other liquids, RTILs will not evaporate at any appreciable rate under ambient conditions. The nonvolatile nature of this liquid phase provides stability for long-term operations and prevents contamination of the gas streams. Furthermore, if the high thermal stability of ILs is taken into account (together with their negligible evaporation rate), it becomes clear that SILM systems can be successfully implemented in separations at elevated temperatures. Higher SILM stability is also attained because of the greater capillary forces associated with high viscosity of ILs, which could reduce displacement of the liquids from the pores under applied pressure gradient. These are the most distinctive advantages of SILMs over conventional SLM membranes. However, the use of RTILs in a SILM system may not be practical if the pressure difference across the membrane is greater than the capillary forces holding the RTIL in the pores of the support. A second approach is to combine RTILs with polymers to make a homogeneous composite membrane. By using cross-linkable polymers, large quantities of RTILs can be incorporated so as to form a homogeneous film. The addition of the polymer can greatly decrease gas permeability because of a decrease in the rate of diffusion. However, as there are no chemical bonds tethering the RTILs to the polymer material, there is also a limit to the pressure difference across the membrane as moderate pressure differentials may compress the membrane and *squeeze* out the RTIL.

In reference to CO_2 capture from flue gases, economic studies on CO_2 capture using membranes operating under vacuum conditions versus compression of the feed gases indicate that membranes that operate with cross membrane pressures of <2 bars and with CO_2/N_2 selectivities >50 can be more efficient than existing systems for CO_2 capture. SILMs can potentially meet this 2 bars and selectivity >50 specification for CO_2 capture.

The developed methodologies for predicting and projecting the upper limits for the permeance of a given gas through a SILM membrane start with the assumption that gas transport through the specific membrane follows a dissolution/diffusion model:

$$P_e = DS \qquad (3.1)$$

where

P_e is the permeability
S is the solubility (in moles per volume per partial pressure)
D is the diffusivity

In a further step, models that describe the components, S and D, in RTILs are taken into account and compared against SILMs data. Kilaru and Scovazzo [45], Camper et al. [49], and Scovazzo et al. [56,77,78] have proposed empirical models for the prediction of gas solubility/selectivity and diffusivity in relation to the RTIL molar volume and viscosity. Eventually, the predicted permeability/selectivity performance of SILMs is compared to the experimental one, usually obtained via the time-lag [79] technique, and the up-to-date results have shown converging of experiment with prediction [80]. A general conclusion of previous studies, encompassing all RTIL families that interact physically with CO_2, is that the critical properties affecting SILM performance are the molar volume and viscosity of the RTILs [80]. These properties have been shown to decisively affect the solubility and diffusivity of gases dissolved in bulk ILs. Specifically, for CO_2/N_2 separation, decreasing the RTIL molar volume increases selectivity and increasing viscosity decreases CO_2 permeability. Apart from the IL physicochemical properties, gas solubility and solubility–selectivity (defined as the ratio of the solubility of the two components in a gas pair) is dependent on the nature of the gas solutes and the ion pair components (primarily the anionic species) that determine the strength of gas/solvent interactions.

The separation performance of the SILM membranes is also dependent on the type of support that is used. In the majority of the publications dealing with SILM systems for CO_2 capture, polymeric porous substrates have been employed as supports for the liquid phase. A small number of studies have been also reported, regarding systems incorporating inorganic (ceramic) membranes as supports. There are also results from a few hybrid or mixed-matrix SILM membranes consisting of polymeric–zeolitic supports with ILs. Besides the combinations of IL and support types, operation conditions also affect the productivity (i.e., permeate flux) and separation efficiency of a given SILM system as well as its stability and capability to be successfully employed in industrial separation process under specific conditions. Feed CO_2 concentration, temperature, and transmembrane pressure determine the obtained permeate flux and selectivity of CO_2 permeation over several other gases. In the following, a reference is provided on the impact of the aforementioned parameters on the separation performance of these membranes.

3.2.1 PHYSICOCHEMICAL PROPERTIES OF THE IL PHASE AND PERMEATING GAS SOLUTES

3.2.1.1 Viscosity of IL Phase

Most RTILs are viscous liquids, their viscosities ranging between 10 and 500 mPa s, being two or three orders of magnitude greater than conventional organic solvents. More and more attentions have been paid to the design of ILs with lower viscosity. It has been found that the viscosity of $[C_n mim][BF_4]$ with n equal to 2, 4, and 6 increases from 66.5 to 314 mPa s [81], whereas for $[C_n mim][Tf_2N]$ [82], with increasing number of alkyl chain carbon atoms from 1 to 4, the viscosity first decreases and then increases again. However, regarding the reported viscosity data for RTILs in the literature, some discrepancy can be noticed between several works. For instance, for the viscosity of [bmim][PF$_6$], Yanfang et al. [83] found a value of 205.57 mPa s at 298.15 K, which was lower than that of a dried sample with 0.019% of water in mass fraction but higher than that of a water-saturated sample containing 2.68% of water, reported by Jacquemin et al. [84]. Zhu et al. [85] have obtained a value of 217.9 mPa s at 298.15 K, which is somewhat higher than that found by the previously mentioned authors, whereas Okoturo and van der Noot [86] reported a lower viscosity of 173 mPa s at 298.15 K. In addition, Harris and Woolf [87] reported a higher value of 273 mPa s at 298.15 K, and up to now, the highest values of viscosity have been obtained by Huddleston et al. [88] of 397 mPa s for a water-equilibrated sample and 450 mPa s for a dried one. Therefore, besides the effect of temperature and experimental method, the presence of trace amounts of impurities such as water and halogen ion in a given RTIL can greatly modify its viscosity values.

It has been found that the influence of temperature on the viscosity of ILs is much greater than that on organic solvents. On the other hand, an Arrhenius-like equation can be used to fit the temperature effect on the viscosity of the IL mixtures:

$$n = n_\infty \exp\left(\frac{E_a}{RT}\right) \qquad (3.2)$$

where
　　n is the viscosity of the IL
　　n_∞ and E_a are the characteristic parameters

The temperature dependency of the viscosity can also be given by the Vogel–Fulcher–Tammann (VFT) equation:

$$n = n_o \exp\left(\frac{B}{T - T_o}\right) \tag{3.3}$$

where
n_o is in mPa s
B (K) and T_o (K) are the constants

Tokuda et al. [44] studied several physicochemical and thermal properties of a series of RTILs prepared from different cationic species combined with the same anion, bis(trifluoromethane sulfonyl)imide ($[(CF_3SO_2)_2N]$), over a wide temperature range. The authors found that the relatively high viscosity of the studied ILs from 40.0 to 76.6 mPa s at 30°C was well justified in terms of their low diffusion coefficients of the sum of the cation and anion. In the temperature range of measurements, the macroscopic viscosity values followed the order of cations: $[(n\text{-}C_4H_9)\text{-}(CH_3)_3N]^+ >$ [bmpro]$^+$ > [bpy]$^+$ > [bmim]$^+$, which coincided with the reverse order of the ionic self-diffusion coefficients. The contrastive features of the viscosity with ionic diffusivity agreed well with a previous report of the same research group for [bmim]-based RTILs with different anionic structures and $[Rmim][(CF_3SO_2)_2N]$ ILs with different alkyl chain length [89], and thus the microscopic ion dynamics reflect the macroscopically observed viscosity.

The viscosity can also affect the solvation properties of RTILs that physically interact with gases. Kilaru and Scovazzo [45] developed a theory relating the low-pressure solubility of gas solutes to the viscosity of several groups of RTILs (with imidazolium, phosphonium, and ammonium cations), based on the Hildebrand solubility parameter, δ_1, of these solvents. The Hildebrand solubility parameter of the solvent is defined as the square root of the cohesive energy density (CED), which is the ratio of the energy of vaporization of the solvent, relative to its molar volume:

$$\delta_1 = \left(\frac{E^{vap}}{V_1^L}\right)^{1/2} \tag{3.4}$$

where
the energy of vaporization (E^{vap}) can be estimated from Eyring's activation energy of viscosity
V_1 is the RTIL molar volume, expressed in cm^3/mol

The energy of vaporization is related to the molar free energy of activation, ΔG_{vis}^0, through a proportionality constant, K_v, giving

$$E^{vap} = K_v \Delta G_{vis}^0 = K_v RT \ln\left(\frac{\mu V_1^L}{hN_A}\right) \tag{3.5}$$

where
  μ is the dynamic viscosity of RTIL (cP)
  h is the Planck constant (J s)
  N_A is Avogadro's number

Substituting Equation 3.5 into Equation 3.4 results in the following form of the solubility parameter equation:

$$\delta_1 = \left\{\frac{K_v RT}{V_1^L} \ln\left[\frac{(1\times10^{-9})\mu V_1^L}{hN_A}\right]\right\}^{1/2} \tag{3.6}$$

Therefore, the Kilaru model states that if the objective is a solubility model applicable to all RTILs (which are nonfunctionalized and physically interact with gas solutes), then viscosity needs to be considered in addition to RTIL molar volume.

The diffusivity of gas solutes is an important parameter, which plays a major role in permeation mechanism for gases such as CO_2 in ILs with high viscosity. By increasing the operation temperature, the viscosities of ILs decrease and the permeability of the gas solutes increases. Figure 3.1 shows CO_2 permeability versus RTIL viscosity in a log–log plot, as given by Scovazzo [80] in a paper dealing with the determination of the upper limits, benchmarks, and critical properties for gas

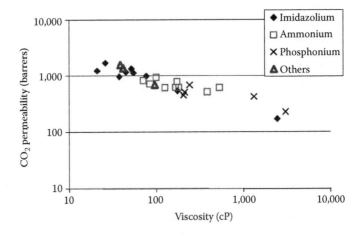

FIGURE 3.1 CO_2 permeability versus RTIL viscosity showing the dominance of viscosity in determining CO_2 gas permeability in SILMs. Data obtained at 30°C. (From Scovazzo, P., *J. Membr. Sci.*, 343, 199, 2009.)

separations using stabilized imidazolium-, phosphonium-, and ammonium-based RTILs. This plot shows the dominance of viscosity in determining CO_2 permeability in SILMs even though molar volume was the dominant factor in determining selectivities. A linear regression for the trend shown in Figure 3.3 has an r^2 value of 0.86 and the following equation:

$$P_{e,CO_2} \text{ (barrers)} = \frac{4953}{(\mu_{RTIL})^{0.388 \pm 0.03}} \tag{3.7}$$

where
 P_{e,CO_2} is the RTIL permeability
 μ_{RTIL} is the RTIL viscosity in cP

Using multivariable linear regression with the independent variables of RTIL molar volume and viscosity produces the following correlation result, which also has an adjusted r^2 value of 0.86:

$$P_{e,CO_2} \text{ (barrers)} = \frac{1323(V_{RTIL})^{0.30 \pm 0.17}}{(\mu_{RTIL})^{0.47 \pm 0.06}} \tag{3.8}$$

One of the conclusions of Scovazzo's report is that permeability in SILMs scales with viscosity, while selectivity scales with molar volume. It is also illustrative to project the permeability of yet unsynthesized RTILs with lower viscosities. A reasonable lower bound on RTIL viscosity (assumed here to be 1 cP) would project a maximum possible CO_2 permeability in RTILs of 5000 barrers. This is only 2.9 times the maximum value already reported in the literature, 1702 barrers for [emim][Tf$_2$N].

3.2.1.2 Solubility

When using an RTIL as either a bulk-fluid absorber or as a membrane, it is necessary to understand the parameters that influence gas solubility. Regular solution theory (RST) has been shown to explain the solubility behavior of many gases in imidazolium-based ILs [54,57,90]. This provides the basis for determining the solubility selectivity for various gas pairs in an IL. This theory can be extended to IL mixtures [91]. In addition, one can use a group contribution approach to provide guidance on structural changes for specific separations [39]. RST is not expected to be an accurate predictor for exact values of gas solubility, especially for polar gases such as SO_2. Camper et al. [49,90] and Scovazzo et al. [54] developed a solubility model for gases in alkyl-imidazolium (or Rmim)-based RTILs. The authors reported that the solubility parameter of the RTIL can be estimated from the lattice energy density of the liquid using the Kapustinskii equation. This equation has been shown to be useful estimating the solubility parameters of ion pairs, such as alkali chlorides [92]. However, accurate determinations of the energy of vaporization for RTILs are difficult, because of their nonvolatile natures. A modified version of the Kapustinskii equation was used to estimate the solubility parameters of imidazolium-based RTILs. This method was only applied to Rmim-based RTILs because the main source of polarizability in these RTILs is the imidazolium ring [49,93]. Since the polarizability is dominated by the imidazolium ring, any error due to ignoring the polarizability by using the Kapustinskii equation would be applied to all the Rmim-based RTILs. Equation 3.9 shows the final result for the derivation by Camper et al. of the solubility parameter of an Rmim-based RTIL with a charge ratio of 1:1:

$$\delta_1 = \left(\frac{2.56 \times 10^6 \, (\text{J/mol})(\text{cm}^3/\text{mol})^{1/3}}{\left(V_1^L\right)^{4/3}} \right)^{1/2} \left(1 - \frac{0.367 \, (\text{cm}^3/\text{mol})^{1/3}}{\left(V_1^L\right)^{1/3}} \right)^{1/2} \quad (3.9)$$

where
δ_1 represents the solubility parameter
V_1^L is the molar volume of the RTIL

Since the molar volumes of the Rmim-based RTILs are fairly large and the charge ratio is 1:1, the solubility parameter of the RTILs can be shown to be related to the molar volume of the RTIL, as shown in the following equation. According to the authors, very small molar volumes push the limit of this assumption, but the solubility parameters would still stay within a 10% error for any reasonable molar volume of an RTIL:

$$\delta_1 \propto \left(\frac{1}{\left(V_1^L\right)^{4/3}} \right)^{1/2} \quad (3.10)$$

Using a simplified version of RST (Equation 3.11) and the preceding equation, a simple relationship between Henry's constant, K_H, in atm units and the molar volume can be shown in Equation 3.12, where a and β (or β^*) are constants that depend only on temperature and the gas being absorbed:

$$\ln\left(K_H\right) = a + \beta \delta_1^2 \quad (3.11)$$

$$\ln\left(K_H\right) = a + \frac{\beta^*}{\left(V_1^L\right)^{4/3}} \quad (3.12)$$

According to the aforementioned equation, the solubility, S_H ($=1/K_H$), of a given gas solute will be higher in ILs with larger molar volume. To directly determine the bulk

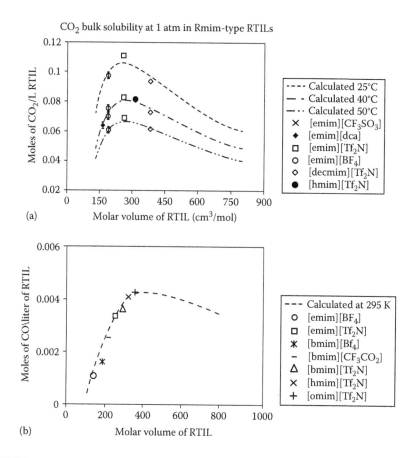

(a)

(b)

FIGURE 3.2 (a) Plot of the moles of CO_2 per liter of RTIL at 1 atm versus the molar volume of the RTIL at various temperatures. (b) Plot of the moles of CO per liter of RTIL at 1 atm versus the molar volume of the RTIL at 295 K. (From Camper, D. et al., *Ind. Eng. Chem. Res.*, 45, 6279, 2006.)

solubility of a certain gas, such as CO_2, this was rearranged so that the solubility could be plotted using moles of gas per liter of RTIL (Equation 3.13) versus molar volume of the RTIL (Figure 3.2a and b):

$$\frac{\text{Mol of gas}}{\text{Lit RTIL}} = \frac{1}{\left[\exp\left(a + \beta^* / \left(V_1^L \right)^{4/3} \right) - 1 \right] V_1^L} \tag{3.13}$$

There is an inherent maximum amount of moles of gas per volume of RTIL that can be absorbed for Rmim-based RTILs, since the molar solubility increases with increasing molar volume of the RTIL. This can be shown mathematically by taking the derivative of the following equation with respect to the molar volume and setting the derivative equal to zero:

$$V_1^L = \left\{ \frac{(4/3)\beta^* \exp\left(a + \beta^* / \left(V_1^L\right)^{4/3}\right)}{\left[\exp\left(a + \beta^* / \left(V_1^L\right)^{4/3}\right) - 1\right]} \right\}^{3/4} = \left[\frac{(4/3)\beta^*}{1 - x_2}\right]^{3/4} \qquad (3.14)$$

This can then be simplified by letting $(1 - x_2) = 1$ (Equation 3.15), which is a good approximation at low pressures:

$$V_1^L = \left(\frac{4\beta^*}{3}\right)^{3/4} \qquad (3.15)$$

Equation 3.16 provides a simple method to determine the molar volume of an Rmim-based RTIL that can absorb the maximum amount of moles of gas per volume of RTIL. All that is needed is the value of β^* from Equation 3.12, which will vary depending on the gas and temperature.

Regarding the CO_2 selectivity with respect to CO_2 permeability for nonfunctionalized RTILs, Scovazzo in a previous paper [80] has stated that modified Camper models can give useable upper bounds for both selectivity and permeability for CO_2 separations using SILMs. The highest selectivities occur at the smallest RTIL molar volumes (e.g., in the case of $[C_n\text{mim}][X]$ type of ILs, $n = 2$), while the highest CO_2 permeability will occur for RTILs with larger molar volumes. Figure 3.3 illustrates

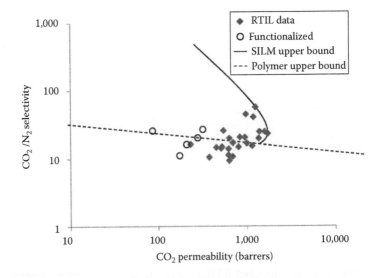

FIGURE 3.3 CO_2/N_2 selectivity versus CO_2 permeability, showing the upper bound or *Robeson plot* for SILMs. (From Scovazzo, P., *J. Membr. Sci.*, 343, 199, 2009.)

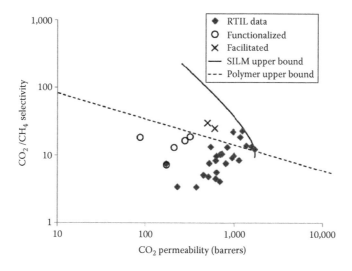

FIGURE 3.4   CO$_2$/CH$_4$ selectivity versus CO$_2$ permeability, showing the upper bound or *Robeson plot* for SILMs. (Taken from Scovazzo, P., *J. Membr. Sci.*, 343, 199, 2009.)

CO$_2$/N$_2$ selectivity versus CO$_2$ permeability, showing the upper bound or *Robeson plot* for SILMs. In addition, Figure 3.4 presents CO$_2$/CH$_4$ selectivity versus CO$_2$ permeability, showing the upper bound or *Robeson plot* for SILMs. Both Figures 3.3 and 3.4 also show that reported attempts at functionalized and facilitated-transport SILMs have resulted in membranes that moved in an unfavorable direction away from the upper bound. Facilitated-transport points calculated for a feed CO$_2$-partial pressure of 20 kPa.

3.2.1.3   Correlation of Gas Solubility to the Cation Nature

Correlations reported in literature have found no evidence of solute/cation interactions by varying cations between polarizable and nonpolarizable. Within an anion class, the gas solubility (in terms of mol/[mol atm]) in the RTILs increases as the length of the cation alkyl chain increases. This may be due to the arrangement of gas molecules into the free volume of the RTILs. [Tf$_2$N]-based modeling has indicated that the solubility of gases in these RTILs follows a different type of molecular ordering from the non-[Tf$_2$N]-based RTILs. This may involve a variation in the arrangement of gas molecules into the free volumes available in the cationic and anionic moieties.

3.2.1.4   Correlation of Gas Solubility to the Anion Nature

The reported correlations also examined solute/anion interactions by varying anions over a large range of electron donor potentials. The non–[Tf$_2$N] anion classification for CO$_2$ solubility, which contained anions with a range of electron donor potential from [Cl]$^-$ to [BETI]$^-$, exhibited no evidence of CO$_2$ interactions with the anions being a significant factor in relative CO$_2$ solubility. This may contradict the widely published statement that the relative CO$_2$ solubility between RTILs is related to solute/anion interactions [40,51].

3.2.1.5   Diffusivity

Scovazzo's group has studied gas diffusivity in RTILs and has developed correlations for ammonium [78] and phosphonium RTILs [77] in addition to imidazolium-based RTILs [56] with the following form:

$$D_{1,RTIL} = A \frac{V_{RTIL}^{a}}{\mu_{RTIL}^{b} V_1^{c}} \tag{3.16}$$

where
   A, a, b, and c are the RTIL-class-specific parameters
   $D_{1,RTIL}$ is the diffusivity of solute, 1, in the RTIL
   V_{RTIL} is the RTIL molar volume
   V_1 is the solute molar volume
   μ_{RTIL} is the RTIL viscosity

While we do not recommend a diffusivity correlation covering all RTILs, we have noted certain *universal* trends that are useful for developing an upper bound for SILM performance. Specifically, diffusivity scales inversely with the square root of viscosity, $b \approx 0.5$, and inversely to the solute molar volume to the power of 1–1.3. This means that diffusivity in RTILs is less dependent on viscosity, and more dependent on solute size, than predicted by the conventional Stokes–Einstein model. Furthermore, for RTILs with cations having long flexible alkyl chains ($R \geq 4$), diffusivity may have a void space dependence. For this reason, the power of a changes with RTIL classification, ranging from 0 for imidazolium RTILs (R generally <4) to 1.57 for ammonium RTILs (R generally >4).

The ratio in Equation 3.16 for two different diffusing solutes will give insights into the role of diffusion selectivity in SILMs:

$$a_{1,2} \text{(diffusivity)} \approx \left(\frac{V_2}{V_1} \right)^{1.15} \tag{3.17}$$

where
   $a_{1,2}$ (diffusivity) is the diffusion selectivity of solute 1 versus solute 2
   V_1 is the molar volume of solute 1
   V_2 is the molar volume of solute 2

Therefore, diffusivity selectivity in a SILM would be approximately equal to the ratio of gas molar volumes. For most gas pairs of interest (CO_2/CH_4, CO_2/N_2, O_2/N_2, etc.), this would be small; for example, the diffusivity selectivity of CO_2 versus CH_4 is 0.94 and the diffusivity selectivity of CO_2 versus N_2 is 1.1. However, diffusivity will have a large impact on the separation productivity in that RTILs with larger viscosities will form SILMs with smaller permeability values. The fact that diffusion selectivity does not play a major role gives a theoretical basis for the experimental observations that SILM mixed-gas selectivities are approximately equal to the ideal selectivity values [75].

Results also show that the diffusion coefficients are very similar so the selectivity is based almost entirely on solubility differences. This is in contrast to gas separations using glassy polymers where the selectivity is based on differences in diffusivity. Regarding the relative contribution of solubility and diffusivity of the gaseous components in their separation efficiency in SILMs, Scovazzo [80] has reported as a conclusion that solubility selectivity dominates the permeability selectivity. The fact that diffusion selectivity does not play a major role gives a theoretical basis for the experimental observations that SILM mixed-gas selectivities are approximately equal to the single-gas or ideal selectivities [75]. For CO_2 separations, there are two critical RTIL properties that affect SILM performance: RTIL molar volume and RTIL viscosity. The permeability selectivity is a function of RTIL molar volume, while the CO_2 permeability is a function of viscosity.

3.2.1.6   Water Content in the IL Phase

It has been reported [94] that essentially no change was found in water content for [bpy][Tf_2N] that was left open to the atmosphere for 24 h. There was also no significant change in viscosity for this IL before and after exposure to the atmosphere. In contrast to these observations with [bpy][Tf_2N], the ILs [bmim][BF_4] and [omim][BF_4] exhibited a 50%–150% increase in water content when they were left open to the air for 24 h. The viscosity of the [BF_4]-bearing ILs was found to decrease by ~10% with this increase in water content. It is important to add that the water content of all of these ILs was <1 wt%, values that are significantly below saturation levels [95,96]. In summary, these previous studies indicate that the addition of water to ILs can cause a decrease of CO_2 solubility because water can limit CO_2–anion interactions as well as increase gas diffusivity and permeability because of the decrease in viscosity resulting from dissolved water.

The effect of water content on CO_2 solubility in the IL [bmim][PF_6] was investigated by Fu et al. [97] who observed a maximum 15% decrease in solubility when water content increased from 0.0067 to 1.6 wt%, with the impact more pronounced at higher pressures (>5 MPa). Little difference in CO_2 solubility was observed when CO_2 mole fraction was defined on a water-free basis. Zhao et al. [98] examined the effect of water on CO_2 and N_2 permeance through SILMs prepared with the IL [bmim][BF_4] supported in a polyethersulfone (PES) membrane. N_2 permeance was independent of water content up to 10 mol% water but was found to increase significantly (>30%) when water content was increased to 20 mol%. CO_2 permeance was found to increase by ~10%–20% when water content increased from 0 to 10 mol% but then decreased as water content was further increased to 20 mol%. A similar maximum in CO_2/N_2 selectivity was observed at ~10 mol% water. The increase in CO_2 and N_2 permeance with added water was attributed to the decrease in IL viscosity that results from the addition of water. The subsequent decrease in CO_2 solubility at higher water content was attributed to hydrogen bonding between water and the IL, apparently limiting the interactions between CO_2 and the anion.

3.2.2 Membrane Supports

A considerable number of reports have shown that the permeation properties of a SILM system are influenced by the support characteristics. The chemical composition and porous structure of the support are of decisive importance for the permeation properties, physicochemical features, and stability of SILM membranes. In general, three types of substrates have been used for fabrication of these systems: polymeric, inorganic, and mixed-matrix (or hybrid) supports.

3.2.2.1 Polymeric Supports

Regarding the polar or nonpolar nature of polymeric supports, Neves et al. [99] compared the stability of SILMs prepared with hydrophilic and hydrophobic polyvinylidene fluoride support membranes with different $[C_n mim]$-bearing ILs by measuring weight changes in the membranes during N_2 permeation measurements. Results revealed smaller weight losses from the membranes prepared with the hydrophobic membranes, which they attributed to better chemical affinity between the immobilized ILs and the support. Regarding the impact of CO_2 permeation on the polymeric-IL systems, Simons et al. [100], in a study of swelling and plasticization with styrene-based polymerized RTIL membranes, found that CO_2 plasticized the polymer in both pure and mixed gases, with enhanced CH_4 permeation rates in the presence of CO_2. The extent of plasticization was dependent on the monomer structure, with plasticization increasing as the length of the RTIL alkyl side chain increased, due to the shifting balance between polymer–polymer ionic interactions and CO_2-induced polymer swelling. Moreover, the critical issue of CO_2 feed pressure on the membrane morphology and subsequent gas separation performance has been evaluated. One important feature of this situation is that the swelling is reversible due to the charged nature of the polymer material. This is in contrast to the irreversible conditioning with conventional polymer membranes for high-pressure CO_2 separations.

3.2.2.2 Inorganic Supports

The physical properties of ILs may also be tuned by means of their immobilization inside porous networks at the nanoscale level. The immobilization may occur either by physical sorption or via chemical grafting, the later method providing better thermal and mechanical stability. Ceramic or porous glass substrates (either symmetric or asymmetric or composite) with parallel channels or tortuous pores can be used as supports. Confinement of ILs within pores comparable to their ion pair dimensions results in drastic alterations in what concerns their structure and physicochemical/thermophysical properties with regard to their bulk phase [101]. Previous experimental and simulation studies have demonstrated that confined ILs undergo changes in phase behavior such as in glass transition temperature and melting point [102–110]. These effects are induced by the microstructural conformation and the possibility of long-range molecular ordering or formation of new phases having different ionic self-diffusivities. The extent of confinement and the resulting ordering of the ionic moieties in specific arrangements depend on the characteristics of the porous network of the supporting material (e.g., specific surface area, pore shape, and pore size distribution) with respect to the dimensions of the ionic components, as well as on the interface chemistry (pore walls with possible charges and groups) of the supporting material. In essence, the microstructure and phase thermal stability of the immobilized or confined ILs is determined by the relative energy of the interactions among the ionic species (e.g., hydrogen bonds and electrostatic interactions) and the interactions at interfaces between ion pairs and solid material, favoring crystallization and ordering.

The molecules of a confined liquid have different dynamics with respect to the bulk fluid. It has been reported that the diffusion coefficient of water decreases by three to four orders of magnitude when it transforms into a frozen state [111,112]. In the bulk phase, RTILs may form a 3D hydrogen bond network as in the case of water molecules, depending on the selected combination of cations and anions. Regarding the stabilizing mechanism, it was speculated that strong cationic/anionic interactions with functional groups present at the nanopores walls within a radial dimension of 1–10 Å may form structured matrix arrangement of the cations/anions within the nanopore networks. Such a formation could give rise to unique gas transport and separation properties different to neat ILs.

The confinement of an IL into the pores can considerably alter the diffusion mechanism of gas solutes in the liquid phase as a result of formation of ordered arrangements, suppression of self-diffusivity, and orientation preferences of the ionic moieties. The long-range ordering and the orientation uniformity of the ions with respect to the pore walls and to their ionic counterparts might create fixed (permanent) diffusion paths along sorption sites throughout the immobilized IL phase, especially when a solid-like frozen phase is formed. As a result, gas solute transport should diverge from the conventional diffusion in a bulk liquid. In this case, the experimental results deviate considerably from predictions based on the solution-diffusion transport. In order to derive the experimental total permeability coefficients of these SILM membranes, the structure morphology and the porous structure of the multilayered membrane's support should be elucidated [101].

3.2.2.3 Mixed-Matrix Supports

An examination of mixed-matrix membranes involving RTILs showed that the addition of zeolites to RTILs, to polymerized RTILs, and to RTIL/polymerized RTIL mixtures increased the permeability of CO$_2$, N$_2$, and CH$_4$ and that CO$_2$/N$_2$ and CO$_2$/CH$_4$ selectivity depended on the membrane system [113].

3.2.3 OPERATION PARAMETERS

3.2.3.1 Temperature

Increasing temperature increases the diffusion rate of the gas solute so that permeance increases. One important property of ILs that can be further exploited is the large temperature range where the material is stable and nonvolatile. One can then use temperature to adjust the properties of the membrane (solubility, diffusivity, reactivity, etc.) to find a suitable range for a given separation.

3.2.3.2   Transmembrane Pressure

In order to successfully implement any SILM system in large-scale gas separation processes, it is important to consider the effect of transmembrane pressure on separation performance and stability, since especially in the case of SILMs with meso- or macroporous supports (either inorganic or polymeric), there may be a pressure limitation, where the liquid is removed from the pores. As a consequence, mixed-gas selectivity values would gradually decrease in the course of a given separation process. While mixed-gas data would give results more applicable to industrial applications, there are considerably more data on pure gas measurements. The permeability of carbon dioxide frequently shows a strong dependence on the CO_2 partial pressure, whereas the N_2 permeability dependence is considerably less. The permeability tends to drop with increasing pressure. If the membrane is vulnerable to plasticization, the permeability will reach a minimum and then rise steeply with increased pressure.

The following section summarizes experimental results and conclusions obtained from some research works reported in the literature that are referred to the development of SILM systems and investigation of their permeation properties and separation performance. Specifically, the results presented next are associated with two of the most important separations involving carbon dioxide in industry that is the separation of CO_2 from nitrogen and from methane.

3.2.3.3 CO_2–N_2 Separations

Jindaratsamee et al. [114] investigated CO_2 separation from N_2 through polymeric SILM membranes, using high feed CO_2 concentrations. The membranes were prepared by impregnation of porous polyvinylidene fluoride (PVDF) supports with [bmim][PF_6] and [bmim][Tf_2N] ILs. The permeability of CO_2 decreased with increasing total pressure difference and the results pointed out that facilitated-transport played a key role for a total pressure at the feed side of 1.05–1.21 atm. The fluorinated anions of the employed ILs acted as the carriers, which were immobilized inside the IL membrane. At higher total pressure difference, the saturation of the carrier was the limiting factor leading to lower efficiency of the carrier, which resulted in a decrease of CO_2 permeability. In addition, at high CO_2 concentration at the feed side, both the permeability of CO_2 and the CO_2/N_2 selectivity were limited.

Cheng et al. [115] studied the effect of the supporting membrane structure on the IL loading, stability (concerning the retaining of the IL within the pores), and performance of SILM membranes, fabricated through deposition of [bmim][BF_4] IL on porous symmetric and asymmetric PVDF supports. It was found that a symmetric PVDF membrane can retain more IL, yet is less stable, than an asymmetric PVDF membrane. The asymmetric support retained approximately 95% IL even at 0.6 MPa. Conversely, the IL loading of the symmetric SILM dropped dramatically when the transmembrane pressure reached 0.3 MPa. At transmembrane pressures of 0.25 MPa and above, the N_2 single-gas permeance through symmetric SILMs increased dramatically, showing again that the IL in these membranes is easier to be removed. Although the asymmetric SILMs showed higher single-gas permeance values, the symmetric SILMs clearly exhibited higher ideal selectivities for both CO_2/N_2 and CO_2/air. The asymmetric SILMs only yielded ideal selectivities no more than 2 even at high transmembrane pressures, whereas the symmetric SILMs exhibited ideal selectivities of CO_2/N_2 and CO_2/air up to 18 and 6, respectively, at transmembrane pressure of 0.25 MPa. According to the authors, the symmetric membrane structure was filled with more IL, reducing the permeance of all gases. Since more CO_2 was dissolved in the supported IL phase, the selectivity of the symmetric membranes was higher.

Neves et al. [99] studied the potential of separating CO_2/N_2 and CO_2/CH_4 gas mixtures using RTILs with imidazolium-based cations (i.e., [bmim][Tf$_2$N], [C$_n$mim] [PF$_6$], where n = 4, 8, and [C$_n$mim][BF$_4$], where n = 4, 10), immobilized in porous hydrophobic (having 220 nm pore size) and hydrophilic (200 nm pores) PVDF membranes. An increase in the permeability values (by an average factor of ~2) with an increase in the alkyl chain length of the RTIL cation was observed for all gases studied (H_2, O_2, N_2, CH_4, and CO_2) and for both RTIL anions ([BF$_4$]$^-$ and [PF$_6$]$^-$). Additionally, when estimating the diffusivities, a decrease by a factor close to 2 was obtained, due to the higher viscosity of the corresponding RTILs. Therefore, the resulting estimated solubility had to increase by a factor of 4, in order to explain the experimentally determined permeability values. These results indicated that even though the gas diffusivity decreases for the more viscous RTILs, a comparatively higher increase in solubility leads to an overall increase in permeability. Therefore, solubility effects seem to play a more important role in the transport of gases, for RTILs with different imidazolium cation alkyl chain lengths, whereas the solubility enhancement was found to be almost identical for all gases studied.

Jindaratsamee et al. [116] studied the effects of temperature and anion species on CO_2 permeability and CO_2/N_2 separation efficiency of SILM membranes, fabricated by impregnation of porous hydrophobic PVDF disc-shaped supports, of 200 nm pore size, with a series of ILs consisting of the [bmim]$^+$ cation and [BF$_4$]$^-$, [PF$_6$]$^-$, [Tf$_2$N]$^-$, [OTf]$^-$, and [dca]$^-$ as counteranions. The authors found that the CO_2 single-gas permeability through all IL membranes increased with temperature. The [bmim] [Tf$_2$N]-bearing membrane gave the highest CO_2 permeability value, while the membrane with [bmim][PF$_6$] yielded the lowest, due to the lowest diffusion coefficient. On the other hand, the ideal selectivity for the CO_2/N_2 gas pair through the studied SILM membranes was reduced with increasing temperature. Unlike the results of CO_2 permeability, the lowest separation efficiency was found for the [bmim][Tf$_2$N]-bearing membrane, in which the permeability of CO_2 was the highest.

Bernardo et al. [117] investigated the structure, mechanical properties, and gas separation performance of elastomeric block copolyamide SILM membranes fabricated by addition of [bmim][CF$_3$SO$_3$] RTIL to Pebax®1657 and Pebax®2533 supports. The membranes were prepared in the form of stable gels of IL and polymer by the solution casting and controlled solvent evaporation method. The gas transport properties of the less permeable Pebax1657 were significantly affected by the presence of the IL resulting in a strong increase of the single-gas permeability of all gases (He, H_2, CO_2, O_2, N_2, and CH_4) and a slight decrease of the ideal selectivity for most gas pairs. The plasticization of the polymer by the IL limited the role of the less permeable hard blocks of the copolymer, reducing the weak size-sieving ability of the neat polymer. The highest selectivity was observed for CO_2/N_2 separation, for which the selectivity decreased from about 60 to about 40, while the CO_2 permeability increased four-fold upon addition of the IL to the support. Both the solubility and the diffusion coefficients of the gases, and as a result also their permeability coefficients, were lower in Pebax1657 system than in Pebax2533. The only exception was the solubility of CO_2, which was similar in both copolymers. For Pebax2533-based membranes, the IL did not influence the transport properties of the polymer matrix. For this polymer, the permeation process is solubility controlled, with or without the presence of IL.

Santos et al. [118] prepared SILM systems incorporating the acetate-based RTILs [emim][Ac], [bmim][Ac], and [Vbtma][Ac], in order to perform selective separation of CO_2 from N_2. The RTILs were supported in PVDF porous membranes. Gas permeabilities increased with temperature, while CO_2/N_2 ideal selectivity was reduced for all the studied RTILs. The permeability activation energies were higher for N_2 than for CO_2. At the highest temperature of measurements (333 K), the highest CO_2 permeability values were equal to 2064.9 barrers for [emim][Ac], 1940.9 barrers for [bmim][Ac], and 2114.2 barrers for [Vbtma][Ac]. The respective CO_2/N_2 ideal selectivities at this temperature were 26.4, 29.7, and 27.3. Comparing the permeability results at increasing temperatures to the upper-bound values for the selectivity versus permeability of polymer membranes for CO_2/N_2 separation given by Robeson, it was concluded that most of these SILMs had better performance than many polymer membranes previously studied, being the experimental results near the upper bound.

Cserjési et al. [119] prepared polymeric SILM membranes by addition of flat sheet porous hydrophobic PVDF supports, having a pore size of 220 nm and a porosity of 75%, on [emim][CF$_3$SO$_3$] and different types of unconventional ILs (Ammoeng™ 100, Ecoeng™ 1111P, Cyphos 102, Cyphos 103, Cyphos 104, and [SEt$_3$][Tf$_2$N]). All SILMs exhibited the highest permeability values for CO_2 and the lowest for N_2. Ammoeng 100 had the lowest permeability and [SEt$_3$][Tf$_2$N] the highest permeability for the four gases tested. Both H_2 single-gas permeability and CO_2/H_2 ideal selectivity changed after the permeation of CO_2. In the case of Ecoeng 1111P and [emim][CF$_3$SO$_3$]-bearing membranes, the CO_2/H_2 separation efficiency increased and in all other cases it decreased, thereby indicating that CO_2 gas mostly has an unfavorable effect on the permeability and selectivity of SILMs, probably due to the plasticization of the supporting polymer membrane. It was also noticed that these SILMs were stable even at as high pressure as 5 bars without a sudden increase in permeability, hence without the IL being pushed out of the membrane pores. Furthermore, at 10 bars, a slight decrease in stability was observed, first for [emim][CF$_3$SO$_3$] and [SEt$_3$][Tf$_2$N], and then at higher pressure, the more hydrophobic Cyphos 102 and Cyphos 103 ILs were removed slowly from the smaller membrane pores making successful gas separation impossible.

Close et al. [120] fabricated SILM membranes by impregnating thin porous anodized alumina (Anodisc®) supports with [emim]-, [bmim]-, and [hmim][Tf$_2$N], [emim][TfA], [C$_3$C$_1$mim][Tf$_2$N], [bmim][Ac], and [C$_4$C$_4$im][Tf$_2$N] ILs. Several N_2 and CO_2 single-gas permeance measurements were performed with such membranes with nominal pore size of 20 and 100 nm. The membrane permeance to both CO_2 and N_2 in the 100 nm pore size membranes was larger (by ~60%) than the values measured in the membranes with the 20 nm pores. However, CO_2/N_2 ideal selectivities were found to be independent of the membrane pore size. It is possible that the properties of the IL (i.e., viscosity or density) were influenced by confinement in the porous substrates. Measured CO_2 permeability values were consistently smaller than values predicted using CO_2 solubility and diffusivity values that were determined by measuring gas uptake into thin IL films. The results indicate that bulk IL properties are different than properties of the same fluids when confined within the pores on the employed alumina supports. However, the difference between measured and predicted permeances was not large for the [emim]-, [bmim]-, and [hmim][Tf_2N] SILM systems and was larger for the [C_3C_1mim][Tf_2N] and [bmim][TfA] systems, and the values differed by a factor of ~4 for the [C_4C_4mim][Tf_2N] system.

Scovazzo et al. [63] studied RTIL membranes made by addition of the following water stable anions: [Tf_2N]−, [CF_3SO_3]−, [Cl]−, and [dca]− to 80% porous hydrophilic PES supports. The authors reported CO_2 permeability values of 350 barrers (for [Cl]−) to 1000 barrers (for [Tf_2N]−) combined with CO_2/N_2 ideal selectivities of 15 (for [Cl]−) to 61 (for [dca]−). These permeability/selectivity values placed the studied RTIL membranes above the upper bound in a CO_2/N_2 Robeson plot of representative polymers. The CO_2/CH_4 ideal selectivities ranged from 4 (for [Cl]−) to 20 (for [dca]−), thereby placing the [dca]-membrane above the upper bound for the CO_2/CH_4 Robeson plot. Moreover, the membranes incorporating [emim][Tf_2N] showed the highest N_2 permeability values, approximately 48 barrers, while [emim][dca] had the lowest N_2 permeability of 10 barrers. In the former case, similar N_2 permeabilities were reported under both low and high humidity conditions. The N_2 permeabilities of [emim][CF_3SO_3] and [thtdp][Cl] were 26 and 24 barrers, respectively. Therefore, the series for the highest N_2 permeability to the lowest was [Tf_2N]− > [CF_3SO_3]− > [Cl]− > [dca]−. There was a 300% difference in the RTIL-membrane CH_4 permeabilities. [emim][Tf_2N] high humidity and [thtdp][Cl] had the highest CH_4 permeabilities with 94 and 89 barrers, respectively, while [emim][dca] has the lowest CH_4 permeability with 31 barrers.

The authors also examined the impact of water on CO_2 solubility in [Tf_2N]-incorporating ILs by measuring CO_2 solubility in dry and water-saturated [emim][Tf_2N]. Henry's law constants were found to be unaffected by the addition of water (39 bars in the dry IL and 38 bars in the water-saturated IL). The CO_2 permeability of a SILM system prepared with this same IL supported in a PES membrane increased by ~10% when the IL was equilibrated at high humidity (85% relative humidity) compared to equilibration at low humidity (10% relative humidity).

3.2.3.4 CO_2–CH_4 Separations

Iarikov et al. [121] prepared SILM membranes by impregnating α-alumina porous supports with [bmpyr][Tf_2N], [hmim][Tf_2N], [mtoa][Tf_2N], [bmim][BF_4], [bmpy][BF_4], [apir][I], and [$P_{6,6,6,14}$][dca] ILs. Two types of inorganic supports were used in their study: tubular α-alumina with a nominal pore size of 5 nm, precoated with a g-alumina layer, and hollow fiber (HF) α-alumina of 100 nm pores. The CO_2 permeance of these membranes was on the order of 10^{-9} mol/(m^2 s Pa) with CO_2/CH_4 selectivity ranging between 5 and 30. The best membranes were prepared using a mixture of [bmpy][BF_4] and [apyr][I] ILs, with a selectivity of 52 at a permeance of 2×10^{-9} mol/(m^2 s Pa). Hence, the properties of the RTIL membranes could be significantly improved by the addition of an IL with an amine group functionality. For membranes prepared with the same anion ([Tf_2N]−), different cations gave approximately the same permeance values but different selectivity in the order [hmim]+ > [bmpy]+ > [mtoa]+. Regarding the effect of pressure on the separation performance of the fabricated membranes, the authors concluded that the SILM membranes are most effective at lower feed pressures. For the membrane bearing [hmim][Tf$_2$N], it was shown that the CO$_2$/CH$_4$ selectivity decreased dramatically with increasing total feed pressure for a 50/50 gas mixture. This trend originated from the dependence of CO$_2$ and CH$_4$ solubility in the ILs on increasing pressure. Hert et al. [122] studied the solubility of CO$_2$/CH$_4$ gas mixtures in [hmim][Tf$_2$N]. They found that although the solubility of pure CO$_2$ was much higher than that of pure CH$_4$, the behavior changed for binary mixtures. In the case of 10/90 and 50/50 CO$_2$/CH$_4$ mixtures, the presence of CO$_2$ actually increased the solubility of CH$_4$ in the IL phase. Since the separation mechanism at work for SILM membranes is solution diffusion, the separation becomes ineffective at relatively mild pressures.

Regarding the temperature dependence of the permeance of CO$_2$ and CH$_4$ gases through a membrane supporting [bmim][BF$_4$], the authors observed that the CH$_4$ permeance increased more significantly (as a result of increasing diffusivity) over a certain temperature range, thereby resulting in lower selectivity at elevated temperatures. The permeance dependence on temperature followed the Arrhenius equation, and the activation energy for CO$_2$ permeance in this case was 7.5 kJ/mol compared to 20 kJ/mol for CH$_4$. Jacquemin et al. [50] determined the solubilities of CO$_2$ and CH$_4$ in [bmim][BF$_4$] as a function of temperature between 283 and 343 K. They found that CO$_2$ solubility in the IL (which was much higher than that of other gases) decreased dramatically with increasing temperature compared to CH$_4$. This explains why membranes that use [bmim][BF$_4$] as the separation medium were not as effective at higher temperatures. In another study by the same research group [123], a similar trend in CO$_2$ and CH$_4$ solubilities was observed for [bmim][PF$_6$].

Shahkaramipour et al. [124] measured the permeability of CO$_2$ and CH$_4$ gases through SILMs incorporating two germinal dicationic ILs, namely, pr[mim]$_2$[Tf$_2$N]$_2$ and h[mim]$_2$[Tf$_2$N]$_2$ stabilized on alumina porous supports with a nominal pore size of 20 nm. Single-gas permeability values of CO$_2$ and CH$_4$ through the investigated SILMs did not significantly change with transmembrane pressure of diffusing gas. The ideal selectivity for CO$_2$/CH$_4$ separation was reduced for both studied ILs with increasing temperature. In fact, the permeability values of both gases through the fabricated SILMs increased with increasing operating temperature, but those of CH$_4$ increased more than CO$_2$ values. Regarding the nature of the supported ILs, the results indicated that CO$_2$ permeabilities (at all the temperatures) and ideal selectivities (at 27°C and 45°C) through a [hmim][Tf$_2$N]-bearing membrane were higher than those of the two diatomic ILs, which had a zwitterionic form. This behavior was explained by the higher viscosity of the two diatomic ILs compared with [hmim][Tf$_2$N], thus demonstrating the major role of the viscosity parameter in permeation mechanism of gas solutes.

3.3 MOF MEMBRANES FOR CO$_2$ SEPARATION

3.3.1 ADVANTAGES OF MOFs COMPARED TO OTHER SOLID ADSORBENTS: ZIFs, A SUBFAMILY OF MOFs

In the past two decades, metal–organic frameworks (MOFs), also known as coordination polymers or coordination networks, have attracted intense research interest as novel functional materials. MOFs are hybrid materials that combine organic ligands

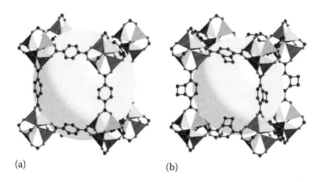

(a)   (b)

FIGURE 3.5   Single-crystal x-ray structures of (a) MOF-5 and (b) IRMOF-6. (From Eddaoudi, M. et al., *Science*, 295, 469, 2002.)

and metal ions or metal-containing clusters and have attracted considerable attention because of their unique structural properties, including superior surface area (up to 6200 m^2/g) relative to those of traditional adsorbents such as activated carbon and zeolites, high porosity (up to 90%), and low crystal density, as well as high thermal and chemical stability. A typical MOF, MOF-5, which is constructed from zinc atoms as the metal centers and terephthalic acid as the organic linker, has its basic structural unit as shown in Figure 3.5 [125]. After removal of the guest molecules, typically solvents in the open pores or channels, the 3D structure of the MOFs can be usually retained and used for other guest adsorption. The major advantage of MOFs over more traditional porous materials, such as zeolites and carbon-based adsorbents, is the ability to tailor pore size and functionality of these materials by choosing appropriate building blocks in order to provide them with the optimal properties for specific host–guest interactions. For instance, by systematically changing the organic ligand in MOF-5, an isoreticular series of isoreticular MOFs (IRMOFs) with similar structures but different pore sizes were obtained, of which IRMOF-6 showed high methane storage capacity [126] (Figure 3.5). To date, there are tens of thousands of MOFs catalogued in the Cambridge Structural Database (CSD), and many of them are porous and stable upon solvent removal. The family of MOFs includes IRMOFs [127], zeolitic imidazolate frameworks (ZIFs) [128], and zeolite-like MOFs (ZMOFs) [129]. Besides changing of the ligands in the mother liquid, the postsynthetic modification of MOFs is a powerful way to tune pore size or surface properties of the pore walls. The combined favorable properties of large surface area, permanent porosity, and tunable pore size/functionality have enabled MOFs as ideal candidates for CO_2 capture. The CO_2 adsorption and desorption in MOFs are explained by a *breathing-type mechanism* or a *gate effect mechanism* [130–132]. Normally, the interaction between adsorbed CO_2 and adsorbents is weak, and the species starts to desorb at temperatures higher than 30°C.

A number of reviews have summarized the work in MOFs for gas adsorption applications including hydrogen storage, methane storage, and CO_2 capture [129,133,134]. Most recently, Liu et al. summarized the CO_2 adsorption both at high pressures and selective adsorption at approximate atmospheric pressures [135,136]. Llewellyn et al. obtained a record capacity of 40 mmol/g for CO_2 capture with MIL-101 at

50 bars and 30°C [136]. Like most porous materials, CO_2 adsorption capacity of MOFs mainly depends on their surface area. Yaghi and coworkers [137] carried out the first systematic study to explore the relationship between surface area and CO_2 capacity. Nine MOFs were selected to examine their structural and porous attributes. Among these MOFs with various topologies (MOF-2, MOF-505 and $Cu_3[BTC]_2$, MOF-74, IRMOF-11, IRMOF-3, and IRMOF-6 and the extra-high-porosity frameworks IRMOF-1 and MOF-177), they found that MOF-177 has the highest surface area among these materials and it also has the highest CO_2 uptake at high pressure, which is 60.0 wt% (33.5 mmol/g) at 35 bars.

MOF-based CO_2 adsorbents should fulfill the following requirements in order to be preferable for industrial applications. High CO_2 capture capacity and, over other components in the flue gas, high thermal and chemical stability. To date, although MOFs have shown exceptionally high CO_2 storage capacity under equilibrium conditions with pure CO_2, most of them show little uptake in the low-pressure regime of 0.1–0.2 bars. Therefore, tremendous efforts have been made to further increase their CO_2 capture selectivity while maintaining or improving their capture capacity under equilibrium conditions. The modifications are usually conducted in two aspects: (1) direct assembly of premodified organic linkers with functional groups and (2) postcovalent functionalization of organic ligands or linker sites.

Examples of direct assembly of new MOFs from particular metal nodes and organic linkers with specific functionalities are ZIFs with gmelinite topology. ZIF is a subcategory of MOF. Different from MOFs, ZIFs consist of transition metal cations connected by imidazolate anions into tetrahedral frameworks, which resemble the zeolite topologies. Unlike MOFs, ZIFs exhibit exceptionally high thermal and chemical stability due to the strong bonding between metal cations and imidazolate anions [138]. In addition, ZIFs offer advantages compared to zeolite due to their properties such as variety of structural and framework diversity [139,140], fine-tunable pore sizes, and chemical functionality [141] and also exhibit high resistance to organic solvents, water, and alkaline solution [142].

In Gmelinite (GME) ZIFs, by changing the imidazole linker, they produced a wide range of pore metrics and functionalities for CO_2 separation processes [143] (Figure 3.6a). The imidazole link functionality was altered from polar (–NO_2, ZIF-78; –CN, ZIF-82; –Br, ZIF-81; –Cl, ZIF-69) to nonpolar (–C_6H_6, ZIF-68; –CH_3, ZIF-79). The order of CO_2 uptake at 1 bar and 298 K was in line with the greater attractions expected between the polar functional groups in the ZIFs and the strongly quadrupolar CO_2, that is, –NO_2 (ZIF-78) > –CN, –Br, –Cl (ZIF-82, -81, -69) > –C_6H_6, –Me (ZIF-68, -79) > –H (ZIF-70) > BPL carbon (widely used in industry for gas separation) (Figure 3.6b, left), while there is no noticeable relationship between CO_2 uptake and pore diameter. This result implies that CO_2 uptake capacity is influenced primarily by functionality effects rather than pore metrics. To this extend, ZIF-78 and ZIF-82 showed higher CO_2/CH_4, CO_2/N_2, and CO_2/O_2 selectivities than the other ZIFs, since the –NO_2 and –CN groups in ZIF-78 and ZIF-82, respectively, have greater dipole moments than the other functionalities, and dipole–quadrupole interactions with CO_2 can be expected (Figure 3.6b, right). These results suggested that highly polar functional groups are helpful for attaining high CO_2 selectivities as well as high CO_2 uptake. ZIFs demonstrate considerable affinity to CO_2 [144]. The CO_2 adsorption isotherms of ZIF-68,

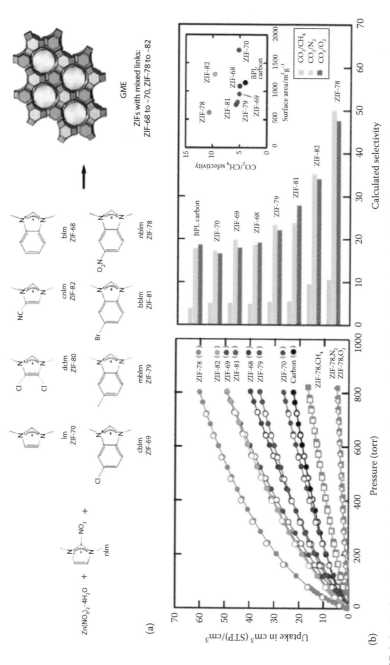

FIGURE 3.6 (a) Reaction of (2-nitroimidazole) nIm plus any other imidazole linker in the *GME*. H atoms have been omitted for clarity. C, black; N, green; O, red; Zn, blue tetrahedra. (b) Left, CO_2 isotherms of the **GME** ZIFs and **BPL** carbon (widely used in industry) at 298 K. CH_4, N_2, and O_2 isotherms for ZIF-78 are also shown. Right, calculated CO_2/CH_4, CO_2/N_2, and CO_2/O_2 selectivities. (Inset) Surface-area dependence of the calculated CO_2/CH_4 selectivities. (From Banerjee, R. et al., *J. Am. Chem. Soc.*, 131, 3875, 2009.)

ZIF-69, and ZIF-70 show steep uptakes in the low-pressure regions indicating a high gas affinity; furthermore, all aforementioned ZIFs also possess high CO_2 uptake capacities [145,146] of 38, 41, 32, 51, 34, 39, and 54 cm³/g for ZIF-68, ZIF-69, ZIF-70, ZIF-78, ZIF-79, ZIF-81, and ZIF-82, respectively. Meanwhile, CO_2 uptakes for ZIF-95 and ZIF-100 were 19.70 and 32.60 cm³/g, respectively [147]. On the other hand, Xie et al. reported that though the BET surface area of the ZIF-7 is smaller than ZIF-8, the CO_2 uptake for ZIF-7 (48 mL/g) was relatively larger than ZIF-8 (18 mL/g).

While both ZIFs and MOFs can hold significantly large amounts of CO_2, the main advantages of ZIFs are based on their physical properties. For example, the relatively high chemical stability of ZIFs compared with MOFs makes them excellent candidates for industrial use. Furthermore, ZIFs have been shown to have a high affinity for CO_2 at low pressures (at 298 K and 1 atm, MOF-177 has a maximum uptake of 7.60 L/L CO_2, while ZIF-69 has a capacity of 82.6 L/L), which is relevant for a pressure swing adsorption-type process for CO_2 capture [137]. In addition, ZIFs show greater selectivity than MOFs for CO_2 from other relevant flue gases (such as CO) as breakthrough experiments using binary gas mixtures such as CO_2/CH_4, CO_2/CO, and CO_2/N_2 (50:50, v/v) further support the affinity of the reported ZIFs for CO_2 by showing complete retention of CO_2 and simultaneous unrestricted passage of CH_4, CO, and N_2 through the pores of the framework [148]. So it is believed that ZIFs are preferable to MOFs for industrial application, especially given the importance of gas selectivity in CO_2 capture, as CO_2 does not come in pure form but rather as in flue gas (mixture of gases).

All the aforementioned characteristics have emerged as a novel type of crystalline porous material with a potential use in a variety of separation applications. Membrane-based technologies, using thin microporous ZIF layers of the order of a few micrometers, are of particular interest in applications such as CO_2 capture, H_2 purification, pervaporation, and alkane/alkene separations.

3.3.2 MOF-Based Membranes for CO_2 Separation

Recent literature reveals that MOF membranes are becoming a major player in the family of membranes for CO_2 separation (polymeric [149], inorganic [150–154], mixed-matrix [155,156], and MOF membranes [157,158]). Apart from the remarkable structure properties, preparation of MOF membranes is simpler compared to zeolite membranes as it does not require high synthesis temperatures as well as the postsynthetic calcination steps. Moreover, by showing great tolerance in high pressures and humid conditions, ZIF membranes offer advantages in gas separation compared to zeolite membranes. ZIF-8 membranes, for example, exhibit high chemical resistance and can sustain temperatures up to 400°C in the air and 550°C in N_2 [138]. In addition, compared to polymeric membranes, ZIFs demonstrate their durability characteristic under high pressure without a CO_2-plasticization effect [159].

3.3.3 CO_2 Separation Mechanism through ZIF Membranes

The permeation mechanism for ZIF membranes is different from zeolite membranes. Theoretically, the transport mechanisms involved in ZIF membranes for gas separation and permeation are adsorption/desorption and diffusion mechanisms [160]. Figure 3.7a shows the permeation mechanism through ZIF-8 membrane. Initially, CO_2 molecules are selectively adsorbed onto the ZIF membrane. Then, the molecules are diffused through the matrix of the membrane with different chemical potentials and, finally, desorbed from the membrane [159]. According to grand canonical Monte Carlo (GCMC) simulations, the CO_2 adsorption site for ZIF-8

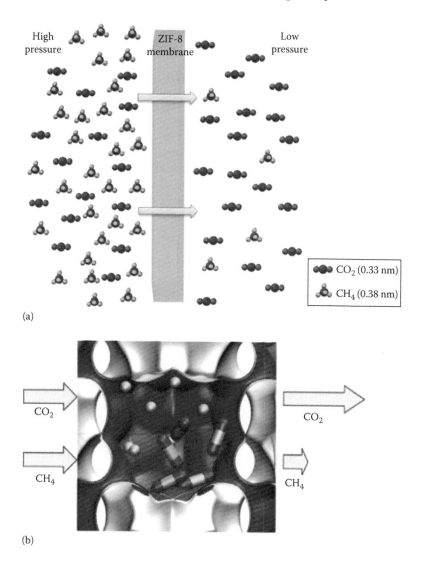

(a)

(b)

FIGURE 3.7 (a) Permeation mechanism of CO_2 and CH_4 through ZIF-8 membrane. (From Lai, L.S. et al., *Sep. Sci. Technol.*) (b) CO_2 molecules lodge preferentially near the windows of ZIF-8. CO_2 hinders the intercage hopping of CH_4. Hindering effects enhance diffusion selectivities in favor of CO_2. (From Chmelik, C. et al., *J. Membr. Sci.*, 397–398, 87, 2012.)

includes the three methyl rings and the six imidazole rings [161]. The C atoms from methyl and N atoms from the imidazole ring interact with electron-deficient carbon from the CO_2 molecules. CO_2 molecules are preferentially permeated through ZIF-8 membrane in the separation of CO_2/CH_4. According to the molecular simulation study, CO_2 molecules were preferentially adsorbed in the window cage of ZIF-8 than the CH_4 molecules (Figure 3.7b). This was because CO_2 molecules are polar with larger quadrupolar moment, while CH_4 molecules are nonpolar with the absence of quadrupolar moment. Hence, electrostatic interaction is found between CO_2 molecules and ZIF frameworks.

3.3.4 ZIF MEMBRANE SYNTHESIS AND PERFORMANCE

Synthesis of ZIF membranes remains a challenging task because the organic ligands do not form bonds with the linkage groups of the support. Different methods have been developed to synthesize ZIF membranes, such as in situ crystallization, stepwise layer-by-layer growth, microwave (MW)-assisted heating, electrospinning, and secondary seeded growth. Some of these methods are adapted from the membrane synthesis of other ordered porous materials such as zeolites. Generally, ZIF membranes are synthesized through in situ growth and secondary seeded growth methods, which will be discussed in the following section [162].

3.3.4.1 In Situ Growth

In situ growth involves one-step process for nucleation and crystallization by immersing the bare support into the reacting solution. Xu et al. [138] reported the one-step synthesis of ZIF-8 membrane on an alumina HF support and repeated the synthesis cycle three times. Such a ZIF-8 membrane exhibited strong CO_2 adsorption properties. Experimental results revealed microcavities inside the continuous ZIF-8 layer. Nevertheless, the H_2/CO_2 separation factor of ~7.1 was still higher than its Knudsen separation index, indicating that membrane exhibited molecular sieving properties.

Liu et al. [163] reported the in situ synthesis of a continuous and c-oriented ZIF-69 membrane on an α-alumina support. ZIF-69 belongs to the group of ZIFs with GME topology and has 12-membered ring (MR) straight channels along the c-axis with a pore size of 0.78 nm. Figure 3.8a is a top view SEM image of the obtained membrane and is consisted of well-intergrown grains, which have fully covered the surface of the substrate. Most of the supported grains have been preferentially aligned with the elongated c-axis vertically to the support forming a crystal layer with an average thickness of 50 μm (cross-sectional view Figure 3.8b).

Single-component permeation results for H_2, CH_4, CO, CO_2, and SF_6 showed a linear relationship between permeance and inversed square root of the molecular weight of the gas molecules, which indicates that their permeation behaviors mainly followed the Knudsen diffusion mechanism (Figure 3.8c). The exception is CO_2, whose permeance was almost three times higher than the Knudsen diffusion rate, and the ideal selectivity of CO_2/CO is 2.5 ± 0.2. This result was attributed to strong affinity of CO_2 with ZIF-69, which has been proved experimentally through CO_2 sorption experiments in ZIF-69 crystals in powder form [143]. Thus, high affinity of CO_2 with ZIF-69 could enhance its diffusion rate through the corresponding

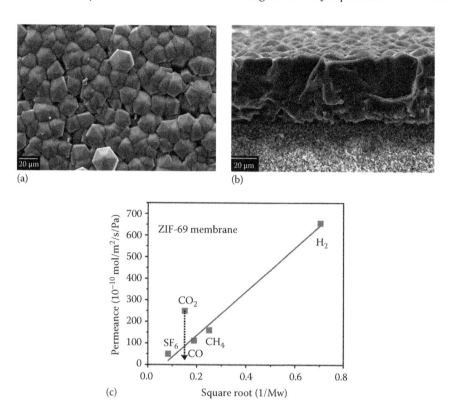

(a)   (b)

(c)

FIGURE 3.8 ZIF-69 membrane: (a) SEM images of top view and (b) cross section. (c) Single-component gas permeation results through ZIF-69 membrane under 1 bar. (From Liu, Y. et al., *J. Membr. Sci.*, 353, 36, 2010.)

membranes by surface diffusion mechanism. The permselectivity of CO_2/CO was found around 3.5 ± 0.1 at room temperature, and the permeance of CO_2 was in the range of $3.6 \pm 0.3 \times 10^{-8}$ mol/(m² s Pa). Both permeance and selectivity were higher in the binary mixture permeation than the single-component mixture. This could be attributed to the selective adsorption of CO_2, which enhanced the transport rate of CO_2. Hence, ZIF-69 membranes proved to be a promising potential candidate in the separation of CO_2 and/or other gases.

3.3.4.2 Secondary Seeded Growth

Seeded growth is another important method, which has been widely used in synthesis of high-performance inorganic zeolite membranes [164,165]. However, very few cases have been reported for ZIF membranes. The seeded growth method decouples the nucleation from growth [164–168] and proceeds into two steps. The first step is to apply a layer of seeds on a support surface. In the second step, the seeded support is exposed to a synthesis solution to carry out secondary growth. During secondary growth, seeds grow and merge forming a continuous membrane. The notable advantage of the seeded growth method is that it can systematically control the membrane orientation, which can be achieved either by the van der Drift competitive growth theory [169] or by preserving the orientation from the seed layer [170]. Other membrane properties such as membrane thickness and grain boundary structure can also be controlled by the seeded growth method, which in many cases determines membrane performance [171]. As a result, this method provided better control for the microstructure of the membrane. The secondary seeded growth method usually produced thinner and less defective membrane because each step involved in this method can be independently manipulated [172].

Secondary seeded growth method has been applied to synthesize a few types of ZIF membranes. Li et al. reported a H_2/CO_2 separation factor of 6.5, which exceeded Knudsen separation factor of 4.7 by a ZIF-7-supported α-alumina membrane [173]. Seeding was performed by dipping of the support in a viscous seeding solution with polyethyleneimine (PEI) as PEI enhanced the linkage between seeds and the support. In another effort to produce a compact ZIF-7 membrane on an asymmetric alumina disc, an aqueous seeding sol with PEI followed by a similar seeding method with small alterations was employed [174]. A clear H_2/CO_2 cutoff was observed at 220°C with a separation factor of 13.6, which exceeded remarkably Knudsen separation factor. The performance of the as-synthesized ZIF-7 membrane without any posttreatment was close to upper bound of micro-/inorganic membranes.

In a next effort, Li et al. [175] improved the seeding PEI technique by following one-pot synthesis in order to form seeds directly in the colloidal solution with PEI as it has a high density of amino groups and it can act as a base to deprotonate benzimidazole, allowing the quick generation of a large number of ZIF-7 nuclei. By altering molar ratio of PEI and reaction duration, the size of ZIF-7 nanoparticles was varied from 40 to 140 nm. In addition, by replacing zinc nitrate with Zn chloride, the existence of chloride ions had a great influence on the growth kinetics of ZIF-7 crystals. The ZIF-7 crystals synthesized with $ZnCl_2$ exhibited prismatic hexagonal shapes with high aspect ratios, promoting the c-out of plane preferred orientation of the ZIF-7 membrane.

Venna et al. [176] reported the seeding synthesis of ZIF-8 membranes by rubbing the inside surface of the tubular α-alumina porous support with dry ZIF-8 seeds. These membranes displayed unprecedented high CO_2 permeances up to ~2.4 × 10^{-5} mol/(m^2 s Pa) and CO_2/CH_4 selectivity values that were ranged from 4 to 7. Separation index of the membranes ranged from 6.5 to 10, values which are comparable with reported alumina-supported SAPO-34 membranes. Despite the low CO_2/CH_4 selectivity most likely due to a high concentration of nonzeolitic pores, high CO_2 permeances contributed to relatively high separation indexes.

A ZIF-8 membrane with a H_2/CO_2 separation factor of 5.2, which exceeded Knudsen constant, was also prepared by Tao et al. [177] using the seeding technique. In this case, crystals used as seeds came from a poorly intergrown film that was produced upon in situ solvothermal reaction and were then rubbed manually by a fine sand paper and finally exposed to secondary growth.

Liu et al. [171] reported growth of a c-oriented ZIF-69 membrane by soaking of an α-alumina support in a seed in DMF suspension for a few seconds. Again by replacing the zinc nitrate source according to Yaghi recipe with zinc acetate, crystal size was manipulated and reduced to 800 nm, in order for these grains to be used as seeds. During secondary growth however, Zn nitrate was used as a zinc source that

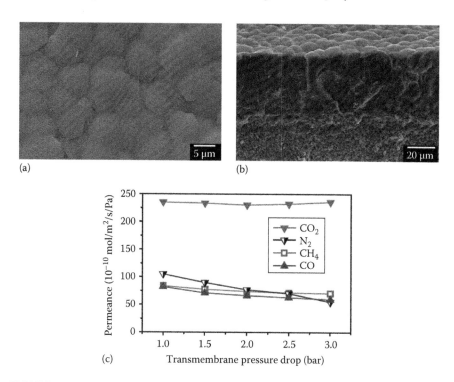

(a)   (b)   (c)   Transmembrane pressure drop (bar)

FIGURE 3.9 (a) Top view of ZIF-69 membrane by secondary growth. (b) Cross-sectional view of ZIF-69 membrane by secondary growth. (c) Single-gas permeances of CO_2, N_2, CH_4, and CO through a ZIF-69 membrane as a function of transmembrane pressure drop at 298 K. (From Liu, Y. et al., *J. Membr. Sci.*, 379, 46, 2011.)

enhanced production of elongated crystal prisms along c-direction, thus promoting the c-preferred orientation of ZIF-69 membrane. Figure 3.9a and b shows the typical SEM images in top view and in cross-sectional view of the ZIF-69 membranes obtained by secondary growth. The top view SEM image shows the membrane was well intergrown with smooth membrane surface. The average grain size inside the membrane was around 10 μm. The cross-sectional view reveals that the ZIF-69 membrane was composed of tightly intergrown crystals highly oriented to the porous alumina support. The membrane thickness was around 40 μm. Figure 3.9c shows the dependence of the single-gas permeance of CO_2, N_2, CO, and CH_4 upon transmembrane pressure drop. The permeance of CO_2 remained almost constant with increased feed pressure, while the permeances of N_2, CO, and CH_4 decreased, indicating negligible defects for the ZIF-69 membrane.

Permeance of CO_2 in this study was higher than those of N_2, CO, and CH_4. This was possibly attributed to the strong selective adsorption of ZIF-69 for CO_2 that caused surface diffusion of CO_2 through ZIF-69 crystals. The ideal selectivities of CO_2/N_2, CO_2/CO, and CO_2/CH_4 were 2.2, 2.9, and 2.7, respectively. The steady-state separation factors of CO_2/N_2, CO_2/CO, and CO_2/CH_4 were around 6.3, 5.0, and 4.6, respectively, and were increased compared to ideal selectivities. This was again attributed to selective adsorption of CO_2 on ZIF-69 crystals, which enhanced the transport rate of CO_2.

Compared to the ZIF-69 membranes reported previously [163], which were fabricated by in situ solvothermal reaction, improvement in both selectivity and permeance was observed for the analogous membranes prepared by the seeding secondary growth as the selectivity for CO_2/CO system was increased from 3.5 to 5.0, while the permeance appeared almost two times higher. The better membrane performance was attributed to the improvement of membrane microstructure, as seeded secondary growth offered a better control over crystal orientation and intergrowth.

3.3.4.3 Microwave-Assisted Heating

In contrast to the conventional heating solvothermal method, which requires a long time (typically half to several days) and high electric power, MW-assisted heating offers the advantage of directly and uniformly heating the contents resulting in a shorter synthesis time with a lower power consumption. Consequently, it emerges as a promising preparation technique in organic synthesis and inorganic hybrid materials and for nanoscale particle preparations, which has advantages in effectively saving reaction time, accelerating the crystallization process, and producing phase-pure products in high yield and large scale.

High-quality membranes have been reported within short synthesis duration [173–175,178]. Ba-SAPO-34 membrane was successfully synthesized by Chew et al. [150] by using MW-assisted synthesis method. A membrane layer with a thickness of ~3–4 μm was grown after a short synthesis duration of 2 h. On the contrary for a conventional hydrothermal synthesis method of a SAPO-34 membrane of the same topology, Li et al. [179] reported the requirement of 24 h synthesis time, while the thickness of the membrane increased to 5 μm.

Bux and coworkers reported the double advantage of applying seeding and subsequent solvothermal MW heating in producing a thin defect-free ZIF-8 membrane [180]. ZIF-8 seeds were dispersed into a PEI in water solution and were attached to porous alumina discs using an automatic dip-coating device. A subsequent solvothermal secondary growth of the seeded support led to the formation of a ZIF-8 layer of around 12 μm thick (Figure 3.10a and b), which was much thinner compared to the well-intergrown ZIF-8 layer, which was prepared by in situ MW heating (~30 μm) (Figure 3.10c and d) [181]. In addition, the ZIF-8 layer produced with the use of seeds exhibited a high degree of (100) orientation indicating that seeding was in fact necessary in order to grow a continuous and defect-free ZIF layer. A H_2/CH_4 separation factor of 15 was observed for the secondary growth membrane, which was slightly higher compared to the ZIF-8 membrane grown by in situ crystallization ($a \sim 11$) due to better intergrowth of grains and a low leak transport. A sharp cutoff for H_2/C_nX_m separation was also observed when increasing the hydrocarbon chain length from C_2 to C_3 for the secondary growth membrane.

Our group demonstrated application of MW heating in order to grow seeds on the support by an in situ solvothermal reaction [182]. The preceding seeding synthesis cycle was repeated twice in order for the seeds to be deposited throughout the support surface. The SEM image of Figure 3.11a indicates crystallinity of seed particles with crystal size of ~1 μm and an approximate interseed separation distance of 1 μm. The subsequent secondary growth took place conventionally forming a compact ZIF layer (Figure 3.11b).

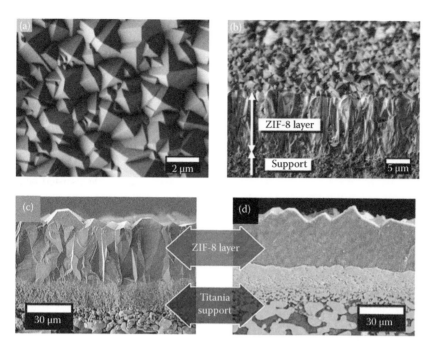

FIGURE 3.10   (a) SEM top view of the well-intergrown ZIF-8 layer after 2 h of secondary growth. (b) SEM top-down view on the corresponding cross section of the broken membrane. (From Bux, H. et al., *Chem. Mater.*, 23(8), 2262, 2011.) (c) Cross-sectional view of ZIF-8 membrane layer prepared by in situ MW heating. (d) EDXS mapping of the ZIF-8 membrane. (From Bux, H. et al., *J. Am. Chem. Soc.*, 131, 16000, 2009.)

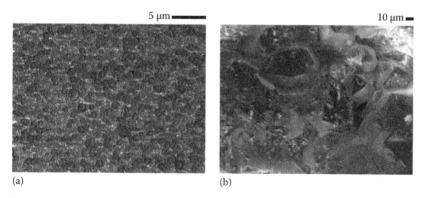

FIGURE 3.11   (a) ZIF-69 seed layer on α-alumina after two-cycle MW seeding. (b) ZIF-69 membrane produced by secondary growth using the seeded support shown in panel (a). (From Tzialla, O. et al., *J. Phys. Chem. C*, 117, 18434, 2013.)

Permeance experimental data of the as-prepared ZIF-69 membranes are shown in Figure 3.8. The as-grown ZIF-69 membrane exhibited a 2–2.5-fold lower permeance compared to the pristine α-alumina support (Figure 3.12a) due to the formation of the ZIF-69 film on the support surface. However, the absence of a CO_2/N_2 separation capability (Figure 3.12b) suggested the existence of defects in the ZIF layer, which

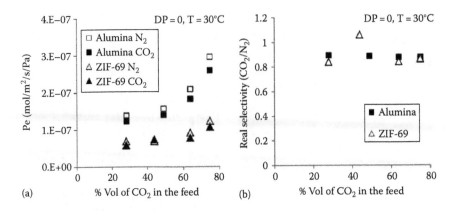

FIGURE 3.12 (a) CO_2 and N_2 permeance as a function of CO_2 concentration in the feed for the α-alumina support and the pristine ZIF-69 membrane. (b) CO_2/N_2 selectivity versus CO_2 concentration in the feed. (From Tzialla, O. et al., *J. Phys. Chem. C*, 117, 18434, 2013.)

originated from intercrystal gaps. Notably, an almost linear correlation between permeance and transmembrane pressure was observed, which is compatible with the Poiseuille flow in macropores, and constituted another piece of evidence that intercrystal, macrosized defects existed.

3.3.5 ENHANCEMENT OF CO_2 ADSORPTION

One of the most attractive properties of MOFs is the ability to tailor the pore structure and functionality due to the high organic functionality and flexibility with which the organic linkers of MOFs can be modified. The search for a new modification method to functionalize pore channel and cavities in MOFs is crucial for extending the application of these new hybrid materials. Besides direct assembly of organic ligands that can pose difficulties as certain functional groups may be hard to incorporate into MOFs, the other strategy for creating desired functionalities in MOFs is the postsynthesis modification of these preconstructed hybrid structures.

Organic links of ZIF-90, for example, which possess aldehyde functionality, have been covalently transformed by two common organic reaction reductions of the aldehyde to an alcohol with $NaBH_4$ and the formation of an imine bond by reaction with ethanolamine in 80% and quantitative yields, respectively (Figure 3.13) [145]. It is noteworthy that upon transformations, the crystallinity of the ZIF material was maintained.

Grafting of amines onto surfaces of porous materials has been proved to enhance adsorption of the acidic CO_2 molecules; several reports have demonstrated the functionalization of MOFs with amine groups. Wang et al. [183] reported a postsynthesis grafting of tetraethylenepentamine (TEPA), which has already been used as an active component in CO_2 adsorbents on the coordinatively unsaturated Cr(III) sites of MIL-101 for the selective adsorption of CO_2 over CO. Upon TEPA grafting, selectivity of MIL-101 increased dramatically from 1.77 to 70.2 at low pressures. The obvious difference between interactions of CO_2 and CO with TEPA led to the high CO_2 selectivity of TEPA-MIL101.

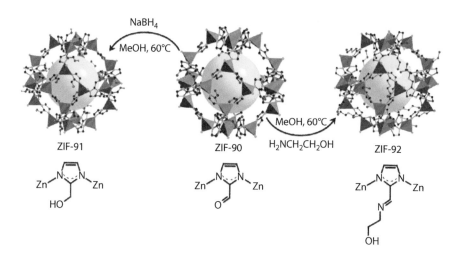

FIGURE 3.13 Isoreticular functionalization of ZIFs: crystal structure of ZIF-90 transformed to ZIF-91 by reduction with $NaBH_4$ and to ZIF-92 by reaction with ethanolamine. The yellow ball indicates space in the cage. H atoms are omitted for clarity, except the H of an alcohol group in ZIF-91 (C, black; N, green; O, red; Cl, pink). (From Phan, A. et al., *Acc. Chem. Res.*, 43, 58, 2009.)

Zhang et al. [184] reported the enhancement of CO_2/N_2 selectivity upon modifying ZIF-8 with ethylenediamine (ED) via a postsynthetic modification technique. The ED-ZIF-8 selectivity for CO_2/N_2 adsorption improved significantly and reached the values of 23 and 13.9 at 0.1 and 0.5 bar, respectively, being almost twice of the achieved selectivity by the unmodified ZIF-8. In addition, Zhang and coworkers reported that ZIF-8 modified by the ammonia impregnation method could adsorb more CO_2 molecules due to an increase of the basicity of ZIF-8 crystals [185]. Chen et al. [186] performed atomistic simulation studies in order to investigate the CO_2 capture of a composite of IL, namely, 1-*n*-butyl-3-methylimidazolium hexafluorophosphate [BMIM][PF$_6$] supported on MOF IRMOF-1. The bulky [BMIM]$^+$ cation resided in the open pore of IRMOF-1, whereas the small [PF$_6$]$^-$ anion preferred to locate in the metal cluster corner and possessed a strong interaction with the framework (Figure 3.14a). Ions in the composite interacted strongly with CO_2, in particular, the [PF$_6$]$^-$ anion, which was the most favorable site for CO_2 adsorption. The composite selectively adsorbed CO_2 from the CO_2/N_2 mixture, with selectivity significantly higher than many other supported ILs. Furthermore, the selectivity increased with increasing IL ratio in the composite (Figure 3.14b). This computational study demonstrated for the first time that IL/MOF composite might be potentially useful for CO_2 capture.

Gupta et al. [187] performed atomistic simulation studies in order to investigate CO_2 capture in IL membranes supported on MOFs. In the 1-*n*-butyl-3-methylimidazolium thiocyanate [BMIM][SCN]–supported Na-rho–ZMOF membrane, the [SCN]$^-$ interacted more strongly than the [BMIM]$^+$ with the MOF support. Compared to polymer membranes and polymer-supported ILs, [BMIM][SCN]/ZMOF possessed higher PCO_2/PN_2 permselectivity as well as higher permeability, which surpassed a

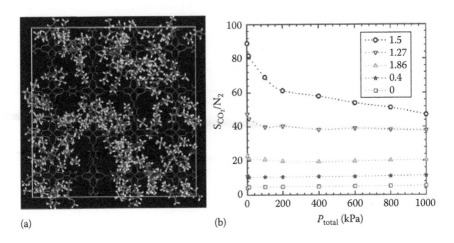

(a)   (b)

FIGURE 3.14   (a) [BMIM][PF$_6$]/IRMOF-1 composite at a weight ratio $W_{IL/IRMOF-1} = 0.4$. N, blue; C in [BMIM]$^+$, green; P, pink; F, cyan; Zn, orange; O, red; C in IRMOF-1, gray; H, white. (b) Selectivity of CO_2/N_2 mixture in IL/IRMOF-1 at $W_{IL/IRMOF-1} = 0$, 0.4, 0.86, 1.27, and 1.5. (From Chen, Y. et al., *J. Phys. Chem. C*, 115, 21736, 2011.)

Robeson upper bound. This computational investigation again suggested the potential use of IL/MOF composite for CO_2 capture.

Although there are several modification methods reported for the enhancement of CO_2 adsorption for MOF in crystal form, there are only few reports on the modification of MOF membranes. Similar to zeolite membranes, most of the MOF layers are polycrystalline with intercrystalline grain boundaries, which compromised membrane quality. Therefore, postmodification techniques are helpful to minimize the nonselective transport through the intercrystalline gaps. Huang et al. [188] reported the covalent postfunctionalization of ZIF-90 membrane by ethanolamine in order to enhance its H_2/CO_2 selectivity. The postfunctionalization strategy offered a double advantage. It caused elimination of invisible intercrystalline defects of the ZIF-90 layer, thus enhancing the molecular sieving performances of the ZIF-90 membrane and constriction of the pore aperture, thus improving molecular sieving for the separation of H_2 from CO_2 and other large gases. The covalent postfunctionalized membrane exhibited an increased H_2/CO_2 selectivity compared to the as-synthesized ZIF-90 membrane from 7.3 to 62.5.

Based on the aforementioned computational investigations, which indicated a significant increase in selectivity of the IL/MOF composite compared to many other supported ILs, our group has recently reported [182] the separation performance toward CO_2 of an IL/ZIF-69 composite membrane, in an attempt to fill the existing intercrystalline gaps of the synthesized ZIF-69 membrane, grown by the seeded technique, and act favorably in enhancing the separation capability due to the synergistic role of the two materials. A CO_2-selective, bulky IL, namely, the 1-octyl-3-methylimidazolium tricyanomethanide ([omim][TCM]), was chosen in order to combine the superior CO_2 performance of ILs with ZIF-69, which is one of the best performing ZIFs for CO_2 capture and separation. The performance of the [omim][TCM]-casted membrane in comparison to the pristine ZIF-69 one is presented in Figure 3.15.

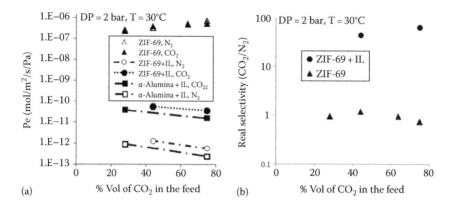

(a)   (b)

FIGURE 3.15 (a) CO_2 and N_2 permeance of the pristine ZIF-69 membrane and of the ZIF-69 membrane after IL casting. Square marks show the respective permeances of an IL-casted α-alumina support. (b) Enhancement of the CO_2/N_2 selectivity of the ZIF-69 membrane after IL casting. (From Tzialla, O. et al., *J. Phys. Chem. C*, 117, 18434, 2013.)

Loading with the IL resulted in significant enhancement of the CO_2/N_2 selectivity, which reached the values of 44 and 64 for CO_2 concentrations of 45% and 75% (vol/vol) (Figure 3.15b) in the feed stream, respectively, accompanied with lower flux through the membrane. These findings indicated that IL plugged and repaired effectively the membrane cracks. Furthermore, for comparison, a blank α-alumina support was loaded by IL, and the resulting IL film exhibited a 2.5 times lower permeance compared to the IL-casted ZIF-69 membrane (Figure 3.15a). This was due to the fact that ZIF crystals exhibit higher permeability than the bulk IL phase and thus enhanced permeability through the membrane.

In addition, we used a sequential flow resistance model in order to interpret the mechanism of permeance through the composite IL/ZIF-69 membrane (Figure 3.16). By applying the experimentally defined permeance values of the support, the IL, and the composite IL/ZIF membrane, we managed to calculate the unknown permeances and CO_2/N_2 selectivities through the ZIF film. These calculated values can also correspond to an ideal, homogeneous, and defect-free ZIF-69 membrane. The calculated CO_2/N_2 selectivity for an ideal ZIF-69 membrane reached the value of 27, which exceeded five times the up-to-date highest-performance ZIF-69 membranes [171].

The aforementioned calculated results demonstrated the exceptional CO_2 separation performance of a potential defect-free ZIF-69 membrane. However, because preparation of an ideal ZIF-69 membrane still remains a challenging issue, plugging of gaps with a CO_2-selective IL has proved to be an efficient method in order to improve membrane performance as the IL/ZIF composite membrane reported here combined the superior CO_2 properties of the two materials and exhibited higher CO_2 selectivity and permeability compared to as-grown ZIF-69 membranes and bulk IL, respectively. Still optimization experiments are currently underway in our group as to restrict IL deposition exclusively in the intercrystalline gaps of the hybrid membranes in order to enhance permeability without sacrificing selectivity by eliminating the additional, slow-diffusivity path caused by the IL layer on the top of the ZIF film.

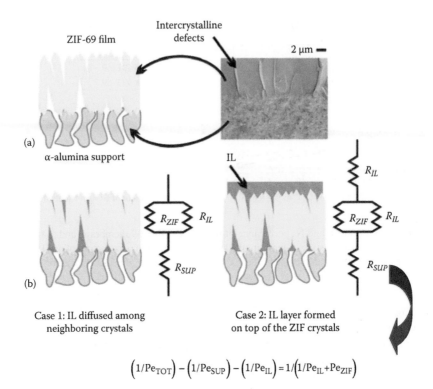

$$\left(1/Pe_{TOT}\right) - \left(1/Pe_{SUP}\right) - \left(1/Pe_{IL}\right) = 1/\left(1/Pe_{IL} + Pe_{ZIF}\right)$$

FIGURE 3.16 Schematic illustration of ZIF-69 membrane morphologies. (a) As-grown membrane where the intercrystalline voids are evident, accompanied by a cross-sectional SEM image of the membrane. (b) IL/ZIF-69 hybrid membrane. Left, IL fills only the intercrystalline gaps. Resistance to the flow is through the ZIF crystals, the intercrystal-located IL, and the support. Right, IL fills the intercrystalline gaps and also forms an additional layer on top of the crystalline film creating an additional resistance to the flow. The hydrophobic nature of the IL does not allow for diffusion and uniform distribution into the pores of the hydrophilic support. (From Tzialla, O. et al., *J. Phys. Chem. C*, 117, 18434, 2013.)

REFERENCES

1. R. Yegani, H. Hirozawa, M. Teramoto, H. Himei, O. Okada, T. Takigawa, N. Ohmura, N. Matsumiya, H. Matsuyama, Selective separation of CO_2 by using novel facilitated transport membrane at elevated temperatures and pressures. *J. Membr. Sci.*, 2007. **291**: 157–164.
2. J. Huang, L. El-Azzami, W.S.W. Ho, Modeling of CO_2-selective water gas shift membrane reactor for fuel cell. *J. Membr. Sci.*, 2005. **261**: 67–75.
3. H. Lin, E. Van wagner, B.D. Freeman, L.G. Toy, R.P. Gupta, Plasticization-enhanced hydrogen purification using polymeric membranes. *Science*, 2006. **311**: 639–642.
4. R.W. Baker, K. Lokhandwala, Natural gas processing with membranes: An overview. *Ind. Eng. Chem. Res.*, 2008. **47**: 2109–2121.
5. S. Basu, A.L. Khan, A. Cano-Odena, C. Liu, I.F.J. Vankelecom, Membrane-based technologies for biogas separations. *Chem. Soc. Rev.*, 2010. **39**: 750–768.
6. K. Okabe, N. Matsuyama, H. Mano, Stability of gel-supported facilitated transport membrane for carbon dioxide separation from model flue gas. *Sep. Purif. Technol.*, 2007. **57**: 242–249.

7. M.K. Barillas, R.M. Enick, M. O'Brien, R. Perry, D.R. Luebke, B.D. Morreale, The CO_2 permeability and mixed gas CO_2/H_2 selectivity of membranes composed of CO_2-philic polymers. *J. Membr. Sci.*, 2011. **372**: 29–39.
8. C. Myers, H. Pennline, D. Luebke, J. Ilconich, J.K. Dixon, E.J. Maginn, J.F. Brennecke, High temperature separation of carbon dioxide/hydrogen mixtures using facilitated supported ionic liquid membranes. *J. Membr. Sci.*, 2008. **322**: 28–31.
9. G.T. Rochelle, Amine scrubbing for CO_2 capture. *Science*, 2009. **325**(5948): 1652–1654.
10. X. Xu et al., Novel polyethylenimine-modified mesoporous molecular sieve of MCM-41 type as high-capacity adsorbent for CO_2 capture. *Energy Fuels*, 2002. **16**(6): 1463–1469.
11. K.T. Chue et al., Comparison of activated carbon and zeolite 13X for CO_2 recovery from flue gas by pressure swing adsorption. *Ind. Eng. Chem. Res.*, 1995. **34**(2): 591–598.
12. E. DÃ-az et al., Enhancement of the CO_2 retention capacity of X zeolites by Na- and Cs-treatments. *Chemosphere*, 2008. **70**(8): 1375–1382.
13. N. Konduru, P. Lindner, N.M. Assaf-Anid, Curbing the greenhouse effect by carbon dioxide adsorption with Zeolite 13X. *AIChE J.*, 2007. **53**(12): 3137–3143.
14. J. Merel, M. Clausse, F. Meunier, Experimental investigation on CO_2 post-combustion capture by indirect thermal swing adsorption using 13X and 5A zeolites. *Ind. Eng. Chem. Res.*, 2007. **47**(1): 209–215.
15. L.M. Robeson, The upper bound revisited. *J. Membr. Sci.*, 2008. **320**: 390–400.
16. W.J. Koros, Evolving beyond the thermal age of separation processes: Membranes can lead the way. *AIChE J.*, 2004. **50**: 2326–2334.
17. T. Matsuura, Progress in membrane science and technology for seawater desalination—A review. *Desalination*, 2001. **134**: 47–54.
18. K.P. Lee, T.C. Arnot, D. Mattia, A review of reverse osmosis membrane materials for desalination–development to date and future potential. *J. Membr. Sci.*, 2011. **370**: 1–22.
19. S.S. Hashim, A.R. Mohamed, S. Bhatia, Oxygen separation from air using ceramic-based membrane technology for sustainable fuel production and power generation. *Renew. Sust. Energy Rev.*, 2011. **15**: 1284–1293.
20. A. Merritt, R. Rajagopalan, H.C. Foley, High performance nanoporous carbon membranes for air separation. *Carbon*, 2007. **45**: 1267–1278.
21. I.S.K. Purnomo, E. Alpay, Membrane column optimisation for the bulk separation of air. *Chem. Eng. Sci.*, 2000. **55**: 3599–3610.
22. A.R. Smith, J. Klosek, A review of air separation technologies and their integration with energy conversion processes. *Fuel Process. Technol.*, 2001. **70**: 115–134.
23. J.K. Adewole, A.L. Ahmad, S. Ismail, C.P. Leo, Current challenges in membrane separation of CO_2 from natural gas: A review. *Int. J. Greenhouse Gas Control*, 2013. **17**: 46–65.
24. T.E. Rufford, S. Smart, G.C.Y. Watson, B.F. Graham, J. Boxall, J.C. Diniz da Costa, E.F. May, The removal of CO_2 and N_2 from natural gas: A review of conventional and emerging process technologies. *J. Petroleum Sci. Eng.*, 2012. **94–95**: 123–154.
25. S.E. Kentish, Polymeric membranes for natural gas processing. *Adv. Membr. Sci. Technol. Sust. Energy Environ. Appl.*, 2011. **11**: 339–360.
26. R.J.R. Uhlhorn, K. Keizer, A.J. Burggraaf, Gas transport and separation with ceramic membranes. Part I. Multilayer diffusion and capillary condensation. *J. Membr. Sci.*, 1992. **66**: 259–269.
27. J. Xiao, J. Wei, Diffusion mechanisms of hydrocarbons in zeolites—I. Theory. *Chem. Eng. Sci.*, 1992. **47**: 1123–1141.
28. J. Xiao, J. Wei, Diffusion mechanisms of hydrocarbons in zeolites—II. Analysis of experimental observations. *Chem. Eng. Sci.*, 1992. **47**: 1143–1159.
29. H. Rhim, S.T. Haynes, Transport of capillary condensate. *J. Colloid. Interface Sci.*, 1975. **52**: 174–181.

30. K.H. Lee, S.T. Hwang, The transport of condensible vapors through a microporous Vycor glass membrane. *J. Colloid. Interface Sci.*, 1986. **110**: 544–555.

31. P. Uchytil, R. Petričkovič, A. Seidel-Morgenstern, Study of capillary condensation of butane in a Vycor glass membrane. *J. Membr. Sci.*, 2005. **264**: 27–36.

32. K.P. Tzevelekos, G.E. Romanosa, E.S. Kikkinides, N.K. Kanellopoulos, V. Kaselouri, Experimental investigation on separations of condensable from non-condensable vapors using mesoporous membranes. *Micropor. Mesopor. Mater.*, 1999. **31**: 151–162.

33. R.S.A. de Lange, J.H.A. Hekkink, K. Keizer, A.J. Burggraaf, Permeation and separation studies on microporous sol-gel modified ceramic membranes. *Micropor. Mater.*, 1995. **4**: 169–186.

34. A.J. Burggraaf, Single gas permeation of thin zeolite (MFI) membranes: Theory and analysis of experimental observations. *J. Membr. Sci.*, 1999. **155**: 45–65.

35. C.E. Powell, G.G. Qiao, Polymeric CO_2/N_2 gas separation membranes for the capture of carbon dioxide from power plant flue gases. *J. Membr. Sci.*, 2006. **279**: 1–49.

36. M.T. Ho, G.W. Allinson, D.E. Wiley, Reducing the cost of CO_2 capture from flue gases using membrane technology. *Ind. Eng. Chem. Res.*, 2008. **47**: 1562–1568.

37. H. Chen, A.S. Kovvali, K.K. Sirkar, Selective CO_2 separation from CO_2-N_2 mixtures by immobilized Glycine-Na-Glycerol membranes. *Ind. Eng. Chem. Res.*, 2000. **39**: 2447–2458.

38. A.S. Kovvali, H. Chen, K.K. Sirkar, Dendrimer membranes: A CO_2-selective molecular gate. *J. Am. Chem. Soc.*, 2000. **122**: 7594–7595.

39. J.E. Bara, T.K. Carlisle, C.J. Gabriel, D. Camper, A. Finotello, D.L. Gin, R.D. Noble, A guide to CO_2 separations in imidazolium-based room-temperature ionic liquids. *Ind. Eng. Chem. Res.*, 2009. **48**: 2739–2751.

40. C. Cadena, J. Anthony, J. Shah, T. Morrow, J. Brennecke, E.J. Maginn, Why is CO_2 so soluble in imidazolium-based ionic liquids? *J. Am. Chem. Soc.*, 2004. **126**: 5300–5308.

41. L.A. Neves, J. Benavente, I.M. Coelhoso, J.G. Crespo, Design and characterization of Nafion membranes with incorporated ionic liquids cations. *J. Membr. Sci.*, 2010. **347**: 42–52.

42. L.A. Neves, I.M. Coelhoso, J.G. Crespo, Methanol and gas crossover through modified Nafion membranes by incorporation of ionic liquid cations. *J. Membr. Sci.*, 2010. **360**: 363–370.

43. M. Guo, J. Fang, H. Xu, W. Li, X. Lu, C. Lan, K. Li, Synthesis and characterization of novel anion exchange membranes based on imidazolium-type ionic liquid for alkaline fuel cells. *J. Membr. Sci.*, 2010. **362**: 97–104.

44. H. Tokuda, K. Ishii, M.A.B.H. Susan, S. Tsuzuki, K. Hayamizu, M. Watanabe, Physicochemical properties and structures of room-temperature ionic liquids. 3. Variation of cationic structures. *J. Phys. Chem. B*, 2006. **110**: 2833–2839.

45. P.K. Kilaru, P. Scovazzo, Correlations of low-pressure carbon dioxide and hydrocarbon solubilities in imidazolium-, phosphonium-, and ammonium-based room-temperature ionic liquids. Part 2. Using activation energy of viscosity. *Ind. Eng. Chem. Res.*, 2008. **47**: 910–919.

46. Y. Yoshida, O. Baba, G. Saito, Ionic liquids based on dicyanamide anion: Influence of structural variations in cationic structures on ionic conductivity. *J. Phys. Chem. B*, 2007. **111**: 4742–4749.

47. S. Feng, G.A. Voth, Molecular dynamics simulations of imidazolium-based ionic liquid/water mixtures: Alkyl side chain length and anion effects. *Fluid Phase Equilib.*, 2010. **294**: 148–156.

48. J.L. Anthony, E.J. Maginn, J.F. Brennecke, Solubilities and thermodynamic properties of gases in the ionic liquid 1-n-butyl-3-methylimidazolium hexafluorophosphate. *J. Phys. Chem. B*, 2002. **106**: 7315–7320.

49. D. Camper, C. Becker, C. Koval, R. Noble, Low pressure hydrocarbon solubility in room temperature ionic liquids containing imidazolium rings interpreted using regular solution theory. *Ind. Eng. Chem. Res.*, 2005. **44**: 1928–1933.

50. J. Jacquemin, M.F. Costa Gomes, P. Husson, V. Majer, Solubility of carbon dioxide, ethane, methane, oxygen, nitrogen, hydrogen, argon, and carbon monoxide in 1-butyl-3-methylimidazolium tetrafluoroborate between temperatures 283 K and 343 K and at pressures close to atmospheric. *J. Chem. Thermodyn.*, 2006. **38**: 490–502.

51. J.L. Anthony, J.L. Anderson, E.J. Maginn, J.F. Brennecke, Anion effects on gas solubility in ionic liquids. *J. Phys. Chem. B*, 2005. **109**: 6366–6374.

52. D. Camper, C. Becker, C. Koval, R. Noble, Diffusion and solubility measurements in room temperature ionic liquids. *Ind. Eng. Chem. Res.*, 2006. **45**: 445–450.

53. L.A. Blanchard, Z. Gu, J.F. Brennecke, High-pressure phase behavior of ionic liquid/CO_2 systems. *J. Phys. Chem. B*, 2001. **105**: 2437–2444.

54. P. Scovazzo, D. Camper, J. Kieft, J. Poshusta, C. Koval, R. Noble, Regular solution theory and CO_2 gas solubility in room-temperature ionic liquids. *Ind. Eng. Chem. Res.*, 2004. **43**: 6855–6860.

55. R.E. Baltus, B.H. Culbertson, S. Dai, H. Luo, D.W. Depaoli, Low-pressure solubility of carbon dioxide in room-temperature ionic liquids measured with a quartz crystal microbalance. *J. Phys. Chem. B*, 2004. **108**: 721–727.

56. D. Morgan, L. Ferguson, P. Scovazzo, Diffusivity of gases in room temperature ionic liquids: Data and correlation obtained using a lag-time technique. *Ind. Eng. Chem. Res.*, 2005. **44**: 4815–4823.

57. D. Camper, J. Bara, C. Koval, R. Noble, Bulk-fluid solubility and membrane feasibility of Rmim-based room-temperature ionic liquids. *Ind. Eng. Chem. Res.*, 2006. **45**: 6279–6283.

58. A. Finotello, J.E. Bara, D. Camper, R.D. Noble, Room-temperature ionic liquids: Temperature dependence of gas solubility selectivity. *Ind. Eng. Chem. Res.*, 2008. **47**: 3453–3459.

59. M.T. Mota-Martinez, M.A.M. Althuluth, M.C. Kroon, C.J. Peters, Solubility of carbon dioxide in the low-viscosity ionic liquid 1-hexyl-3-methylimidazolium tetracyanoborate. *Fluid Phase Equilib.*, 2012. **332**: 35–39.

60. M.C. Kroon, A. Shariati, M. Costantini, J. van Spronsen, G.J. Witkamp, R.A. Sheldon, C.J. Peters, High-pressure phase behavior of systems with ionic liquids: Part V. The binary system carbon dioxide + 1-butyl-3-methylimidazolium tetrafluoroborate. *J. Chem. Eng. Data*, 2005. **50**: 173–176.

61. E.D. Bates, R.D. Mayton, I. Ntai, J.H. Davis Jr., CO_2 capture by a task-specific ionic liquid. *J. Am. Chem. Soc.*, 2002. **124**: 926–927.

62. B.E. Gurkan, J.C. de la Fuente, E.M. Mindrup, L.E. Ficke, B.F. Goodrich, E.A. Price, W.F. Schneider, J.F. Brennecke, Equimolar CO_2 absorption by anion functionalized ionic liquids. *J. Am. Chem. Soc.*, 2010. **132**: 2116–2117.

63. P. Scovazzo, J. Kieft, D.A. Finan, C. Koval, D. DuBois, R. Noble, Gas separations using non-hexafluorophosphate [PF_6](−) anion supported ionic liquid membranes. *J. Membr. Sci.*, 2004. **238**: 57–63.

64. Y.-Y Jiang, Z. Zhou, Z. Jiao, L. Li, Y.-T. Wu, Z.-B. Zhang, SO_2 gas separation using supported ionic liquid membranes. *J. Phys. Chem. B*, 2007. **111**: 5058–5061.

65. J. Ilconich, C. Myers, H. Pennline, D. Luebke, Experimental investigation of the permeability and selectivity of supported ionic liquid membranes for CO_2/He separation at temperatures up to 125°C. *J. Membr. Sci.*, 2007. **298**: 41–47.

66. X. Han, D.W. Armstrong, Ionic liquids in separations. *Acc. Chem. Res.*, 2007. **40**: 1079–1086.

67. S. Hanioka, T. Maruyama, T. Sotani, M. Teramoto, H. Matsuyama, K. Nakashima, M. Hanaki, F. Kubota, M. Goto, CO_2 separation facilitated by task-specific ionic liquids using a supported liquid membrane. *J. Membr. Sci.*, 2008. **314**: 1–4.

68. L.A. Neves, N. Nemestóthy, V.D. Alves, P. Cserjesi, K. Belafi-Bako, I.M. Coelhoso, Separation of biohydrogen by supported ionic liquid membranes. *Desalination*, 2009. **240**: 311–315.

69. F.J. Hernández-Fernández, A.P. De los Rios, F. Tomás-Alonso, J.M. Palacios, G. Víllora, Preparation of supported ionic liquid membranes: Influence of the ionic liquid immobilization method on their operational stability. *J. Membr. Sci.*, 2009. **341**: 172–177.

70. T.K. Carlisle, J.E. Bara, A.L. Lafrate, D.L. Gin, R.D. Noble, Main-chain imidazolium polymer membranes for CO_2 separations: An initial study of a new ionic liquid inspired platform. *J. Membr. Sci.*, 2010. **359**: 37–43.

71. Y.-I. Park, B.-S. Kim, Y.-H. Byun, S.-H. Lee, E.-W. Lee, J.-M. Lee, Preparation of supported ionic liquid membranes (SILMs) for the removal of acidic gases from crude natural gas. *Desalination*, 2009. **236**: 342–348.

72. A.P. de los Rios, F.J. Hernández-Fernándeza, F. Tomas-Alonso, J.M. Palacios, D. Gomez, M. Rubio, G. Villora, A SEM-EDX study of highly stable supported liquid membranes based on ionic liquids. *J. Membr. Sci.*, 2007. **300**: 88–94.

73. F.F. Krull, M. Hechinger, W. Kloeckner, M. Verhuelsdonk, F. Buchbender, H. Giese, T. Melin, Ionic liquid imbibition of ceramic nanofiltration membranes. *Colloids Surf. A Physicochem. Eng. Aspects*, 2009. **345**: 182–190.

74. O.C. Vangeli, G.E. Romanos, K.G. Beltsios, D. Fokas, Ch.P. Athanasekou, N.K. Kanellopoulos, Development and characterization of chemically stabilized ionic liquid membranes—Part I: Nanoporous ceramic supports. *J. Membr. Sci.*, 2010. **365**: 366–377.

75. P. Scovazzo, D. Havard, M. McShea, S. Mixon, D. Morgan, Long-term, continuous mixed-gas dry fed CO_2/CH_4 and CO_2/N_2 separation performance and selectivities for room temperature ionic liquid membranes. *J. Membr. Sci.*, 2009. **327**: 41–48.

76. L.M. Robeson, Correlation of separation factor versus permeability for polymeric membranes. *J. Membr. Sci.*, 1991. **62**(2): 165–185.

77. L. Ferguson, P. Scovazzo, Solubility, diffusivity, and permeability of gases in phosphonium-based room temperature ionic liquids: Data and correlations. *Ind. Eng. Chem. Res.*, 2007. **46**: 1369–1374.

78. R. Condemarin, P. Scovazzo, Gas permeabilities, solubilities, diffusivities, and diffusivity correlations for ammonium-based room temperature ionic liquids with comparison to imidazolium and phosphonium RTIL data. *Chem. Eng. J.*, 2009. **147**: 51–57.

79. Y. Hou, R.E. Baltus, Experimental measurement of the solubility and diffusivity of CO_2 in room-temperature ionic liquids using a transient thin-liquid-film method. *Ind. Eng. Chem. Res.*, 2007. **46**: 8166–8175.

80. P. Scovazzo, Determination of the upper limits, benchmarks, and critical properties for gas separations using stabilized room temperature ionic liquid membranes (SILMs) for the purpose of guiding future research. *J. Membr. Sci.*, 2009. **343**: 199–211.

81. K.N. Marsh, J.A. Boxall, R. Lichtenthaler, Room temperature ionic liquids and their mixtures—A review. *Fluid Phase Equilib.*, 2004. **219**: 93–98.

82. P. Bonhôte, A.P. Dias, N. Papageorgiou, K. Kalyanasundaram, M. Grätzel, Hydrophobic, highly conductive ambient-temperature molten salts. *Inorg. Chem.*, 1996. **35**: 1168–1178.

83. G. Yanfang, W. Tengfang, Y. Dahong, P. Changjun, L. Honglai, H. Ying, Densities and viscosities of the ionic liquid [C_4mim][PF_6] + N,N-dimethylformamide binary mixtures at 293.15 K to 318.15 K. *Chin. J. Chem. Eng.*, 2008. **16**: 256–262.

84. J. Jacquemin, P. Husson, A.A.H. Padua, V. Majer, Density and viscosity of several pure and water-saturated ionic liquids. *Green Chem.*, 2006. **8**: 172–180.

85. J.O. Zhu, J. Chen, C.Y. Li, W.Y. Fei, Viscosities and interfacial properties of 1-methyl-3-butylimidazolium hexafluorophosphate and 1-isobutenyl-3-methylimidazolium tetrafluoroborate ionic liquids. *J. Chem. Eng. Data*, 2007. **52**: 812–816.

86. O.O. Okoturo, T.J. Vander-Noot, Temperature dependence of viscosity for room temperature ionic liquids. *J. Electroanal. Chem.*, 2004. **568**: 167–181.

87. K.R. Harris, L.A. Woolf, Temperature and pressure dependence of the viscosity of the ionic liquid 1-butyl-3-methylimidazolium hexafluorophosphate. *J. Chem. Eng. Data*, 2005. **50**: 1777–1782.

88. J.G. Huddleston, A.E. Visser, W.M. Reichert, H.D. Willauer, G.A. Broker, R.D. Rogers, Characterization and comparison of hydrophilic and hydrophobic room temperature ionic liquids incorporating the imidazolium cation. *Green Chem.*, 2001. **3**: 156–164.

89. K.H. Tokuda, K. Ishii, M.A.B.H. Susan, M. Watanabe, Physicochemical properties and structures of room temperature ionic liquids. 1. Variation of anionic species. *J. Phys. Chem. B*, 2004. **108**: 16593–16600.

90. D. Camper, P. Scovazzo, R.D. Noble, Gas solubilities in room temperature ionic liquids. *Ind. Eng. Chem. Res.*, 2004. **43**: 3049–3054.

91. A. Finotello, J.E. Bara, D. Camper, R.D. Noble, Ideal gas solubilities and solubility selectivities in a binary mixture of ionic liquids. *J. Phys. Chem. B*, 2008. **112**: 2335–2339.

92. T. Takamatsu, The application of the regular solution theory to the ion-pair systems. *Bull. Chem. Soc. Jpn.*, 1974. **47**: 1287.

93. G. Giraud, C.M. Gordon, I.R. Dunkin, K. Wynne, The effects of anion and cation substitution on the ultrafast solvent dynamics of ionic liquids: A time-resolved optical Kerr-effect spectroscopic study. *J. Chem. Phys.*, 2003. **119**: 464.

94. B. Buchheit, The effect of water and light alcohols on the viscosity of tetrafluoroborate containing ionic liquids, undergraduate Honors thesis. Department of Chemical & Biomolecular Engineering, Clarkson University, Potsdam, NY, 2008.

95. M.G. Freire, P.J. Carvalho, R.L. Gardas, I.M. Marrucho, L.M.N.B.F. Santos, J.A.P. Coutinho, Mutual solubilities of water and the [Cnmim][Tf$_2$N] hydrophobic ionic liquids. *J. Phys. Chem. B*, 2008. **112**: 1604–1610.

96. M.G. Freire, C.M.S.S. Neves, P.J. Carvalho, R.L. Gardas, A.M. Fernandes, I.M. Marrucho, L.M.N.B.F. Santos, J.A.P. Coutinho, Mutual solubilities of water and hydrophobic ionic liquids. *J. Phys. Chem. B*, 2007. **111**: 13082–13089.

97. D.B. Fu, X.W. Sun, J.J. Pu, S.Q. Zhao, Effect of water content on the solubility of CO_2 in the ionic liquid [bmim][PF$_6$]. *J. Chem. Eng. Data*, 2006. **51**: 371–375.

98. W. Zhao, G.H. He, L.L. Zhang, J. Ju, H. Dou, F. Nie, C.N. Li, H.J. Liu, Effect of water in ionic liquid on the separation performance of supported ionic liquid membrane for CO_2/N_2. *J. Membr. Sci.*, 2010. **350**: 279–285.

99. L.A. Neves, J.G. Crespo, I.M. Coelhoso, Gas permeation studies in supported ionic liquid membranes. *J. Membr. Sci.*, 2010. **357**: 160–170.

100. K. Simons, K. Nijmeijer, J.E. Bara, R.D. Noble, M. Wessling, How do polymerized room-temperature ionic liquid membranes plasticize during high pressure CO_2 permeation? *J. Membr. Sci.*, 2010. **360**: 202–209.

101. A.I. Labropoulos, G.E. Romanos, E. Kouvelos, P. Falaras, V. Likodimos, M. Francisco, M.C. Kroon, B. Iliev, G. Adamova, T.J.S. Schubert, Alkyl-methylimidazolium tricyanomethanide ionic liquids under extreme confinement onto nanoporous ceramic membranes. *J. Phys. Chem. C*, 2013. **117**: 10114–10127.

102. S. Chen, G. Wu, M. Sha, S. Huang, Transition of ionic liquid [bmim][PF$_6$] from liquid to high-melting-point crystal when confined in multiwalled carbon nanotubes. *J. Am. Chem. Soc.*, 2007. **129**: 2416–2417.

103. Y.L. Verma, M.P. Singh, R.K. Singh, Ionic liquid assisted synthesis of nano-porous TiO$_2$ and studies on confined ionic liquid. *Mater. Lett.*, 2012. **86**: 73–76.

104. M.P. Singh, R.K. Singh, S. Chandra, Studies on Imidazolium-based ionic liquids having a large anion confined in a nanoporous silica gel matrix. *J. Phys. Chem. B*, 2011. **115**: 7505–7514.

105. O.C. Vangeli, G.E. Romanos, K.G. Beltsios, D. Fokas, E.P. Kouvelos, K.L. Stefanopoulos, N.K. Kanellopoulos, Grafting of imidazolium based ionic liquid on the pore surface of nanoporous materials—Study of physicochemical and thermodynamic properties. *J. Phys. Chem. B*, 2010. **114**: 6480–6491.

106. M.-A. Néouze, J.L. Bideau, P. Gaveau, S. Bellayer, A. Vioux, Ionogels, new materials arising from the confinement of ionic liquids within silica-derived networks. *Chem. Mater.*, 2006. **18**: 3931–3936.

107. K.L. Stefanopoulos, G.E. Romanos, O.C. Vangeli, K. Mergia, N.K. Kanellopoulos, A. Koutsioubas, D. Lairez, Investigation of confined ionic liquid in nanostructured materials by a combination of SANS, contrast-matching SANS, and nitrogen adsorption. *Langmuir*, 2011. **27**: 7980–7985.

108. M. Sha, G. Wu, H. Fang, G. Zhu, Y. Liu, Liquid-to-solid phase transition of a 1,3-dimethylimidazolium chloride ionic liquid monolayer confined between graphite walls. *J. Phys. Chem. C*, 2008. **112**: 18584–18587.

109. M. Sha, G. Wu, Y. Liu, Z. Tang, H. Fang, Drastic phase transition in ionic liquid [Dmim] [Cl] confined between graphite walls: New phase formation. *J. Phys. Chem. C*, 2009. **113**: 4618–4622.

110. Q. Dou, M. Sha, H. Fu, G. Wu, Melting transition of ionic liquid [bmim][PF_6] crystal confined in nanopores: A molecular dynamics simulation. *J. Phys. Chem. C*, 2011. **115**: 18946–18951.

111. Y.X. Zhu, S. Granick, Viscosity of interfacial water. *Phys. Rev. Lett.*, 2001. **87**: 096104.

112. R. Zangi, A.E. Mark, Bilayer ice and alternate liquid phases of confined water. *J. Chem. Phys.*, 2003. **119**: 1694.

113. Y.C. Hudiono, T.K. Carlisle, J.E. Bara, Y.F. Zhang, D.L. Gin, R.D. Noble, A three-component mixed-matrix membrane with enhanced CO_2 separation properties based on zeolites and ionic liquid materials. *J. Membr. Sci.*, 2010. **350**: 117–123.

114. P. Jindaratsamee, A. Ito, S. Komuro, Y. Shimoyama, Separation of CO_2 from the CO_2/N_2 mixed gas through ionic liquid membranes at the high feed concentration. *J. Membr. Sci.*, 2012. **423–424**: 27–32.

115. L.-H. Cheng, M.S.A. Rahaman, R. Yao, L. Zhang, X.-H. Xu, H.-L. Chen, J.-Y. Lai, K.-L. Tung, Study on microporous supported ionic liquid membranes for carbon dioxide capture. *Int. J. Greenhouse Gas Control*, 2014. **21**: 82–90.

116. P. Jindaratsamee, Y. Shimoyama, H. Morizaki, A. Ito, Effects of temperature and anion species on CO_2 permeability and CO_2/N_2 separation coefficient through ionic liquid membranes. *J. Chem. Thermodyn.*, 2011. **43**: 311–314.

117. P. Bernardo, J.C. Jansen, F. Bazzarelli, F. Tasselli, A. Fuoco, K. Friess, R. Izák, V. Jarmarová, M. Kačírková, G. Clarizia, Gas transport properties of Pebax®/room temperature ionic liquid gel membranes. *Sep. Purif. Technol.*, 2012. **97**: 73–82.

118. E. Santos, J. Albo, A. Irabien, Acetate based Supported Ionic Liquid Membranes (SILMs) for CO_2 separation: Influence of the temperature. *J. Membr. Sci.*, 2014. **452**: 277–283.

119. P. Cserjési, N. Nemestóthy, K. Bélafi-Bakó, Gas separation properties of supported liquid membranes prepared with unconventional ionic liquids. *J. Membr. Sci.*, 2010. **349**: 6–11.

120. J.J. Close, K. Farmer, S.S. Moganty, R.E. Baltus, CO_2/N_2 separations using nanoporous alumina supported ionic liquid membranes: Effect of the support on separation performance. *J. Membr. Sci.*, 2012. **390–391**: 201–210.

121. D.D. Iarikov, P. Hacarlioglu, S.T. Oyama, Supported room temperature ionic liquid membranes for CO_2/CH_4 separation. *Chem. Eng. J.*, 2011. **166**: 401–406.

122. D.G. Hert, J.L. Anderson, S.N.V. Aki, J.F. Brennecke, Enhancement of oxygen and methane solubility in 1-hexyl-3-methylimidazolium bis(trifluoromethylsulfonyl) imide using carbon dioxide. *Chem. Commun.*, 2005: 2603.

123. J. Jacquemin, P. Husson, V. Majer, M.F. Costa Gomes, Low-pressure solubilities and thermodynamics of solvation of eight gases in 1-butyl-3-methylimidazolium hexafluorophosphate. *Fluid Phase Equilib.*, 2006. **240**: 87–95.

124. N. Shahkaramipour, M. Adibi, A.A. Seifkordi, Y. Fazli, Separation of CO_2/CH_4 through alumina supported geminal ionic liquid membranes. *J. Membr. Sci.*, 2014. **455**: 229–235.

125. H. Li et al., Design and synthesis of an exceptionally stable and highly porous metal–organic framework. *Nature*, 1999. **402**: 276–279.

126. M. Eddaoudi et al., Systematic design of pore size and functionality in isoreticular MOFs and their application in methane storage. *Science*, 2002. **295**: 469–472.

127. J. Lee et al., Metal–organic framework materials as catalysts. *Chem. Soc. Rev.*, 2009. **38**: 1450–1459.

128. R. Zou et al., Storage and separation applications of nanoporous metal–organic frameworks. *CrystEngComm.*, **12**: 1337–1353.

129. G. Ferey et al., Why hybrid porous solids capture greenhouse gases? *Chem. Soc. Rev.*, 2011. **40**: 550–562.

130. S. Bourrelly et al., Different adsorption behaviors of methane and carbon dioxide in the isotypic nanoporous metal terephthalates MIL-53 and MIL-47. *J. Am. Chem. Soc.*, 2005. **127**: 13519–13521.

131. D. Li, K. Kaneko, Hydrogen bond-regulated microporous nature of copper complex-assembled microcrystals. *Chem. Phys. Lett.*, 2001. **335**: 50–56.

132. K.S. Walton et al., Understanding inflections and steps in carbon dioxide adsorption isotherms in metal–organic frameworks. *J. Am. Chem. Soc.*, 2007. **130**: 406–407.

133. L.J. Murray, M. Dinca, J.R. Long, Hydrogen storage in metal–organic frameworks. *Chem. Soc. Rev.*, 2009. **38**: 1294–1314.

134. Y. Liu, Z.U. Wang, H.-C. Zhou, Recent advances in carbon dioxide capture with metal–organic frameworks. *Greenhouse Gases Sci. Technol.*, 2012. **2**: 239–259.

135. J. Liu et al., Progress in adsorption-based CO_2 capture by metal–organic frameworks. *Chem. Soc. Rev.*, 2012. **41**: 2308–2322.

136. P.L. Llewellyn et al., High uptakes of CO_2 and CH_4 in mesoporous metal organic frameworks MIL-100 and MIL-101. *Langmuir*, 2008. **24**: 7245–7250.

137. A.R. Millward, O.M. Yaghi, Metal organic frameworks with exceptionally high capacity for storage of carbon dioxide at room temperature. *J. Am. Chem. Soc.*, 2005. **127**: 17998–17999.

138. G. Xu et al., Preparation of ZIF-8 membranes supported on ceramic hollow fibers from a concentrated synthesis gel. *J. Membr. Sci.*, 2011. **385–386**: 187–193.

139. D. Fairen-Jimenez et al., Opening the gate: Framework flexibility in ZIF-8 explored by experiments and simulations. *J. Am. Chem. Soc.*, 2011. **133**: 8900–8902.

140. B. Assfour, S. Leoni, G. Seifert, Hydrogen adsorption sites in zeolite imidazolate frameworks ZIF-8 and ZIF-11. *J. Phys. Chem. C*, 2010. **114**(31): 13381–13384.

141. H. Huang et al., Effect of temperature on gas adsorption and separation in ZIF-8: A combined experimental and molecular simulation study. *Chem. Eng. Sci.*, 2011. **66**: 6297–6305.

142. K.S. Park et al., Exceptional chemical and thermal stability of zeolitic imidazolate frameworks. *Proc. Natl. Acad. Sci. USA*, 2006. **103**: 10186–10191.

143. R. Banerjee et al., Control of pore size and functionality in isoreticular zeolitic imidazolate frameworks and their carbon dioxide selective capture properties. *J. Am. Chem. Soc.*, 2009. **131**: 3875–3877.

144. A. Battisti, S. Taioli, G. Garberoglio, Zeolitic imidazolate frameworks for separation of binary mixtures of CO_2, CH_4, N_2 and H_2: A computer simulation investigation. *Micropor. Mesopor. Mater.*, 2011. **143**: 46–53.

145. A. Phan et al., Synthesis, structure, and carbon dioxide capture properties of zeolitic imidazolate frameworks. *Acc. Chem. Res.*, 2009. **43**: 58–67.

146. J.-R. Li et al., Carbon dioxide capture-related gas adsorption and separation in metal–organic frameworks. *Coord. Chem. Rev.*, 2011. **255**: 1791–1823.

147. B. Wang et al., Colossal cages in zeolitic imidazolate frameworks as selective carbon dioxide reservoirs. *Nature*, 2008. **453**: 207–211.

148. R. Banerjee et al., High-throughput synthesis of zeolitic imidazolate frameworks and application to CO_2 capture. *Science*, 2008. **319**: 939–943.

149. K.-J. Kim et al., CO_2 separation performances of composite membranes of 6FDA-based polyimides with a polar group. *J. Membr. Sci.*, 2003. **211**: 41–49.

150. T.L. Chew, A.L. Ahmad, S. Bhatia, Ba-SAPO-34 membrane synthesized from microwave heating and its performance for CO_2/CH_4 gas separation. *Chem. Eng. J.*, 2011. **171**: 1053–1059.

151. S. Li et al., High-pressure CO_2/CH_4 separation using SAPO-34 membranes. *Ind. Eng. Chem. Res.*, 2005. **44**: 3220–3228.

152. S.R. Venna, M.A. Carreon, Amino-functionalized SAPO-34 membranes for CO_2/CH_4 and CO_2/N_2 separation. *Langmuir*, 2011. **27**: 2888–2894.

153. Y. Zhang et al., Blocking defects in SAPO-34 membranes with cyclodextrin. *J. Membr. Sci.*, 2010. **358**: 7–12.

154. M.L. Carreon, S. Li, M.A. Carreon, AlPO-18 membranes for CO_2/CH_4 separation. *Chem. Commun.*, 2012. **48**: 2310–2312.

155. T. Li et al., Carbon dioxide selective mixed-matrix composite membrane containing ZIF-7 nano-fillers. *J. Membr. Sci.*, 2013. **425–426**: 235–242.

156. M.J.C. Ordoñez et al., Molecular sieving realized with ZIF-8/Matrimid® mixed-matrix membranes. *J. Membr. Sci.*, 2010. **361**: 28–37.

157. J.A. Bohrman, M.A. Carreon, Synthesis and CO_2/CH_4 separation performance of Bio-MOF-1 membranes. *Chem. Commun.*, 2012. **48**: 5130–5132.

158. Z. Xie et al., Alumina-supported cobalt-adeninate MOF membranes for CO_2/CH_4 separation. *J. Mater. Chem. A*, 2014. **2**: 1239–1241.

159. L.S. Lai et al., Zeolitic Imidazolate Frameworks (ZIF): A potential membrane for CO_2/CH_4 separation. *Sep. Sci. Technol*, 2014. **49**: 1490–1508.

160. C. Chmelik, J. van Baten, R. Krishna, Hindering effects in diffusion of CO_2/CH_4 mixtures in ZIF-8 crystals. *J. Membr. Sci.*, 2012. **397–398**: 87–91.

161. D. Liu et al., Experimental and molecular simulation studies of CO_2 adsorption on zeolitic imidazolate frameworks: ZIF-8 and amine-modified ZIF-8. *Adsorption*, 2013. **19**: 25–37.

162. M. Shah et al., Current status of metal organic framework membranes for gas separations: Promises and challenges. *Ind. Eng. Chem. Res.*, 2012. **51**: 2179–2199.

163. Y. Liu et al., Synthesis and characterization of ZIF-69 membranes and separation for CO_2/CO mixture. *J. Membr. Sci.*, 2010. **353**: 36–40.

164. G.N. Karanikolos et al., Continuous c-Oriented AlPO4-5 films by tertiary growth. *Chem. Mater.*, 2007. **19**: 792–797.

165. C.M. Veziri et al., Toward submicrometer c-Oriented nanoporous films with unidimensional pore network: AFI film morphology control by precursor mixture manipulation. *Chem. Mater.*, 2010. **22**: 1492–1502.

166. G.N. Karanikolos et al., Growth of $AlPO_4$-5 and CoAPO-5 films from amorphous seeds. *Micropor. Mesopor. Mater.*, 2008. **115**: 11–22.

167. J.A. Stoeger et al., Oriented CoSAPO-5 membranes by microwave-enhanced growth on TiO_2-coated porous alumina. *Angew. Chem. Int. Ed.*, 2012. **51**: 2470–2473.

168. J.A. Stoeger et al., On stability and performance of highly c-oriented columnar $AlPO_4$–5 and CoAPO-5 membranes. *Micropor. Mesopor. Mater.*, 2012. **147**: 286–294.

169. G. Bonilla, D.G. Vlachos, M. Tsapatsis, Simulations and experiments on the growth and microstructure of zeolite MFI films and membranes made by secondary growth. *Micropor. Mesopor. Mater.*, 2001. **42**: 191–203.

170. Z. Lai, M. Tsapatsis, J.P. Nicolich, Siliceous ZSM-5 membranes by secondary growth of b-oriented seed layers. *Adv. Funct. Mater.*, 2004. **14**: 716–729.

171. Y. Liu et al., Synthesis of highly c-oriented ZIF-69 membranes by secondary growth and their gas permeation properties. *J. Membr. Sci.*, 2011. **379**: 46–51.

172. C. Algieri et al., Preparation of thin supported MFI membranes by in situ nucleation and secondary growth. *Micropor. Mesopor. Mater.*, 2001. **47**: 127–134.

173. Y.-S. Li et al., Molecular sieve membrane: Supported metal–organic framework with high hydrogen selectivity. *Angew. Chem. Int. Ed.*, 2010. **49**: 548–551.

174. Y. Li et al., Zeolitic imidazolate framework ZIF-7 based molecular sieve membrane for hydrogen separation. *J. Membr. Sci.*, 2010. **354**: 48–54.

175. Y.-S. Li et al., Controllable synthesis of metal–organic frameworks: From MOF nanorods to oriented MOF membranes. *Adv. Mater.*, 2010. **22**: 3322–3326.

176. S.R. Venna, M.A. Carreon, Highly permeable zeolite imidazolate framework-8 membranes for CO$_2$/CH$_4$ separation. *J. Am. Chem. Soc.*, 2009. **132**: 76–78.

177. K. Tao, C. Kong, L. Chen, High performance ZIF-8 molecular sieve membrane on hollow ceramic fiber via crystallizing-rubbing seed deposition. *Chem. Eng. J.*, 2013. **220**: 1–5.

178. H. Bux et al., Ethene/ethane separation by the MOF membrane ZIF-8: Molecular correlation of permeation, adsorption, diffusion. *J. Membr. Sci.*, 2011. **369**: 284–289.

179. S. Li, J.L. Falconer, R.D. Noble, SAPO-34 membranes for CO$_2$/CH$_4$ separations: Effect of Si/Al ratio. *Micropor. Mesopor. Mater.*, 2008. **110**: 310–317.

180. H. Bux et al., Oriented zeolitic imidazolate framework-8 membrane with sharp H$_2$/C$_3$H$_8$ molecular sieve separation. *Chem. Mater.*, 2011. **23**(8): 2262–2269.

181. H. Bux et al., Zeolitic imidazolate framework membrane with molecular sieving properties by microwave-assisted solvothermal synthesis. *J. Am. Chem. Soc.*, 2009. **131**: 16000–16001.

182. O. Tzialla et al., Zeolite imidazolate framework ionic liquid hybrid membranes for highly selective CO$_2$ separation. *J. Phys. Chem. C*, 2013. **117**: 18434–18440.

183. X. Wang, H. Li, X.-J. Hou, Amine-functionalized metal organic framework as a highly selective adsorbent for CO$_2$ over CO. *J. Phys. Chem. C*, 2012. **116**: 19814–19821.

184. Z. Zhang et al., Enhancement of CO$_2$ adsorption and CO$_2$/N$_2$ selectivity on ZIF-8 via postsynthetic modification. *AIChE J.*, 2013. **59**: 2195–2206.

185. Z. Zhang et al., Improvement of CO$_2$ adsorption on ZIF-8 crystals modified by enhancing basicity of surface. *Chem. Eng. Sci.*, 2011. **66**: 4878–4888.

186. Y. Chen et al., Ionic liquid/metal organic framework composite for CO$_2$ capture: A computational investigation. *J. Phys. Chem. C*, 2011. **115**: 21736–21742.

187. K.M. Gupta, Y. Chen, J. Jiang, Ionic liquid membranes supported by hydrophobic and hydrophilic metal organic frameworks for CO$_2$ capture. *J. Phys. Chem. C*, 2013. **117**: 5792–5799.

188. A. Huang, J. Caro, Covalent post-functionalization of zeolitic imidazolate framework ZIF-90 membrane for enhanced hydrogen selectivity. *Angew. Chem. Int. Ed.*, 2011. **50**: 4979–4982.

4 H$_2$ and Olefin/Paraffin Separation with Porous Materials

State of the Art and New Developments

Canan Gücüyener, Jorge Gascon, and Freek Kapteijn

CONTENTS

4.1 INTRODUCTION

Coal, petroleum oil, and natural gas are the primary raw materials for the production of most bulk chemicals [1]. The finite nature of these fossil reserves represents one of the biggest challenges for today's society. Petrochemical industries are obliged to improve their current processes and/or invest more in the development of new methods for producing these bulk chemicals [2,3].

However, as long as there are sufficient natural resources available [4], current processes can only be improved by

- Introducing better materials for catalysis and separation [5–8]
- Introducing more energy and atom efficient processes available on the market [9–12]
- Intensifying the current processes by combining the reactive and separation processes [13,14]

In addition, the production of bulk chemicals is also quite energy intensive. The main reactions involved in the production of bulk chemicals either rely on irreversible reactions or suffer from formation of side products. While catalysis engineers work on the development of better catalysts and combining them with better reactors to intensify and increase the selectivity of these reactions [15–22], efficiency of current processes is still highly dependent on separation processes and their improvement. According to Dutch records of 2007, 40% of the energy use in the (petro)chemical industry is spent in the separation of products, which in turn means more than 100 PJ annum^{-1} [23]. According to another report presented to the U.S. Department of Energy in 2005, 22% of total industrial energy consumption by chemical and petroleum industries was due to separation processes [24].

The major industrial separation technologies and the relative energy consumption of these technologies are highlighted in Figure 4.1. Distillation, evaporation, and drying are highly energy-intensive (due to the phase change character) but mature and well-established processes. These separations are based on the differences in boiling points of the components to be separated. These technologies represent 80% of the energy consumption of the industrial separations. In the case of gaseous mixtures, separations can be achieved by cryogenic distillation, adsorption, and membrane

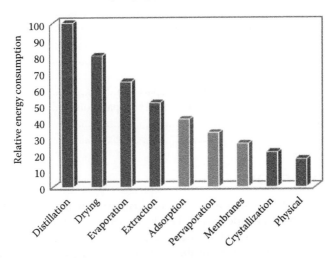

FIGURE 4.1 Relative energy consumption depending on the separation method used. (Reproduced from Materials for separation technologies: Energy and emission reduction opportunities, http://www.eere.energy.gov/, last accessed: August 2013.)

separation, either via gas or vapor permeation [25]. However, adsorption and membrane technologies, even though relatively much less energy consuming, currently do not account for more than 3% of the industrial energy consumption [24]. The latter stresses the importance of research into new adsorbents and membranes that can provide a technology push to increase this share in industrial separations.

Processes in the petrochemical industries are highly complex and divided into many different sections for the production of different base chemicals [1]. In this chapter, the main focus is divided on the production of lower olefins by steam cracking and their state-of-the-art separation processes, as it is reported to be the most energy consuming process in the chemical industry [26]. Further discussion is devoted to H$_2$ production and separation technology. Both separation processes are the core of this chapter.

4.2 STATE OF THE ART

4.2.1 OLEFIN PRODUCTION

Production of light olefins is mainly carried out by steam cracking. In steam cracking, hydrocarbon feedstocks are converted to light olefins such as ethene, propene, butene, and other products in the presence of steam. Steam is mainly added as a diluent but also reacts with the carbonaceous deposits formed during the process. Feedstock can be liquid (such as gas oil and naphtha) or gas (mainly ethane and propane). Depending upon the distribution of the natural resources throughout the world, it is no surprise that the hydrocarbon feedstock is mainly naphtha in Europe and Japan, while ethane is the main feedstock in the United States. Approximately 10 wt% of the crude oil is naphtha, while 1–14 wt% of the natural gas is ethane. Both ethane and naphtha cracking follows the same thermal cracking production route, though operation conditions need to be optimized for each feedstock [1,26]. When naphtha is used as the starting material, ethene yields are in the range of 19–24 wt%, whereas it is in between 42 and 51 wt% when ethane is used [1].

Steam cracking involves three main sections: pyrolysis, primary fractionation/compression, and product recovery/fractionation (Figure 4.2). In the pyrolysis section, feedstock is vaporized in the presence of a superheated steam and cracked into smaller hydrocarbons via free-radical mechanism in the absence of a catalyst. Due to the higher stability of ethane, higher temperatures and longer residence times than in the case of naphtha are required for acceptable pyrolysis conversion. Next, the products are quenched to 550°C through transfer line exchangers before they enter the primary fractionation/compression section. Here, the gas fraction is compressed and quenched to temperatures up to 15°C while the acid gases, carbon dioxide, and water are separated. Finally, the products are recovered in the recovery/fractionation section, which requires drying, precooling, recompression, and refrigeration [1,26].

Fractionation of C$_2$, C$_3$, and C$_4$ components is dependent upon the feedstock, desired purity, and desired range of products. In the conventional case, the dried gases are cooled to −120°C and sent through the demethanizer. The process continues with deethanizer, depropanizer, and debutanizer, which are used for prefractionation of the desired products. Finally, the leftover stream is sent to C$_2$, C$_3$, and C$_4$

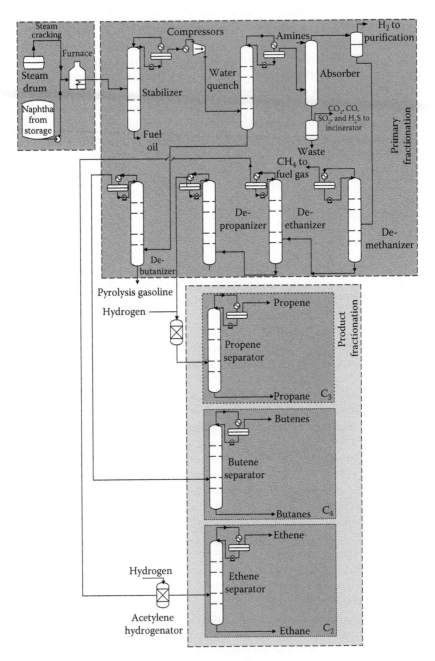

FIGURE 4.2 Production of light olefins from naphtha. (Reproduced from Chaudhuri, U.R., *Chemical Industries*, CRC Press, San Diego, CA, 2011.)

splitters in order to produce ethene from ethane (mainly), propene from propane (mainly), and for C_4 hydrocarbons. Production scheme of light olefins starting from naphtha is depicted in Figure 4.2 [27]. However, it should be pointed out that the separation of butanes and butenes is much more complicated than as highlighted in this basic sketch: depending upon the desired C_4 product, the process conditions (cracking method and severity) and the contents of the starting materials used at the steam cracker, the C_4 fraction, and the separation process can change [1]. In some cases, the C_4 fraction would be hydrogenated to butanes and directed back to the pyrolysis unit in order to produce higher yields of lighter olefins.

Distillation is still the primary process used for separating most of the steam cracking products to achieve the desired product purity [25]. The columns for separating olefins from paraffins of the C_2, C_3, and C_4 components are among the most energy-intensive distillation applications in oil refining [1]. In case of C_2 and C_3 splitter, separation is either achieved by cryogenic distillation where low temperatures are employed or by high-pressure distillation [28,29]. In case of C_4 hydrocarbon separation, in many cases, extractive distillation is the choice due to the close proximity of their boiling points and azeotrope formation [30]. When the relative volatilities of the components are <1.2, distillation is very inefficient [9,31]. The relative volatility of ethene to ethane is moderately small, ranging from about 1.13 for high-pressure mixtures rich in ethene to 2.34 for low-pressure mixtures rich in ethane [32]. Ethane/ethene separation takes place in a cryogenic distillation tower consisting of over 150 trays [26]. The relative volatility of propane/propene mixture is between 1.0 and 1.1 at temperatures in the range of 244–327 K and total pressure of 1.7–22 bar [33]. A typical propene/propane fractionator in steam crackers requires a large number of stages up to 200 and high reflux ratios from 12 to 20 due to the close boiling points of propene and propane [34]. Olefin/paraffin separation makes up almost 7% of the total distillation energy demand of the United States [26]. The energy demand of this distillation represents up to 85% of the total cost of the entire steam cracking process [35].

Ethene and propene have almost no end use (they can be used as refrigerant—in fact in cryogenic distillation, propene is the choice of refrigerant [9]), but they are important building blocks for the production of many intermediate and end chemicals of industrial interest (Figure 4.3). For example, ethene is used for the production of polyethene (PE), ethene oxide (EO), ethene glycol (EG), ethene dichloride (EDC), ethyl benzene (EBz), and styrene monomer [9,10,36]. It is expected that the demand for ethene will reach 160 million tons by 2015 [37] (Table 4.1).

Propene is mainly used for the production of polypropene ([PP] two thirds of the whole demand [42]), although it is also used for the production of isopropanol, acrylic acid (AA), oxo-alcohols (OA), cumene (C), propene oxide (PO), acrylonitrile (AN), and other intermediate and end products (Figure 4.3). Increase in demand for the polypropene industry [43] makes propene production extended not only to the side product of ethene production from steam crackers but also production through other refinery sources, mainly as a by-product of fluidized catalytic cracking plants [42–46]. Certainly when the demands are even higher, dehydrogenation of propane is the main choice [47].

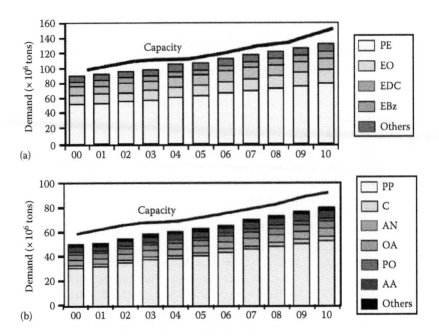

FIGURE 4.3 Overall demand for (a) ethene and (b) propene in years 2000–2010. (Reproduced from World light olefins analysis, http://www.cmaiglobal.com/worldanalysis/wpoabook. aspx, last accessed: August 2013.)

TABLE 4.1
Olefin/Paraffin Ratio of Different Olefin Production Streams

Olefin Production Streams	$C_2=/C_2$	$C_3=/C_3$	References
Naphtha cracking	83–88/12–17	96.6–97.2/2.8–3.4	[1]
	92/8	98/2	[39]
Ethane cracking	48–62.7/37.3–52	88.3–89.8/10.2–11.7	[1]
Fluidized-bed catalytic cracking	73/27	87/13	[39]
	45/55	73/17	[40]
Propane dehydrogenation	—	99.5/0.5	[40]
MTO	93.8–99.3/0.7–6.2	88.1–99.5/0.5–11.9	[41]

Other sources of olefinic carbons with respect to their mixtures with paraffins in these streams are given in Section 4.2.2 (Table 4.2). Although other technologies such as metathesis and direct methane conversion to olefins exist, their applicability are rather limited or still in the urge of commercialization.

C_4 hydrocarbon fraction mainly consists of butenes (but-1-ene, but-2-ene [*cis*- and *trans*-], 2-methylpropene), butadiene, and butane (iso- and normal-), although some butyls (normal-, iso-, sec-, and tert-), and negligible amounts of C_3 and C_5 hydrocarbons might also be present [30,49–51]. If naphtha is used as the feedstock, butenes are highest in content (50–56 wt%), whereas butadiene (61–76 wt%) takes the lead

TABLE 4.2

H$_2$ Production Streams

H$_2$ Production Streams	Amount of H$_2$ (vol%)	Other Gases
Naphtha cracker off-gas	45–80	N$_2$, He, CO, CO$_2$, CH$_4$
Ethene cracker off-gas	70–85	CH$_4$, He, CO, C$_2$H$_4$
Styrene monomer off-gas	90–95	CH$_4$, CO$_2$, C$_2$H$_4$, C$_6$H$_6$, C$_8$H$_{10}$
C$_3$/C$_4$ dehydrogenation	90–95	CH$_4$, C$_3$, C$_4$
Chlor-alkali off-gas	99.5+	Cl$_2$, O$_2$, N$_2$
Ammonia synthesis purge	60–70	N$_2$, CH$_4$, Ar
Catalytic reformer off-gas	70–85	CH$_4$, C$_2$–C$_{10}$
Catalytic cracker purge	10–20	N$_2$, O$_2$, CH$_4$, CO, CO$_2$, H$_2$S, H$_2$O, C$_2$–C$_8$
Hydrocracker purge	75–85	CH$_4$, H$_2$S, H$_2$O, C$_2$–C$_6$
Hydrotreater purge	75–85	CH$_4$
MTBE off-gas	80–85	N$_2$, CO, CO$_2$, CH$_4$
Toluene hydrodealkylation purge	50–60	CH$_4$, C$_2$H$_6$
Coke oven gas	55–65	CH$_4$, CO, CO$_2$, N$_2$, Ar, O$_2$, C$_2$H$_4$

Source: Gunardson, H.H., *Industrial Gases in Petrochemical Processing: Chemical Industries*, Taylor & Francis Group, 1997.

in case of ethane. In many cases, firstly, butadiene is isolated from the rest of the C$_4$ hydrocarbons [50] by extractive distillation. In modern technologies while solvents such as dimethylformamide, dimethylacetamide, and *N*-methylpyrrolidone (NMP) are used, in older technologies where the alkyne fraction of the C$_4$ hydrocarbons are carried out prior to extractive distillation, solvents such as acetone, furfural alcohol, and acetonitrile are used [51]. A simple sketch for the C$_4$ butadiene separation is depicted in Figure 4.4, using the BASF's NMP process as the basis [48]. Recovery of butadiene with such a process approaches yields above 98 wt% [52,53]. In the

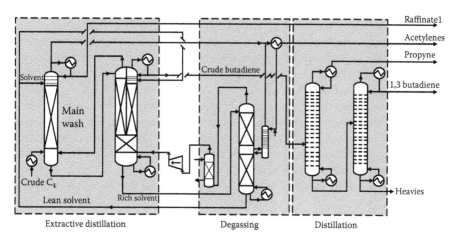

FIGURE 4.4 BASF's NMP process for Butadiene Extraction. (From NMP process, http://www.intermediates.basf.com/chemicals/butadiene-extraction/features, last accessed: August 2013.)

extraction zone (the second column from the left in the given figure), a column of 40–80 theoretical plates is operated most preferably at 4.5 bar and at 40°C–60°C. However, the cost of such a separation method is estimated to require 9.75 GJ per ton of butadiene [23]. Raffinate 1, the overhead product of the first column (main wash), containing mainly butenes, isobutylene, and butanes and negligible amounts of butadiene, still needs further separation: for example, reacting raffinate 1 with methanol, methyl tert-butyl ether can be produced, while the butenes/butane could be recovered through a series of hydroisomerization, simple and extractive fractionation, and isomerization processes [50]. C_4 olefins are important base materials for the production of many blended polymers that are desirable for their resilience at high temperatures [54]. For example, but-2-enes could be blended with ethene to achieve polymers with different properties: ethene and *cis*-but-2-ene alternating copolymers are crystalline, whereas the alternating copolymers composed of ethene and *trans*-but-2-ene are amorphous, which changes their application area [55].

Ethane/ethene and propane/propene separations have long been considered as highly energy consuming separations that can be replaced by low-energy alternatives [56–58], and with the increasing demand for these two building blocks, a great deal of effort has been and still is being devoted to the development of more efficient separation methods [9,19,26,28,29,33,35,37,44,45,56–66]. On the other hand, 1,3-butadiene separation from butenes and butanes would be highly interesting if an alternative process allowed obtaining these valuable products at minimum possible energy consumption [49].

4.2.2 H_2 PRODUCTION

In a world of finite fossil energy resources and increasing greenhouse gas emissions, H_2 is long considered as the new and clean energy carrier. However, there is a diversity of opinions about how to produce H_2 efficiently, store and deliver the produced H_2, if it can be safely used, and whether the required infrastructure to have such an economy is viable enough [67–71]. One of the ideas toward the solution of the problems is using the infrastructure available for natural gas for H_2 as well, by simply creating a mixture. This further indicates that produced mixture should be separated at the point of use, and discussions on the best separation technology (membranes, cryogenic distillation, or pressure swing adsorption [PSA]) have already started [72]. In spite of these discussions, global investment in hydrogen-related processes has accelerated dramatically over the past few years and is now in the range of U.S. billion dollars [73]. Commercially, hydrogen is used in petroleum and chemical processes extensively: in producing ammonia, hydrodesulfurization, Fischer–Tropsch synthesis, upgrading petroleum products and chemicals (hydrogenation, hydrocracking, production of methanol, higher alcohols, urea, and hydrochloric acid), and as a reducing agent in metallurgical industry [73,74]. Thus, production and separation of H_2 will always be an important issue.

Hydrogen is one of the most abundant elements in the universe, but in nature, it is mostly bound to carbon or oxygen [69]. Although possibilities of H_2 production from electrolysis of water, nuclear energy, or renewable resources (electrically derived) such as solar, biomass, or wind exist, still the 80% of the mass production of H_2 currently is based on fossil fuels [75]: mostly by steam reforming as a product of the syngas production [73,76–78] (Figure 4.5) and from the dehydrogenation of

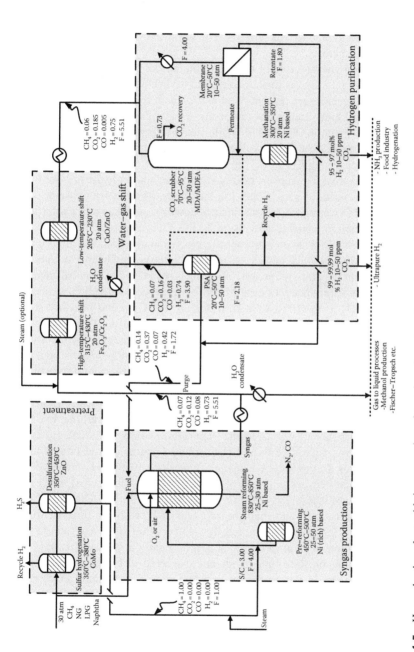

FIGURE 4.5 H₂ production plant with incorporation of a series of reactors for pretreatment, WGS, and H₂ purification. *F* is the flow rate defined in arbitrary units with the feed flow rate being 1.0. The numbers to the right of the molecular symbols are stream composition numbers in mol or vol fraction. (Reproduced from Ritter, J.A. and Ebner, A.D., *Sep. Sci. Technol.*, 42, 1123, 2007.)

alkanes [47]. Other industrial streams that require further separation of H_2 are given in Table 4.2. These streams mainly consist of H_2 (55–99.5+ vol%) but also gases such as CO, CO_2, CH_4, and C_2H_4 [79]. *In situ* removal technologies like membrane reactors have already proven to be beneficial for *shifting* the thermodynamic equilibrium for many dehydrogenation reactions [80–89]. The same approach could be followed for other industrial streams.

Steam reforming is the first step in the gas to liquid fuel processing [1]. Syngas, produced after steam reformer, is further converted to fuels via the Fischer–Tropsch process. A series of reactors are incorporated in the steam reformer: A prereformer, a steam reformer, and high- and/or low-temperature water gas shift reactors. A flow sheet of the plant is given in Figure 4.5. While steam reformer is the main reactor for the conversion of CH_4 to CO, the following water gas shift unit enables further H_2 production by converting CO to CO_2 by steam. In some cases, a prereformer is used as shown in Figure 4.5 to convert mostly ethane and heavier hydrocarbons into methane that is present in the natural gas. In other cases, a secondary reformer that is basically a partial oxidation (PO) unit is located downstream the primary reformer unit to achieve maximum methane conversion. In PO, O_2 is reacted with methane.

Reactions occurring in a steam reformer are reversible and occurs at high temperatures (450°C–850°C), which makes this process highly energy intensive. Energy efficiency of H_2 production through steam reforming is really low: in terms of energy, 0.66 MJ of hydrogen is produced for each MJ of fossil fuel consumed [90].

Purification of H_2 from mixtures with CO, CO_2, N_2, CH_4, and possibly H_2O is industrially carried out by amine absorption in combination with methanation (elimination of CO), or PSA, or membranes combined with methanation [74]. Amine adsorption (wet scrubbing) is employed when 95–97 vol% purity of H_2 is enough. For high-purity H_2, currently, PSA and fractional/cryogenic distillation dominate the market but they are not as cost effective [74,75,91]. In PSA, pressures reaching 10 MPa are required to achieve the desired purity [73,91]. In addition, these PSA units utilize layered beds containing two to four different adsorbents to optimize the separation. Many different systems are employed in PSA; such as Polybed (UOP) and Gemini (Air Products and Chemicals) processes. In the Polybed system, typically a 10-bed 11-step process is applied where each bed is composed of 2 layers (thus 2 adsorbents), whereas in the Gemini process, these 2 layers are further divided into 2, enabling first the removal of CO_2 (6 beds) and then the removal of H_2 (3 beds) thus yielding also 2 pure effluents [92]. It is estimated that 450 trillion Btu year^{-1} could be saved with a 20% improvement in the H_2 separation and purification train [73]. Membrane separation, on the other hand, is considered to be the most promising method due to its low energy consumption, lower investment cost, and ease of (continuous) operation [75]. Polysep (UOP) and Prism (Air Products and Chemicals) are two technologies that are already available in the market. In both cases, polymeric membranes are applied; the only difference lies on the fact that the polymeric layer in case of the Prism system is supported on hollow fibers, whereas in Polysep technology, it is based on a spiral wound sheet–type contractor [73]. The H_2 purity achieved by these membrane technologies is lower than that of the PSA operation ($\approx$98%); however, the recovery rates are much higher (up to 97% H_2 recovery with Polysep compared to up to 90% recovery with Polybed [93]). During the depressurization steps of PSA, a waste gas—commonly referred to as tail gas—is produced containing CO, CO_2, CH_4, N_2, and some H_2, which is burnt in the reformer as fuel. However, a high content of CO_2 in the tail gas can cause instability in the flame and is kept under 85 vol% within the reformer fuel [94]. In addition, late regulations regarding the emission of CO_2 put further limitations to the content that could be added to the fuel [95]. Lately, quite a lot of effort is also invested on the direct aromatization of the methane, to avoid production of CO_2 [18,96,97]. In this scenario, in order to activate the methane, high temperatures are needed: at 1 bar, the equilibrium conversion is about 5% at 600°C, 11.4% at 700°C, and 16.2% at 800°C [96]. As the temperature is increased, however, coke formation is an issue. The removal of hydrogen is one of the most efficient ways to overcome the thermodynamic limits of this reaction, enabling lower temperatures to be applied. Direct methane aromatization, however, can also yield high benzene in the fuel. This is not desirable since benzene is a carcinogenic substance. Lately, also effort has been devoted to benzene alkylation with paraffins [98–107], which could be achieved in a membrane reactor, where H_2 is removed through the membrane to overcome equilibrium limitations. In this way, paraffins, which are less expensive than olefins, can be used directly. In fact, dehydrogenation and alkylation are combined in one process with separation.

4.3 ALTERNATIVE SEPARATION PROCESSES

Adsorption appears to be the most attractive alternative to be implemented on olefin production because of the maturity of the basic technology [28]. Most of the existing adsorbents display a stronger affinity for the unsaturated hydrocarbon [58,63,66,108] due to differences in the strength of adsorption attributed to the olefin double bond [34]. This is also the reason why this technology is not yet economically favorable [65]. Discovery of adsorbents that show a higher selectivity for paraffin would represent a real improvement, as the whole separation scheme would become much simpler [28].

Another industrial activity where potentially much improvement can be obtained is the separation of H_2. Large-scale production of H_2 often requires large capital investments. On the other hand, H_2 is in many cases not the main product of these reactions and could be integrated with *in situ* separation technologies. In the actual production schemes of H_2, the best alternative seems to be the membrane reactor processes since their use allows at the same time the reduction in the separation/purification load and recycle volume after the reaction, increasing in the yield of the process. This will imply a lower energy consumption and reduced raw material consumption and offers a compact system with lower footprint area [13,47,80–82,84,86,88]. In some cases, even these two separation challenges could be coupled when olefin is produced from the counterpart paraffin and the only side product is the H_2 [80].

4.3.1 ADSORPTIVE PROCESSES

4.3.1.1 Adsorption and Adsorptive Separations

Adsorption is the attachment of molecules from the gaseous or liquid phase onto the surface of a solid substrate. The adsorbing phase is the adsorbent, and the material concentrated or adsorbed at the surface of that phase is the adsorbate. Adsorption is a spontaneous process and can be physical and chemical. Physical adsorption is caused mainly by dispersive van der Waals forces and electrostatic forces between the adsorbate molecules and the adsorbent surface. It is a reversible process. In chemisorption, a reaction occurs with the surface groups, yielding a chemical bond, which is either ionic or covalent. In the remaining text, unless otherwise stated, all the information given will be referring to the physical adsorption of a gas.

Although adsorption can occur at any surface, adsorbents used in adsorptive processes are porous solids, preferably having a high surface area per unit of volume. The most widely used commercial adsorbents are zeolites, activated carbons, activated aluminas, and silica gels, owing to their significant surface area and intrinsic properties. However, it is clear that for selective adsorption of a certain compound using a solid, there are trivial requirements from the adsorbent to be fulfilled: high selectivity, high internal volume (high capacity), regenerability, and high mechanical resistance (attrition resistance). On top of these properties, the adsorbent is expected to be cheap for industrial applications.

Separation of different molecules over the same adsorbent can be based on thermodynamics or kinetics. In case of thermodynamics, enthalpic and/or entropic factors rule the separation. Due to limited pore volume and surface area, every adsorbent has an equilibrium loading once in contact with an adsorbent that is reached after a certain time. This equilibrium loading is dependent upon adsorbent–adsorbate interactions, temperature, and pressure. In case of enthalpy-driven separation, one molecule adsorbs stronger than the other, causing the strongly adsorbed molecule to be preferentially retained in the structure. When the adsorption enthalpies of different molecules are quite similar, one molecule might have access to a larger number of configurations in the adsorbed phase, enabling entropy-based separation. The time required to achieve thermodynamic equilibrium may be also important, particularly when the size of the adsorbent is close to the size of the molecules to be separated [31,109,110]. In these cases, diffusion rules the separation, that is, kinetic separation. However, kinetic selectivity might be transitory, and if enough time is given, the thermodynamic selectivity might take over. In industry, for the separation of gases, mostly enthalpy-driven separations are preferred, although there are a few examples (such as separation of O_2 and N_2) where kinetic selectivity rules the separation [111].

In adsorptive separation, the feed stream is put in contact with the adsorbent that is normally packed in a fixed bed. The less (thermodynamics) or slower (kinetics) adsorbing component in the feed stream exits (breaks through) the column faster than the more (thermodynamics)/faster (kinetics) adsorbing compound. Before the retained compound also starts breaking through, the feed should be stopped and the adsorbent should be regenerated by desorbing the retained compound. By changing one of the process parameters, it is possible to regenerate the adsorbent, since adsorption equilibrium is given by specific operating conditions (composition, T and P).

When regeneration of the adsorbent is performed by reducing the total pressure of the system, the process is termed PSA, since the total pressure of the system *swings* between high pressure in feed and low pressure in regeneration [109]. When the pressure swings between atmospheric and vacuum pressures, it is named as vacuum swing adsorption (VSA). Typical operating conditions for VSA are 100 kPa for the adsorption stage and 1 kPa for the regeneration stage. In some cases, vacuum swing and pressure swing can also be coupled, making the process vacuum pressure swing adsorption (VPSA). When temperature is varied for regeneration, the process is called temperature swing adsorption (TSA). Mechanical work is always more expensive than heat; thus, it is desirable to have the TSA rather than PSA processes. However, since it is faster to control the pressure, industrially, the PSA process is more convenient due to much faster cycles enabling higher throughputs for the same volume of adsorbent [31,112]. One can also conclude that, for similar reasons, VSA is also less preferred.

In lab scale, adsorptive separation is studied in discontinuous mode, usually adsorption followed by regeneration, referred to as breakthrough experiments (Figure 4.6). In industry, in order to increase the output of the process, many beds are used in series, making this process semicontinuous. To even further increase the output of the operation and to keep the raffinate (output stream obtained during regeneration) or extract (output stream obtained during adsorption) pure, additional steps

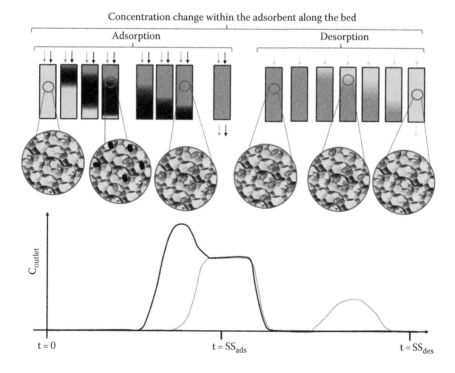

FIGURE 4.6 Binary adsorption followed by desorption over the adsorbent bed with respect to concentration change in the outlet. SS refers to the steady state.

might be added. For example, in an industrial PSA operation, the following steps can be combined in operations with two or more beds: (1) Pressurization, (2) high-pressure feed, (3) cocurrent depressurization, (4) countercurrent depressurization, (5) countercurrent purge, and (6) equalization. The number of beds and sequencing in between beds are arranged in a way to maximize selectivity and productivity of the operation.

There have been many approaches to reach continuity in this semicontinuous process. One of the ways is the utilization of continuous countercurrent systems involving circulation of the adsorbent (true moving bed [TMB]) or design of a fluid flow system that simulates the adsorbent circulation (simulated moving bed [SMB]) [31]. In the latter, through changing a rotary valve that is connected to different parts of the fixed bed, the direction of different flows (extract, feed, raffinate, and desorbent) could be controlled precisely to maximize the separation selectivity and capacity. Although these processes enable higher throughputs, their applications are limited to the separations where selectivities are low (minimum acceptable separation between the extract and the raffinate product is about 3 for SMB [31]) since they are also really expensive: TMB involves circulation of the adsorbent that decreases the sorbent life, whereas SMB is accompanied with distillation to further separate the obtained different streams from the used desorbent. However, both processes have found their place in industrially important applications, such as Sorbex, Parex, Molex, and Olex.

4.3.1.2 Adsorbents for Olefin Separation

Many different adsorbents, such as activated carbon [113], silica gel [114], titanosilicate [37], mesoporous silica [115,116], microporous silica [114,117], and clay [118], are suggested in literature for the adsorptive separation of light olefins from paraffins (such as ethene/ethane and propene/propane separation), but most of the reports concentrated on the use of zeolites.

Zeolites are hydrated aluminosilicate crystals with alkali or earth alkaline metals in their framework. The silicon and aluminum atoms are tetrahedrally coordinated with each other through shared oxygen atoms. The charge difference between silicon (+4) and aluminum (+3) atoms causes a negatively charged framework, which is balanced by a cation of alkaline or alkaline earth metals. These cations, which are located in the pores of zeolite, are mobile and exchangeable. The Si/Al ratio of the zeolite framework varies between one and infinity depending on the type of zeolite. Zeolite A (LTA) with a Si/Al ratio of one has the highest Al content among all zeolites. On the other hand, silicalite-1 (MFI) is a pure silica zeolite with a Si/Al ratio of infinity [119]. Pore sizes and BET surface areas range from 0.3 to 0.8 nm and 200–900 $m^2\ g^{-1}$, respectively, depending on the framework. Thus, different types of zeolite frameworks enable a palette for the selection of the best adsorbent for a given separation.

For olefin/paraffin separation, alumina-rich zeolites are mainly selected because of their high equilibrium selectivity. Due to electrostatic forces arising from the exchangeable cations, olefins are adsorbed much more strongly than the corresponding paraffins [28]. On the other hand, all-silica zeolites with 8-membered-ring (8MR) pore openings have also been suggested for these separations since they allow kinetic separation of light olefins with their small pore sizes. Nonetheless, various zeolite framework topologies have been selected for this purpose, such as LTA [28,63,120–123], FAU [28,35,44,45,58,65,66,124,125], CHA [33,126], ITW [127], ITE [126], IHW [60], AFN [128], and DDR [108,126]. From these zeolite structures, CHA and DDR showed the best potential for a kinetically driven separation, whereas FAU and LTA have shown the best capacity for thermodynamic separation [28]. On the other hand, CHA, a zeolite with one of the lowest framework densities, small pore opening, and 3D porous structure, provides a perfect kinetic separation between olefin and paraffin. Although most of these works focused on propane/propene separation, recently by Kim et al. [129], many other zeolites (existing and hypothetical) are shown to have potential for ethene/ethane separation. This potential is based on the performance curve (where selectivity times loading higher than 3 is regarded as a *high-performing* region) obtained by the ideal adsorbed solution theory (IAST), although many of these zeolites have not been considered before in the literature.

Other attempts regarding the separation of olefins focused on modified zeolites that would allow π-complexation between the adsorbent and the adsorbate. During adsorption, these metals and their ions can form σ bonds with their s-orbitals. Moreover, d-orbitals of the metals can back-donate electrons to the antibonding π-orbitals of olefin. Chemical complexation bonds are generally stronger than van der Waals interactions, so for adsorptive separations, they should be weak enough to enable regeneration. Although many d-orbital metals are available for this complexation, many efforts focused on the Ag- and Cu-modified zeolites [130] since the complexes created are easily reversible under adsorptive processes without destroying the original olefin. Another key advantage of π-complexation sorbents is that they are capable of separating multiple olefins (C_2's + C_3's + C_4's) from paraffins [131], which might be desirable in certain cases, such as separation of olefins in the raffinate of the butadiene separation unit (see Section 4.2.1) or removal of olefins from natural gas.

Van Miltenburg et al. [66] studied pure and Cu-modified NaY zeolites for the binary separation of ethane/ethene and propene/propene and showed higher selectivities for the Cu-modified NaY zeolite: approaching to 20 times higher selectivity for ethene and 6 times higher selectivity for propene compared to that of the unmodified NaY, depending upon the temperature and the partial pressure. On the other hand, they have also reported diffusion limitations for the modified zeolite and concluded that it could be avoided by the use of smaller crystals, though no further investigation is carried out to confirm this hypothesis. Aguado et al. [132] studied Ag-exchanged CaA zeolite for the binary separation of ethene/ethane mixtures. They have reported exclusion of ethane over Ag-exchanged zeolite while still some adsorption occurred for nonexchanged CaA zeolite. By the exchange of Ag with Ca in the framework, the pore opening is reduced to the range between the ethane and ethene kinetic diameters. The adsorption capacity of ethene for single adsorption isotherms (57 cm³ g⁻¹ at total column pressure of 100 kPa and temperature of 30°C) is in the same range as in the case of NaY [66]. Many other examples regarding the use of Ag-/Cu-modified zeolites can be found in the literature [128,133,134].

Zeolites and modified zeolites have also been investigated using semicontinuous adsorptive process schemes. Grande et al. [58,63] reported propane/propene separation by VPSA using zeolite A and Li-exchanged zeolite X. Campo et al. [45] reported as well the VPSA on FAU particles, in addition to SMB technology [44] for improving the throughput of the process for the same separation. Purities exceeding 97% can be achieved in the effluent for propene with these adsorptive separations. Recently, a comprehensive summary on PSA processes for propane/propene separation is highlighted by Khalighi et al. [33], where modified and unmodified adsorbents are compared to each other in terms of capacity and selectivity. Si-CHA is concluded to be suitable for four-step VPSA operation, when olefin/paraffin feed of 85–15 mol% in composition is used, yielding 99% pure propene and 90% pure propane with a 96/98 recovery for propane/propene. This is the best result obtained so far for propane/propene adsorptive separation.

Even though these separations are long considered as *on the way to commercialization* [34], the only available commercial adsorptive separation for olefin remains the Olex process [135], in which normal olefins are separated from cyclic and/or branched olefins and not olefins from their corresponding paraffins. It is believed that the main problem for applicability of all these adsorbents lies in the reverse selectivity of the adsorbent, that is, olefins are adsorbed rather than paraffins, causing energy-intensive regeneration methods for the removal of a desired compound from the adsorbent [28]. In most cases, vacuum (mechanical work) is applied during regeneration since temperature increase with sorbents might lead to oligomerization and even to cracking, causing blockage of the pores of the zeolite [108]. Although some materials have been suggested to have paraffin selectivity [29,136,137], these investigations are only based on simulations and further experiments are needed to confirm the behavior.

4.3.1.3 Metal Organic Frameworks as Adsorbents for Olefin Production

In the last two decades, a new class of porous materials has been proposed for different adsorptive separations: metal organic frameworks (MOFs), or more widely speaking coordination polymers, are known from the late 1950s [138] and early 1960s [139–143], although it was not until the end of the last century when the field was relaunched thanks to the efforts, among others, of Robson and coworkers [144,145], Kitagawa et al. [146,147], Yaghi and Li [148], Gardner et al. [149], and Riou and Férey [150].

MOFs are formed by metal clusters as inorganic units and ligands as organic units that are connected through coordination bonds forming crystalline porous solids. The possibility of selecting different clusters and combining them with different linkers gave the researchers a playground for designing new structures that could be targeted for different applications [151]. In addition, MOF structures show exceptionally high porosity (the highest BET surface area recently reported is $\approx$7000 m^2 g^{-1} [152]), which attracted quite some attention regarding their application in different gas separations. Apart from the obtained enormous surface areas, organic functional groups and coordinatively unsaturated metal sites can be seen as an advantage for separation by tailoring specific interactions with specific molecules.

The main concern from the application point of view has always been the chemical, thermal, and mechanical stability of these MOFs, since their uniform pore structure created an unavoidable comparison with zeolites. In addition, the first reports regarding MOFs were mostly related to MOF-5, which was not stable under atmospheric conditions [153,154]. Today, there are multiple MOFs that can be used even under harsh conditions [155–157], pelletized [62,158,159], and used for different applications, though the range may be still away from that of zeolites [160]. Many reviews have been published regarding potential applications, including separation processes [151,159,161–166]. Another disadvantage regarding these materials is the expense of synthesis compared to that of zeolites due to the use of expensive building blocks and low yields. However, lately, more and more materials are reported that could use water and/or lower content of organic solvent in the synthesis [167–170] and could be produced through electrochemical synthesis [171] or synthesis at room temperature [172] rather than hydrothermal synthesis, enabling high yields for the synthesis [170,173] that could overcome price issues.

One of the first MOFs reported for olefin/paraffin separation was CuBTC (HKUST-1). CuBTC is formed by linking Cu(II) pairs through benzene-1,3,5-tircarboxylate and has two types of cages, connected to each other through 0.9 and 0.46 nm windows. In its as-synthesized form, H$_2$O is connected to Cu(II), which leaves a coordinatively unsaturated site (CUS) after solvent removal. These unsaturated sites act as electron acceptors [174] and thus may have π-bond interaction with the olefin double bond, increasing selectivity toward olefin (in a way, these CUSs act similarly as the extra framework cations present in zeolites). First adsorption experiments regarding this material supported this hypothesis, showing a preferential uptake toward olefin [175,176]. Simulations followed to explain the location of the preferential adsorption sites on CuBTC [177,178]. Both studies proved the preferential uptake of paraffin at the cage centers, whereas for olefin, the CUS is the main adsorption site until saturation. Later investigations focused on the adsorptive separation of isobutane/isobutene [179,180], propane/propene [181,182], and even octane/octene [110] in terms of breakthrough experiments, mostly in the low-pressure region. A separation factor of 2.1 toward isobutene (30°C, 1 kPa, equimolar) [180] and 2 for propene (20°C, 5 kPa, equimolar) [181] is reported. In further studies, commercially available pellets (tablets) are used [183]. Ferreira et al. [62] studied these tablets for propane/propene separation in terms of pseudo-binary/pseudo-ternary breakthrough experiments where the desorbent potential of propane–propene and isobutane is also compared. They have reported a separation factor of 30 at 10 kPa and 100°C using a 75/25 vol% mixture of propene/propane. Increasing further the pressure toward 150 kPa resulted in a decrease of this separation factor to 2. Isobutane, on the other hand, is proven to be a good desorbent that could displace propane from the column and can be displaced by propene. They have also observed a decrease in capacity when using CuBTC extrudates, which was related to the partial destruction of the material by extrusion. Further studies by the same group focused on propane/propene separation by breakthrough experiments to assess the possible application of this adsorbent in SMB operation, where isobutane as the desorbent using an improved CuBTC tablet provided from the same supplier [158]. In this case, more concentrated mixtures of propane/propene are studied to mimic the effluent of the C$_3$ splitter of the light olefin production unit (Figure 4.2). Although an increased adsorption capacity is reported, separation factors varied in the same region as in previous experiments (1.4–3) under operation at 100°C and

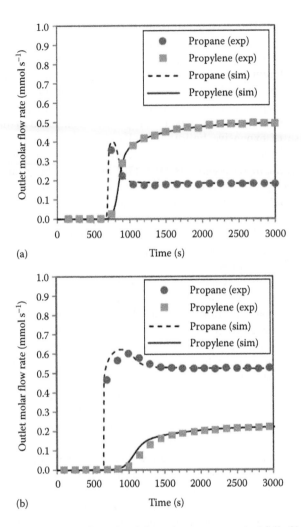

FIGURE 4.7 Propane/propene binary breakthrough curves over a bed full of helium at 373 K and 150 kPa: (a) 25:20:55 helium/propane/propene and (b) 70:30 propane/propene. (Taken from Plaza, M.G. et al., *Sep. Purif. Technol.*, 90, 109, 2012.)

total pressure of 150–250 kPa. Later, the material is also used under VSA conditions (at 373 K, 250 kPa–adsorption, and 10 kPa–regeneration, a mixture of 30/70 propane/propene used in adsorption and pure propane is used as the purge), and selectivity ≈2.3 is obtained (Figure 4.7) [184].

Further reports regarding olefin/paraffin separation, where propene was selectively adsorbed, concern the use of M-DOBDC (also referred to as MOF-74) series [185,186], containing Mg, Co, Mn, and Fe as the metal (M), where DOBDC represents the 2,5-dioxido-1,4-benzene-dicarboxylate. Highest mixture adsorption selectivities are calculated for Co (ca. 24 for Mn-MOF-74, ca. 46 for Co-MOF-74, and 4.5 for Mg-MOF-74) using IAST (based on single-component isotherms), which are also the highest selectivities reported so far for olefin/paraffin separation.

This higher selectivity for Co is arising from the large difference in binding energies between propene and propane caused by the specific interactions with the metal center. Binding energies calculated for propane/propene molecules with Co-, Mn-, and Mg-MOF-74 are 49/24.4, 45.1/24.5, and 44.8/23.6 kJ mol^{-1}, respectively. Breakthrough analysis using Co-MOF-74 and initially 100% propane-saturated column confirmed propene selectivity yielding 6.8 mol kg^{-1} of capacity at room temperature (neither pressure of the mixture nor the selectivity data were given for this breakthrough experiment). He et al. [187] illustrated also the highly selective behavior of Co-MOF-74 for C_2H_4/C_2H_6, C_3H_6/C_3H_8, and $CH_4/C_2H_2/C_2H_4/C_2H_6$, $CH_4/C_2H_2/C_2H_4/C_2H_6/C_3H_6/C_3H_8$ mixtures by performing breakthrough simulations, which in each case showed to have a better capacity compared to that of zeolites such as NaY and Na-ETS-10. Further breakthrough analysis on this material at 22°C and total column pressure of 1 kPa yielded separation factors of 1.7 and 2.9 for equimolar mixtures of ethane/ethene and propane/propene, respectively [57].

Like zeolites, the aforementioned materials are selective toward olefin and not paraffin, and it is likely that they will still not be considered for adsorptive separation in industry. Compared to zeolites, MOFs are not formed by covalent bonds but rather from coordination bonds and/or weak cooperative bonds (π–π stacking, H bonding, van der Waals interactions), making them not as robust as zeolites. In addition, some MOFs can further have flexibility in response to an external stimulus (i.e., pressure, temperature, light, electric field, guest molecules). These dynamic and flexible changes can cause a shift in the structure by exchange or removal of guest molecules [188]. Effects of these dynamic changes on selective gas adsorption have been reviewed by Li et al. [161], dividing these changes into four groups:

1. Selective adsorption based on size/shape exclusion accompanied by pore size/shape change
2. Selective adsorption based on adsorbate–surface interactions accompanied by pore size/shape change
3. Selective adsorption based on gate opening or structural rearrangement induced by adsorbate–surface interactions
4. Selective adsorption based on adsorbate-specific gate-opening pressures

The first group could be viewed as the molecular sieving effect that is also observed by zeolites. At certain temperatures and pressures, certain molecules can adsorb to the structure, whereas others are sieved. However, once the adsorbate is in the structure, the structure changes in order to host further molecules, creating an (observed) hysteresis in the adsorption isotherm. In case of the second group, adsorption of different gases might occur without a sieving effect. Depending on the interactions of a specific adsorbate, the structure might change shape as the amount of adsorbate increases, yielding a step in the isotherm (which is called the breathing effect). Dynamic changes classified for 3 and 4 have a subtle difference. In case of group 3, the structure might be amorphous and, due to specific interactions induced by the adsorbate, could then transform and adsorb certain molecules, although no adsorption might occur for others. Else, although the material might be adsorbing any molecule, certain adsorbate interactions can induce a gate opening for smaller cages or at increasing pressures, increasing the available adsorption sites and thus yielding a higher capacity for certain molecules. In case of group 4, in order for any adsorption to occur, a gate opening is required that is achieved at different pressures depending upon the interactions of the adsorbent with the adsorbate gate entrance.

Such a material showing group 4–type flexibility was firstly observed by Li and Kaneko [189] when studying the adsorption characteristics of Cu-complex-assembled microcrystals, that is, $\{Cu(bpy)(BF_4)_2(H_2O)_2 \cdot (bpy)\}_n$ where bpy represents 4,4'-bipyridine. This material has a 2D sheet structure where bpy molecules are stacked through π–π interactions. Cu bonds to the bpy ligand from two sides and connect to two other bpy through H bridges via water molecules and two BF_4^- anions surrounding the Cu from two opposite sides in the same square planar plane with H_2O molecules. N_2 (at $-196°C$), CO_2 (at $0°C$), and Ar (at $-186°C$) adsorption using materials pretreated at different temperatures ($25°C–200°C$) indicated a gate-opening effect in this material, which was previously assumed to be nonporous, placing this material as an example for group 4. For example, for the sample that was pretreated at the same temperature although Ar adsorption started at a relative pressure of ≈0.65 and N_2 adsorption started at a relative pressure of ≈0.02. Although such a trend can also be observed for the samples pretreated at different temperatures using the same adsorbate, this trend is more related to changes in the pore volume due to removal of H_2O and BF_4^-. The gate opening occurring in this material is related to the possibility of breakage of the hydrogen bonding in between the sheets due to the interactions of different molecules with the structure, enabling rearrangement in the framework to have a 1D channel.

Further reports continued for this gate-opening effect in zeolite imidazolate frameworks (ZIFs) [156], which are named after their resemblance in their IM–M–IM angles to that of O–Si–O ($145°$). Their exceptional thermal and chemical stability, possibility of synthesizing large single crystals [190,191] for various characterization techniques [192,193], and first reports regarding their separation performance [191,192,194–197] placed them in the center of attention.

ZIF-7 was the first material to be recognized as a possible candidate for olefin/paraffin separation due to its flexibility. Although the first separation report was regarding their performance for H_2 separation from hydrocarbons [196,198], later studies focused on olefin/paraffin separation [199,200]. ZIF-7 is formed by connecting Zn ions through benzimidazole (BIM) linkers. It has a sodalite topology with a crystallographic six-membered-ring (6MR) pore opening of 0.3 nm although it can adsorb even *trans*-2-butene and isobutane, indicative of group 4 flexibility. Further, ZIF-7 has an anomalous adsorption behavior toward paraffins and olefins: at the same temperature, gate opening occurs at lower pressures for paraffins compared to olefins, leaving an operation window that could be used in adsorptive separation. Both ethene/ethane and propene/propane could be separated over the material within this operation window, enabling a pure olefin effluent from the adsorption. Moreover, regeneration over the material could be attained almost within the same timescale as the adsorption, making this unique material highly preferable for industrial-scale operation. Selectivities up to 3.8 (at $30°C$ and 200 kPa) and 2.7 (at $100°C$ and 120 kPa) are reported for ethane/ethene and propane/propene separation, respectively. Further DFT studies indicated that the selectivity toward paraffin is attained within the operation window due to a higher energy required for olefin to pass from the gate. However, if the breakthrough analysis is carried out at a pressure where both olefin and paraffin adsorb, then separation is attained through kinetics, which could yield a higher selectivity toward olefin.

ZIF-8, formed by Zn ions and methyl imidazolate ligands, also has a sodalite topology and a pore opening of 0.34 nm. Early on after the first synthesis of this material, N$_2$ (−196°C) and Ar (−186°C) adsorption measurements [156] indicated a two-step hysteresis (both hystereses occur at different relative pressures for these gases) in the isotherm, which was suggested to be the consequence of rearrangement of adsorbed molecules (increase in the packing efficiency at certain pressures). Further studies indicated that the changes occurring in ZIF-8 is related to the rearrangement of the methylimidazole linkers (Figure 4.8) and result in an overall increase in the unit cell of the structure [193,201–203]. Zheng et al. observed by performing diffusion studies that, although many adsorbents could be adsorbed on the structure, the effective pore diameter should be considered as ≈0.4 nm rather than the crystallographic pore diameter of 0.34 nm [204], although adsorption of 2,3-dimethylbutane with a kinetic diameter of 0.58 nm can also occur [205]. These studies indicate that ZIF-8 has type 3 flexibility, and due to possible different interactions, olefin/paraffin separation could be achieved.

First studies on ZIF-8 related to olefin/paraffin separation reported differences in uptake rates of propane and propene. A higher diffusivity of propane compared to that of propene ($D_{C_3H_8}/D_{C_3H_6} = 125$ at 30°C) [64] indicated a possible kinetic separation. Follow-up breakthrough analysis [57,206] revealed a separation factor of only 1.49 ($S_{C_3H_8/C_3H_6}$) for an equimolar mixture of propane and propene at 22°C at a total column pressure of 1 bar. On the other hand, ethane and ethene separation yielded at the same conditions a separation factor of 2.01 despite the predicted lower diffusivity ratio ($D_{C_2H_6}/D_{C_2H_4} \approx 5$ [207]). At these conditions, although the uptake of C$_3$ compounds is much higher than C$_2$ compounds (≈4.5 mol kg^{-1} for propane and propene compared to 1.5 mol kg^{-1} for ethane and 0.8 mol kg^{-1} for ethene, using

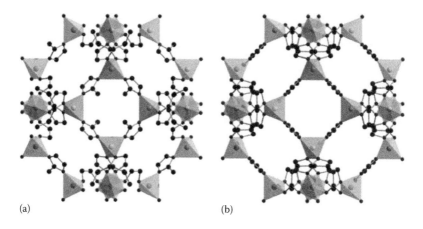

(a) (b)

FIGURE 4.8 Rearrangement of methylimidazolate ligands of ZIF-8 at different pressures under methanol/ethanol hydrostatic medium (a) at ambient pressure and (b) at 1.47 GPa. (Taken from Moggach, S.A. et al., *Angew. Chem. Int. Ed.*, 48, 7087, 2009.)

single-component adsorption isotherms), the energy required for C_3 compounds to pass through the gate is much higher than for the C_2 compounds [204], yielding a much lower selectivity in dynamic adsorptive separation. Further studies regarding the effect of temperature on the adsorptive separation also revealed that selectivity increases as the temperature is decreased, in line with the increase in the ideal adsorption selectivities of C_2H_8 and C_2H_6; the breakthrough adsorption selectivity also increased to 4 at 2°C [57,206].

4.3.2 MEMBRANE PROCESSES

4.3.2.1 Membrane Separation

A membrane is a solid or liquid phase (film), which separates two fluids. In a more detailed fashion, a membrane is a semipermeable film, allowing passage of molecules or particles, depending on their size and chemical nature (polarity, charge, etc.) under the influence of a driving force [208,209].

Important nomenclature for membranes and membrane processes can be found elsewhere [210]. The main separation mechanisms behind different types of membranes are depicted in Figure 4.9. In case of dense membranes, separation is based on the solution-diffusion mechanism, that is, the targeted compound first needs to adsorb or dissolve in the membrane, diffuse through the membrane, and desorb on the other side of the membrane. This type of mechanism is observed for metallic, dense ceramic, and dense polymeric (nonporous) membranes. In this mechanism, usually the diffusion is the rate-limiting step (especially for ceramic and polymeric membranes), but the selectivity is high, approaching infinity for metallic membranes. On the other hand, in case of porous membranes, separation can be based on different mechanisms, depending on the pore sizes of the membrane and the interactions of the molecules with the surface of the membrane.

In case of macroporous and mesoporous membranes, the main transport mechanism involved is nonselective viscous flow, due to the bigger size of the pores compared to that of the mean free path of the molecules. Knudsen diffusion applies when the molecular free path of the molecule is smaller than the pore size of the membrane. If the pore size of the membrane is comparable with the molecular sizes of the molecules, surface diffusion and activated diffusion takes over. On the other hand, if the sizes of the molecules are bigger than the pore size of the membrane, then

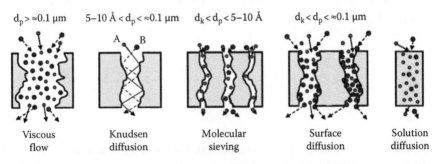

FIGURE 4.9   Different gas separation mechanisms involved in membranes.

molecular sieving is the main mechanism. From the molecular separation point of view, Knudsen diffusion is not favorable since it results in low separation factors for the desired compound, whereas molecular sieving is the most desirable since it may yield infinite selectivity. Considering that not every membrane is perfect, these mechanisms can also play their share in one membrane, and permeation of one compound through the membrane can also be affected by the compound, due to inhibition/blocking or competition for the same site.

Gas separations are of industrial interest since they allow in-line separation of the outlet streams without extensive heating/cooling cycles, compared to many other separation methods (such as distillation). Many membranes have been introduced in recent years with different specifications for several industrial separations. Problems with scale-up [211–213] are the reasons that application of membranes in many industrial streams is disregarded, especially in the case of inorganic membranes. On the other hand, considering their efficiency and operational simplicity, high selectivity and permeability for the transport of specific compounds, low energy requirement, and good thermal and chemical stability in line with their environmental compatibility, membrane operations are in many cases the answer to process intensification [11,13]. From 1981 onward, more and more attention is given to membrane separation, such as H$_2$ recovery in hydrotreater purge, in the purge of NH$_3$ synthesis, and removal of CO$_2$ from natural gas [1] (which in fact is the biggest membrane industrial separation activity [214]), which also indicates the presence of light at the end of a tunnel.

Polymeric membranes are the most commonly applied membranes for industrial applications due to the fact that they are relatively cheap, flexible, easily reproducible, and scalable, which are key factors for industrial application [74,75,215–217]. On the other hand, the two quality criteria that are important for the application of these membranes (selectivity and permeability), as in the case of most membranes, show a trade-off: the higher the selectivity, the lower the permeation and vice versa. Robeson evaluated different types of membranes and mixtures to define the operation limit of the current polymeric membranes available, or in more accepted term the *upper bound* [218,219]. Two important conclusions could be drawn from these studies: (1) Although it is easier to prepare polymeric membranes from low glass transition temperature polymers, the upper bound limit is mainly formed by glassy polymers [219,220], and (2) to make the polymeric membranes more industrially attractive, new strategies should be considered [220,221].

Such strategies defined the main theme of polymer membrane research over the last 25 years, reshaping itself on two main pillars: improving the sorption selectivity of the membrane material and increasing the sieving capability of the membrane [56,221] since the separation is the product of the solution-diffusion mechanism. The new strategies applied so far, but not limited to, are the synthesis of new and high-performance polymers; polymer blending; chemical cross linking [220,222]; doping with carrier agents (such as metal salts, halogen ions) [223,224]; addition of highly selective materials (such as zeolites, MOFs, carbon molecular sieves [CMS]) to the polymer matrix, so called mixed-matrix membranes (MMMs) [225–229]; and pyrolysis of the polymeric membranes under controlled conditions, resulting in CMS membranes [227]. However, from all these strategies, only a limited part has the potential of finding a place in actual industrial applications. Koros and coworkers made a remarkable comparison showing the limits of such strategies for the separation of olefins from paraffins.

Most of the materials reported in the literature for ethene/ethane or propene/propane is carried out using glassy high-performance polyimides such as 6FDA (4,4'-(hexafluoroisopropylidene)diphthalic anhydride)-based polymers. However, the ideal performances of these membranes are limited to a maximum selectivity of 8 (for a permeability of <2 Barrer) and permeability of 100 Barrer (for a selectivity of 6) in case of ethene/ethane separations, which slightly increases in case of propene/propane separation: a maximum selectivity of ≈30 (for a permeability of <1 Barrer) and permeability of 200 Barrer (for a selectivity of 2) [56,230]. These limits, on the other hand, can be pushed further, for example, with facilitated transport membranes. For such separations, membranes containing metal salts (Cu or Ag based, with the same idea presented in Section 4.3.1.2) show exceptional selectivities (selectivity of 40 with a permeance of 2.9×10^{-8} mol m^{-2} s^{-1} Pa^{-1} for ethene/ethane separation [223] and selectivity of 10.2 with a permeability of 188 Barrer for propene/propane separation [231]), far beyond the upper bound. However, this starting performance declined rapidly, making their long-term applicability questionable. MMMs fabricated by dispersing MOFs in a 6FDA-based polymer matrix, on the other hand, proved to be a promising alternative for propene/propane separation. Compared to zeolites that require in most cases an additional compatibilizer [232–234], MOFs show an excellent compatibility with the polymer matrix [235–240] due to organic groups present in their structure. By using ZIF-8 nanoparticles and increasing the loading in the polymer matrix to 48 wt%, propene/propane selectivity reached up to ≈30 while the permeability increased to ≈55 Barrer in pure gas permeation, whereas it was ≈17 and ≈41 Barrer in mixed gas [241]. On the other hand, the same approach did not turn out to be as favorable for ethene/ethane separation. Using the same matrix and loading, selectivities only increased slightly (≈3.2 ethene/ethane selectivity for a ≈220 Barrer permeability of ethene) resulting only in a slight improvement relative to the upper bound. In this case, the best strategy for improvement of the polymeric membranes was to employ CMS [56].

CMS membranes are formed by hexagonal graphite-like sheets with slit-like pores, exhibiting a molecular sieving effect and allowing discrimination of small molecules (such as ethene and ethane) with subtle shape differences. Pyrolysis of 6FDA-based membranes at 675°C under controlled conditions (oxygen/argon environment) results in an ethene/ethane selectivity up to ≈8 and the permeability of ≈51 Barrer for pure gas. These membranes also showed a selectivity of 7.4 and 22 Barrer, for the separation of an ethene/ethane mixture obtained from an ethane cracker (63.2 mol% ethene/23.9 mol% ethane), values far above the upper bound but still insufficient for their direct industrial application [56].

Although in many cases current gas separation membranes may not satisfy industrial requirements to directly replace existing units, their integration in existing separation units to a hybrid system [242] may be attractive to overcome capacity limitations of, for example, distillation units (debottlenecking), to decrease the energy consumption of the whole separation scheme [243], or to demonstrate the viability of the membrane separation.

4.3.2.2 Membranes for H$_2$ Separation

Diffusivity of H$_2$ is usually higher than of other gases such as CO$_2$, CH$_4$, and N$_2$, making polymeric membranes an important candidate for the removal of H$_2$ from streams containing these gases [74]. However, most of the reported membranes show Knudsen range H$_2$/CO$_2$ selectivities or a high selectivity in combination with a low permeability. Recently, by chemical cross-linking of polyimide membranes (6FDA based) with diamine groups (1,3-diaminopropane), a drastic improvement in pure gas H$_2$/CO$_2$ selectivity (1–101) is obtained [244]. Further strategies for performance improvement also comprise development of MMMs, especially by using MOFs as fillers. Perez et al. [245] prepared MMMs using 30 wt% loading of MOF-5 in Matrimid® matrix, achieving pure H$_2$ gas permeability of 53.8 Barrer and 120 H$_2$/CH$_4$ selectivity, which is just on top of the 2008 upper bound, whereas the same membrane yielded an ideal selectivity of 2.7 for H$_2$/CO$_2$, which dropped to 2.3 for equimolar H$_2$/CO$_2$ mixture. Ordoñez et al. [246] used ZIF-8 in the same polymer matrix, having the best ideal selectivities and permeabilities with 50–60 wt% loading at 35°C. In case of 50 wt% loaded ZIF-8/Matrimid membranes, pure H$_2$ permeability of 20 Barrer resulted in a selectivity of 472 and 3.8 for H$_2$/CH$_4$ and H$_2$/CO$_2$, respectively. Further increase in loading to 60 wt% declined the H$_2$/CH$_4$ to 357 but increased the H$_2$/CO$_2$ selectivity up to 4.4 for a 36 Barrer pure gas permeability of H$_2$. These results remain still below the Robeson upper limit. For 50 and 60 wt% loaded ZIF-8/Matrimid membranes, separation factors of 3.5 and 7 were obtained with equimolar H$_2$/CO$_2$ mixtures, respectively. Yang et al. studied both ZIF-7 and ZIF-8 in a polybenzimidazole (PBI) matrix and ZIF-8/PBI MMM showed better selectivities [247,248]. Permeabilities and selectivities obtained by 30 wt% loaded ZIF-8 MMMs exhibited 105 Barrer pure H$_2$ permeability and 12.3 ideal selectivity of H$_2$/CO$_2$, thus exceeding the present upper bound. Unfortunately, no separation performance was reported for this membrane.

The first studies related to pure MOF membranes were done on CuBTC (HKUST-1) [249–251]. Ni-MOF-74 membranes yielded ideal H$_2$/N$_2$, H$_2$/CO$_2$, and H$_2$/CH$_4$ selectivities of 3, 9.1, and 2.9, respectively, with a single H$_2$ permeance of 10^{-6} mol m^{-2} s^{-1} Pa^{-1} although increasing transmembrane pressure showed a linear increase in permeation for all the components, indicative of defective membranes [252]. Later studies focused on increasing the interaction of the MOF layer with the support by using polymeric linkers, such as PEI and APTES. This approach yielded membranes with much higher selectivities [253]. Caro and coworkers published on different types of MOF membranes, in many cases using the same approach [59,191,195–197,254–257]. For instance, Li et al. [196] reported ZIF-7 membranes of 1.5 μm thick using PEI, yielding a selectivity of 18, 13.6, and 14 at 220°C for equimolar H$_2$/N$_2$, H$_2$/CO$_2$, and H$_2$/CH$_4$ mixtures with 4.42, 4.55, and 4.36 × 10^{-8} mol m^{-2} s^{-1} Pa^{-1}, respectively. Huang et al. [195] studied ZIF-90 membranes using APTES as the polymeric linker showing selectivities of 7.3 for equimolar H$_2$/CO$_2$ mixture at 200°C. By covalent functionalization of ZIF-90 membranes after synthesis with ethanolamine mixture, selectivity enhanced to 16 at the same temperature with a H$_2$ permeance of 2.02 × 10^{-7} mol m^{-2} s^{-1} Pa^{-1}, which was also six times higher than the premodified membrane [256]. The best pure gas performance reported so far is with APTES-modified ZIF-8 membranes, yielding ideal selectivities of 17 with a H_2 permeance of 5.73×10^{-5} mol m^{-2} s^{-1} Pa^{-1} at 25°C [258], although no mixture separation data are reported for this membrane. NH$_2$-MIL-53(Al) membranes synthesized over polished glass frit yielded a separation factor of 30.9, 23.9, and 20.7 with a H_2 permeance of 19.8, 17.9, and 15.1×10^{-7} mol m^{-2} s^{-1} Pa^{-1} for equimolar H_2/N_2, H_2/CO_2, and H_2/CH_4 mixtures at 15°C [259]. The highest temperature this membrane was tested was 80°C and resulted in a decrease in the H_2/CO_2 separation factor to ≈23.5. The stability of the membrane is studied by following the separation of H_2/CO_2 during operation over 120 h, showing no significant change in permeance and selectivity at 15°C.

The thermal stability of polymeric membranes is limited [74], restricting their use in *in situ* applications in hydrogen production plants or dehydrogenation units. Moreover, their relatively high sensitivity to swelling and compaction and low chemical resistance for compounds that could be present in petrochemical streams, such as sulfur, add other constraints for their industrial application [74,215,216]. Also hybrid membranes, such as mixed-matrix and MOF types, may suffer from these problems, and more studies are needed regarding their chemical and thermal stability for any specific application.

Nonpolymeric membranes considered for hydrogen separation at high temperatures are carbon-based membranes, microporous silica, Pd and Pd alloys, ceramics (dense, perovskite), and zeolite membranes [74,215,216,260–266]. Each and every one of them has advantages and disadvantages. Carbon membranes, rather than self-standing structures [267], require multiple polymer deposition and carbonization cycles [75] in order to obtain defect-free membranes. In most cases, such membranes have so low permeances that they could never be used in industrial separations (H_2 permeability of 2.78×10^{-17} mol m m^{-2} s^{-1} Pa^{-1}—measured under the pressure difference of 0.1 MPa—and an ideal H_2/N_2 selectivity of 46 are reported recently for an ordered nanoporous membrane [268]). Moreover, these types of membranes are susceptible to O_2-containing streams at high temperatures, and strongly adsorbing vapors might clog the pores [269].

Ceramic membranes are fairly inert and can be used at high temperatures. Excellent separation factors (up to 1000) and fluxes (up to 10^{-7} mol m^{-2} s^{-1} Pa^{-1}) are reported for SiO_2-type membranes regarding H_2/N_2 separation [270]. However, pure silica membranes are hydrothermally unstable, hampering their application in many industrial separations. Later studies focused on the preparation of membranes with improved stability: by the addition of a fluorinated silica source to the membrane preparation procedure [271,272] and by doping metals/earth metals to the membrane [273,274]. However, although these modifications enabled higher hydrothermal stability, obtained performances are not comparable to previously obtained results (selectivities around 7 for H_2/CO_2). By the use of organic–inorganic hybrid alkoxide as a starting raw material (such as BTESE, $(C_2H_4O)_3Si–CH_2CH_2–Si(OC_2H_4)_3$, and BTESM, $(C_2H_4O)_3Si–CH_2–Si(OC_2H_4)_3$) [275–278], quantitative improvements are obtained in terms of hydrothermal stability (less thermal Si–OH groups)—theoretically stable until 300°C [276]—and after exposure to steam, no significant change was observed regarding separation performance, in the order of 2×10^{-5} mol m^{-2} s^{-1} Pa^{-1} for H_2, with an ideal selectivity of 500 for H_2/SF_6 [276]. Unfortunately, no mixture separation performance is reported. Further separation application mainly focused on pervaporation [279] and reverse osmosis [277], although some membrane reactor experiments indicated the applicability for H$_2$ separation. Other ceramic membranes available, such as TiO$_2$ or ZrO$_2$, show low selectivity toward H$_2$ (only just above Knudsen selectivity) [269,273].

Perovskite-type oxides (proton-conducting ceramics), including BaZrO$_3$, SrZrO$_3$, SrCeO$_3$, and BaCeO$_3$, doped with a rare earth oxide have been reported in the literature for H$_2$ separation [280–282]. Among these membranes, SrCeO$_3$ and BaCeO3 doped with Y, Yb, and Gb own the market as high-temperature proton conductors [280]. However, their low electron conductivity usually yields a low H$_2$ permeability. A solution to this problem was the addition of metals (or alloys) such as Pd, Pd–Ag, Pd–Cu, Nb, Ta, and Zr, yielding the so-called cermet membranes [283]. These membranes are formed by dispersion of metal (40–50 vol%) species in a ceramic matrix. This approach does not only increase electron transfer but also enhances the adsorption and ionization of hydrogen on the membrane surface and mechanical stability [281]. Although these membranes offer relatively high hydrogen fluxes at temperatures of about 900°C–1000°C (Balachandran et al. [283] reported H$_2$ flux of 1.5×10^{-5} mol m^{-2} s^{-1} Pa^{-1} at 900°C and 0.9 bar H$_2$ feed pressure for 50 vol% Pd with Y$_2$O$_3$-stabilized ZrO$_2$), these temperatures are quite high for a sustainable process.

Pd membranes have been widely studied for H$_2$ separation in different petrochemical processes, in spite of their high price [74,75,215,261,284]. Separation is based on the solution-diffusion transport and the achievable selectivity is infinite in case of defect-free membranes. There are, however, several drawbacks associated with these membranes. In actual mixture streams, the nonhydrogen species present within the mixture can cause severe poisoning problems. For example, CO binds strongly to Pd at temperatures below 200°C [285], blocking the dissociative adsorption of hydrogen and decreasing the permeability of H$_2$ to a great extent. Adsorbed sulfur has a similar irreversible impact [215,286]. Another limitation of pure palladium membranes is the formation of undesired phases. When applied at operating conditions, below 20 atm and 300°C, increasing hydrogen concentration causes the formation of the β-phase, which coexists with the α-phase. The remarkable difference between the expansion coefficients of these two phases results in distortion, dislocation, multiplication, and hardening of the membrane layer. This can further cause splitting of the membrane after a few hydrogenation/dehydrogenation cycles [263,264]. To overcome these drawbacks, Pd-based alloys are used. Metals commonly applied in combination with Pd are Ag, Au, Cu, and Ni [286–290]. However, this can further increase the cost of the membrane and the number of preparation steps: expansion coefficient differences of the composite membrane and the support can become a further problem. Recently, a report suggested Pt as an excellent metallic glue, considering that the thermal expansion coefficient of Pt is in between Pd and Al. Low fluxes are attributed to the decrease in solubility with the addition of Pt. However, tests that varied at temperatures of 150°C–350°C yielded stability for many heating/cooling cycles [291]. Along the same line, Pt–Pd–Y$_2$O$_3$-stabilized ZrO$_2$ (YSZ) membranes tested under water gas shift conditions showed higher fluxes compared to pure Pd membranes due to less inhibition of the membrane by the other components in the mixture [292]. The highest fluxes yet obtained for these composite membranes still lie in the range of 10^{-3} to 1 mol m^{-2} s^{-1}, even though selectivities ranged from 3.7 to infinity [262].

4.3.2.3   Zeolite Membranes for H_2 Separation

Zeolites are recognized as attractive materials to prepare membranes because of their microporous crystalline structures with monodispersed pores. Zeolite membranes are generally synthesized as composites of thin zeolitic films on thick macroporous supports, usually by hydrothermal treatment of the substrate in the presence of aluminosilicate precursors (Si/Al = 0–∞). The thin zeolitic layer will be responsible for the separation, while the support provides mechanical strength [293]. The pore diameter of the separating zeolite layer is chosen in such a way that the range of the kinetic diameter of the molecules to be separated is such to enforce molecular sieving as the determining diffusion regime. In addition to this molecular exclusion effect, due to the mutual effect of mixture adsorption and diffusion, differences on the removal of one component from the mixture compared to single permeation can be observed. Usually a trade-off is observed from the view of selectivity vs. permeance: as the selectivity increases, the permeance decreases and vice versa. To enhance fluxes through the membrane while keeping high selectivities, thin zeolite films are preferred [74,216,293,294].

For the removal of H_2 (with a kinetic diameter of 0.289 nm) from the main production streams, which contain also chemicals having fairly similar kinetic diameters (like CO, CO_2, CH_4, and C_2H_4 with kinetic diameters of molecules being 0.38, 0.33, 0.38, and 0.39 nm, respectively), the selected zeolite membrane should have a small pore size that will allow high selectivity in terms of H_2. This is why most of the studies focused on 8MR structures, such as AlPO-4 (LTA), SAPO-34 (CHA), and DDR. Besides separation performance, another desired property is the chemical and structural stability of the membrane in high-temperature media, including water vapor, acidic compounds, and possible corrosive impurities [260]. Zeolite membranes are considered as highly thermally stable with respect to their counterparts; however, their application at temperatures above 300°C has been hardly reported in the open literature. These aluminosilicate crystals tend to show lower stability with increasing temperature as the Al content within the framework increases. Moreover, the synthesis of high-quality membranes also depends on the Si/Al ratio in the framework, as shown by Caro and Noack et al. [295–298] for MFI and FAU membranes. Another issue that has to be addressed is the membrane thickness. While keeping the selectivity toward the desired component, it is also desired to keep the permeances as high as possible, which could only be achieved by decreasing the thickness of the zeolite layer. Most of the zeolite membranes reported for this separation have thickness values varying from 5 to 30 μm. Synthesis of thinner membranes is desirable and still challenging [74,293,294,299]. As the thickness decreases, inevitably formation of defects increases [260]. Moreover, during the calcination step of the membranes, cracks and nonzeolitic pores may be formed [74,75,260]. To overcome these issues, many advanced techniques, such as chemical vapor deposition (CVD), atomic layer deposition (ALD), catalytic cracking deposition (CCD), or silylation, are applied to modify or repair the membranes after calcination [300–302]. On the other hand, it is more desirable to avoid crack formation in the first place by replacing the calcination procedure with a less harsh method.

Recently, Zhang et al. [303] reported the decomposition of di-*n*-propyl amine and tetraethyl ammonium hydroxide (used as templates) from SAPO-34 powder and membranes in a vacuum chamber that is held at a pressure of 10^{-4} kPa and temperature of 400°C under N$_2$ atmosphere. With this treatment, they were able to obtain SAPO-34 membranes showing twice higher permeation toward CO$_2$ compared to that of the membranes where the template is removed under flowing air. Kuhn et al. [304,305] showed the application potential of ozonication using zeolite powders and membranes, yielding crack-free layers and complete detemplation. In ozonication, O$_3$, which is a stronger oxidizing agent, is used in combination with O$_2$, enabling lower temperatures to be applied for the removal of the template. In addition to changes made in the calcination step, it is as important to increase the attachment of the layer to the support, to avoid formation of cracks in between the zeolite and the support layer. This was achieved by introducing covalent/cationic linkers [306–311] that change the surface properties of the support layer, enabling better attachment of the zeolite crystals during the synthesis. An additional advantage of this method is the formation of uniform layers and even sometimes the formation of layers with oriented crystals [311], which is preferable for anisotropic zeolites, where orientation can influence positively the gas permeation behavior.

The current status of zeolite membranes applied for H$_2$ separation is summarized in Table 4.3. It should be noted that many of the literature studies just give ideal selectivity data. However, to understand their applicability in real separations, emphasis in the following sections is given to mixture behavior.

Zheng et al. [312] reported the hydrogen separation behavior of low-quality DD3R membranes at elevated temperatures. DD3R has a 2D structure with 0.36×0.44 nm ellipsoidal pore opening. The membranes were synthesized by both *in situ* and secondary growth methods. After calcination, the membranes were postsynthetically modified by CVD. These modified membranes presented a H$_2$/CO$_2$ ideal selectivity of 32. Regrettably, no mixture separation data were provided. Using the same modification technique, Kanezashi et al. [300] also reported the performance change of DDR membranes. Accordingly, the membranes prepared by secondary growth showed Knudsen behavior for He and H$_2$ below 100°C, with activated diffusion being the separation mechanism at higher temperatures. The permeance of CO$_2$ increased drastically with decreasing temperature in a manner similar to surface diffusion, and it was even higher than H$_2$ and He at temperatures below 200°C. After CVD modification, permeances of H$_2$ and He were always higher than that of CO$_2$. Although the modified membranes show higher values for selectivity and permeance [75,300], long-term utilization of a modified membrane is still an issue to be addressed (this repair is based on utilization of amorphous silica and the problems regarding the hydrothermal stability of amorphous silica have been discussed in Section 4.3.2.2). Van den Bergh et al. [313] reported on high-temperature single and mixture permeation behavior of DD3R membranes. A commercial membrane, provided by NGK Insulators, was employed and mixture data up to 500°C were reported. The permeation behavior of H$_2$ was not affected significantly by increasing temperature. Separation selectivities of 400, 2.2, and 12 were achieved for H$_2$/*i*-butane, H$_2$/CO$_2$, and H$_2$/N$_2$ mixtures at 400°C, whereas permeances ranged between 10^{-7} and 10^{-8} mol m^{-2} s^{-1} Pa^{-1}.

TABLE 4.3

Comparison of Binary Gas Mixture Separation Performances of Different Zeolite Membranes Reported in the Literature for H₂ Separation

Membrane	T (°C)	Feed Mixture Composition (%/%)	Feed Pressure (kPa)	Permeate Pressure (kPa)	Sweep[a] (+/−)	Permeance of H₂ in the Mixture (10^{-7} mol m⁻² s⁻¹ Pa⁻¹)	Separation Factor			References
							H_2/CO_2	H_2/N_2	H_2/CH_4	
Silicalite/ZSM-5 bilayer	150	50/50	101.3	NA	+	0.7	23	—	—	[314]
MFI—silylated	500	50/50	NA	NA	+	0.12	45	—	—	[315]
ITQ-29	200	48.5/48.5/3[b]	101.3	NA	+	2.3	—	—	5.8	[316]
Na-LTA	25	50/50	101.3	NA	+	4.6	6.4	5.1	4	[307]
FAU	100	50/50	101.3	NA	+	4	6	6	4	[311]
DDR	30	50/50	201.3	101.3	+	0.71	0	—	—	[313]
	400	50/50	201.3	101.3	+	0.52	4.5	—	—	
	30	50/50	201.3	101.3	+	0.92	—	5	—	
	400	50/50	201.3	101.3	+	0.5	—	12.2	—	
B-ZSM-5	250	50/50	222	84	−	1.3	1.8	2.3	—	[301]
silylated	250	50/50	222	84	−	1.4	—	—	1.6	

Material										Ref.
SAPO-34— silylated	25	50/50	222	84	—	0.48	—	—	59	
SAPO-34/Al₂O₃ composite	200	50/50	350	84	—	0.7	23	600	—	[317]
	200	50/50	900	84	—	0.6	—	970	—	
AlPO₄	35	50/50	101.3	1	+	2.7	—	6	—	[318]
	35	50/50	101.3	1	+	2.1	9.7	—	—	
SAPO-34	30	50/50	100	NA	—	90	16.66	20.91	—	[319]
SAPO-34	−20	43/57	1600	84	—	0.002	0.0073	—	—	[320]
	35	43/57	1600	84	—	0.027	0.0625	—	—	
	−20	54/46	1600	84	—	1.5	—	—	18	
	250	54/46	1600	84	—	1.7	—	—	19	
SSZ-13	25	50/50	223	85	—	0.81	—	—	6.2	[321]
	200	50/50	223	85	—	1.4	—	—	6.1	

a Measurements with sweep (+) and without sweep (−).
b Steam was added to check the stability of the membrane.

Huang et al. [309] synthesized a new type of membrane using LTA-type $AlPO_4$ zeolite material by using a secondary growth method. Due to its cation-free nature, $AlPO_4$ has an estimated pore size of 0.4 nm. The membrane was only tested at 20°C and 50°C and no data on the hydrothermal stability were given. However, the reported permeance values are very promising. The binary mixture behavior of the membrane toward H_2 was tested in the presence of CO_2, O_2, CH_4, and C_3H_8, and selectivities were 7.6, 6.1, 4.3, and 143, respectively, while the H_2 permeances ranged from 2.1 to 2.5×10^{-7} mol m^{-2} s^{-1} Pa^{-1}. In fact, a H_2 selective composite $AlPO_4$-5/ $AlPO_4$-34 membrane was previously proposed by Guan et al. [318]. The membrane showed a 9.7 separation factor for H_2/CO_2 at 35°C, but the addition of water to the stream caused a decrease in the permeance values.

Hong et al. [320] used SAPO-34 membranes to remove H_2 from CH_4 or CO_2 to simulate the hydrogen production stream. SAPO-34 has a structure analogous to natural zeolite chabazite with a pore size of 0.38 nm. Due to the hydrophobicity of the structure, the adsorption of CO_2 is stronger than that of CH_4 and much stronger than that of H_2. In the study, the membrane had a Si/Al ratio of 0.15 and the highest temperature applied during gas separation was 250°C. CO_2/H_2 ideal selectivities ~140 are reported at 0°C. As the temperature increases, diffusion becomes more important than adsorption and the CO_2/H_2 ideal selectivity decreases to 2 at 220°C. H_2 permeances of 2×10^{-9} mol m^{-2} s^{-1} Pa^{-1} are reported for CO_2–H_2 mixture separation at 220°C. The permeances of the components were also affected by their presence: while H_2 permeance decreases, methane permeance increases in the presence of H_2. This phenomenon is experienced for many membranes: CH_4 adsorbs much stronger than H_2, but H_2 diffuses much faster and it speeds up the CH_4 as well, yielding fluxes even higher than single-component fluxes [322].

Although the best results achieved so far are for AlPO- and SAPO-type membranes, and thermally, these zeolites are proven to be stable until 1000°C [323], their hydrothermal stability is generally lower than for Al-containing structures (AlO_2^- and SiO_2), especially at high temperatures. Poshusta et al. [324] showed that after exposure of the membrane to humid conditions, membrane quality has deviated from the original, depending upon the membrane and exposure conditions. For instance, a membrane that was left in laboratory conditions for 4 months in a vial showed a decrease in CO_2/CH_4 ideal selectivity from 14 to 2.8.

SSZ-13, having the same CHA-type framework as SAPO-34, was also studied by Kalipcilar et al. [321]. Membranes with a Si/Al ratio of 20 were synthesized over stainless steel supports. Synthesis was repeated for 4–5 times to yield defect-free membranes before calcination. H_2 permeances of $\approx 1.4 \times 10^{-7}$ mol m^{-2} s^{-1} Pa^{-1} in single-component measurements dropped to 8.2×10^{-8} mol m^{-2} s^{-1} Pa^{-1} at 25°C in an equimolar H_2/CH_4 mixture (pressure drop maintained across the membrane was 138 kPa), which was recovered at 200°C. The highest separation factor of 7.8 was obtained at 100°C for the best membrane, and at other temperatures, the selectivity ranged between 6.2 and 7.8. Separation of equimolar mixtures of CO_2/N_2, CO_2/ CH_4, and H_2/n-C_4H_{10} was also studied. The highest H_2/n-C_4H_{10} selectivity of 6.9 was obtained at 147°C, indicating that part of the permeation took place through nonzeolitic pores since n-butane is not expected to fit in the pores of this zeolite. This membrane was further studied for pervaporation of water to determine the stability

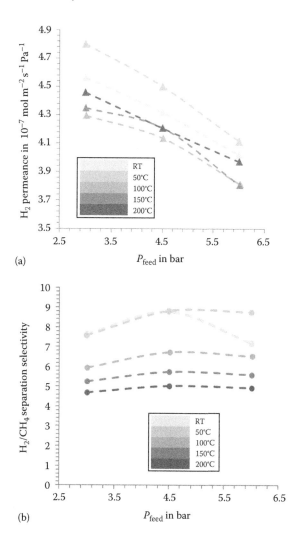

FIGURE 4.10 (a) H_2 permeance and (b) separation selectivities obtained over high-silica SSZ-13 membrane for an equimolar H_2/CH_4 mixture by varying feed pressure and temperature. Permeate was atmospheric and no sweep was used. (From Gucuyener, C. et al., *J. Memb. Sci.*, in preparation.)

of the membrane under 69.5 wt% HNO_3 at 298 K. Forty one percent HNO_3 mixture was obtained on the permeate side after 13 days, indicating the possibility of breaking the azeotrope and also proving the membranes high acid stability (Figure 4.10).

Further studies regarding SSZ-13 -type membranes focused on high-silica structures. Membranes with a Si/Al ratio of ≈60 were intentionally selected to improve the hydrothermal stability. H_2/CH_4 ideal selectivity reaches up to 60, which is more than 20 times that of the Knudsen selectivity. Pure H_2 permeances (1.1×10^{-6} mol m^{-2} s^{-1} Pa^{-1}) obtained for these membranes are much higher, reaching to 62 times higher compared with that of SAPO-34 [320] and 8 times higher compared to the

TABLE 4.4

Comparison of Water–Gas Shift Membrane Reactor Studies Using Different MFI Membranes for H_2 Removal

	Zhang et al.[a] [326]		Wang et al. [314]		Kim et al. [327]
Membrane	Modified MFI		Modified bilayer MFI		Modified MFI
Support	α-Alumina		Y_2O_3-ZrO_2-α-alumina		α-Alumina
Catalyst	CuO/ZnO/Al_2O_3		CeO_2/Fe_2O_3	No catalyst	CeO_2/Fe_2O_3
H_2O/CO	1.5	5	3	1	3.5
Space velocity	1,500	1,500	60,000	48,000	7,500
	L (NTP) kg_{cat}^{-1} h^{-1}		**h^{-1} (GHSV)**		**h^{-1} (WHSV)**
Temperature (°C)	200	200	500	500	550
Feed pressure (kPa)	200	200	202	202	202
Sweep gas		N_2		He	N_2
Sweep rate (mL min^{-1})		20		20	20
H_2 permeance (10^{-7} mol m^{-2} s^{-1} Pa^{-1})	0.06	0.0273	1.04	1.18	1.5
Selectivity H_2/CO_2			25	23	31
H_2/CO				29	25

[a] For this reference, in the given space velocities, NTP refers to normal temperature and pressure.

SSZ-13 membrane [321]. Under mixture conditions, a H_2 permeance of 4.6 10^{-7} mol m^{-2} s^{-1} Pa^{-1} and a separation selectivity of 7.6 are attained for equimolar mixtures, at a pressure drop of 2 bar at 25°C. As the temperature is increased, although there is no significant decrease in the permeance of H_2, due to the activated permeation of CH_4, selectivities further drop to 4.7 at 200°C [325].

Recently, more studies regarding membrane reactor performance of zeolite membranes under industrially relevant conditions during the water gas shift reaction have been published. The membranes studied under these conditions were always MFI type [314,326,327]. This choice, although it might be related to the expectation of high permeances, is more likely the consequence of the number of studies regarding the behavior of MFI-type membranes. Almost in all studies reported, MFI membranes were modified with CCD, indicating that these membranes were not of a high quality. Nevertheless, CO conversion rates up to 92% and H_2 recovery of 20% are reported under both high- and low-temperature water gas shift conditions (Table 4.4).

4.4 CONCLUSIONS

Separation processes are among the most energy-intensive operations in the chemical industry, with olefin/paraffin and H_2 separation being among the most outstanding examples. In this chapter, we have summarized the main advantages and limitations of the use of porous materials for the separation of these important bulk chemical production streams.

Currently, H$_2$ separation from gas to liquid streams is based either on cryogenic distillation or on adsorptive separation that is handled by a 10-bed 11-step process. On the other hand, *in situ* removal of H$_2$ or in-line separation of H$_2$ could potentially decrease the cost of such processes. For H$_2$ separation, although Pd membranes are proven to have a better selectivity and permeability compared to many other membrane types, the price and possible poisoning of such membranes seem to be very important issues. In addition, current methods to overcome their chemical stability are based on upgrading the membranes with additional precious metals, which increases the cost further. In case of silica, MMM, and MOF membranes, although showing promising results, (hydro)thermal stability during *in situ* operation could potentially become an important problem, unless further studies prove otherwise. Currently, although most of the membrane reactor studies for H$_2$ separation are based on 10MR zeolites due to the impressive amount of information provided for these types of membranes in the literature, 8MR zeolites are more desirable to yield higher selectivities and permeances. Although SAPO- and AlPO-type membranes are the most studied 8MR membranes from this separation, DDR and SSZ-13 have much better chemical stability. Further studies should focus more on the membrane reactor performance and scale-up of such membranes to unravel their actual potential. Once the issues associated with zeolite membrane preparation have been solved, their large-scale application, especially for the *in situ* removal of hydrogen in membrane reactors, should become a very attractive technology.

When it comes to olefin/paraffin separation, a materials' issue has been identified. In order to efficiently (from an energetic point of view) separate small olefin/paraffin mixtures, we lack materials with the desired properties. For adsorption-based processes, materials displaying a preferential adsorption of paraffin over olefin offer ideal properties for the direct production of pure olefin (the desired product). Although the first materials displaying such properties have already been reported (such as ZIF-7 and ZIF-8), a great deal of effort should be devoted now to better understand and control this reverse selectivity and to develop new adsorption-based processes for this separation. Regarding membranes for this separation, a higher degree of control over membrane synthesis (based on polymers, on structured materials, or on CMSs) is required before real application can be considered.

For the next few decades, we foresee intensive research in these two key technological areas that should eventually result in new energy-efficient and environmentally friendly technologies. At this stage, it is already clear that implementation of such technologies is going to be slow and that maturity of membrane- and adsorbent-based processes, especially in the case of olefin/paraffin mixture separation, will be reached via implementation of hybrid separation systems where these new technologies will be combined with more mature processes such as distillation.

REFERENCES

1. J.A. Moulijn, M. Makkee, and A. van Diepen, *Chemical Process Technology*. John Wiley & Sons Ltd., England, U.K., 2013.
2. P. Kaiser, R.B. Unde, C. Kern, and A. Jess, Production of liquid hydrocarbons with CO$_2$ as carbon source based on reverse water-gas shift and fischer-tropsch synthesis, *Chemie-Ingenieur-Technik* 85(2013) 489–499.

3. L. Han, C. Ding, and H. Lui, Studies on olefin production by steam cracking of waste oil blended with naphtha, *Applied Mechanics and Materials* 291–294 (2013) 738–743.

4. M. Tagliabue, D. Farrusseng, S. Valencia, S. Aguado, U. Ravon, C. Rizzo, A. Corma, and C. Mirodatos, Natural gas treating by selective adsorption: Material science and chemical engineering interplay, *Chemical Engineering Journal* 155 (2009) 553–566.

5. C.J. Gonzalez, J. Mazurelle, and W. Vermeiren, Upgrading light naphtas for increased olefins production, EP2243814 A1, 2010.

6. C.L. Kibby, K. Jothimurugesan, T.K. Das, R.J. Saxton, and A.W. Burton, Zeolite supported cobalt hybrid fischer-tropsch catalyst, US7943674 B1, 2011.

7. N.C. Schiødt, Chromium-free water gas shift catalyst, US8119099 B2, 2012.

8. W. Dolan, B. Speronello, A. Maglio, D. Reinersten, and M.D. Rehms, Lower reactivity adsorbent and higher oxygenate capacity for removal of oxygenates from olefin streams, WO2011043974 A2, 2012.

9. M. Benali and B. Aydin, Ethane/ethylene and propane/propylene separation in hybrid membrane distillation systems: Optimization and economic analysis, *Separation and Purification Technology* 73 (2010) 377–390.

10. D.B. Manley, Thermodynamically efficient distillation: Ethylene recovery, *Latin American Applied Research* 28 (1998) 1–6.

11. M.G. Buonomenna, Membrane processes for a sustainable industrial growth, *RSC Advances* 3 (2013) 5694–5740.

12. R.A. Sheldon, Atom efficiency and catalysis in organic synthesis, *Pure and Applied Chemistry* 72 (2000) 1233–1246.

13. E. Drioli, A. Brunetti, G. Di Profio, and G. Barbieri, Process intensification strategies and membrane engineering, *Green Chemistry* 14 (2012) 1561–1572.

14. J.R. Hufton, S. Mayorga, and S. Sircar, Sorption-enhanced reaction process for hydrogen production, *AIChE Journal* 45 (1999) 248–256.

15. R. Henry, M. Tayakout-Fayolle, P. Afanasiev, C. Lorentz, G. Lapisardi, and G. Pirngruber, Vacuum gas oil hydrocracking performance of bifunctional Mo/Y zeolite catalysts in a semi-batch reactor, *Catalysis Today* 220–222 (2013) 159–167.

16. A. Nakhaei Pour and M.R. Housaindokht, Study of activity, products selectivity and physico-chemical properties of bifunctional Fe/HZSM-5 Fischer-Tropsch catalyst: Effect of catalyst shaping, *Journal of Natural Gas Science and Engineering* 14 (2013) 29–33.

17. Y. Zhang, D. Yu, W. Li, S. Gao, and G. Xu, Bifunctional catalyst for petroleum residue cracking gasification, *Fuel* 117 (2013) 1196–1203.

18. A.K. Aboul-Gheit, A.E. Awadallah, S.M. Abdel-Hamid, A.A. Aboul-Enein, and D.S. El-Desouki, Direct conversion of natural gas to petrochemicals using monofunctional Mo/SiO$_2$ and H-ZSM-5 zeolite catalysts and bifunctional Mo/H-ZSM-5 zeolite catalyst, *Petroleum Science and Technology* 30 (2012) 893–903.

19. J. Weitkamp, A. Raichle, and Y. Traa, Novel zeolite catalysis to create value from surplus aromatics: Preparation of C^{2+}-n-alkanes, a high-quality synthetic steamcracker feedstock, *Applied Catalysis A: General* 222 (2001) 277–297.

20. S. Sartipi, K. Parashar, M. Makkee, J. Gascon, and F. Kapteijn, Breaking the Fischer-Tropsch synthesis selectivity: Direct conversion of syngas to gasoline over hierarchical Co/H-ZSM-5 catalysts, *Catalysis Science and Technology* 3 (2013) 572–575.

21. S. Sartipi, K. Parashar, M.J. Valero-Romero, V.P. Santos, B. Van Der Linden, M. Makkee, F. Kapteijn, and J. Gascon, Hierarchical H-ZSM-5-supported cobalt for the direct synthesis of gasoline-range hydrocarbons from syngas: Advantages, limitations, and mechanistic insight, *Journal of Catalysis* 305 (2013) 179–190.

22. S. Sartipi, J.E. Van Dijk, J. Gascon, and F. Kapteijn, Toward bifunctional catalysts for the direct conversion of syngas to gasoline range hydrocarbons: H-ZSM-5 coated Co versus H-ZSM-5 supported Co, *Applied Catalysis A: General* 456 (2013) 11–22.

23. http://www.ecn.nl/ (Last accessed: August 2013).
24. Materials for Separation Technologies: Energy and Emission Reduction Opportunities, http://www.eere.energy.gov/ (Last accessed: August 2013).
25. A. Mersmann, B. Fill, R. Hartmann, and S. Maurer, The potential of energy saving by gas-phase adsorption processes, *Chemical Engineering & Technology* 23 (2000) 937–944.
26. T. Ren, M. Patel, and K. Blok, Olefins from conventional and heavy feedstocks: Energy use in steam cracking and alternative processes, *Energy* 31 (2006) 425–451.
27. U.R. Chaudhuri, *Fundamentals of Petroleum and Petrochemical Engineering*. CRC Press, Boca Raton, FL (Taylor & Francis-distributor, London, U.K.), 2011.
28. D.M. Ruthven and S.C. Reyes, Adsorptive separation of light olefins from paraffins, *Microporous and Mesoporous Materials* 104 (2007) 59–66.
29. M.C. Kroon and L.F. Vega, Selective paraffin removal from ethane/ethylene mixtures by adsorption into aluminum methylphosphonate-α: A molecular simulation study, *Langmuir* 25 (2009) 2148–2152.
30. Y.A. Sangalov, K.S. Minsker, and G.E. Zaikov, *Polymers Derived from Isobutylene: Synthesis, Properties, Application*. VSP Internatioanl Science Publishers, Utrecht, the Netherlands, 2001.
31. D.M. Ruthven, *Principles of Adsorption and Adsorption Processes*. John Wiley & Sons, Mississauga, Ontario, Canada, 1984.
32. D.A. Barclay, J.L. Flebbe, and D.B. Manley, Relative volatilities of the ethane-ethylene system from total pressure measurements, *Journal of Chemical & Engineering Data* 27 (1982) 135–142.
33. M. Khalighi, Y.F. Chen, S. Farooq, I.A. Karimi, and J.W. Jiang, Propylene/propane separation using SiCHA, *Industrial & Engineering Chemistry Research* 52 (2013) 3877–3892.
34. P.F. Bryan, Removal of propylene from fuel-grade propane, *Separation and Purification Reviews* 33 (2004) 157–182.
35. A. Van Miltenburg, W. Zhu, F. Kapteijn, and J.A. Moulijn, Adsorptive separation of light olefin/paraffin mixtures, *Chemical Engineering Research and Design* 84 (2006) 350–354.
36. Ethylene product overview, http://www.shell.com (Last accessed: August 2013).
37. M. Shi, A.M. Avila, F. Yang, T.M. Kuznicki, and S.M. Kuznicki, High pressure adsorptive separation of ethylene and ethane on Na-ETS-10, *Chemical Engineering Science* 66 (2011) 2817–2822.
38. World Light Olefins Analysis, http://www.cmaiglobal.com/worldanalysis/wpoabook. aspx (Last accessed: August 2013).
39. S. Choi, Y.S. Kim, D.S. Park, S.J. Kim, and I.M. Yang, Process for increasing production of light olefins from hydrocarbon feedstock in catalytic cracking, WO2007043738 A1, 2007.
40. A. Chauvel and G. Lefebvre, *Synthesis-Gas Derivatives and Major Hydrocarbons*. Editions Technip, Paris, France, 1989.
41. J. Li, Y. Wei, G. Liu, Y. Qi, P. Tian, B. Li, Y. He, and Z. Liu, Comparative study of MTO conversion over SAPO-34, H-ZSM-5 and H-ZSM-22: Correlating catalytic performance and reaction mechanism to zeolite topology, *Catalysis Today* 171 (2011) 221–228.
42. A.M. Ribeiro, M.C. Campo, G. Narin, J.C. Santos, A. Ferreira, J.S. Chang, Y.K. Hwang et al., Pressure swing adsorption process for the separation of nitrogen and propylene with a MOF adsorbent MIL-100(Fe), *Separation and Purification Technology* 110 (2013) 101–111.
43. J.S. Plotkin, The changing dynamics of olefin supply/demand, *Catalysis Today* 106 (2005) 10–14.
44. M.C. Campo, M.C. Baptista, A.M. Ribeiro, A. Ferreira, J.C. Santos, C. Lutz, J.M. Loureiro, and A.E. Rodrigues, Gas phase SMB for propane/propylene separation using enhanced 13X zeolite beads, *Adsorption* 20 (2013) 1–15.

45. M.C. Campo, A.M. Ribeiro, A. Ferreira, J.C. Santos, C. Lutz, J.M. Loureiro, and A.E. Rodrigues, New 13X zeolite for propylene/propane separation by vacuum swing adsorption, *Separation and Purification Technology* 103 (2013) 60–70.
46. Naphtha cracking light olefins production, www.eptq.com (Last accessed: August 2013).
47. U. Illgen, R. Schafer, M. Noack, P. Kolsch, A. Kuhnle, and J. Caro, Membrane supported catalytic dehydrogenation of iso-butane using an MFI zeolite membrane reactor, *Catalysis Communications* 2 (2001) 339–345.
48. NMP process, http://www.intermediates.basf.com/chemicals/butadiene-extraction/features (Last accessed: August 2013).
49. L. Babyak, O. Matsyak, and V. Shevchuk, Conversion of C4 fraction of hydrocarbon pyrolysis products over ZVM + 2% Zn high-silica zeolite catalyst, *Chemistry & Chemical Technology* 5 (2011) 95–99.
50. D.N. Andreevskii, G.Y. Kabo, G.N. Roganov, Z.N. Polyakov, A.D. Peshchenko, and I.I. Pis'man, Isolation of individual olefins from the C4 fraction, *Chemistry and Technology of Fuels and Oils* 8 (1972) 487–491.
51. K. Weissermel and H.-J. Arpe, *Industrial Organic Chemistry*. Wiley VCH Publishers, Weinheim, Germany, 1997.
52. S.D. Barnicki, Synthetic organic chemicals, in: J.A. Kent, (ed.), *Handbook of Industrial Chemistry and Biotechnology*, Springer, London, U.K.
53. K. Kindler and H. Puhl, Method for separating a C4-hydrocarbon mixture, US 6337429 B1, 2002.
54. W.C. White, Butadiene production process overview, *Chemico-Biological Interactions* 166 (2007) 10–14.
55. N.Y. Chen, C. Hill, and S.J. Lucki, Method for separating trans from cis isomers, 3524895, 1970.
56. M. Rungta, C. Zhang, W.J. Koros, and L. Xu, Membrane-based ethylene/ethane separation: The upper bound and beyond, *AIChE Journal* 59 (2013) 3475–3489.
57. U. Böhme, B. Barth, C. Paula, A. Kuhnt, W. Schwieger, A. Mundstock, J. Caro, and M. Hartmann, Ethene/ethane and propene/propane separation via the olefin and paraffin selective metal-organic framework adsorbents CPO-27 and ZIF-8, *Langmuir* 29 (2013) 8592–8600.
58. C.A. Grande, J. Gascon, F. Kapteijn, and A.E. Rodrigues, Propane/propylene separation with Li-exchanged zeolite 13X, *Chemical Engineering Journal* 160 (2010) 207–214.
59. H. Bux, C. Chmelik, R. Krishna, and J. Caro, Ethene/ethane separation by the MOF membrane ZIF-8: Molecular correlation of permeation, adsorption, diffusion, *Journal of Membrane Science* 369 (2011) 284–289.
60. M. Palomino, A. Cantin, A. Corma, S. Leiva, F. Rey, and S. Valencia, Pure silica ITQ-32 zeolite allows separation of linear olefins from paraffins, *Chemical Communications* 0 (2007) 1233–1235.
61. A. Van Miltenburg, W. Zhu, F. Kapteijn, and J.A. Moulijn, Zeolite based separation of light olefin and paraffin mixtures, in: J. Čejka, N. Žilková, and P. Nachtigall, (eds.), *Studies in Surface Science and Catalysis*, Elsevier B.V., Amsterdam, the Netherlands. 2005. pp. 979–986.
62. A.F.P. Ferreira, J.C. Santos, M.G. Plaza, N. Lamia, J.M. Loureiro, and A.E. Rodrigues, Suitability of Cu-BTC extrudates for propane-propylene separation by adsorption processes, *Chemical Engineering Journal* 167 (2011) 1–12.
63. C.A. Grande, F. Poplow, and A.E. Rodrigues, Vacuum pressure swing adsorption to produce polymer-grade propylene, *Sep. Sci. Technol.* 45 (2010) 1252–1259.
64. K. Li, D.H. Olson, J. Seidel, T.J. Emge, H. Gong, H. Zeng, and J. Li, Zeolitic imidazolate frameworks for kinetic separation of propane and propene, *Journal of the American Chemical Society* 131 (2009) 10368–10369.

65. F.A.D. Silva and A.E. Rodrigues, Propylene/propane separation by vacuum swing adsorption using 13X zeolite, *AIChE Journal* 47 (2001) 341–357.

66. A. Van Miltenburg, J. Gascon, W. Zhu, F. Kapteijn, and J. Moulijn, Propylene/propane mixture adsorption on faujasite sorbents, *Adsorption* 14 (2008) 309–321.

67. R. Shinnar, The hydrogen economy, fuel cells, and electric cars, *Technology in Society* 25 (2003) 455–476.

68. J.A. Turner, Sustainable hydrogen production, *Science* 305 (2004) 972–974.

69. Y.S.H. Najjar, Hydrogen safety: The road toward green technology, *International Journal of Hydrogen Energy* 38 (2013) 10716–10728.

70. F. Rigas and P. Amyotte, Myths and facts about hydrogen hazards, *Chemical Engineering Transactions* 31 (2013) 913–918.

71. G.W. Crabtree, M.S. Dresselhaus, and M.V. Buchanan, The hydrogen economy, *Physics Today* 57 (2004) 39–44.

72. B. Ibeh, C. Gardner, and M. Ternan, Separation of hydrogen from a hydrogen/methane mixture using a PEM fuel cell, *International Journal of Hydrogen Energy* 32 (2007) 908–914.

73. J.A. Ritter and A.D. Ebner, State of the art adsorption and membrane separation processes for hydrogen production in the chemical and petrochemical industries, *Separation Science and Technology* 42 (2007) 1123–1193.

74. G.Q. Lu, J.C. Diniz da Costa, M. Duke, S. Giessler, R. Socolow, R.H. Williams, and T. Kreutz, Inorganic membranes for hydrogen production and purification: A critical review and perspective, *Journal of Colloid and Interface Science* 314 (2007) 589–603.

75. N.W. Ockwig and T.M. Nenoff, Membranes for hydrogen separation, *Chemical Reviews* 107 (2007) 4078–4110.

76. T.V. Choudhary and D.W. Goodman, CO-free production of hydrogen via stepwise steam reforming of methane, *Journal of Catalysis* 192 (2000) 316–321.

77. C.G. Maciel, T.D.F. Silva, E.M. Assaf, and J.M. Assaf, Hydrogen production and purification from the water-gas shift reaction on CuO/CeO2-TiO2 catalysts, *Applied Energy* 112 (2013) 52–59.

78. Y.H. Zhang, S.Z. Xin, T.C. Li, Q.L. Xu, S.P. Zhang, Z.W. Ren, and Y.J. Yan, A study on the integrated process for hydrogen-rich gas production from biomass, *Energy Sources, Part A: Recovery, Utilization and Environmental Effects* 35 (2013) 1905–1913.

79. H.H. Gunardson, *Industrial Gases in Petrochemical Processing: Chemical Industries.* CRC Press, Taylor & Francis, New York, 1997.

80. M. Sheintuch, and D.A. Simakov, Alkanes Dehydrogenation in: M. De Falco, L. Marrelli, and G. Iaquaniello, (eds.), *Membrane Reactors for Hydrogen Production Processes*, Springer, London, U.K. pp. 183–200.

81. H.A. Al-Megren, G. Barbieri, I. Mirabelli, A. Brunetti, E. Drioli, and M.C. Al-Kinany, Direct conversion of n-butane to isobutene in a membrane reactor: Thermodynamic analysis, *Industrial and Engineering Chemistry Research* 52 (2013) 10380–10386.

82. D. Casanave, P. Ciavarella, K. Fiaty, and J.A. Dalmon, Zeolite membrane reactor for isobutane dehydrogenation: Experimental results and theoretical modelling, *Chemical Engineering Science* 54 (1999) 2807–2815.

83. D. Casanave, A. Giroir-Fendler, J. Sanchez, R. Loutaty, and J.A. Dalmon, Control of transport properties with a microporous membrane reactor to enhance yields in dehydrogenation reactions, *Catalysis Today* 25 (1995) 309–314.

84. P. Ciavarella, D. Casanave, H. Moueddeb, S. Miachon, K. Fiaty, and J.A. Dalmon, Isobutane dehydrogenation in a membrane reactor: Influence of the operating conditions on the performance, *Catalysis Today* 67 (2001) 177–184.

85. P. Ciavarella, H. Moueddeb, S. Miachon, K. Fiaty, and J.A. Dalmon, Experimental study and numerical simulation of hydrogen/isobutane permeation and separation using MFI-zeolite membrane reactor, *Catalysis Today* 56 (2000) 253–264.

86. J.P. Collins, R.W. Schwartz, R. Sehgal, T.L. Ward, C.J. Brinker, G.P. Hagen, and C.A. Udovich, Catalytic dehydrogenation of propane in hydrogen permselective membrane reactors, *Industrial and Engineering Chemistry Research* 35 (1996) 4398–4405.

87. R. Schafer, M. Noack, P. Kölsch, S. Thomas, A. Seidel-Morgenstern, and J. Caro, Development of a H2-selective SiO2-membrane for the catalytic dehydrogenation of propane, *Separation and Purification Technology* 25 (2001) 3–9.

88. J. van den Bergh, C. Gücüyener, J. Gascon, and F. Kapteijn, Isobutane dehydrogenation in a DD3R zeolite membrane reactor, *Chemical Engineering Journal* 166 (2010) 368–377.

89. J. van den Bergh, N. Nishiyama, and F. Kapteijn, Zeolite membranes in catalysis: What is new and how bright is the future? in: A. Cybulski, J.A. Moulijn, and A. Stankiewicz, (eds.), *Novel Concepts in Catalysis and Chemical Reactors*, Wiley-VCH Verlag GmbH & Co. KGaA, Weinheim, Germany. pp. 211–237.

90. D. Capoferri, B. Cucchiella, G. Iaquaniello, A. Mangiapane, S. Abate, and G. Centi, Catalytic partial oxidation and membrane separation to optimize the conversion of natural gas to syngas and hydrogen, *ChemSusChem* 4 (2011) 1787–1795.

91. B. Zornoza, C. Casado, and A. Navajas, Chapter 11—Advances in hydrogen separation and purification with membrane technology, in: M.G. Luis, A. Gurutze, and P.M. Dieguez, (eds.), *Renewable Hydrogen Technologies*, Elsevier, Amsterdam, the Netherlands. pp. 245–268.

92. S. Sircar and T.C. Golden, Purification of hydrogen by pressure swing adsorption, *Separation Science and Technology* 35 (2000) 667–687.

93. UOP Hydrogen Selection Matrix, http://www.uop.com/processing-solutions/gas-processing/hydrogen/ (Last accessed: December 2013).

94. Recovery of CO_2 from tail gas in industrial facilities to substitute the use of fossil fuels for production of CO_2, http://cdm.unfccc.int/ (Last accessed: August 2013).

95. Kyoto protocol reference manual: On accounting of emissions and assigned amount, http://unfccc.int/ (Last accessed: August 2013).

96. K. Skutil and M. Taniewski, Some technological aspects of methane aromatization (direct and via oxidative coupling), *Fuel Processing Technology* 87 (2006) 511–521.

97. L.M. Petkovic and D.M. Ginosar, Direct production of hydrogen and aromatics from methane or natural gas: Review of recent U.S. patents, *Recent Patents on Chemical Engineering* 5 (2012) 2–10.

98. S.I. Abasov, F.A. Babayeva, R.R. Zarbaliyev, G.G. Abbasova, D.B. Tagiyev, and M.I. Rustamov, Low-temperature catalytic alkylation of benzene by propane, *Applied Catalysis A: General* 251 (2003) 267–274.

99. F.A. Babaeva, S.I. Abasov, and M.I. Rustamov, Conversions of mixtures of propane and benzene on Pt, Re/Al₂O₃ + H-zeolite systems, *Petroleum Chemistry* 50 (2010) 42–46.

100. C. Bigey, and B.L. Su, Propane as alkylating agent for alkylation of benzene on HZSM-5 and Ga-modified HZSM-5 zeolites, *Journal of Molecular Catalysis A: Chemical* 209 (2004) 179–187.

101. X. Huang, X. Sun, S. Zhu, and Z. Liu, Benzene alkylation with propane over Mo modified HZSM-5, *Catalysis Letters* 119 (2007) 332–338.

102. X. Huang, X. Sun, S. Zhu, and Z. Liu, Benzene alkylation with propane over metal modified HZSM-5, *Reaction Kinetics and Catalysis Letters* 91 (2007) 385–390.

103. D.B. Lukyanov and T. Vazhnova, A kinetic study of benzene alkylation with ethane into ethylbenzene over bifunctional PtH-MFI catalyst, *Journal of Catalysis* 257 (2008) 382–389.

104. D.B. Lukyanov and T. Vazhnova, Highly selective and stable alkylation of benzene with ethane into ethylbenzene over bifunctional PtH-MFI catalysts, *Journal of Molecular Catalysis A: Chemical* 279 (2008) 128–132.

105. V.V. Ordomskiy, L.I. Rodionova, I.I. Ivanova, and F. Luck, Dehydroalkylation of benzene with ethane over Pt/H-MFI in the presence of hydrogen scavengers, *ChemCatChem* 4 (2012) 681–686.
106. A.V. Smirnov, E.V. Mazin, V.V. Yuschenko, E.E. Knyazeva, S.N. Nesterenko, I.I. Ivanova, L. Galperin, R. Jensen, and S. Bradley, Benzene alkylation with propane over Pt-modified MFI zeolites, *Journal of Catalysis* 194 (2000) 266–277.
107. S. Todorova and B.L. Su, Propane as alkylating agent for benzene alkylation on bimetal Ga and Pt modified H-ZSM-5 catalysts: FTIR study of effect of pre-treatment conditions and the benzene adsorption, *Journal of Molecular Catalysis A: Chemical* 201 (2003) 223–235.
108. J. Gascon, W. Blom, A. van Miltenburg, A. Ferreira, R. Berger, and F. Kapteijn, Accelerated synthesis of all-silica DD3R and its performance in the separation of propylene/propane mixtures, *Microporous and Mesoporous Materials* 115 (2008) 585–593.
109. C.A. Grande, Advances in pressure swing adsorption for gas separation *ISRN, Chemical Engineering* 2012 (2012) 13.
110. D. Peralta, G. Chaplais, A. Simon-Masseron, K. Barthelet, C. Chizallet, A.A. Quoineaud, and G.D. Pirngruber, Comparison of the behavior of metal-organic frameworks and zeolites for hydrocarbon separations, *Journal of the American Chemical Society* 134 (2012) 8115–8126.
111. W.J.C.B.D. Thomas, *Adsorption Technology and Design*. Butterworth-Heinemann, Boston, MA, 1998.
112. C.L. Cavalcante Jr., Industrial adsorption separation processes: Fundamentals, modeling and applications, *Latin American Applied Research* 30 (2000) 357–364.
113. W. Zhu, J.C. Groen, A. van Miltenburg, F. Kapteijn, and J.A. Moulijn, Kureha activated carbon characterized by the adsorption of light hydrocarbons, in: F.R.-R.J.R. P.L. Llewellyn, and N. Seaton, (eds.), *Studies in Surface Science and Catalysis*, Elsevier, Amsterdam, the Netherlands. pp. 287–294.
114. S.U. Rege, J. Padin, and R.T. Yang, Olefin/paraffin separations by adsorption: π-complexation vs. kinetic separation, *AIChE Journal* 44 (1998) 799–809.
115. C.A. Grande, N. Firpo, E. Basaldella, and A.E. Rodrigues, Propane/propene separation by SBA-15 and π-complexed Ag-SBA-15, *Adsorption* 11 (2005) 775–780.
116. L. Chen and X. Liu, π-complexation mesoporous adsorbents Cu-MCM-48 for ethylene-ethane separation, *Chinese Journal of Chemical Engineering* 16 (2008) 570–574.
117. J. Padin and R.T. Yang, New sorbents for olefin/paraffin separations by adsorption via π-complexation: Synthesis and effects of substrates, *Chemical Engineering Science* 55 (2000) 2607–2616.
118. S.H. Cho, S.S. Han, J.N. Kim, N.V. Choudary, P. Kumar, and S.G.T. Bhat, Adsorbents, method for the preparation and method for the separation of unsaturated hydrocarbons for gas mixtures, US6315816, 2001.
119. D.W. Breck, *Zeolite Molecular Sieves: Structure, Chemistry and Use*. John Wiley & Sons, New York, 1974.
120. M. Mofarahi and S.M. Salehi, Pure and binary adsorption isotherms of ethylene and ethane on zeolite 5A, *Adsorption* 19 (2013) 101–110.
121. M. Bulow, C.J. Guo, D. Shen, F.R. Fitch, A.I. Shirley, and V.A. Malik, Separation of alkenes and alkanes, US6200366, 2001.
122. C.A. Grande and A.E. Rodrigues, Adsorption of binary mixtures of propane–propylene in carbon molecular sieve 4A, *Industrial & Engineering Chemistry Research* 43 (2004) 8057–8065.
123. C.A. Grande, S. Cavenati, P. Barcia, J. Hammer, H.G. Fritz, and A.E. Rodrigues, Adsorption of propane and propylene in zeolite 4A honeycomb monolith, *Chemical Engineering Science* 61 (2006) 3053–3063.
124. C.J. Leslie, L.P. Joseph, and E.R. Nightingale Jr., Olefin separation with strontium and cadmium molecular sieves, US3355509, 1967.

125. J. Gascon and F. Kapteijn, Use of a li-faujasite for separation of olefin/paraffin mixtures, WO/2009/104960, 2009.

126. D.H. Olson, M.A. Camblor, L.A. Villaescusa, and G.H. Kuehl, Light hydrocarbon sorption properties of pure silica Si-CHA and ITQ-3 and high silica ZSM-58, *Microporous and Mesoporous Materials* 67 (2004) 27–33.

127. J.J. Gutiérrez-Sevillano, D. Dubbeldam, F. Rey, S. Valencia, M. Palomino, A. Martín-Calvo, and S. Calero, Analysis of the ITQ-12 zeolite performance in propane—Propylene separations using a combination of experiments and molecular simulations, *Journal of Physical Chemistry C* 114 (2010) 14907–14914.

128. S.U. Rege and R.T. Yang, Propane/propylene separation by pressure swing adsorption: Sorbent comparison and multiplicity of cyclic steady states, *Chemical Engineering Science* 57 (2002) 1139–1149.

129. J. Kim, L.C. Lin, R.L. Martin, J.A. Swisher, M. Haranczyk, and B. Smit, Large-scale computational screening of zeolites for ethane/ethene separation, *Langmuir* 28 (2012) 11914–11919.

130. T.Y. Ralph, *Adsorbents: Fundamentals and Applications*. John Wiley & Sons, Hoboken, NJ, 2003.

131. J. Padin, S.U. Rege, R.T. Yang, and L.S. Cheng, Molecular sieve sorbents for kinetic separation of propane/propylene, *Chemical Engineering Science* 55 (2000) 4525–4535.

132. S. Aguado, G. Bergeret, C. Daniel, and D. Farrusseng, Absolute molecular sieve separation of ethylene/ethane mixtures with silver zeolite A, *Journal of the American Chemical Society* 134 (2012) 14635–14637.

133. S. Hosseinpour, S. Fatemi, Y. Mortazavi, M. Gholamhoseini, and M.T. Ravanchi, Performance of cax zeolite for separation of C2H6, C2H4, and CH4 by adsorption process; capacity, selectivity, and dynamic adsorption measurements, *Separation Science and Technology* 46 (2011) 349–355.

134. Y. Xie, N. Bu, J. Liu, G. Yang, J. Qui, N. Yang, and Y. Tang, Adsorbents for use in the separation of carbon monoxide and/or unsaturated hydrocarbons from mixed gases, US4917711, 1990.

135. B. Mcculloch and J.R. Lansbarkis, Process for separating normal olefins from non-normal olefins, US5276246, 1994.

136. A.G. Albesa, M. Rafti, D.S. Rawat, J.L. Vicente, and A.D. Migone, Ethane/ethylene adsorption on carbon nanotubes: Temperature and size effects on separation capacity, *Langmuir* 28 (2012) 1824–1832.

137. L. Huang and D. Cao, Selective adsorption of olefin-paraffin on diamond-like frameworks: Diamondyne and PAF-302, *Journal of Materials Chemistry A* 1 (2013) 9433–9439.

138. Y. Kinoshita, I. Matsubara, T. Higuchi, and Y. Saito, The crystal structure of bis(adiponitrilo)copper(I) nitrate, *Bulletin of the Chemical Society of Japan* 32 (1959) 1221–1226.

139. E.A. Tomic, Thermal stability of coordination polymers, *Journal of Applied Polymer Science* 9 (1965) 3745–3752.

140. A.B. Aleksandr and N.G. Matveeva, Polymeric chelate compounds, *Russian Chemical Reviews* 29 (1960) 119.

141. B.P. Block, S.H. Rose, C.W. Schauman, E.S. Roth, and J. Simkin, Coordination polymers with inorganic backbones formed by double-bridging of tetrahedral elements [14], *Journal of the American Chemical Society* 84 (1962) 3200–3201.

142. F.W. Knobloch and W.H. Rauscher, Coordination polymers of copper(II) prepared at liquid-liquid interfaces, *Journal of Polymer Science* 38 (1959) 261–262.

143. M. Kubo, M. Kishita, and Y. Kuroda, Polymer molecules involving coordination links in the crystals of cupric oxalate and related compounds, *Journal of Polymer Science* 48 (1960) 467–471.

144. S.R. Batten, B.F. Hoskins, and R. Robson, Two interpenetrating 3D networks which generate spacious sealed-off compartments enclosing of the order of 20 solvent molecules in the structures of Zn(CN)(NO₃)(tpt)2/3·solv (tpt = 2,4,6-tri(4-pyridyl)-1,3,5-triazine, solv = ~3/4C₂H₂C₁₄·3/4CH₃OH or ~3/2CHCl₃·1/3CH₃OH), *Journal of the American Chemical Society* 117 (1995) 5385–5386.

145. B.F. Hoskins and R. Robson, Design and construction of a new class of scaffolding-like materials comprising infinite polymeric frameworks of 3D-linked molecular rods. A reappraisal of the Zn(CN)₂ and Cd(CN)₂ structures and the synthesis and structure of the diamond-related frameworks [N(CH₃)₄] [CuIZnIIKCN₄], *Journal of the American Chemical Society* 112 (1990) 1546–1554.

146. S. Kitagawa, S. Kawata, Y. Nozaka, and M. Munakata, Synthesis and crystal structures of novel copper(I) co-ordination polymers and a hexacopper(I) cluster of quinoline-2-thione, *Journal of the Chemical Society, Dalton Transactions* 9 (1993) 1399–1404.

147. S. Kitagawa, S. Matsuyama, M. Munakata, and T. Emori, Synthesis and crystal structures of novel one-dimensional polymers, [{M(bpen)X}∞] [M = CuI, X = PF6-; M = AgI, X = ClO₄ -; bpen = trans-1,2-bis(2-pyridyl)ethylene] and [{Cu(bpen)(CO)(CH₃CN)(PF6)}∞], *Journal of the Chemical Society, Dalton Transactions* (1991) 2869–2874.

148. O.M. Yaghi and H. Li, Hydrothermal synthesis of a metal—Organic framework containing large rectangular channels, *Journal of the American Chemical Society* 117 (1995) 10401–10402.

149. G.B. Gardner, D. Venkataraman, J.S. Moore, and S. Lee, Spontaneous assembly of a hinged coordination network, *Nature* 374 (1995) 792–795.

150. D. Riou and G. Férey, Hybrid open frameworks (MIL-n). Part 3: Crystal structures of the HT and LT forms of MIL-7: A new vanadium propylenediphosphonate with an open-framework. Influence of the synthesis temperature on the oxidation state of vanadium within the same structural type, *Journal of Materials Chemistry* 8 (1998) 2733–2735.

151. J. Gascon and F. Kapteijn, Metal-organic framework membranes-high potential, bright future? *Angewandte Chemie—International Edition* 49 (2010) 1530–1532.

152. O.K. Farha, I. Eryazici, N.C. Jeong, B.G. Hauser, C.E. Wilmer, A.A. Sarjeant, R.Q. Snurr, S.T. Nguyen, A.Ö. Yazaydın, and J.T. Hupp, Metal–organic framework materials with ultrahigh surface areas: Is the sky the limit? *Journal of the American Chemical Society* 134 (2012) 15016–15021.

153. J.A. Greathouse and M.D. Allendorf, The interaction of water with MOF-5 simulated by molecular dynamics, *Journal of the American Chemical Society* 128 (2006) 10678–10679.

154. S.S. Kaye, A. Dailly, O.M. Yaghi, and J.R. Long, Impact of preparation and handling on the hydrogen storage properties of Zn₄O(1,4-benzenedicarboxylate)3 (MOF-5), *Journal of the American Chemical Society* 129 (2007) 14176–14177.

155. V. Colombo, S. Galli, H.J. Choi, G.D. Han, A. Maspero, G. Palmisano, N. Masciocchi, and J.R. Long, High thermal and chemical stability in pyrazolate-bridged metal-organic frameworks with exposed metal sites, *Chemical Science* 2 (2011) 1311–1319.

156. K.S. Park, Z. Ni, A.P. Cote, J.Y. Choi, R. Huang, F.J. Uribe-Romo, H.K. Chae, M. O'Keeffe, and O.M. Yaghi, Exceptional chemical and thermal stability of zeolitic imidazolate frameworks, *Proceedings of the National Academy of Sciences of the United States of America* 103 (2006) 10186–10191.

157. A. Czaja, E. Leung, N. Trukhan, and U. Müller, Industrial MOF synthesis, in: D. Farrusseng, (ed.), *Metal-Organic Frameworks*, Wiley-VCH Verlag GmbH & Co. KGaA, Weinheim, Germany. pp. 337–352.

158. M.G. Plaza, A.F.P. Ferreira, J.C. Santos, A.M. Ribeiro, U. Müller, N. Trukhan, J.M. Loureiro, and A.E. Rodrigues, Propane/propylene separation by adsorption using shaped copper trimesate MOF, *Microporous and Mesoporous Materials* 157 (2012) 101–111.

159. U. Mueller, M. Schubert, F. Teich, H. Puetter, K. Schierle-Arndt, and J. Pastré, Metal-organic frameworks—Prospective industrial applications, *Journal of Materials Chemistry* 16 (2006) 626–636.

160. F.X. Llabres i Xamena and J. Gascon, Chapter 14 Towards future MOF catalytic applications, in: F.L.i. Xamena and J. Gascon, (eds.), *Metal Organic Frameworks as Heterogeneous Catalysts*, The Royal Society of Chemistry. pp. 406–424.

161. J.-R. Li, R.J. Kuppler, and H.-C. Zhou, Selective gas adsorption and separation in metal-organic frameworks, *Chemical Society Reviews* 38 (2009) 1477–1504.

162. J.-R. Li, J. Sculley, and H.-C. Zhou, Metal–organic frameworks for separations, *Chemical Reviews* 112 (2011) 869–932.

163. R.Q. Snurr, J.T. Hupp, and S.T. Nguyen, Prospects for nanoporous metal-organic materials in advanced separations processes, *AIChE Journal* 50 (2004) 1090–1095.

164. A.U. Czaja, N. Trukhan, and U. Muller, Industrial applications of metal-organic frameworks, *Chemical Society Reviews* 38 (2009) 1284–1293.

165. G. Fèrey, Hybrid porous solids: Past, present, future, *Chemical Society Reviews* 37 (2008) 191–214.

166. H. Wu, Q. Gong, D.H. Olson, and J. Li, Commensurate adsorption of hydrocarbons and alcohols in microporous metal organic frameworks, *Chemical Reviews* 112 (2012) 836–868.

167. M. He, J. Yao, L. Li, K. Wang, F. Chen, and H. Wang, Synthesis of zeolitic imidazolate framework-7 in a water/ethanol mixture and its ethanol-induced reversible phase transition, *ChemPlusChem* 78 (2013) 1222–1225.

168. K. Kida, M. Okita, K. Fujita, S. Tanaka, and Y. Miyake, Formation of high crystalline ZIF-8 in an aqueous solution, *CrystEngComm* 15 (2013) 1794–1801.

169. J. Liu, F. Zhang, X. Zou, G. Yu, N. Zhao, S. Fan, and G. Zhu, Environmentally friendly synthesis of highly hydrophobic and stable MIL-53 MOF nanomaterials, *Chemical Communications* 49 (2013) 7430–7432.

170. E. Stavitski, M. Goesten, J. Juan-Alcañiz, A. Martinez-Joaristi, P. Serra-Crespo, A.V. Petukhov, J. Gascon, and F. Kapteijn, Kinetic control of metal–organic framework crystallization investigated by time-resolved in situ x-ray scattering, *Angewandte Chemie International Edition* 50 (2011) 9624–9628.

171. A. Martinez Joaristi, J. Juan-Alcañiz, P. Serra-Crespo, F. Kapteijn, and J. Gascon, Electrochemical synthesis of some archetypical Zn^{2+}, Cu^{2+}, and Al^{3+} metal organic frameworks *Crystal Growth and Design* 12 (2012) 3489–3498.

172. D.J. Tranchemontagne, J.R. Hunt, and O.M. Yaghi, Room temperature synthesis of metal-organic frameworks: MOF-5, MOF-74, MOF-177, MOF-199, and IRMOF-0 *Tetrahedron* 64 (2008) 8553–8557.

173. C.E. Wilmer, O.K. Farha, T. Yildirim, I. Eryazici, V. Krungleviciute, A.A. Sarjeant, R.Q. Snurr, and J.T. Hupp, Gram-scale, high-yield synthesis of a robust metal-organic framework for storing methane and other gases, *Energy & Environmental Science* 6 (2013) 1158–1163.

174. J.R. Karra and K.S. Walton, Effect of open metal sites on adsorption of polar and nonpolar molecules in metal-organic framework Cu-BTC, *Langmuir* 24 (2008) 8620–8626.

175. Q. Min Wang, D. Shen, M. Bülow, M. Ling Lau, S. Deng, F.R. Fitch, N.O. Lemcoff, and J. Semanscin, Metallo-organic molecular sieve for gas separation and purification, *Microporous and Mesoporous Materials* 55 (2002) 217–230.

176. N. Lamia, M. Jorge, M.A. Granato, F.A. Almeida Paz, H. Chevreau, and A.E. Rodrigues, Adsorption of propane, propylene and isobutane on a metal–organic framework: Molecular simulation and experiment, *Chemical Engineering Science* 64 (2009) 3246–3259.

177. E. García-Pérez, J. Gascón, V. Morales-Flórez, J.M. Castillo, F. Kapteijn, and S. Calero, Identification of adsorption sites in Cu-BTC by experimentation and molecular simulation, *Langmuir* 25 (2009) 1725–1731.

178. M. Rubeš, A.D. Wiersum, P.L. Llewellyn, L. Grajciar, O. Bludský, and P. Nachtigall, Adsorption of propane and propylene on CuBTC metal–Organic framework: Combined theoretical and experimental investigation, *The Journal of Physical Chemistry C* 117 (2013) 11159–11167.

179. M. Hartmann, D. Himsl, S. Kunz, and O. Tangermann, Olefin/paraffin separation over the metal organic framework material Cu$_3$(BTC)2, in: P.M. Antoine Gédéon, and B. Florence, (eds.), *Studies in Surface Science and Catalysis*, Elsevier B. V., Amsterdam, the Netherlands. pp. 615–618.

180. M. Hartmann, S. Kunz, D. Himsl, O. Tangermann, S. Ernst, and A. Wagener, Adsorptive separation of isobutene and isobutane on Cu$_3$(BTC)2, *Langmuir* 24 (2008) 8634–8642.

181. A. Wagener, M. Schindler, F. Rudolphi, and S. Ernst, Metallorganische Koordinationspolymere zur adsorptiven Trennung von Propan/Propen-Gemischen, *Chemie Ingenieur Technik* 79 (2007) 851–855.

182. J.W. Yoon, I.T. Jang, K.Y. Lee, Y.K. Hwang, and J.S. Chang, Adsorptive separation of propylene and propane on a porous metal-organic framework, copper trimesate, *Bulletin of the Korean Chemical Society* 31 (2010) 220–223.

183. Metal Organic Framework Samples, http://www.catalysts.basf.com/p02/USWeb-Internet/catalysts/en/content/microsites/catalysts/prods-inds/energy-storage/MOF-samples (Last accessed: November 2013).

184. M.G. Plaza, A.M. Ribeiro, A. Ferreira, J.C. Santos, U.H. Lee, J.-S. Chang, J.M. Loureiro, and A.E. Rodrigues, Propylene/propane separation by vacuum swing adsorption using Cu-BTC spheres, *Separation and Purification Technology* 90 (2012) 109–119.

185. Y.-S. Bae, C.Y. Lee, K.C. Kim, O.K. Farha, P. Nickias, J.T. Hupp, S.T. Nguyen, and R.Q. Snurr, High Propene/Propane Selectivity in Isostructural Metal–Organic Frameworks with High Densities of Open Metal Sites, *Angewandte Chemie International Edition* 51 (2012) 1857–1860.

186. Z. Bao, L. Yu, Q. Ren, X. Lu, and S. Deng, Adsorption of CO$_2$ and CH$_4$ on a magnesium-based metal organic framework, *Journal of Colloid and Interface Science* 353 (2011) 549–556.

187. Y. He, R. Krishna, and B. Chen, Metal-organic frameworks with potential for energy-efficient adsorptive separation of light hydrocarbons, *Energy & Environmental Science* 5 (2012) 9107–9120.

188. S. Kitagawa and K. Uemura, Dynamic porous properties of coordination polymers inspired by hydrogen bonds, *Chemical Society Reviews* 34 (2005) 109–119.

189. D. Li and K. Kaneko, Hydrogen bond-regulated microporous nature of copper complex-assembled microcrystals, *Chemical Physics Letters* 335 (2001) 50–56.

190. S.R. Venna, J.B. Jasinski, and M.A. Carreon, Structural evolution of zeolitic imidazolate framework-8, *Journal of the American Chemical Society* 132 (2010) 18030–18033.

191. H. Bux, F. Liang, Y. Li, J. Cravillon, M. Wiebcke, and J. Caro, Zeolitic imidazolate framework membrane with molecular sieving properties by microwave-assisted solvothermal synthesis, *Journal of the American Chemical Society* 131 (2009) 16000–16001.

192. H. Bux, C. Chmelik, J.M. Van Baten, R. Krishna, and J. Caro, Novel MOF-membrane for molecular sieving predicted by IR-diffusion studies and molecular modeling, *Advanced Materials* 22 (2010) 4741–4743.

193. S.A. Moggach, T.D. Bennett, and A.K. Cheetham, The effect of pressure on ZIF-8: Increasing pore size with pressure and the formation of a high-pressure phase at 1.47 GPa *Angewandte Chemie—International Edition* 48 (2009) 7087–7089.

194. T.H. Bae, J.S. Lee, W. Qiu, W.J. Koros, C.W. Jones, and S. Nair, A high-performance gas-separation membrane containing submicrometer-sized metal-organic framework crystals, *Angewandte Chemie—International Edition* 49 (2010) 9863–9866.

195. A. Huang, W. Dou, and J. Caro, Steam-stable zeolitic imidazolate framework ZIF-90 membrane with hydrogen selectivity through covalent functionalization, *Journal of the American Chemical Society* 132 (2010) 15562–15564.

196. Y. Li, F. Liang, H. Bux, W. Yang, and J. Caro, Zeolitic imidazolate framework ZIF-7 based molecular sieve membrane for hydrogen separation, *Journal of Membrane Science* 354 (2010) 48–54.

197. Y.S. Li, F.Y. Liang, H. Bux, A. Feldhoff, W.S. Yang, and J. Caro, Molecular sieve membrane: Supported metal-organic framework with high hydrogen selectivity *Angewandte Chemie—International Edition* 49 (2010) 548–551.

198. S.C.S. Reyes, Jose G. , Z. Ni, C.S. Paur, P. Kortunov, J. Zengel, and H.W. Deckman, Separation of hydrogen from hydrocarbons utilizing zeolitic imidazolate framework materials, US20090211440, 2009.

199. C. Gücüyener, J. Van Den Bergh, J. Gascon, and F. Kapteijn, Ethane/ethene separation turned on its head: Selective ethane adsorption on the metal-organic framework ZIF-7 through a gate-opening mechanism, *Journal of the American Chemical Society* 132 (2010) 17704–17706.

200. J. Van Den Bergh, C. Gücüyener, E.A. Pidko, E.J.M. Hensen, J. Gascon, and F. Kapteijn, Understanding the anomalous alkane selectivity of ZIF-7 in the separation of light alkane/alkene mixtures, *Chemistry—A European Journal* 17 (2011) 8832–8840.

201. D. Fairen-Jimenez, S.A. Moggach, M.T. Wharmby, P.A. Wright, S. Parsons, and T. Düren, Opening the Gate: Framework flexibility in ZIF-8 explored by experiments and simulations, *Journal of the American Chemical Society* 133 (2011) 8900–8902.

202. D. Fairen-Jimenez, R. Galvelis, A. Torrisi, A.D. Gellan, M.T. Wharmby, P.A. Wright, C. Mellot-Draznieks, and T. Düren, Flexibility and swing effect on the adsorption of energy-related gases on ZIF-8: Combined experimental and simulation study, *Dalton Transactions* 41 (2012) 10752–10762.

203. L. Zhang, Z. Hu, and J. Jiang, Sorption-induced structural transition of zeolitic imidazolate framework-8: A hybrid molecular simulation study, *Journal of the American Chemical Society* 135 (2013) 3722–3728.

204. B. Zheng, Y. Pan, Z. Lai, and K.W. Huang, Molecular dynamics simulations on gate opening in ZIF-8: Identification of factors for ethane and propane separation, *Langmuir* 29 (2013) 8865–8872.

205. A.F.P. Ferreira, M.C. Mittelmeijer-Hazeleger, M.A. Granato, V.F.D. Martins, A.E. Rodrigues, and G. Rothenberg, Sieving di-branched from mono-branched and linear alkanes using ZIF-8: Experimental proof and theoretical explanation, *Physical Chemistry Chemical Physics* 15 (2013) 8795–8804.

206. U. Böhme, C. Paula, V.R. Reddy Marthala, J. Caro, and M. Hartmann, Exceptional adsorption and separation properties of the molecular sieve ZIF-8, *Ungewöhnliche Adsorptions- und Trenneigenschaften des Molekularsiebs ZIF-8* 85 (2013) 1707–1713.

207. C. Chmelik, D. Freude, H. Bux, and J. Haase, Ethene/ethane mixture diffusion in the MOF sieve ZIF-8 studied by MAS PFG NMR diffusometry, *Microporous and Mesoporous Materials* 147 (2012) 135–141.

208. A. Tavolaro and E. Drioli, Zeolite membranes, *Advanced Materials* 11 (1999) 975–996.

209. M. Mulder, *Basic Principles of Membrane Technology*. Springer, Dordrecht, the Netherlands, 1996.

210. P. Aptel, J. Armor, R. Audinos, R.W. Baker, R. Bakish, G. Belfort, B. Bikson et al., Terminology for membranes and membrane processes (IUPAC Recommendations 1996), *Journal of Membrane Science* 120 (1996) 149–159.

211. W.J. Koros and R. Mahajan, Pushing the limits on possibilities for large scale gas separation: Which strategies? *Journal of Membrane Science* 175 (2000) 181–196.
212. M.C. Den Exter. 1996. Explatory study of the synthesis and properties of 6-, 8- and 10-ring tectosilicates and their potential application in zeolite membranes. In Chemical Engineering. Technical University of Delft, Delft, the Netherlands. pp. 1–223.
213. P. Gorgojo, Ó. de la Iglesia, and J. Coronas, Preparation and characterization of zeolite membranes, in: M. Reyes and M. Miguel, (eds.), *Membrane Science and Technology*, Elsevier B. V., Amsterdam, the Netherlands. pp. 135–175.
214. R.W. Baker and K. Lokhandwala, Natural gas processing with membranes: An overview, *Industrial and Engineering Chemistry Research* 47 (2008) 2109–2121.
215. S. Adhikari and S. Fernando, Hydrogen membrane separation techniques, *Industrial and Engineering Chemistry Research* 45 (2006) 875–881.
216. T.M. Nenoff and J. Dong, Highly selective zeolite membranes, in: V. Valtchev, S. Mintova, and M. Tsapatsis, (Eds.), *Ordered Porous Solids*, Elsevier B. V., Amsterdam. 365–386.
217. J.D. Perry, K. Nagai, and W.J. Koros, Polymer membranes for hydrogen separations, *MRS Bulletin* 31 (2006) 745–749.
218. L.M. Robeson, Correlation of separation factor versus permeability for polymeric membranes, *Journal of Membrane Science* 62 (1991) 165–185.
219. L.M. Robeson, The upper bound revisited, *Journal of Membrane Science* 320 (2008) 390–400.
220. D.F. Sanders, Z.P. Smith, R. Guo, L.M. Robeson, J.E. McGrath, D.R. Paul, and B.D. Freeman, Energy-efficient polymeric gas separation membranes for a sustainable future: A review, *Polymer (United Kingdom)* 54 (2013) 4729–4761.
221. L. Shao, B.T. Low, T.S. Chung, and A.R. Greenberg, Polymeric membranes for the hydrogen economy: Contemporary approaches and prospects for the future, *Journal of Membrane Science* 327 (2009) 18–31.
222. N. Du, H.B. Park, M.M. Dal-Cin, and M.D. Guiver, Advances in high permeability polymeric membrane materials for CO$_2$ separations, *Energy & Environmental Science* 5 (2012) 7306–7322.
223. T.C. Merkel, R. Blanc, I. Ciobanu, B. Firat, A. Suwarlim, and J. Zeid, Silver salt facilitated transport membranes for olefin/paraffin separations: Carrier instability and a novel regeneration method, *Journal of Membrane Science* 447 (2013) 177–189.
224. L.C. Tomé, D. Mecerreyes, C.S.R. Freire, L.P.N. Rebelo, and I.M. Marrucho, Polymeric ionic liquid membranes containing IL-Ag+ for ethylene/ethane separation via olefin-facilitated transport, *Journal of Materials Chemistry A* 2 (2014) 5631–5639.
225. R. Nasir, H. Mukhtar, Z. Man, and D.F. Mohshim, Material advancements in fabrication of mixed-matrix membranes, *Chemical Engineering and Technology* 36 (2013) 717–727.
226. H.B. Tanh Jeazet, C. Staudt, and C. Janiak, Metal-organic frameworks in mixed-matrix membranes for gas separation, *Dalton Transactions* 41 (2012) 14003–14027.
227. M.G. Buonomenna, W. Yave, and G. Golemme, Some approaches for high performance polymer based membranes for gas separation: Block copolymers, carbon molecular sieves and mixed matrix membranes, *RSC Advances* 2 (2012) 10745–10773.
228. J. Gascon, F. Kapteijn, B. Zornoza, V. Sebastián, C. Casado, and J. Coronas, Practical approach to zeolitic membranes and coatings: State of the art, opportunities, barriers, and future perspectives, *Chemistry of Materials* 24 (2012) 2829–2844.
229. M. Shah, M.C. McCarthy, S. Sachdeva, A.K. Lee, and H.-K. Jeong, Current status of metal–organic framework membranes for gas separations: Promises and challenges, *Industrial & Engineering Chemistry Research* 51 (2011) 2179–2199.
230. R.L. Burns, and W.J. Koros, Defining the challenges for C$_3$H$_6$/C$_3$H$_8$ separation using polymeric membranes, *Journal of Membrane Science* 211 (2003) 299–309.

231. L.D. Pollo, L.T. Duarte, M. Anacleto, A.C. Habert, and C.P. Borges, Polymeric membranes containing silver salts for propylene/propane separation *Brazilian, Journal of Chemical Engineering* 29 (2012) 307–314.

232. U. Cakal, L. Yilmaz, and H. Kalipcilar, Effect of feed gas composition on the separation of CO_2/CH_4 mixtures by PES-SAPO 34-HMA mixed matrix membranes, *Journal of Membrane Science* 417–418 (2012) 45–51.

233. E. Karatay, H. Kalipçilar, and L. Yilmaz, Preparation and performance assessment of binary and ternary PES-SAPO 34-HMA based gas separation membranes, *Journal of Membrane Science* 364 (2010) 75–81.

234. E.E. Oral, L. Yilmaz, and H. Kalipcilar, Effect of gas permeation temperature and annealing procedure on the performance of binary and ternary mixed matrix membranes of polyethersulfone, SAPO-34, and 2-hydroxy 5-methyl aniline. *Journal of Applied Polymer Science* 131 (2014) 8498–8505.

235. P. Burmann, B. Zornoza, C. Téllez, and J. Coronas, Mixed matrix membranes comprising MOFs and porous silicate fillers prepared via spin coating for gas separation, *Chemical Engineering Science* 107 (2014) 66–75.

236. T. Rodenas, M. Van Dalen, E. García-Pérez, P. Serra-Crespo, B. Zornoza, F. Kapteijn, and J. Gascon, Visualizing MOF mixed matrix membranes at the nanoscale: Towards structure-performance relationships in CO_2/CH_4 separation over NH_2-MIL-53(Al)@PI, *Advanced Functional Materials* 24 (2014) 249–256.

237. M. Valero, B. Zornoza, C. Téllez, and J. Coronas, Mixed matrix membranes for gas separation by combination of silica MCM-41 and MOF NH2-MIL-53(Al) in glassy polymers, *Microporous and Mesoporous Materials* 192 (2013) 23–28.

238. B. Zornoza, A. Martinez-Joaristi, P. Serra-Crespo, C. Tellez, J. Coronas, J. Gascon, and F. Kapteijn, Functionalized flexible MOFs as fillers in mixed matrix membranes for highly selective separation of CO_2 from CH_4 at elevated pressures, *Chemical Communications* 47 (2011) 9522–9524.

239. B. Zornoza, C. Tellez, J. Coronas, J. Gascon, and F. Kapteijn, Metal organic framework based mixed matrix membranes: An increasingly important field of research with a large application potential, *Microporous and Mesoporous Materials* 166 (2013) 67–78.

240. T. Rodenas, M. van Dalen, P. Serra-Crespo, F. Kapteijn, and J. Gascon, Mixed matrix membranes based on NH_2-functionalized MIL-type MOFs: Influence of structural and operational parameters on the CO_2/CH_4 separation performance, *Microporous and Mesoporous Materials* 192 (2013) 35–42.

241. C. Zhang, Y. Dai, J.R. Johnson, O. Karvan, and W.J. Koros, High performance ZIF-8/6FDA-DAM mixed matrix membrane for propylene/propane separations, *Journal of Membrane Science* 389 (2012) 34–42.

242. L. Xu, M. Rungta, M.K. Brayden, M.V. Martinez, B.A. Stears, G.A. Barbay, and W.J. Koros, Olefins-selective asymmetric carbon molecular sieve hollow fiber membranes for hybrid membrane-distillation processes for olefin/paraffin separations, *Journal of Membrane Science* 423–424 (2012) 314–323.

243. A. Motelica, O.S.L. Bruinsma, R. Kreiter, M. den Exter, and J.F. Vente, Membrane retrofit option for paraffin/olefin separation—A technoeconomic evaluation, *Industrial & Engineering Chemistry Research* 51 (2012) 6977–6986.

244. M. Pera-Titus, Porous inorganic membranes for CO_2 capture: Present and prospects, *Chemical Reviews* 114 (2013) 1413–1492.

245. E.V. Perez, K.J. Balkus Jr, J.P. Ferraris, and I.H. Musselman, Mixed-matrix membranes containing MOF-5 for gas separations, *Journal of Membrane Science* 328 (2009) 165–173.

246. M.J.C. Ordoñez, K.J. Balkus Jr, J.P. Ferraris, and I.H. Musselman, Molecular sieving realized with ZIF-8/Matrimid® mixed-matrix membranes, *Journal of Membrane Science* 361 (2010) 28–37.

247. T. Yang, G.M. Shi, and T.S.C. Chung, Symmetric and asymmetric zeolitic imidazolate frameworks (ZIFs)/Polybenzimidazole (PBI) nanocomposite membranes for hydrogen purifi cation at high temperatures, *Advanced Energy Materials* 2 (2012) 1358–1367.

248. T. Yang, Y. Xiao, and T.S. Chung, Poly-/metal-benzimidazole nano-composite membranes for hydrogen purification, *Energy and Environmental Science* 4 (2011) 4171–4180.

249. H. Guo, G. Zhu, I.J. Hewitt, and S. Qiu, "Twin Copper Source" growth of metal–organic framework membrane: $Cu_3(BTC)_2$ with high permeability and selectivity for recycling H_2, *Journal of the American Chemical Society* 131 (2009) 1646–1647.

250. J. Gascon, S. Aguado, and F. Kapteijn, Manufacture of dense coatings of $Cu_3(BTC)_2$ (HKUST-1) on α-alumina, *Microporous and Mesoporous Materials* 113 (2008) 132–138.

251. S. Zhou, X. Zou, F. Sun, F. Zhang, S. Fan, H. Zhao, T. Schiestel, and G. Zhu, Challenging fabrication of hollow ceramic fiber supported $Cu_3(BTC)_2$ membrane for hydrogen separation, *Journal of Materials Chemistry* 22 (2012) 10322–10328.

252. D.J. Lee, Q. Li, H. Kim, and K. Lee, Preparation of Ni-MOF-74 membrane for CO_2 separation by layer-by-layer seeding technique, *Microporous and Mesoporous Materials* 163 (2012) 169–177.

253. R. Ranjan, and M. Tsapatsis, Microporous metal organic framework membrane on porous support using the seeded growth method, *Chemistry of Materials* 21 (2009) 4920–4924.

254. A. Huang, H. Bux, F. Steinbach, and J. Caro, Molecular-sieve membrane with hydrogen permselectivity: ZIF-22 in LTA topology prepared with 3-aminopropyltriethoxysilane as covalent linker, *Angewandte Chemie—International Edition* 49 (2010) 4958–4961.

255. Y.S. Li, H. Bux, A. Feldhoff, G.N. Li, W.S. Yang, and J. Caro, Controllable synthesis of metal-organic frameworks: From MOF nanorods to oriented MOF membranes, *Advanced Materials* 22 (2010) 3322–3326.

256. A. Huang, and J. Caro, Covalent post-functionalization of zeolitic imidazolate framework ZIF-90 membrane for enhanced hydrogen selectivity, *Angewandte Chemie International Edition* 50 (2011) 4979–4982.

257. A. Huang, Y. Chen, Q. Liu, N. Wang, J. Jiang, and J. Caro, Synthesis of highly hydrophobic and permselective metal–organic framework Zn(BDC)(TED)0.5 membranes for H_2/CO_2 separation, *Journal of Membrane Science* 454 (2014) 126–132.

258. Z. Xie, J. Yang, J. Wang, J. Bai, H. Yin, B. Yuan, J. Lu, Y. Zhang, L. Zhou, and C. Duan, Deposition of chemically modified [small alpha]-Al_2O_3 particles for high performance ZIF-8 membrane on a macroporous tube. *Chemical Communications* 48 (2012) 5977–5979.

259. F. Zhang, X. Zou, X. Gao, S. Fan, F. Sun, H. Ren, and G. Zhu, Hydrogen selective NH2-MIL-53(Al) MOF membranes with high permeability, *Advanced Functional Materials* 22 (2012) 3583–3590.

260. J. Dong, Y.S. Lin, M. Kanezashi, and Z. Tang, Microporous inorganic membranes for high temperature hydrogen purification, *Journal of Applied Physics* 104 (2008) 121301.

261. J.W. Phair, and S.P.S. Badwal, Materials for separation membranes in hydrogen and oxygen production and future power generation, *Science and Technology of Advanced Materials* 7 (2006) 792–805.

262. A. Basile, A. Iulianelli, T. Longo, S. Liguori, and M. Falco, Pd-based selective membrane state-of-the-art, in: M. De Falco, L. Marrelli, and G. Iaquaniello, (Eds.), *Membrane Reactors for Hydrogen Production Processes*, Springer, London, U.K. pp. 21–55.

263. O. Hatlevik, S.K. Gade, M.K. Keeling, P.M. Thoen, A.P. Davidson, and J.D. Way, Palladium and palladium alloy membranes for hydrogen separation and production: History, fabrication strategies, and current performance, *Separation and Purification Technology* 73 (2010) 59–64.

264. A.G. Knapton, Palladium alloys for hydrogen diffusion membranes, *Platinum Metals Review* 21 (1977) 44–50.

265. S.N. Paglieri, and J.D. Way, Innovations in palladium membrane research, *Separation and Purification Methods* 31 (2002) 1–169.

266. T. Norby, and R. Haugsrud, Dense ceramic membranes for hydrogen separation, in: A.F. Sammells, and M.V. Mundschau, (Eds.), *Nonporous Inorganic Membranes*, Wiley-VCH Verlag GmbH & Co. KGaA, Weinheim, Germany. pp. 1–48.

267. T. Pietraß, Carbon-based membranes, *MRS Bulletin* 31 (2006) 765–769.

268. B. Zhang, Y. Shi, Y. Wu, T. Wang, and J. Qiu, Preparation and characterization of supported ordered nanoporous carbon membranes for gas separation, *Journal of Applied Polymer Science* 131 (2013) 39925.

269. J. Caro, and M. Noack, Chapter 1—Zeolite membranes—Status and prospective, in: E. Stefan, (Ed.), *Advances in Nanoporous Materials*, Elsevier B. V., Amsterdam, the Netherlands. pp. 1–96.

270. S.J. Khatib and S.T. Oyama, Silica membranes for hydrogen separation prepared by chemical vapor deposition (CVD), *Separation and Purification Technology* 111 (2013) 20–42.

271. Q. Wei, F. Wang, Z.R. Nie, C.L. Song, Y.L. Wang, and Q.Y. Li, Highly hydrothermally stable microporous silica membranes for hydrogen separation, *Journal of Physical Chemistry B* 112 (2008) 9354–9359.

272. Z.F. Hong, Q. Wei, G.H. Li, X.W. Wang, Z.R. Nie, and Q.Y. Li, Preparation, H_2 separation and hydrothermal stability of trifluoropropyl-modified silica membranes, *Chinese Journal of Inorganic Chemistry* 29 (2013) 941–947.

273. H. Qi, H. Chen, L. Li, G. Zhu, and N. Xu, Effect of Nb content on hydrothermal stability of a novel ethylene-bridged silsesquioxane molecular sieving membrane for H_2/CO_2 separation, *Journal of Membrane Science* 421–422 (2012) 190–200.

274. V. Boffa, G. Magnacca, L.B. Jørgensen, A. Wehner, A. Dörnhöfer, and Y. Yue, Toward the effective design of steam-stable silica-based membranes, *Microporous and Mesoporous Materials* 179 (2013) 242–249.

275. H.L. Castricum, A. Sah, R. Kreiter, D.H.A. Blank, J.F. Vente, and J.E. ten Elshof, Hybrid ceramic nanosieves: Stabilizing nanopores with organic links, *Chemical Communications* (2008) 1103–1105.

276. M. Kanezashi, K. Yada, T. Yoshioka, and T. Tsuru, Design of silica networks for development of highly permeable hydrogen separation membranes with hydrothermal stability, *Journal of the American Chemical Society* 131 (2009) 414–415.

277. R. Xu, J. Wang, M. Kanezashi, T. Yoshioka, and T. Tsuru, Development of robust organosilica membranes for reverse osmosis, *Langmuir* 27 (2011) 13996–13999.

278. http://www.hybsi.com/home/ (Last accessed: December 2013).

279. H.M. van Veen, M.D.A. Rietkerk, D.P. Shanahan, M.M.A. van Tuel, R. Kreiter, H.L. Castricum, J.E. ten Elshof, and J.F. Vente, Pushing membrane stability boundaries with HybSi® pervaporation membranes, *Journal of Membrane Science* 380 (2011) 124–131.

280. M. Matsuka, R. Braddock, and I. Agranovski, Numerical study of hydrogen permeation flux in SrCe0.95Yb0.05O3-α and SrCe0.95Tm0.05O3-α, *Solid State Ionics* 178 (2007) 1011–1019.

281. X. Meng, J. Song, N. Yang, B. Meng, X. Tan, Z.-F. Ma, and K. Li, Ni–BaCe0.95Tb0.05O3–δ cermet membranes for hydrogen permeation, *Journal of Membrane Science* 401–402 (2012) 300–305.

282. S. Zhan, X. Zhu, W. Wang, F. Huang, and W. Yang, Permeation properties and stability of Ni-BaCe 0.4Zr 0.4Nd 0.2O 3-δ membrane for hydrogen separation 512–515 (2012) 1422–1425.

283. U. Balachandran, T.H. Lee, L. Chen, S.J. Song, J.J. Picciolo, and S.E. Dorris, Hydrogen separation by dense cermet membranes, *Fuel* 85 (2006) 150–155.

284. O. Hatlevik, S.K. Gade, M.K. Keeling, P.M. Thoen, A.P. Davidson, and J.D. Way, Palladium and palladium alloy membranes for hydrogen separation and production: History, fabrication strategies, and current performance, *Separation and Purification Technology* 73 (2010) 59–64.

285. P. Kamakoti, and D.S. Sholl, A comparison of hydrogen diffusivities in Pd and CuPd alloys using density functional theory, *Journal of Membrane Science* 225 (2003) 145–154.

286. A. Kulprathipanja, G.O. Alptekin, J.L. Falconer, and J.D. Way, Pd and Pd–Cu membranes: Inhibition of H2 permeation by H2S, *Journal of Membrane Science* 254 (2005) 49–62.

287. H. Gharibi, M. Saadatinasab, and A. Zolfaghari, Hydrogen permeability and sulfur tolerance of a novel dual membrane of PdAg/PdCu layers deposited on porous stainless steel, *Journal of Membrane Science* 447 (2013) 355–361.

288. F. Gallucci, S. Tosti, and A. Basile, Pd-Ag tubular membrane reactors for methane dry reforming: A reactive method for CO$_2$ consumption and H$_2$ production, *Journal of Membrane Science* 317 (2008) 96–105.

289. A.M. Tarditi, and L.M. Cornaglia, Novel PdAgCu ternary alloy as promising materials for hydrogen separation membranes: Synthesis and characterization, *Surface Science* 605 (2011) 62–71.

290. A.M.A. El Naggar, and G. Akay, Novel intensified catalytic nano-structured nickel–zirconia supported palladium based membrane for high temperature hydrogen production from biomass generated syngas, *International Journal of Hydrogen Energy* 38 (2013) 6618–6632.

291. N. Itoh, E. Suga, and T. Sato, Composite palladium membrane prepared by introducing metallic glue and its high durability below the critical temperature, *Separation and Purification Technology* 121 (2013) 46–53.

292. A.E. Lewis, D.C. Kershner, S.N. Paglieri, M.J. Slepicka, and J.D. Way, Pd–Pt/YSZ composite membranes for hydrogen separation from synthetic water–gas shift streams, *Journal of Membrane Science* 437 (2013) 257–264.

293. A.S.T. Chiang, and K.J. Chao, Membranes and films of zeolite and zeolite-like materials, *Journal of Physics and Chemistry of Solids* 62 (2001) 1899–1910.

294. E.E. McLeary, J.C. Jansen, and F. Kapteijn, Zeolite based films, membranes and membrane reactors: Progress and prospects, *Microporous and Mesoporous Materials* 90 (2006) 198–220.

295. M. Noack, P. Kölsch, J. Caro, M. Schneider, P. Toussaint, and I. Sieber, MFI membranes of different Si/Al ratios for pervaporation and steam permeation, *Microporous and Mesoporous Materials* 35–36 (2000) 253–265.

296. M. Noack, P. Kölsch, V. Seefeld, P. Toussaint, G. Georgi, and J. Caro, Influence of the Si/Al-ratio on the permeation properties of MFI-membranes, *Microporous and Mesoporous Materials* 79 (2005) 329–337.

297. M. Noack, G.T.P. Mabande, J. Caro, G. Georgi, W. Schwieger, P. Kölsch, and A. Avhale, Influence of Si/Al ratio, pre-treatment and measurement conditions on permeation properties of MFI membranes on metallic and ceramic supports, *Microporous and Mesoporous Materials* 82 (2005) 147–157.

298. J. Caro, D. Albrecht, and M. Noack, Why is it so extremely difficult to prepare shape-selective Al-rich zeolite membranes like LTA and FAU for gas separation? *Separation and Purification Technology* 66 (2009) 143–147.

299. H. Verweij, Y.S. Lin, and J. Dong, Microporous silica and zeolite membranes for hydrogen purification, *MRS Bulletin* 31 (2006) 756–764.

300. M. Kanezashi, J. O'Brien-Abraham, Y.S. Lin, and K. Suzuki, Gas permeation through DDR-type zeolite membranes at high temperatures, *AIChE Journal* 54 (2008) 1478–1486.
301. M. Hong, J.L. Falconer, and R.D. Noble, Modification of zeolite membranes for H_2 separation by catalytic cracking of methyldiethoxysilane, *Industrial and Engineering Chemistry Research* 44 (2005) 4035–4041.
302. Z. Tang, J. Dong, and T.M. Nenoff, Internal surface modification of MFI-type zeolite membranes for high selectivity and high flux for hydrogen, *Langmuir* 25 (2009) 4848–4852.
303. Y. Zhang, B. Tokay, H.H. Funke, J.L. Falconer, and R.D. Noble, Template removal from SAPO-34 crystals and membranes, *Journal of Membrane Science* 363 (2010) 29–35.
304. J. Kuhn, J. Gascon, J. Gross, and F. Kapteijn, Detemplation of DDR type zeolites by ozonication, *Microporous and Mesoporous Materials* 120 (2009) 12–18.
305. J. Kuhn, M. Motegh, J. Gross, and F. Kapteijn, Detemplation of [B]MFI zeolite crystals by ozonication, *Microporous and Mesoporous Materials* 120 (2009) 35–38.
306. S. Aguado, J. Gascon, D. Farrusseng, J.C. Jansen, and F. Kapteijn, Simple modification of macroporous alumina supports for the fabrication of dense NaA zeolite coatings: Interplay of electrostatic and chemical interactions, *Microporous and Mesoporous Materials* 146 (2011) 69–75.
307. A. Huang, and J. Caro, Cationic polymer used to capture zeolite precursor particles for the facile synthesis of oriented zeolite LTA molecular sieve membrane, *Chemistry of Materials* 22 (2010) 4353–4355.
308. A. Huang, F. Liang, F. Steinbach, and J. Caro, Preparation and separation properties of LTA membranes by using 3-aminopropyltriethoxysilane as covalent linker, *Journal of Membrane Science* 350 (2010) 5–9.
309. A. Huang, F. Liang, F. Steinbach, T.M. Gesing, and J. Caro, Neutral and cation-free LTA-type aluminophosphate ($AlPO_4$) molecular sieve membrane with high hydrogen permselectivity, *Journal of the American Chemical Society* 132 (2010) 2140–2141.
310. A. Huang, N. Wang, and J. Caro, Synthesis of multi-layer zeolite LTA membranes with enhanced gas separation performance by using 3-aminopropyltriethoxysilane as interlayer, *Microporous and Mesoporous Materials* 164 (2012) 294–301.
311. A. Huang, N. Wang, and J. Caro, Seeding-free synthesis of dense zeolite FAU membranes on 3-aminopropyltriethoxysilane-functionalized alumina supports, *Journal of Membrane Science* 389 (2012) 272–279.
312. Z. Zheng, A.S. Hall, and V.V. Guliants, Synthesis, characterization and modification of DDR membranes grown on α-alumina supports, *Journal of Materials Science* (2008) 1–4.
313. J. van den Bergh, A. Tihaya, and F. Kapteijn, High temperature permeation and separation characteristics of an all-silica DDR zeolite membrane, *Microporous and Mesoporous Materials* 132 (2010) 137–147.
314. H. Wang, X. Dong, and Y.S. Lin, Highly stable bilayer MFI zeolite membranes for high temperature hydrogen separation, *Journal of Membrane Science* 450 (2014) 425–432.
315. Z. Hong, F. Sun, D. Chen, C. Zhang, X. Gu, and N. Xu, Improvement of hydrogen-separating performance by on-stream catalytic cracking of silane over hollow fiber MFI zeolite membrane, *International Journal of Hydrogen Energy* 38 (2013) 8409–8414.
316. A. Huang and J. Caro, Steam-stable hydrophobic ITQ-29 molecular sieve membrane with H2 selectivity prepared by secondary growth using Kryptofix 222 as SDA, *Chemical Communications* 46 (2010) 7748–7750.
317. M. Yu, H.H. Funke, R.D. Noble, and J.L. Falconer, H_2 separation using defect-free, inorganic composite membranes, *Journal of the American Chemical Society* 133 (2011) 1748–1750.

318. G. Guan, T. Tanaka, K. Kusakabe, K.I. Sotowa, and S. Morooka, Characterization of AlPO4-type molecular sieving membranes formed on a porous α-alumina tube, *Journal of Membrane Science* 214 (2003) 191–198.

319. J.K. Das, N. Das, and S. Bandyopadhyay, Highly oriented improved SAPO 34 membrane on low cost support for hydrogen gas separation, *Journal of Materials Chemistry A* 1 (2013) 4966–4973.

320. M. Hong, S. Li, J.L. Falconer, and R.D. Noble, Hydrogen purification using a SAPO-34 membrane, *Journal of Membrane Science* 307 (2008) 277–283.

321. H. Kalipcilar, T.C. Bowen, R.D. Noble, and J.L. Falconer, Synthesis and separation performance of SSZ-13 zeolite membranes on tubular supports, *Chemistry of Materials* 14 (2002) 3458–3464.

322. J.M. Van De Graaf, F. Kapteijn, and J.A. Moulijn, Modeling permeation of binary mixtures through zeolite membranes, *AIChE Journal* 45 (1999) 497–511.

323. L. Wondraczek, G. Gao, D. Möncke, T. Selvam, A. Kuhnt, W. Schwieger, D. Palles, and E.I. Kamitsos, Thermal collapse of SAPO-34 molecular sieve towards a perfect glass, *Journal of Non-Crystalline Solids* 360 (2013) 36–40.

324. J.C. Poshusta, R.D. Noble, and J.L. Falconer, Characterization of SAPO-34 membranes by water adsorption, *Journal of Membrane Science* 186 (2001) 25–40.

325. C. Gucuyener, M.F. De Lange, N. Kosinov, E.J.M. Hensen, J. Gascon, and F. Kapteijn, Modeling permeation of binary gas mixtures across an SSZ-13 zeolite membrane, *Journal of Membrane Science* (in preparation).

326. Y. Zhang, Z. Wu, Z. Hong, X. Gu, and N. Xu, Hydrogen-selective zeolite membrane reactor for low temperature water gas shift reaction, *Chemical Engineering Journal* 197 (2012) 314–321.

327. S.J. Kim, S. Yang, G.K. Reddy, P. Smirniotis, and J. Dong, Zeolite membrane reactor for high-temperature water-gas shift reaction: Effects of membrane properties and operating conditions, *Energy and Fuels* 27 (2013) 4471–4480.

5 Separation of Ethylene/ Ethane Mixtures with Silver Zeolites

Sonia Aguado and David Farrusseng

CONTENTS

5.1 OPPORTUNITIES AND MECHANISMS FOR PARAFFIN/OLEFIN SEPARATION

The continuing rise in crude oil prices and the growing responsibility to reduce the emissions of greenhouse gases resulted in many research projects to find alternative separation processes to fulfill the growing demand of olefins and to reduce the related energy demands. Most attention is paid to absorption-, adsorption-, and membrane-based processes, often aided by a bonding of the olefin, for example, via π-complexation.

Ethylene–ethane separation by adsorption on a solid adsorbent is an energy-efficient alternative. Adsorbents can be classified into two categories depending on separation mechanisms involved: (1) selective adsorption and (2) molecular sieving.

The *selective adsorbents*, which are by far the most studied, rely on the preferential uptake of the olefin. Adsorbents for the separation of an olefin from a paraffin often include high surface area as zeolites, carbons, and resins clays. The selectivity of adsorbents can be improved by the introduction of metal ions that selectively bind to the olefins. An example is the introduction of Cu^+ or Ag^+ ions on the support. Among them, silver-exchanged faujasites, that is, AgX and AgY, have already been extensively studied.[1-8] They form a relatively stable π-complex with the olefin, while the paraffin only adsorbs physically via weak van der Waals forces. To obtain a large number of adsorption sites on the supports, a high dispersion of the metal ions is required.

On the other hand, the *molecular sieving* mechanism relies on the exclusion of one component based on size criteria.[9,10] A porous material, which is only accessible for one of the components in an olefin/paraffin mixture, is not commercially available yet. The development of an adsorbent with molecular sieving property is a challenge for this application due to the small molecular diameter difference that exists between C_2H_4 and C_2H_6, that is, 4.163 and 4.443 Å, respectively.[11]

Many patented processes for olefin/paraffin separation by pressure swing adsorption (PSA) and vacuum swing adsorption (VSA) exist.[12-19] Xie et al. described the use of supports such as zeolite 4A, zeolite X, zeolite Y, alumina, and silica, treated with a copper salt to remove ethane and/or carbon monoxide from ethylene.[1] Yang et al. observed and modeled the separation behavior of ethylene from gaseous mixtures, using copper salts and silver compounds supported alternatively on silica, alumina, MCM-41, zeolite 4A, carbon molecular sieves, and resin Amberlyst-35.[20] Cho et al. reported clay-based adsorbents treated with silver salts for ethylene/ethane separation. However, up to 20% of the olefin is adsorbed in an irreversible manner.[2] Kapteijn and coworkers investigated the use of Li faujasite and CuCl dispersed in NaX crystals for olefin/paraffin separation.[3-5] Copper-modified zeolites 3A and 4A have been used also for the this target separation.[6] Carter et al. found that cadmium- or strontium-substituted type X and type Y zeolites are better than other metal-substituted type X zeolites for the olefin/paraffin separation.[13]

An evolution in pore size control for crystalline molecular sieves began with the discovery of Engelhard titanosilicate-4 (ETS-4).[21] The crystal lattice of ETS-4 systematically contracts upon dehydration at elevated temperatures. These structural changes can be used to control the lattice dimensions and the channel apertures of ETS-4 to *tune* the effective size of the pores. This phenomenon, known as the molecular gate effect, has achieved commercial success in natural gas purification.[22] For example, Sr-ETS-4 contracted titanosilicates (CTS) adsorbents have been applied to difficult size-based separations including N_2/CH_4 (3.64 and 3.76 Å, respectively) on an industrial scale.[22] In 2010, this separation with Na-ETS-10 zeolite modified with a mono-, di-, or trivalent cation and mixtures thereof was patented. Several of the modified ETS-10 materials show improved adsorption capacity, but the selectivity decreased compared with the raw Na-ETS-10 material. Nevertheless, modification with Ba^{2+} and Ba^{2+}/H^+ provided a good balance of selectivity and capacity for the separation of ethylene/ethane mixtures.[23]

However, the contraction process damages the ETS-4 framework and reduces the adsorption capacity.[24] Hence, the control exerted on pore size fine-tuning of robust zeolites by cation exchanges and/or thermal treatment is limited.[21]

5.2 SILVER ZEOLITE 5A AS A PROMISING SEPARATING AGENT

We show that the exceptional paraffin/olefin separation behavior of AgA arises from the combination of preferential adsorption of the olefin over the paraffin and the steric size exclusion of ethane.

Ag ion exchange is performed by exposing the as-received commercial zeolite powders, that is, NaX, NaY, and 5A, to an excess of silver nitrate aqueous solution. One gram of zeolite is treated with 100 mL of a solution of 0.5 M of silver nitrate in water for 5 h, with stirring, at 353 K.

5.2.1 Effect of the Pretreatment Temperature

First, we evaluate the effect of the pretreatment temperature on the subsequent adsorption performance. Howard et al. attributed the discrepancies in ethylene uptake on various silver zeolite A samples to differences in pretreatment conditions.[25] As demonstrated by Gellens et al., the position and oxidation state of the cations depend upon the severity of the dehydration conditions.[26,27] In order to assess these results, the adsorption uptake of ethylene and ethane was evaluated for samples that had been subject to four different pretreatment conditions. Samples were heated up to either 423, 548, and 673 K under a nitrogen flow or 423 K under vacuum. As shown in Figure 5.1, the ethylene uptake is similar for the four selected pretreatment temperatures, with the sample treated under vacuum performing slightly better. In contrast to this similar performance for ethylene uptake, more pronounced differences were observed for the ethane uptake. An increase of the latter for samples pretreated at 673 K occurred. Taking into account these results and considering possible future industrial applications, for successive experiments, heating at 423 K under nitrogen flow was selected as pretreatment condition.

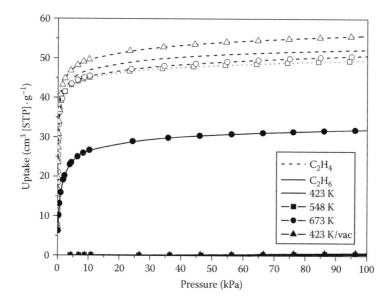

FIGURE 5.1 Adsorption isotherms of ethylene (open symbols) and ethane (close symbols) measured by volumetric method at 303 K on Ag5A at different pretreatment temperatures.

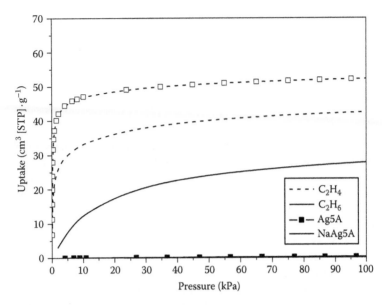

FIGURE 5.2 Adsorption isotherms of ethylene and ethane measured by volumetric method at 303 K on 5A, partially (NaAg5A) and completely Ag exchanged (Ag5A).

5.2.2 EFFECT OF THE Ag LOADING

In similar way, the adsorption performance on partially Ag-exchanged zeolite 5A was assessed, to investigate the disruption of the presence of alkali metal, as claimed by several authors.[28,29] Figure 5.2 illustrates the uptake of ethylene and ethane on zeolite 5A partially and completely Ag exchanged. Clearly, the ethylene uptake is reduced in the case of a partially Ag-exchanged zeolite, associated with an increase of the ethane uptake, leading to an obvious reduction of the paraffin/olefin separation selectivity.

5.2.3 COMPARISON WITH AgX AND AgY

Ethylene and ethane adsorption isotherms on Ag zeolites Ag5A, AgX, and AgY are presented in Figure 5.3. Clearly, ethane adsorption does not occur on Ag5A at the pressure ranges investigated, whereas the gas is substantially adsorbed on AgX and AgY. In contrast, the adsorption of ethylene on silver-exchanged zeolites Ag5A and AgX is quite similar, being a little bit higher on AgY. The adsorption capacity at 100 kPa reaches 2 mol·kg⁻¹, which is among the highest values found for zeolite-based adsorbents and comparable to CuCl dispersed in NaX crystals.[4] Silver-ETS-10 reaches about 1 mol·kg⁻¹ in the same pressure range.[20,22,23] Based on adsorption isotherms measured at different temperatures, we can estimate the enthalpy of ethylene adsorption to be $\Delta H_{303 K} = -95$ kJ·mol⁻¹, which corresponds to other Ag zeolites.[30] The ethylene loading at low pressures is attributed to strong interactions between ethylene and silver clusters inside the zeolite.[31,32] Ethylene adsorption occurs in a similar manner in Ag5A, AgX, and AgY. On the other hand, ethane can diffuse in AgX and AgY, whereas it cannot in Ag5A.

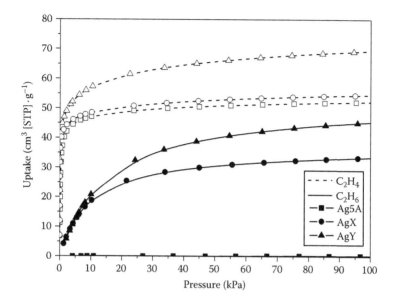

FIGURE 5.3   Adsorption isotherms of ethylene and ethane measured by volumetric method at 303 K on Ag5A, AgX, and AgY.

The ethylene/ethane separation selectivity is infinite for Ag5A, while it is limited to 2.5 for AgX and 2.85 for AgY based on the capacity ratios determined at 10 kPa. Reproducibility experiments were carried out to confirm the extremely limited ethane adsorption on Ag5A. Consecutive single adsorption–desorption isotherms of ethylene and ethane were performed without removal of sample from the measurement cell and without additional thermal treatment between the consecutive measurements. These cycling tests show no changes in the ethylene adsorption capacity of Ag5A and, more importantly, confirm the absence of ethane adsorption; see Figure 5.4.

5.2.4   Performance in Breakthrough Experiments

The outstanding adsorption properties of Ag5A are confirmed with breakthrough experiments carried out using an ethane–ethylene mixture. Figure 5.5 shows the breakthrough curves of a mixture ethylene/ethane/nitrogen (10:10:80) through a column filled with 0.9 g of Ag5A at 303 K and 1 atm. After the first seconds, the mixture is fed; ethane breaks through the Ag5A column. The exclusion of ethane in the Ag5A column is consistent with the absence of ethane uptake as observed in the volumetric measurements. During the first 15 min, the adsorbent could retain ethylene and the outlet stream contained pure ethane. After 25 min, the concentration of the outlet gas reflected the feed gas composition, indicating the saturation of Ag5A. By integration, the amount of ethylene adsorbed was found to be 57 $cm^3 \cdot g^{-1}$, which is in agreement to volumetric measurements.

The lower ethylene/ethane selectivity for AgX and AgY has two consequences for the separation results in the breakthrough experiment. Firstly, we can observe that

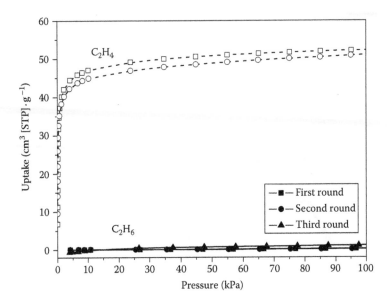

FIGURE 5.4 Adsorption isotherms of ethylene (open symbols) and ethane (close symbols) measured by volumetric method at 303 K on Ag5A for three rounds of experiments.

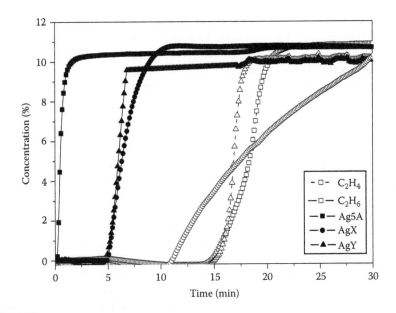

FIGURE 5.5 Breakthrough curves of an ethane (close symbols) and ethylene (open symbols) mixture in nitrogen (10:10:80 kPa), 25 mL (STP)·min^{-1} at 303 K on 0.9 g of Ag5A (ν), AgX (λ), and AgY (π) zeolites.

TABLE 5.1

Free Diameter of 5A and Ag5A Zeolites

	Space Group	Unit Cell Parameter [Å]		Distance [Å]	Free Diameter [nm]
5A	Fm-3c	24.64674(15)	O1–O1	6.939(7)	0.424
			O2–O2	7.545(12)	0.484
Ag5A	Pm-3m	12.29944(16)	O1–O1	7.05(2)	0.435
			O2–O2	7.059(14)	0.436

ethane and ethylene are adsorbed simultaneously in the first minutes. Ethane breaks through the column after 4 min. Secondly, the ethylene capacity is lower since ethylene breaks earlier due to the lower capacity remaining.

The molecular sieving mechanism, for example, size exclusion of ethane, was investigated by powder x-ray diffraction and Ar adsorption on the starting 5A zeolite and silver-exchanged Ag5A.[26,29] Note that, in contrast to Ag5A, ethane and ethylene are adsorbed on zeolite 5A. Rietveld refinement performed on powder x-ray diffractograms of both Ag5A and 5A solids provides free diameter of the channel apertures: 4.24 and 4.84 Å for zeolite 5A and 4.35 and 4.36 Å for zeolite Ag5A. Table 5.1 gives the crystallographic free diameters of the channels obtained considering an oxygen radius of 1.35 Å (hard sphere model).

The free diameter data for Ag5A are clearly larger than the kinetic diameter of ethylene (4.163 Å) and smaller than those of ethane (4.443 Å). As a consequence, ethylene can penetrate in the porous structure of Ag5A, whereas ethane is excluded in perfect agreement with adsorption and separation observations. Furthermore, ethane and ethylene can both diffuse into the porous structure of 5A since their kinetic diameters are smaller than the free diameters measured for 5A. The smaller pore size for Ag5A with respect to 5A has been confirmed by Ar adsorption isotherms at 77 K (Figure 5.6), which shows a decrease of approximately 0.6 Å in the mean diameter. Although silver exchanged on 5A results in a slight decrease of free diameter, it can account for the observed cutoff between ethylene and ethane.

5.2.5 SHAPED ADSORBENTS

Commercial zeolites are generally available in bound forms where the zeolite crystals with dimensions of 1–5 Å are incorporated in regular particle shapes, such as beads, pellets, and extrudates using a binder material such as clay, alumina, and polymers. The purpose of the bound forms, with particle diameters commonly ranging between 0.5 and 6.0 mm, is to reduce the pressure drop in adsorbent industrial columns. The binder phase of the bound particles generally contains a network of meso- and macropores to facilitate transport of the adsorbate molecules from the external gas phase to the mouths of the zeolite crystal pores. Typically, the binder material comprises 10%–20% by weight of the pellet mass. Adsorption of gases on the binder material is generally weak or negligible compared to adsorption on the zeolite.[14] The ion exchange procedure was performed on commercial zeolite 5A

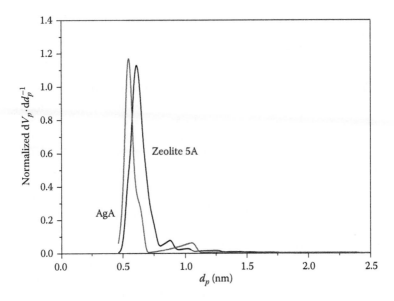

FIGURE 5.6   Pore size distribution.

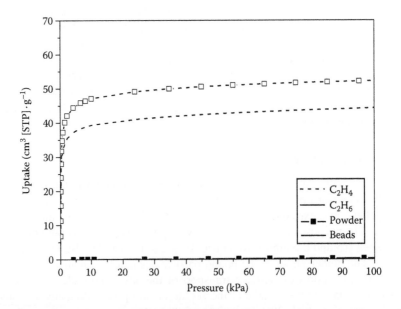

FIGURE 5.7   Adsorption isotherms of ethylene and ethane measured by volumetric method at 303 K on Ag5A in powder and bead shape.

beads, and subsequent adsorption experiments allowed to investigate the impact on ethylene adsorption and prove the total exclusion of ethane, even with the existence of a binder material. Figure 5.7 illustrates the adsorption uptakes of ethylene and ethane on Ag5A beads. It can be seen that, as could be expected due to the presence of a binder material, the ethylene uptake suffers a decrease that is proportional to the binder content. The ethane uptake experiences no change, preserving the total exclusion of ethane from the framework.

5.3 CONCLUSIONS

The molecular sieving separation of an ethane–ethylene mixture has been shown for the first time. It results in an absolute ethylene selectivity and higher ethylene capacity compared to AgX and AgY. The Ag5A adsorbent can be applied in VSA-/TSA-type processes similar to AgX.[13,16] Ag5A could be a valuable adsorbent for the capture of ethylene from streams involved in the oxidative coupling of methane where diluted ethane/ethylene mixtures are present.[33]

REFERENCES

1. Xie, Y., Bu, N., Liu, J., Yang, G., Qiu, J., Yang, N., Tang, Y. US Patent 4917711, 1990.
2. Cho, S., Han, S., Kim, J., Choudary, N., Kumar, P., Garadi, S., Bhat, T. US Patent 6315816, 2001.
3. van Miltenburg, A., Gascon, J., Zhu, W., Kapteijn, F., Moulijn, J. A. *Adsorption* 2008, 14, 309.
4. van Miltenburg, A., Zhu, W., Kapteijn, F., Moulijn, J. A. *Chem. Eng. Res. Des.* 2006, 84, 350.
5. Gascon, J., Kapteijn, F. WO Patent 2009104960, 2009.
6. Bulow, M., Guo, C., Shen, D., Fitch, F., Shirley, A., Malik, V. US Patent 6200366, 2001.
7. Hosseinpour, S., Fatemi, S., Mortazavi, Y., Gholamhoseini, M., Ravanchi, M. T. *Sep. Sci. Technol.* 2011, 46, 349.
8. Rege, S. U., Padin, J., Yang, R. T. *AIChE J.* 1998, 44, 799.
9. Chudasama, C. D., Sebastian, J., Jasra, R. V. *Ind. Eng. Chem. Res.* 2005, 44, 1780.
10. Lee, Y., Reisner, B. A., Hanson, J. C., Jones, G. A., Parise, J. B., Corbin, D. R., Toby, B. H., Freitag, A., Larese, J. Z. *J. Phys. Chem. B* 2001, 105, 7188.
11. Sircar, S., Myers, A. L. Gas separation by zeolites. In *Handbook of Zeolite Science and Technology*, Anesbach, S. M., Carrado, K. A., Dutta, P. K., eds. New York: Marcel Dekker, Inc., 1063, 2003.
12. US Patent 1104208, 1968.
13. Carter, J., Lucchesi, P., Nightingale, E. US Patent 3355509, 1967.
14. Kimberlin, C., Mattox, W. US Patent 2866835, 1958.
15. Kling, M. US Patent 4992601, 1991.
16. Mitariten, M. US Patent 5245099, 1993.
17. Rosback, D. US Patent 3755153, 1973.
18. Rosback, D., Neuzil, R. US Patent 4036744, 1977.
19. Cheng, L., Wilson, S. US Patent 6296688, 2001.
20. Yang, R., Padin, J., Rege, S. US Patent 6423881, 2002.
21. Lin, C. C. H., Sawada, J. A., Wu, L., Haastrup, T., Kuznicki, S. M. *J. Am. Chem. Soc.* 2009, 131, 609.
22. Kucnicki, S., Bell, V. US Patent 6517611, 2003.
23. Kuznicki, S., Anson, A., Segin, T., Lin, C. WO Patent 20090187053, 2009.

24. Anson, A., Lin, C. C. H., Kuznicki, T. M., Kuznicki, S. M. *Chem. Eng. Sci.* 2010, 65, 807.
25. Howard, J., Kadir, Z. A., Robson, K. *Zeolites* 1983, 3, 113.
26. Gellens, L. R., Mortier, W. J., Schoonheydt, R. A., Uytterhoeven, J. B. *J. Phys. Chem.* 1981, 85, 2783.
27. Gellens, L. R., Mortier, W. J., Uytterhoeven, J. B. *Zeolites* 1981, 1, 11.
28. Grobet, P. J., Schoonheydt, R. A. *Surf. Sci.* 1985, 156, 893.
29. Sun, T., Seff, K. *Chem. Rev.* 1994, 94, 857.
30. Huang, Y. *J. Catal.* 1980, 61, 476.
31. Choi, E. Y., Kim, S. Y., Kim, Y., Seff, K. *Micropor. Mesopor. Mater.* 2003, 62, 201.
32. Kim, Y., Seff, K. *J. Am. Chem. Soc.* 1977, 99, 7055.
33. Salerno, D., Arellano-Garcia, H., Wozny, G. *Energy* 2011, 36, 4518.

6 Selective Removal of CH$_4$ from CH$_4$/CO/H$_2$ Mixtures

*Carlos A. Grande, Giorgia Mondino, Anna Lind,
Ørnulv Vistad, and Duncan Akporiaye*

CONTENTS

6.1 INTRODUCTION

The classical route to produce hydrogen is steam methane reforming (Rostrup-Nielsen, 1984; Twigg, 1989). The reforming reactions are extremely endothermic, and to achieve high conversions and CO selectivities, reforming is typically performed at high temperatures in presence of a nickel catalyst (Rostrup-Nielsen, 1984; Twigg, 1989; Xu and Froment, 1989). If the objective is to obtain syngas, that is, a mixture containing H$_2$ and CO, the water–gas shift reaction, which converts CO into CO$_2$, has to be avoided.

The major reforming products are H_2, CO, and CO_2 contained in a mixture with unreacted methane and water. Pressure swing adsorption (PSA) is a standard technology for purification of hydrogen generated in a reformer reactor (Ruthven et al., 1994; Stöcker et al., 1998; Waldron and Sircar, 2000; Sircar and Golden, 2010). PSA units operate at low temperature, so the stream should be cold prior to the separation process. A typical feed stream contains over 70% of H_2 and 15%–20% CO_2, less than 4% CO and CH_4, and is saturated with water vapor.

When the objective of the plant is to produce syngas, the required separation is quite different due to the different reactor effluent composition. The amount of hydrogen is much smaller, while that of carbon monoxide is significantly higher to establish H_2/CO ratios between 1 and 2. The amount of CO_2 and unreacted methane are typically very small. However, if the reaction temperature is reduced, the amount of unconverted methane increases and may become considerable.

Under such conditions, methane should be separated from the syngas mixture and recycled to the reformer reactor in order to improve the overall hydrogen recovery. State-of-the-art CH_4 separation from CH_4/CO/H_2 mixtures is cryogenic distillation (Kerry, 2007). Given the low temperatures used in cryogenic distillation and considering that the obtained syngas will also be used in further steps again at higher temperatures, the energetic consumption is high. As an alternative to distillation, methane removal by stripping with propane has already been considered (Dang, 1983).

The approach in the present chapter is to consider an adsorption process for selective methane removal from a ternary CH_4/CO/H_2 mixture. In order to demonstrate the concept, a mixture with a fixed gas composition was employed: 20% CH_4/40% CO/40% H_2.

The possibility of carrying out this separation by adsorption was evaluated taking into account that in PSA units for H_2 purification, CH_4 is preferentially retained in the column over CO. In that sense, the first target was to identify an adsorbent that is as selective as possible towards methane.

It was observed that in most zeolites, carbon monoxide is preferentially adsorbed over methane (Kyriacos and Boord, 1957; Lopes et al., 2009a; Onyestyák, 2011). The same trend is exhibited by some metal–organic frameworks (Karra and Walton, 2008) and some aluminophosphates (ALPOs) (Predescu et al., 1996).

However, in some other adsorbents, CH_4 is preferentially adsorbed over CO. This is the case for some activated carbons (Wilson and Danner, 1983; Grande et al., 2008; Lopes et al., 2009a,b) and some ALPOs (Predescu et al., 1996). Furthermore, in a previous study, ZSM-5 was identified as the adsorbent with higher CH_4/CO ratio (Heymans et al., 2011). Screening of this separation was done using activated carbon, but the selectivity obtained was not satisfactory. The selectivities exhibited by ALPOs are interesting, but their capacity is generally rather limited.

Based on previously reported data, ZSM-5 was selected as the initial adsorbent. It was verified that the selectivity CH_4/CO increased with the Si/Al ratio. Adsorption equilibrium and kinetics were measured on a commercial sample of ZSM-5 crystals with Si/Al = 200 (Süd-Chemie, Germany). Further testing of adsorption equilibria and kinetics was performed on a *shaped* material, that is, a commercial sample of ZSM-5 extrudates also with Si/Al = 200 (Clariant, Germany). Additionally, in order to evaluate the effect of aluminum in the adsorbents, crystals of a pure silica zeolite beta synthesized in-house were also tested. Adsorption equilibria of pure gases, that is, CH$_4$, CO, and H$_2$, at three different temperatures, that is, 283, 313, and 343 K, were measured on all the samples. Diffusion experiments in the ZSM-5 crystals were carried out by pulse chromatography at 313 and 343 K. Furthermore, the parameters obtained for the ZSM-5 extrudates were verified by measuring ternary breakthrough curves in order to ensure that the mathematical model based on pure-component data can describe the fixed-bed behavior of the gas mixture. This served as a basis to model a two-column PSA process for selective methane removal. A modified Skarstrom cycle including pressure equalization step was simulated, and the effect of operating variables was analyzed.

6.2 MATERIALS AND METHODS

6.2.1 ZEOLITE BETA SYNTHESIS

The as-synthesized pure silica zeolite beta was made according to a procedure previously reported (Camblor et al., 1996). First, the structure directing agent (tetraethyl-ammonium hydroxide [TEAOH]) was diluted with water, and the silica source (tetraethyl orthosilicate [TEOS]) was added. The mixture was stirred for about 7 h at room temperature to allow evaporation of all the EtOH formed during hydrolysis of the silica source. After mixing, hydrofluoric acid (40 wt%) was added to the transparent solution, and a white solid was formed. The molar ratio of the gel after evaporation was 1.0 SiO$_2$:0.27 TEA$_2$O:0.54 HF:7.25 H$_2$O. The solid was transferred to a Teflon-lined stainless steel autoclave and heated with rotation at 140°C for 39 or 140 h. After synthesis, a fine white powder and clear solution was formed. The crystals were cleaned several times by deionized water and centrifugation. The as-synthesized samples were heated to 550°C in the presence of air to remove the organics. The zeolite beta structure of the produced powder was confirmed by x-ray diffraction (XRD).

6.2.2 ADSORBENT CHARACTERIZATION

The ZSM-5 and beta zeolite were analyzed by XRD. The measurements were performed on a PANalytical Empyrean diffractometer. The system is equipped with a PIXcel^{3D} solid-state detector. The measurements were carried out in reflection geometry with a step size of 0.013° and an accumulation time of 0.54 s/step, using CuK$_\alpha$ radiation (λ = 1.54187 Å).

The textural properties of the materials were determined by nitrogen adsorption measurements at 77 K using a BELSORP mini II (BEL, Japan) instrument. The samples were outgassed at 423 K for 18 h before measurement.

6.2.3 ADSORPTION EQUILIBRIUM MEASUREMENTS

Adsorption equilibrium data of pure gases, that is, CH$_4$, CO, and H$_2$, on all the adsorbents were measured in a BELSORP HP instrument (Japan). The measurements were carried out at 283, 313, and 343 K. Isotherms were measured with CH$_4$

(purity > 99.995%), CO (purity > 99.95%), and H_2 (purity > 99.999%) provided by Yara, Norway, and used without additional purification. Activation (degassing) of the samples was performed at 423 K under vacuum for 12 h with a heating ramp of 1 K/min. After the initial degassing, regeneration of the sample was only carried out using vacuum. The same adsorbent sample was used in all the measurements. Due to safety issues regarding measurements of carbon monoxide, the isotherms of this gas were only measured up to 5 bar, while measurements for H_2 and CH_4 were carried out until 30 bar. Adsorption and desorption were recorded multiple times, proving that adsorption is reversible.

6.2.4 DIFFUSION MEASUREMENTS

The experiments for diffusion measurements were performed by injecting a small amount (~20 µL) of pure gas in a stream of carrier gas, that is, helium, which was considered as an inert gas, passing through the adsorbent. For this purpose, a fixed-bed installation was employed where the column containing the adsorbent was placed inside a gas chromatograph (Agilent 6890 Gas Chromatograph, United States). The fixed-bed effluent was directly send to a thermal conductivity detector (TCD), where the gas concentration could be measured. The gas flow rate was regulated with a mass flow controller (Bronkhörst, the Netherlands) and measured at the outlet of the TCD.

The activation, c.q., degassing, of the samples was performed at 423 K under a continuous helium flow for 12 h. After regeneration, a helium gas stream flowed continuously through the system to prevent any moisture adsorption that, otherwise, might occur from the atmosphere.

Responses to pulses of pure CH_4 (purity > 99.995%), CO (purity > 99.95%), and H_2 (purity > 99.999%) provided by Yara, Norway, on the adsorbents were measured at 313 and 343 K employing different flow rates, in the range of 20–84 mL/min (measured at room temperature). Each pulse was repeated in order to test reproducibility. Experimental conditions used in the experiments are listed in Table 6.1.

6.2.5 TERNARY BREAKTHROUGH CURVE MEASUREMENTS

Breakthrough curves of a ternary mixture composed of CH_4 (20%), CO (40%), and H_2 (40%) were measured at 298 and 343 K at a total pressure of 10 bar and using two different flow rates, that is, 30 and 50 mL/min. Degassing of the sample was carried out at 423 K overnight passing a flow of helium of 15 mL/min. The operating conditions used in measuring the experimental data are listed in Table 6.2. Pressure in the column was controlled with a back-pressure regulator from Omega (max. pressure of 12 bar), and products were analyzed using mass spectroscopy (Thermo, United States). Gases were analyzed at different masses: CH_4 (15), CO (28), H_2 (2), and He (4). The system was leak-tight and, hence, was *air-free* so that no effect of N_2 (also 28) needs to be accounted for. Additional variations of H_2O (18), O_2 (32), and CO_2 (44) were analyzed.

TABLE 6.1
Experimental Conditions Used on the Pulse Experiments of CH_4, CO, and H_2 in ZSM-5 Crystals and Extrudates

Adsorbate	CH_4, CO, H_2
He flow rate (mL/min)[a]	20, 30, 40, 60, 84
Pressure (bar)	1
Temperature (K)	313, 343
Bed length (cm)	7.71
Bed diameter (cm)	0.72
Adsorbent	ZSM-5 (Si/Al = 200)
Type	Crystals
Adsorbent mass (g)	1.96
Type	Extrudates
Adsorbent mass (g)	1.76
Average pellet diameter (mm)	1.7
Adsorbent density (kg/m³)[a]	995.2
Pellet porosity, ε_p	0.64

[a] Measured with Hg pycnometry.

TABLE 6.2
Experimental Conditions Used on the Ternary Breakthrough Curves Using CH_4, CO, and H_2 in ZSM-5 Extrudates

Variable	Value
Column length (m)	0.0975
Column diameter (m)	0.0091
Column porosity	0.355
Column thickness (m)	0.0018
Temperature (K)	295 and 313
Pressure (bar)	10
Flow rate (cm³/min)	26.6 and 48.3

6.3 MODELING PROCEDURES

6.3.1 ADSORPTION EQUILIBRIUM

Pure-gas adsorption equilibria were regressed using a Langmuir model. This model is based on kinetic principles assuming that the rate of adsorption equals the rate of desorption (Do, 1998). The main assumptions of the model are as follows:

1. The surface is homogeneous implying that the adsorption energy is constant.
2. Adsorption is localized, that is, no mobility is retained, and each adsorption site can exactly accommodate one molecule.

One of the main advantages of the model is that adsorption of multicomponent mixtures can be estimated based on the parameters obtained from pure-gas measurements. The Langmuir model is one of the simplest models to describe adsorption of gases on porous solids and is still one of the most employed in the design of adsorption units. Its mathematical expression for multicomponent mixtures is

$$q_i = \frac{q_{m,i}K_{eq,i}P_i}{1 + \sum_{i=1}^{N} K_{eq,i}P_i} \tag{6.1}$$

where
$q_{m,i}$ is the maximum amount adsorbed of component i
$K_{eq,i}$ is the adsorption constant that has an exponential dependence of temperature according to

$$K_{eq,i} = K_i^{\infty} \exp\left(\frac{-\Delta H_i}{RT}\right) \tag{6.2}$$

where
K_i^{∞} is the adsorption constant at infinite temperature
$(-\Delta H_i)$ is the heat of adsorption

6.3.2 PULSE CHROMATOGRAPHY

Kinetic and equilibrium adsorption parameters of the pure gases were estimated applying the method of moments to the experimental pulse responses of a packed bed. The first and second moments (μ and σ^2, respectively) were calculated from the experimental data as follows:

$$\mu = \frac{\int_0^{\infty} Ct\,dt}{\int_0^{\infty} C\,dt} \tag{6.3}$$

$$\sigma^2 = \frac{\int_0^{\infty} C(t-\mu)^2\,dt}{\int_0^{\infty} C\,dt} \tag{6.4}$$

where
C is the gas concentration
t is the time

Theoretically, the first moment, also known as stoichiometric time, is given by the following expression:

$$\mu = \frac{L}{u}\left(1+\left(\frac{1-\varepsilon}{\varepsilon}\right)K_H\right) \tag{6.5}$$

where
  L is the bed length
  μ is the interstitial velocity of the gas
  ε is the bed porosity
  K_H is the dimensionless Henry's law constant

For systems of pure-gas adsorption at low concentrations, the following equation expresses the relation of the ratio of the variance and the double of the square of the first moment square of a pulse ($\sigma^2/2\mu^2$) with the sum of all diffusional resistances (Ruthven, 1984):

$$\frac{\sigma^2}{2\mu^2} = \frac{D_{ax}}{uL}+\left(\frac{u}{L}\right)\left(\frac{\varepsilon}{1-\varepsilon}\right)\left(\frac{R_p}{3K_f}+\frac{R_p^2}{15\varepsilon_p D_p}+\frac{r_c^2}{15K_H D_c}\right)\left(1+\frac{\varepsilon}{(1-\varepsilon)K_H}\right)^{-2} \tag{6.6}$$

where
  D_{ax} is the axial dispersion coefficient
  D_p is the pore diffusivity
  K_f is the film mass transfer coefficient
  D_c is the crystal diffusivity
  R_p and r_c are the particle and crystal radius, respectively
  L is the bed length
  ε_p is the particle porosity

Applying Equations 6.5 and 6.6 to the first and second moment, it is possible to calculate the equilibrium, that is, the Henry coefficient, and kinetic, that is, diffusivities, parameters.

By carrying out pulse experiments at different temperatures, the limiting diffusivity at infinite temperature, D_c^0, and the activation energy, E_a, can be obtained by fitting the results according to

$$\frac{D_c}{r_c^2} = \frac{D_c^0}{r_c^2}\exp\left(-\frac{E_a}{R_g T}\right) \tag{6.7}$$

where
  r_c is the crystal radius
  R_g is the universal gas constant
  T is the temperature

In Equation 6.6, the contributions of macropore resistance, film mass transfer resistance, and axial dispersion were evaluated numerically by applying literature correlations for the calculation of pore diffusivity (D_p), axial dispersion (D_{ax}), and film mass transfer (k_f) coefficients, respectively (Da Silva and Rodrigues, 2001).

The axial dispersion contribution can be determined by plotting $(\sigma^2/2\mu^2)(L/u)$ versus $1/u^2$ (Ruthven, 1984). At low Reynolds numbers, where the D_{ax} is not a strong function of velocity, these plots should be linear with the slope being equal to D_{ax} and the intercept corresponding to the total mass transfer resistance. Such a plot offers a convenient way of separating the axial dispersion term from the mass transfer term.

Note that only crystal resistance is considered when analyzing the ZSM-5 crystal data, while both macropore and crystal resistances are considered for extrudates.

As adsorbate concentrations employed in the experiments are sufficiently low to describe adsorption equilibrium by the Henry law, that is, a linear isotherm, and to minimize heat generation ensuring isothermal operation, isothermal process was considered, and the following linear isotherm equation employed:

$$q^* = K C_m \tag{6.8}$$

where
  q^* is the adsorbed phase concentration on the crystal surface
  K is the equilibrium coefficient
  C_m is the molar concentration of the gas in the macropore

The low amount of gas adsorbed compared to the total amount of gas injected also allows to assume constant velocity of the gas along the bed. Furthermore, axial dispersed plug flow was assumed. The mass balance in a differential element of the bed is

$$\varepsilon \frac{\partial C_b}{\partial t} + (1-\varepsilon)\rho_p \frac{\partial \overline{q}}{\partial t} = \varepsilon D_c \frac{\partial^2 C_b}{\partial z^2} - u_0 \frac{\partial C_b}{\partial z} \tag{6.9}$$

where
  C_b is the molar concentration of the adsorbate in the bulk phase
  $\overline{q}$ is the particle averaged adsorbed concentration
  ρ_p is the adsorbent particle density
  z is the axial coordinate

The particle averaged adsorbed phase concentration $\overline{q}$ is defined by

$$\overline{q} = \frac{2}{R_p^2} \int_0^{R_p} \overline{q} R \, dR \tag{6.10}$$

where
  R_p is the radius of the adsorbent particle
  R is the radial coordinate of the particle
  $\overline{q}$ is the crystal averaged adsorbed phase concentration

The latter is defined as

$$\bar{q} = \frac{2}{r_c^2} \int_0^{r_c} qr\,dr \tag{6.11}$$

where
    r_c is the radius of the crystal
    r is the radial coordinate of the crystal

For the diffusion of the adsorbate in a crystal, the Fickian description was used:

$$\frac{\partial q}{\partial t} = \frac{1}{r_c^2}\frac{\partial q}{\partial t}\left(D_c r^2 \frac{\partial q}{\partial r}\right) \tag{6.12}$$

where D_c is the crystal diffusivity.
    The diffusion of the adsorbate in macropores is described by

$$\varepsilon_p \frac{\partial C_m}{\partial t} + (1-\varepsilon_p)\rho_p \frac{\partial \bar{q}}{\partial t} = \varepsilon_p D_p \frac{\partial^2 C_m}{\partial R^2} + \frac{2}{R}\frac{\partial C_m}{\partial R} \tag{6.13}$$

where
    D_p is the macropore diffusivity
    k_f is the film mass transfer coefficient

The model was solved numerically by gPROMS (PSE, United Kingdom), using central finite difference method (CFDM) discretization of the spatial domains.

6.3.3 FIXED-BED AND PSA MATHEMATICAL MODEL

A 1D model was employed, and the mass transfer within the macropores and micropores was simplified to a bilinear driving force (bi-LDF) model in the simulation of the ternary breakthrough curves. Also, since the bed performs in a nonadiabatic and nonisothermal regime, energy balances in the gas, solid, and column wall were accounted for. In lab-scale units, the column wall has a considerably high heat capacity. The mathematical model for the fixed bed is described in Table 6.3 and was previously tested for a diverse number of gas separations (Da Silva and Rodrigues, 2001; Grande et al., 2010; Santos et al., 2011; Lopes et al., 2012).

The fixed-bed model is the basis for the simulation of a PSA process. A modified Skarstrom cycle was used (Skarstrom, 1960; Berlin, 1966) for initial evaluation of the principle. The cycle has six different steps:

1. Feed: The feed stream is fed to the column from the feed column end.
2. Depressurization: In order to recycle some gas rich in CO, the column has an initial cocurrent depressurization to an intermediate pressure.

TABLE 6.3
Mathematical Model Used in the PSA Simulations

Mass balance of component i

$$\varepsilon_c \frac{\partial C_i}{\partial t} = \varepsilon_c \frac{\partial}{\partial z}\left(C_T D_{ax} \frac{\partial y_i}{\partial z}\right) - \frac{\partial(uC_i)}{\partial z} - \frac{(1-\varepsilon_c)a'k_{fi}}{1+Bi_i/5}\left(C_i - \overline{C_{P,i}}\right)$$

Momentum balance (Ergun equation)

$$\frac{\partial P}{\partial z} = -\frac{150\mu_g(1-\varepsilon_c)^2}{\varepsilon_c^3 d_p^2}u + \frac{1.75(1-\varepsilon_c)\rho_g}{\varepsilon_c^3 d_p}|u|u$$

Linear driving force (LDF) equation for macropores

$$\varepsilon_p \frac{\partial \overline{C_{P,i}}}{\partial t} + \rho_p \frac{\partial \langle \overline{q_i} \rangle}{\partial t} = \varepsilon_p \frac{15D_{p,i}}{R_p^2}\frac{Bi_i}{5+Bi_i}\left(C_i - \overline{C_{P,i}}\right)$$

LDF equation for micropores

$$\frac{\partial \langle \overline{q_i} \rangle}{\partial t} = K_{LDF,i}\left(q_i^* - \langle \overline{q_i} \rangle\right)$$

Energy balance in the gas phase

$$\varepsilon_c C_T C_v \frac{\partial T_g}{\partial t} = \frac{\partial}{\partial z}\left(\lambda \frac{\partial T_g}{\partial z}\right) - uC_T C_p \frac{\partial T_g}{\partial z} + \varepsilon_c R_g T_g \frac{\partial C_T}{\partial t} - (1-\varepsilon_c)a'h_f\left(T_g - T_s\right) - \frac{2h_w}{R_w}\left(T_g - T_w\right)$$

Energy balance in the sold phase (adsorbent)

$$(1-\varepsilon_c)\left[\varepsilon_p \sum_{i=1}^{n}\overline{C_{P,i}}C_{vi} + \rho_p \sum_{i=1}^{n}\langle \overline{q_i} \rangle C_{v,ads,i} + \rho_p C_{ps}\right]\frac{\partial T_s}{\partial t} = (1-\varepsilon_c)\varepsilon_p R_g T_s \frac{\partial C_i}{\partial t}$$

$$+(1-\varepsilon_c)\rho_p \sum_{i=1}^{n}(-\Delta H_i)\frac{\partial \langle \overline{q_i} \rangle}{\partial t} + (1-\varepsilon_c)a'h_f\left(T_g - T_s\right)$$

Energy balance in the wall (heat capacitance)

$$\rho_{wall}C_{p,wall}\frac{\partial T_w}{\partial t} = \frac{D_w h_w}{e(D_w + e)}\left(T_g - T_w\right) - \frac{U(T_w - T_\infty)}{(D_w + e)\ln\left(\dfrac{D_w + e}{D_w}\right)}$$

The ideal gas law

$$P = C_t R_g T_g \quad C_t = \sum_{i=1}^{n}C_i$$

3. Blowdown or evacuation: Gas is allowed to exit the column countercurrently to the feed stream. In this step, the lower pressure is attained.
4. Purge: Part of the purified gas is recycled countercurrently to additionally remove methane from the column (both gas and adsorbed phase).
5. Pressure equalization: The gas exiting from step 2 is introduced in the column countercurrently.
6. Pressurization: Part of the feed stream is also used to cocurrently pressurize the bed to the feed pressure in order to start a new cycle.

Simulations were carried out using a two-column PSA unit shown in Figure 6.1 (Liu et al., 2011). A reference simulation was carried out using the operating

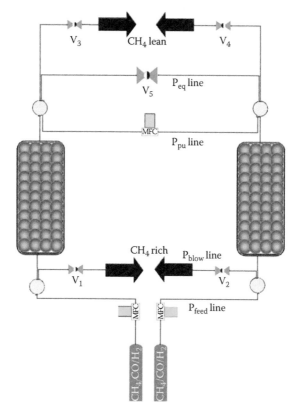

FIGURE 6.1 Scheme of the PSA unit used in the simulations.

conditions listed in Table 6.4. Based on the results of the simulation, the performance parameters of the unit were product (H_2/CO) purity and recovery and H_2/CO ratio. These parameters are defined by:

$$H_2 \text{ purity} = \frac{\int_0^{t\,\text{feed}} C_{H2}u \mid_{z=L} dt}{\sum \int_0^{t\,\text{feed}} C_i u \mid_{z=L} dt} \tag{6.14}$$

$$H_2 \text{ recovery} = \frac{\int_0^{t\,\text{feed}} C_{H2}u \mid_{z=L} dt}{\sum \int_0^{t\,\text{press}} C_{H2}u \mid_{z=0} dt + \int_0^{t\,\text{feed}} C_{H2}u \mid_{z=0} dt} \tag{6.15}$$

$$\frac{H_2}{CO} \text{ ratio} = \frac{\int_0^{t\,\text{feed}} C_{H2}u \mid_{z=L} dt}{\sum \int_0^{t\,\text{feed}} C_{CO}u \mid_{z=L} dt} \tag{6.16}$$

TABLE 6.4

Operating Conditions Used to Simulate the PSA Unit to Separate CH$_4$/CO/H$_2$ Mixture with ZSM-5 Adsorbent

Feed pressure (bar)	10	Feed slow rate (SLPM)	1000
Feed temperature (K)	303	Purge flow rate (SLPM)	125
$y_{CH_4,feed}$	0.20	t_{feed} (s)	350
$y_{CO,feed}$	0.40	t_{depres} (s)	50[b]
$y_{H_2,feed}$	0.20	t_{blow} (s)	300
Column diameter (m)	0.50	t_{purge} (s)	150
Column length (m)	2.50[a]	t_{press} (s)	100

[a] Value for the reference case. Length also evaluated as 2.3 and 2.1 m.
[b] Value for pressure equalization step also.

The model was solved in gPROMS (PSE, United Kingdom) using CFDM (350 points in the axial dimension of the column).

6.4 RESULTS AND DISCUSSION

6.4.1 MATERIALS

XRD patterns of the ZSM-5 powder, ZSM-5 extrudates, and zeolite beta are shown in Figure 6.2. The XRD patterns are typical for these zeolites.

The surface area and micropore volume of the ZSM-5 samples were measured by N$_2$ adsorption at 77 K. The results are shown in Figure 6.3. The results of the crystals

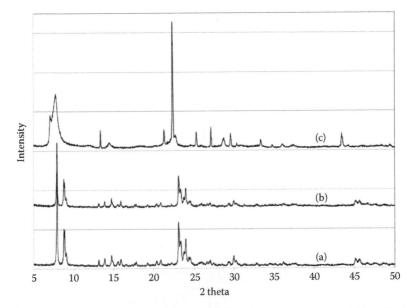

FIGURE 6.2   XRD patterns of (a) ZSM-5 powder, (b) ZSM-5 extrudate, and (c) zeolite beta.

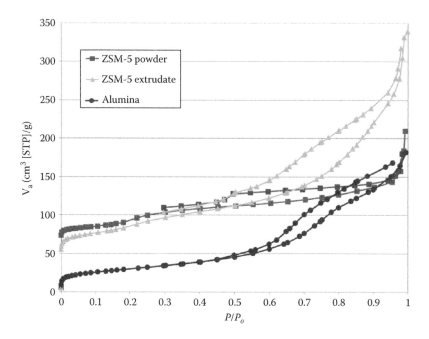

FIGURE 6.3 N$_2$ adsorption–desorption isotherms at 77 K for ZSM-5 powder, ZSM-5 extrudate, and alumina.

TABLE 6.5

Textural Properties of ZSM-5 Powder and ZSM-5 Extrudate Determined with N$_2$ Adsorption at 77 K

Material	Micropore Volume (cm^3/g)	BET Area (m^2/g)
ZSM-5 powder	0.111	343
ZSM-5 extrudates	0.077	298
Alumina	0.007	104

indicate a microporous sample practically without meso-/macropores and show the typical pore filling at high P/P_o. In the case of the extrudates, the sample showed a smaller pore volume and surface area per unit mass due to the presence of the binder. The alumina content in the zeolite extrudates amounted to 30%. For comparison, in Figure 6.3, the results of alumina are shown, showing a similar behavior at high P/P_o.

A summary of the results, that is, micropore volume as well as the BET surface area, for the ZSM-5 powder, ZSM-5 extrudates, and alumina is listed in Table 6.5.

6.4.2 Adsorption Equilibrium

Adsorption equilibrium of pure CH$_4$, CO, and H$_2$ at three different temperatures, that is, 283, 313, and 343 K, on ZSM-5 crystals with Si/Al = 200 is shown in Figure 6.4.

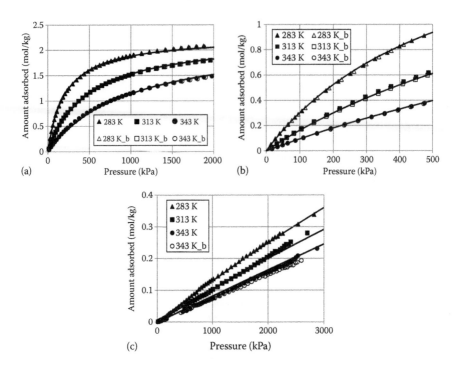

FIGURE 6.4 Adsorption equilibrium of CH_4 (a), CO (b), and H_2 (c) on ZSM-5 (Si/Al = 200) crystals. Solid lines are fitting from the Langmuir isotherm.

For comparison, similar data measured on ZSM-5 extrudates with Si/Al = 200 are shown in Figure 6.5. Adsorption data measured on crystals of pure silica zeolite beta are shown in Figure 6.6. Solid lines in all the figures correspond to the fitting of the Langmuir model with the parameters reported in Table 6.6. All measurements could be reproduced indicating that the adsorption is reversible.

The CH_4/CO selectivity was calculated based on the adsorption equilibria and is shown in Figure 6.7. At low pressures, which are important for preferential adsorption from multicomponent mixtures, the selectivity exhibited by the ZSM-5 crystals exceeded that of the zeolite beta. Since the amount adsorbed by both materials was similar, it was decided to continue the studies with ZSM-5.

The adsorption capacity of the ZSM-5 extrudates was significantly lower than that obtained with the crystals. This is due to the utilization of 30% binder, which reduces the adsorption on a weight basis. Furthermore, alumina preferentially adsorbs CO rather than CH_4, explaining why the lower selectivity of the extrudates at low pressures compared to the pure crystals cannot entirely be related to a simple dilution effect by the binder.

The pure-gas adsorption equilibrium results indicated that both ZSM-5 and zeolite beta can be used to selectively remove CH_4 from a ternary mixture consisting of CH_4/CO/H_2.

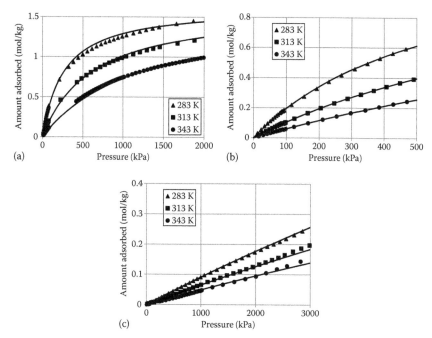

FIGURE 6.5 Adsorption equilibrium of CH$_4$ (a), CO (b), and H$_2$ (c) on extrudates of ZSM-5 (Si/Al = 200). Solid lines are fitting from the Langmuir isotherm.

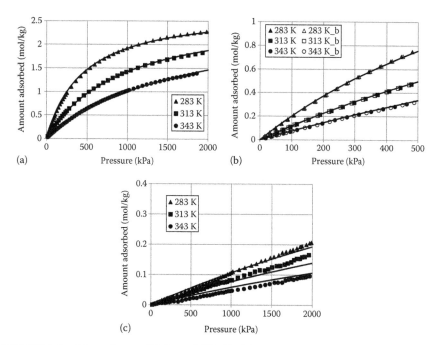

FIGURE 6.6 Adsorption equilibrium of CH$_4$ (a), CO (b), and H$_2$ (c) on crystals of pure silica zeolite beta. Solid lines are fitting from the Langmuir isotherm.

TABLE 6.6

Langmuir Parameters of CH₄, CO, and H₂ Adsorption Equilibrium on ZSM-5 (Crystal and Extrudates) and Zeolite Beta

System	CH_4	CO	H_2
ZSM-5 powder (Si/Al = 200)			
q_{max}, mol/kg	2.262	1.973	3.348
K_∞, kPa⁻¹	6.01×10^{-7}	1.21×10^{-6}	3.50×10^{-7}
$(-\Delta H)$, J/mol	21,212	17,191	5,742
ZSM-5 extrudates (Si/Al = 200)			
q_{max}, mol/kg	1.634	1.308	2.866
K_∞, kPa⁻¹	5.81×10^{-7}	1.13×10^{-6}	9.78×10^{-7}
$(-\Delta H)$, J/mol	20,656	17,312	8,271
Zeolite beta (pure silica)			
q_{max}, mol/kg	2.770	2.400	1.000
K_∞, kPa⁻¹	8.00×10^{-7}	2.50×10^{-6}	4.00×10^{-6}
$(-\Delta H)$, J/mol	18,700	13,900	8,000

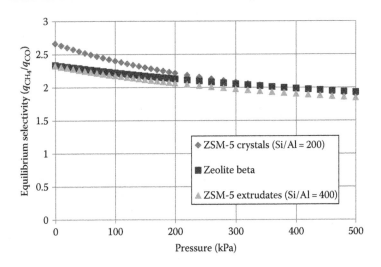

FIGURE 6.7 Equilibrium selectivity (methane/carbon monoxide) on the different samples at 313 K.

6.4.3 DIFFUSION

Adsorption kinetic parameters of pure gases, that is, CH_4, CO, and H_2, diluted in helium, in ZSM-5 (with Si/Al = 200) crystals and ZSM-5 (with Si/Al = 200) extrudates were obtained by two sets of pulse experiments at 1 bar and at two different temperatures, that is, 303 and 343 K, for different carrier gas flow rates in the range of 20–85 mL/min.

The CH_4 and CO Henry coefficients as determined from pulse chromatography are comparable with the results obtained from adsorption equilibrium. However, there are large differences for H_2. It should be noted that the results of pulse chromatography of H_2 are considered to be less accurate due to the small amount of gas used and, hence, the small amount adsorbed. Also, the experimental error associated to the measurement of the hydrogen peak was considerable because of the small difference between the thermal conductivities of hydrogen and the carrier gas used, that is, helium, which causes low sensitiveness of the TCD.

The results obtained for the ZSM-5 crystals are shown in Figure 6.8, and the ones for the extrudates are shown in Figure 6.9. Note that the experiments were performed using different flow rates to evaluate the kinetic coefficients under different conditions. It was possible to simulate the data acquired on the crystals with a single diffusion parameter. The results are listed in Table 6.7. The diffusion coefficients follow the molecular size of the molecules: diffusion is slower for methane followed by CO and H_2 diffusion is very fast. It should be noted that the temperature dependence of the diffusion coefficient was only calculated using two temperatures and, hence, is only reliable within a small temperature interval.

The diffusion coefficients for ZSM-5 extrudates are listed in Table 6.7. In the extrudates, the film mass transfer and pore diffusivity could be neglected, and most of the resistance could be attributed to crystal diffusion. Note that the results of the crystal diffusion of the samples are different and this is due to the different morphology of the crystals that were provided by different suppliers. Although the difference is considerable, the diffusion in the zeolite crystals and extrudates is quite fast.

In order to verify the reliability of experimental data, for each gas-adsorbent pair at the same temperature at different carrier flow rate, the values of $(\sigma^2/2\mu^2)(L/u)$ were plotted versus $1/u^2$. The plots were linear, and their slopes, corresponding to the axial dispersion coefficient D_{ax}, coincided with the theoretical D_{ax}, within the experimental error. An example of these plots is shown in Figure 6.10.

6.4.4 TERNARY BREAKTHROUGH CURVES

The behavior of the multicomponent mixture was assessed by measuring ternary breakthrough curves in the ZSM-5 extrudates. The experimental conditions are listed in Table 6.3. The results at 295 K using the smaller flow rate, that is, 26.6 cm³/min, are shown in Figure 6.11. The temperature at $z = 0.05$ m from feed inlet was continuously monitored and is also shown since the transient, while adsorption is taking place, is not isothermal. The breakthrough curve at a higher flow rate, that is, 48.3 cm³/min, is shown in Figure 6.12, together with temperature evolution. The breakthrough curve at 313 K with the low flow rate is shown in Figure 6.13.

The solid lines in the figures indicate that the mathematical model can successfully reproduce the behavior of the ternary mixture at two different flow rates and two temperatures. In all these breakthrough measurements, hydrogen was the first gas to break through the column followed by CO and finally CH_4. In the experiments, the gas velocity was quite small, and the mass transfer zone is mainly due to axial dispersion. Furthermore, it was observed that the temperature increase due to adsorption was not very intense, that is, below 10 K, and is due to CO and CH_4 adsorption. This means that in a cyclic process, that is, when CO is previously adsorbed in the column and subsequently displaced by CH_4, the temperature variations can be quite small.

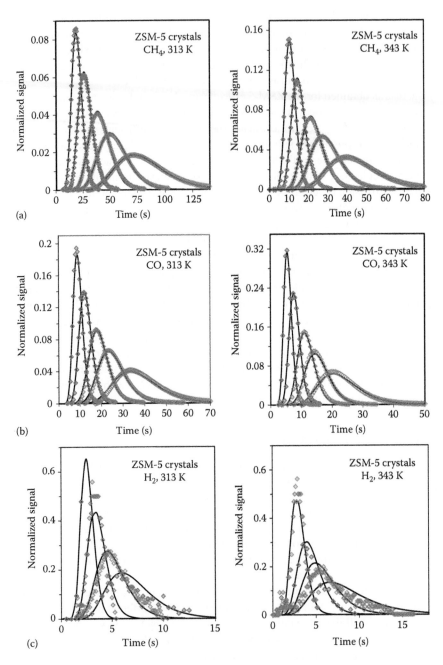

FIGURE 6.8 Pulse responses of ZSM-5 crystals at 313 K (left) and 343 K (right) for (a) CH_4, (b) CO, and (c) H_2: experimental (dotted line) and simulations (solid line). The carrier flow rate (20, 30, 40, 60, and 84 cm^3/min) increases from right to left.

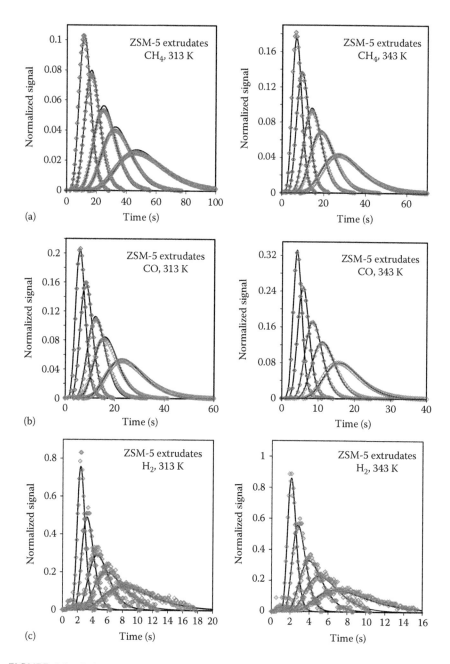

FIGURE 6.9 Pulse responses of ZSM-5 extrudates at 313 K (left) and 343 K (right) for (a) CH₄, (b) CO, and (c) H₂: experimental (dotted line) and simulations (solid line). The carrier flow rate (20, 30, 40, 60, and 84 cm³/min) increases from right to left.

TABLE 6.7

Crystal Diffusivity and Henry Constant of CH$_4$, CO, and H$_2$ in ZMS-5 Crystals Determined from Pulse Chromatography Experiments

Adsorbate	CH$_4$		CO		H$_2$	
Temperature (K)	313	343	313	343	313	343
$D_{ax} \times 10^5$ (m^2/s)	5.27	6.14	5.51	6.42	12.23	14.23
$\left(D_c/r_c^2\right) \times 10^2$ (1/s)	25	44	58	104	260	300
$\left(D_c^0/r_c^0\right)$ (1/s)	160.31		460.24		206.64	
E_a (kJ/mol)	16.82		17.38		12.07	
$K_H^a \times 10^3$ (kPa^{-1})	2.43	1.31	1.22	0.71	0.14	0.11
$K_H^b \times 10^3$ (kPa^{-1})	2.08	1.02	0.89	0.50	0.003	0.003

Other diffusional resistances used in the model are shown.

a Obtained from pulse experiments.

b Obtained from equilibrium measurements.

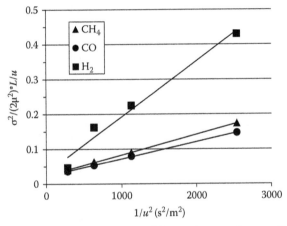

FIGURE 6.10 $(\sigma^2/2\,\mu^2)(L/u)$ plotted against $1/u^2$ from pulse experiments of CH$_4$, CO, and H$_2$ in ZSM-5 extrudates performed at 343 K.

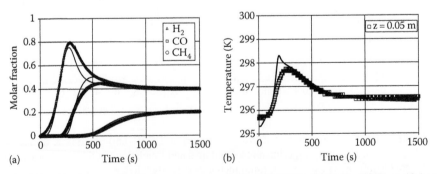

(a) (b)

FIGURE 6.11 Ternary breakthrough curve of CH$_4$ (20%)/CO (40%)/H$_2$ (40%) at 10 bar and 295 K using a flow rate of 26.8 mL/min. Experimental conditions are listed in Table 6.2.

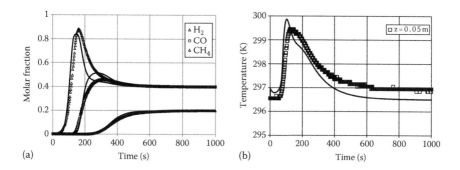

FIGURE 6.12   Ternary breakthrough curve of CH$_4$ (20%)/CO (40%)/H$_2$ (40%) at 10 bar and 295 K using a flow rate of 48.6 mL/min. Experimental conditions are listed in Table 6.2.

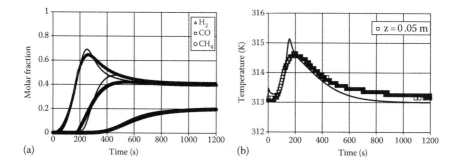

FIGURE 6.13   Ternary breakthrough curve of CH$_4$ (20%)/CO (40%)/H$_2$ (40%) at 10 bar and 313 K using a flow rate of 26.8 mL/min. Experimental conditions are listed in Table 6.2.

The ternary breakthrough curves have also confirmed that the selective separation of methane from this ternary mixture is possible, at least within the investigated range of operating conditions.

6.5   PSA SIMULATION

All experiments discussed in the previous sections have shown that the selected material, ZSM-5 zeolite, with Si/Al = 200 can be used for selective separation of CH$_4$ from a ternary mixture CH$_4$/CO/H$_2$. However, in order to perform this industrially, the separation has to be performed cyclically and efficiently in a PSA process. For this reason, we have evaluated the performance of a PSA unit to carry out this separation. The cycle used is a modified Skarstrom cycle including a pressure equalization step. Two columns were modeled, and hence, the feed is only interrupted while the columns are in pressure equalization, that is, twice per cycle. The real unit will have more than two columns to continuously process the feed, but the simulation of two columns only allows us already to assess the importance of recycling H$_2$ and CO, with different ratios, in the purge, equalization, and pressurization steps.

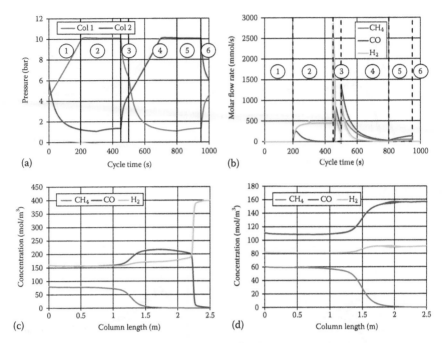

FIGURE 6.14 Simulation results of a PSA process for selective removal of CH_4 from $CH_4/$ CO/H_2 mixture (see conditions in Table 6.4). (a) Pressure history, (b) exiting molar flow, (c) concentration at the end of feed step, and (d) concentration at the end of depressurization step. Numbers in (a) and (b) indicate cycle steps: 1 = pressurization; 2 = feed; 3 = depressurization; 4 = blowdown; 5 = purge; 6 = equalization.

It should be pointed that a good PSA cycle should result only in minor changes to the feed H_2/CO ratio of one, thus minimizing the difference of adsorption equilibrium of CO and H_2.

The reference case (see Table 6.4 for conditions) considers that the column is sufficiently long to remove all methane from the feed. The results of a single cycle under cyclic steady state (CSS) of this simulation are displayed in Figure 6.14. CSS was achieved in less than 20 cycles since the diffusion is quite fast and also the temperature variations within the column are below 10 K.

Looking at the molar flow rate evolution, one can observe that no methane leaves the column during the entire feed step. Moreover, no methane is leaving the column during the depressurization either and, hence, is not recycled to the column in any step of the cycle. For this reason, the methane removal can be considered as complete. However, the bed is so large that also CO and H_2 separation takes place: CO is also not breaking through the column during the feed step. CO is being recycled only during the depressurization step, which explains why only a small portion of CO leaves the column during the feed step. Thus, the H_2/CO ratio in this simulation is 7.21, which is far away from the desired value. However, looking at the methane concentration at the end of the depressurization step, it can be concluded that the column can be shortened. If that happens, CO will break through the column during the feed step and thus reduce the H_2/CO ratio value.

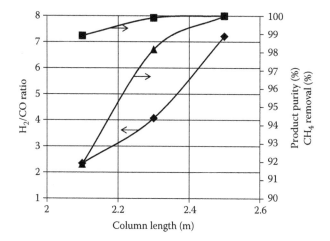

FIGURE 6.15 Performance parameters of PSA simulation as a function of column length (and thus CO recycling). H_2/CO read on the left axis, while methane removal (triangles) and product purity (squares) read on the right axis.

Simulations were carried out using the same step times, that is, taking into account that $t_{press} + t_{feed} = 450$ since pressurization will be faster in a smaller column, but changing the column length. A change in the column length will result in a change of the amount of CO obtained in the product and also recycled in the pressure equalization step. The performance indicators as a function of the column length are shown in Figure 6.15. As mentioned earlier, reducing the column length, more CO is recycled internally and thus reduces the obtained H_2/CO ratio.

The simulations demonstrated that the selective separation of CH_4 from the ternary mixture $CH_4/CO/H_2$ is possible. However, in order to compensate for the difference in CO and H_2 capacity, another more specific cycle has to be used, which allows more CO internal recycle.

6.6 CONCLUSIONS

The selective separation of CH_4 by PSA from a ternary mixture of $CH_4/CO/H_2$ was evaluated. It was observed that materials with low aluminum content such as ZSM-5 (Si/Al = 200) and pure silica zeolite beta were more selective to CH_4 than to CO and H_2. Adsorption equilibrium of pure gases on this adsorbent obeys a Langmuir description between $283 > T > 313$ K and $0 > P > 3000$ kPa. Using the parameters obtained from pure gases, the loading of the ternary mixture in a single point could be predicted. Diffusion experiments were also performed to determine crystal diffusivity coefficients to be used in a mathematical model to simulate a PSA process.

Ternary breakthrough curves at 10 bar and with CH_4 (20%)/CO (40%)/H_2 (40%) were performed serving as a final validation tool. The selectivity of the ZSM-5 adsorbent was slightly affected by the binder material (alumina).

A modified Skarstrom cycle (pressurization, feed, depressurization, blowdown, purge, and equalization) was used in the initial simulations. Using this cycle, it was possible to completely remove CH_4 from the ternary mixture, with the expense of changing the H_2/CO ratio. Increasing the internal recycle of CO (and avoiding CO to be removed in the blowdown step), it is possible to diminish the impact of the PSA process in the H_2/CO ratio while selectively removing CH_4. A more targeted PSA cycle should be used for this purpose.

ACKNOWLEDGMENTS

This paper reports work undertaken in the context of the project *OCMOL (Oxidative Coupling of Methane followed by Oligomerization to Liquids)*. OCMOL is a large-scale collaborative project supported by the European Commission in the 7th Framework Programme (GA no. 228953). For further information about OCMOL, see http://www.ocmol.eu

REFERENCES

Berlin, N. H. Method for providing an oxygen-enriched environment. U.S. Patent 3,280,536, 1966.

Camblor, M. A., Corma, A., Valencia, S. Spontaneous nucleation of pure silica zeolite-beta free of connectivity defects. *Chem. Commun.* 1996, 20, 2365–2366.

Dang, V.-D. Separation of methane from hydrogen and carbon monoxide by an absorption/stripping process. *Industrial Gas Separation*, White, T. E., Yon, C. M., Wagener, E. H. (eds.). ACS Symposium Series 223, ACS Publications, Washington, DC, 1983, 235–253.

Da Silva, F. A., Rodrigues, A. E. Propylene/propane separation by VSA using commercial 13X zeolite pellets. *AIChE J.* 2001, 47, 341–357.

Do, D. D. *Adsorption Analysis: Equilibria and Kinetics.* Imperial College Press, London, U.K., 1998.

Grande, C. A, Lopes, F. V. S., Ribeiro, A. M., Loureiro, J. M., Rodrigues, A. E. Adsorption of off-gases from steam methane reforming (H_2, CO_2, CH_4, CO and N_2) on activated carbon. *Sep. Sci. Technol.* 2008, 43, 1338–1364.

Grande, C. A., Poplow, F., Rodrigues, A. E. Vacuum pressure swing adsorption to produce polymer-grade propylene. *Sep. Sci. Technol.* 2010, 45, 1252–1259.

Heymans, N., Alban, B., Moreau, S., De Weireld, G. Experimental and theoretical study of the adsorption of pure molecules and binary systems containing methane, carbon monoxide, carbon dioxide and nitrogen. application to the syngas generation. *Chem. Eng. Sci.* 2011, 66, 3850–3858.

Karra, J. R., Walton, K. S. Effect of open metal sites on adsorption of polar and nonpolar molecules in metal-organic framework Cu-BTC. *Langmuir* 2008, 24, 8620–8626.

Kerry, F. G. *Industrial Gas Handbook. Gas Separation and Purification.* CRC Press, Boca Raton, FL, 2007.

Kyryacos, G., Boord, C. E. Separation of hydrogen, oxygen, nitrogen, methane and carbon monoxide by gas adsorption chromatography. *Anal. Chem.* 1957, 29, 787–788.

Liu, Z., Grande, C. A., Li, P., Yu, J., Rodrigues, A. E. Multi-bed vacuum pressure swing adsorption for carbon dioxide capture from flue gases. *Sep. Purif. Technol.* 2011, 81, 307–317.

Lopes, F. V. S., Grande, C. A, Ribeiro, A. M., Loureiro, J. M., Rodrigues, A. E. Adsorption of H_2, CO_2, CH_4, CO and N_2 in activated carbon and zeolite for hydrogen production. *Sep. Sci. Technol.* 2009a, 44, 1045–1073.

Lopes, F. V. S., Grande, C. A, Ribeiro, A. M., Loureiro, J. M., Rodrigues, A. E. Enhancing capacity of activated carbons for hydrogen purification. *Ind. Eng. Chem. Res.* 2009b, 48, 3978–3990.

Lopes, F. V. S., Grande, C. A., Rodrigues, A. E. Fast-cycling VPSA for hydrogen purification. *FUEL* 2012, 93, 510–523.

Onyestyák, G. Comparison of dinitrogen, methane, carbon monoxide, and carbon dioxide mass-transport dynamics in carbon and zeolite molecular sieves. *Helvetica Chim. Acta* 2011, 94, 206–217.

Predescu, L., Tezel, F. H., Chopra, S. Adsorption of nitrogen, methane, carbon monoxide and their binary mixtures on aluminophosphate molecular sieves. *Adsorption* 1996, 3, 7–25.

Rostrup-Nielsen, J. R. *Catalytic Steam Reforming*, 1st edn. Springer-Verlag, Berlin, Germany, 1984.

Ruthven, D. M. Principles of adsorption and adsorption processes. Wiley, New York,1984.

Ruthven, D. M., Farooq, S., Knaebel, K. S. *Pressure Swing Adsorption*. VCH Publishers, New York, 1994.

Santos, M. P. S., Grande, C. A., Rodrigues, A. E. Pressure swing adsorption for biogas upgrading. Effect of recycling streams in PSA design. *Ind. Eng. Chem. Res.* 2011, 50, 974–985.

Sircar, S., Golden, T. C. *Hydrogen and Syngas Production and Purification Technologies*, Liu, K., Song, C., Subranami, V. (eds.). Wiley, Hoboken, NJ, Chapter 10, 2010.

Skarstrom, C. W. Method and apparatus for fractionating gas mixtures by adsorption. US Patent 2,944,627, 1960.

Stöcker, J., Whysall, M., Miller, G. Q. *30 Years of PSA Technology for Hydrogen Purification.* UOP LLC, Des Plaines, IL, 1998.

Waldron, W. E., Sircar, S. Parametric study of a pressure swing adsorption process. *Adsorption* 2000, 6, 179–188.

Wilson, R. J., Danner, R. P. Adsorption of synthesis gas-mixture components on activated carbon. *J. Chem. Eng. Data* 1983, 28, 14–18.

Xu, J., Froment, G. F. Methane steam reforming, methanation and water-gas shift: I. Intrinsic kinetics. *AIChE J.* 1989, 35, 88–96.

Section II

Integration of Innovative Catalysts with the GTL Process

7 Combining Catalyst Formulation and Microkinetic Methodologies in the Detailed Understanding and Optimization of Methane Oxidative Coupling

P.N. Kechagiopoulos, L. Olivier, C. Daniel,
A.C. van Veen, Joris W. Thybaut,
Guy B. Marin, and C. Mirodatos

CONTENTS

7.1   INTRODUCTION

Ethylene is the most widely produced organic commodity in chemical industry with an annual global demand of over 140×10^6 tons and a growth rate of 3.5% per year [1]. It is mainly used as a monomer for making polyethylene. Ethylene could also potentially serve to make high octane number fuels via oligomerization in the near future provided a cost-efficient high-capacity access is retained. However, it is at present mainly produced via steam cracking of naphtha or ethane.

Even though crude oil resources will deplete within a few decades, the remaining actual span of availability is much debated [2,3]. Accessing alternative, abundant, and sustainable sources for petrochemicals would be of utmost interest, if that technology allowed the production of ethylene at competitive costs compared to those of present cracking processes. Methane, a major constituent of natural gas, but also increasingly available in biogas, landfill, and shale gas, was identified as auspicious alternative raw material for ethylene production via oxidative coupling of methane (OCM).

7.1.1 Methane Conversion to Ethylene

Conversion of methane to ethylene can be achieved either via direct or indirect routes. Indirect routes involve the production of synthesis gas, a mixture of CO and H_2, also shortly denoted as syngas, via three main catalytic processes:

1. Highly endothermic methane reforming with H_2O, that is, steam reforming, or with CO_2, that is, dry reforming, both requiring extensive heat input
2. Slightly exothermic methane partial oxidation that requires, however, costly pure oxygen as feedstock
3. Autothermal reforming being essentially a combination of alternatives (1) and (2)

Subsequent conversion of syngas to ethylene may proceed via methanol synthesis, involving the water–gas shift reaction, followed by conversion of methanol to ethylene [4].

In contrast, a direct conversion route of methane to ethylene avoids the expensive syngas production step. As such, OCM would conceptually be economically more favorable than indirect routes. Prerequisite for commercialization is appropriate reaction performance [5,6]. Nonetheless, even at current performance levels and feedstock costs, direct methane conversion routes are found economically attractive in remote, gas-rich regions, where natural gas is available at negligible costs [7].

The transformation of methane to ethane and ethylene by OCM can be described via the following two global reactions, respectively:

$$CH_4 + \tfrac{1}{4}O_2 \rightarrow \tfrac{1}{2}C_2H_6 + \tfrac{1}{2}H_2O \quad (\Delta H^{\circ}_{298} = -88 \text{ kJ mol}^{-1}) \quad (7.1)$$

$$CH_4 + \tfrac{1}{2}O_2 \rightarrow \tfrac{1}{2}C_2H_4 + H_2O \quad (\Delta H^{\circ}_{298} = -140 \text{ kJ mol}^{-1}) \quad (7.2)$$

Both transformations are exothermic and, hence, ideally speaking, require no external energy supply for maintaining sufficiently high temperatures to sustain the reaction. However, the stability of methane, as evident from the C–H binding energy in Table 7.1, implies that its activation requires temperatures exceeding 600°C. Only recently reported catalytic materials enable high selectivity in a temperature range

---

TABLE 7.1

Cumulative C–H Binding Energies for Various Alkanes

Molecule	C–H Binding Energy (kJ mol^{-1})
Methane	438.4
Ethane	419.5
Ethylene	444.0

Source: Zavyalova, U. et al., *ChemCatChem*, 3, 1935, 2011.

---

from 450°C to 600°C [8]. However, the performance of these rare-earth oxycarbonate-based materials drops at higher temperatures due to structural degradation. The major challenge in OCM relates to the facile consecutive oxidation of desired products, that is, ethane and ethylene, into carbon oxides caused by the presence of oxygen at high temperatures. Even the total oxidation of methane may be favored at these conditions.

7.1.2 METHANE COUPLING TO C_2

First reports claiming a direct catalytic formation of higher hydrocarbons from methane were a U.S. patent from Mitchell and Waghorne [9] and an article by Fang and Yeh [10]. The latter work reported a methane conversion to ethane in the presence of a ThO_2/SiO_2 catalyst. Keller and Bhasin [11] conducted a first screening of potential OCM catalysts using air as oxidant. Tested catalysts consisted of various metals supported on α-alumina. The authors proposed a mechanism involving surface oxygen cycling between two different valence states and the formation of ethane by coupling of adsorbed CH_3 radicals after their formation on the surface. This work in the early 1980s was then followed by a vast variety of investigations until the 1990s. About one thousand articles were published on the subject during that period.

First attempts to propose more elaborate mechanisms have been made in the work by Stone [12], using UV-Vis and diffuse reflectance infrared Fourier transform spectroscopy (DRIFTS) on SrO, MgO, and CaO microcrystalline surfaces. Those authors put in evidence the heterolytic dissociative adsorption of gas molecules such as water and CH_4 on these highly ionic and basic surfaces and the formation of adsorbed carbonates from CO and CO_2 adsorption. They also highlighted the variation of basicity of such oxides by forming solid solutions. Lunsford worked extensively on Li/MgO catalysts [13–15]. These studies confirmed the formation of methyl radicals on the surface, followed by their coupling in the gas phase and not on the surface, as initially proposed by Keller and Bhasin [11]. The authors highlighted the major role of [$Li^+ O^-$] centers in the CH_3 radical formation. The proposed mechanism [16] was the basis for more complex reaction networks elaborated since then and is now commonly accepted by most authors. Actually, the work of Lunsford is also the basis for the kinetic modeling presented in the current work, the elaboration of which has already been published elsewhere [17–19].

The most important steps on the catalyst surface (see Equations 7.3 through 7.7) are presented in the following. The main and rate-determining step (RDS) is the homolytic activation of one C–H bond in the methane molecule on an O^- oxygen surface species, leading to the formation of a methyl radical:

$$CH_4 + (O^-)_s \rightarrow CH_3 \bullet + (OH^-)_s \qquad (7.3)$$

This RDS can be kinetically described as an Eley–Rideal reaction [20]. Other electrophilic species such as O_2^- have tentatively been proposed as well for describing the active surface oxygen species as presented in more detail later. However, we will consider at this stage only the most commonly agreed O^- dissociated adspecies.

Almost all active OCM catalysts reported in literature are metal oxides, able to generate the electrophilic surface oxygen $(O^-)_s$. Numerous authors studied OCM over various combinations of mixed metal oxides with isotopic exchange techniques and transient kinetics [21–25]. The pair $\{O^{2-} + \square\}$ was identified as the active site on the catalyst surface [26], where $\square$ represents a lattice oxygen vacancy. The site activates gaseous oxygen into O^- species, the latter being responsible for the C–H bond activation, according to the following equilibrium [27,28]:

$$\tfrac{1}{2}O_2 + \{O^{2-} + \square\} \leftrightarrow 2\,(O^-)_s \qquad (7.4)$$

After hydrogen abstraction from the hydrocarbon, the active site is regenerated by producing water:

$$2(OH^-)_s \leftrightarrow H_2O + \{O^{2-} + \square\} \qquad (7.5)$$

Considering a primary reaction pathway in OCM to C_2 constituted only by the afore-mentioned steps, many studies [29–38] as well as a review [39] have shown that suitable materials possess lattice oxygen being able to activate methane homolytically. The latter materials consist generally of ionic, strongly basic, mixed alkaline- and rare-earth metal oxides.

During the coupling step, two methyl radicals recombine in the gas phase to form an ethane molecule:

$$2CH_3{}^\bullet \rightarrow C_2H_6 \qquad (7.6)$$

In general, this reaction requires a collision with a third body to accommodate and evacuate the excess energy and, hence, stabilize the product ethane.

Subsequent ethane dehydrogenation leads to ethylene formation:

$$C_2H_6 \leftrightarrow C_2H_4 + H_2 \quad (\Delta H^{\circ}_{298} = +137\ kJ\ mol^{-1}) \qquad (7.7)$$

The mechanism for ethane dehydrogenation was extensively studied in the literature and involved essentially a sequence of heterogeneously catalyzed as well as homogeneous oxidation and reaction steps [40–46]. The reaction is assumed to proceed on the same oxygen activated sites as that for the methane activation [47].

Figure 7.1 summarizes the main surface and gas-phase steps assumed for OCM.

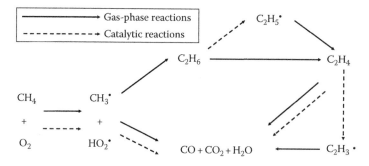

FIGURE 7.1   General scheme of the reaction network in OCM.

7.1.3   PARALLEL AND CONSECUTIVE REACTIONS IN OCM

As can also be seen in Figure 7.1, apart from the desired coupling reactions, the OCM reaction network also includes parallel and consecutive reactions. The most important effect of these reactions is the decrease in C_2 selectivity with increasing oxygen partial pressure in the feedstock caused by the promotion of deep oxidation of methane and coupling products. This deep oxidation of hydrocarbons into CO_x has been shown [48,49] to proceed preferentially via oxidation of vinyl radicals by oxygen in the gas phase to CO, followed by an oxidation of CO to CO_2 on the surface of the catalyst. Moreover, as mentioned earlier, ethane dehydrogenation occurs both in the gas phase and on the catalyst surface. Finally, the role of adsorbed CO_x, forming surface carbonates, the stability of which depends on surface basicity, has also been perceived as a key feature of performing catalysts, with contradictory roles of these adspecies. On the one hand, carbonates poison reversibly the active sites for OCM, and on the other hand, they stabilize sites against permanent deactivation [38,50–52]. Thus, many steps accounting for parallel and consecutive, heterogeneous, as well as homogeneous reaction steps were included to the previously described primary reaction network:

- C_2H_4 and C_2H_6 activation (Mars–van-Krevelen and Eley–Rideal mechanisms)
- CO and CO_2 adsorption/desorption, with a formation of surface carbonate species
- Water adsorption/desorption, involving surface hydroxyl species
- Possible radical quenching at the surface
- Oxidation of CO to CO_2 at the surface of the catalyst

Table 7.2 summarizes the main reaction steps proposed in the literature for given catalysts, by Stansch et al. over La_2O_3/CaO [53], Sohrabi et al. over $CaTiO_3$ [54], Lacombe et al. over La_2O_3 [55], Olsbye et al. over $BaCO_3/La_2O_n(CO_3)_{3-n}$ (n > 1.5) [56], Traykova et al. over La_2O_3/MgO [57], and Shahri and Alavi [58] over $Mn/Na_2WO_4/SiO_2$.

The formal reactions selected by these authors are accounting for

- Methane conversion to ethane
- Methane, ethane, and ethylene oxidation to CO_x
- Carbon monoxide oxidation to carbon dioxide
- Ethylene reforming
- Ethane dehydrogenation to ethylene
- Water–gas shift reaction

Daneshpayeh et al. [59] found that the model best fitting their experimental data on $Mn/Na_2WO_4/SiO_2$ catalyst was Model 1 from Stansch et al. [53], which is among the most complete global models published in the literature. It includes both heterogeneous and homogeneous steps and primary and consecutive ones. On the contrary, the models of Lacombe et al. [55] and Sohrabi et al. [54] do not consider the thermal dehydrogenation of ethane to ethylene nor the water–gas shift reaction, while

TABLE 7.2

Stoichiometric Equations of Reaction Network Kinetic Models

Model	1	2	3	4	5	6
Reactions \ Catalysts	$La_2O_3/$ CaO	$CaTiO_3$	La_2O_3	$BaCO_3/$ $La_2O_n(CO_3)_{3-n}$ $(n > 1.5)$	$La_2O_3/$ MgO	$Mn/$ $Na_2WO_4/$ SiO_2
1 $2CH_4 + \frac{1}{2}O_2 \rightarrow C_2H_6 + H_2O$	✓	✓	✓	✓	✓	✓
2 $CH_4 + O_2 \rightarrow CO + H_2O + H_2$	✓			✓		
3 $CH_4 + (3/2)O_2 \rightarrow CO + 2H_2O$		✓			✓	✓
4 $CH_4 + 2O_2 \rightarrow CO_2 + 2H_2O$	✓	✓	✓	✓		✓
5 $2CH_4 + O_2 \rightarrow C_2H_4 + 2H_2O$		✓				
6 $CO + \frac{1}{2}O_2 \rightarrow CO_2$	✓		✓			
7 $C_2H_6 + \frac{1}{2}O_2 \rightarrow C_2H_4 + H_2O$	✓		✓		✓	✓
8 $C_2H_6 + O_2 \rightarrow 2CO + 3H_2$			✓			
9 $C_2H_6 + (5/2)O_2 \rightarrow 2CO + 3H_2O$				✓		
10 $C_2H_6 + (7/2)O_2 \rightarrow 2CO_2 + 3H_2O$			✓	✓		
11 $C_2H_6 \rightarrow C_2H_4 + H_2$	✓			✓	✓	
12 $C_2H_4 + O_2 \rightarrow 2CO + 2H_2$			✓			
13 $C_2H_4 + 2O_2 \rightarrow 2CO + 2H_2O$	✓			✓		✓
14 $C_2H_4 + 3O_2 \rightarrow 2CO_2 + 2H_2O$				✓		
15 $C_2H_4 + 2H_2O \rightarrow 2CO + 4H_2$	✓					
16 $CO_2 + H_2 \rightarrow CO + H_2O$	✓				✓	✓
17 $CO + H_2O \rightarrow CO_2 + H_2$	✓				✓	✓

Source: From Model 1, Stansch et al. [53]; Model 2, Sohrabi et al. [54]; Model 3, Lacombe et al. [55]; Model 4, Olsbye et al. [56]; Model 5, Traykova et al. [57]; Model 6, Shahri and Alavi [58].

catalytic oxidative dehydrogenation of ethane is not included in the reaction network of Olsbye et al. [56]. Similarly, the reaction network of Traykova et al. [57] does not account for the total oxidation of hydrocarbons, whereas the thermal dehydrogenation of ethane was not included in the work of Shahri and Alavi [58]. Nonetheless, all the models discussed earlier describe well specific OCM features on the corresponding catalysts and operating conditions.

Thus, one major difficulty in trying to propose a *universal* mechanism is that heterogeneous reaction steps may strongly depend on the nature of the surface, whereas the gas-phase reaction network is reasonably well agreed and catalyst independent. Interactions between the catalyst surface and gas-phase species depend on binding energies and adsorption enthalpies, which, in turn, depend on the nature and the environment of the active sites present on the catalyst surface. As will be elaborated in Section 7.3.6.1, the development of a microkinetic model based on fundamental principles that considers all relevant elementary steps and, at the same time, includes catalyst descriptors that account for the catalyst intrinsic properties on the reaction kinetics will be the challenge and guideline of the present work.

7.1.4 Key Catalyst Characteristics

Most research effort on OCM materials started with simple oxides and evolved toward complex, multipromoted mixed oxides exhibiting enhanced performances, such as the formulas based on Na-W-Mn/SiO$_2$ [60–62]. These developments allowed elaborating progressively more complex reaction mechanisms accounting better and better for the different kinetic features of OCM [17,18,63–65]. Even if many surface phenomena remain under debate, several key features are now commonly accepted:

- Basic sites combined with surface defects activate gas-phase oxygen on the catalyst surface.
- C–H bond activation proceeds on these activated oxygen sites.
- CO, CO$_2$, and H$_2$O surface species are in quasi-equilibrium with their gas-phase counterparts.
- CO is catalytically oxidized toward CO$_2$.

Therefore, all catalysts share some properties that can potentially be captured by selected catalyst descriptors. Identifying these descriptors requires a more precise description of interactions between surface and gas phase [17,18,63,64,66].

7.1.5 Ionic Oxygen Species on the Catalyst Surface

According to the surface mechanism of methane activation presented earlier, a good performing material should exhibit active ionic oxygen species on its surface. This property is shared by alkaline-earth, rare-earth, some lanthanides, and transition metal oxides, making the latter, or more precisely their combination in solid mixtures, the best performing materials [39,49]. Several works [51,67–78] investigated the relationships between the basicity of materials and their performances as OCM catalysts and, hence, allowed a progressive understanding in the role of the oxygen surface species on the investigated catalysts. Among all ionic oxygen species that can be present on the catalytic surface according to the equilibrium

$$O_{2(gas)} \leftrightarrow O_{2(ads)} \leftrightarrow O_2^- \leftrightarrow O_2^{2-} \leftrightarrow O^- \leftrightarrow O^{2-} \tag{7.8}$$

the most reactive ones are O$^-$ and O^{2-}, which are also the most basic and the least electrophilic [32,35,79]. Reactivity and Lewis-type basicity of these sites depend on the coordination state of the oxygen, that is, on the associated metal cation playing a key role as stabilizer [49].

Moreover, the lower the coordination of the oxygen species, the higher its basicity. The strongly basic species O^{2-} associated to a vacancy is able to activate reversibly O$_2$ and CO$_2$ leading to either O$^-$ or a carbonate on surface. In the case of a very basic oxide, the pair {O^{2-} + □} would even perform a heterolytic methane dissociation [80]. As stressed previously, O$^-$ and O$_2^-$, as less basic sites, are assumed to be responsible for methyl and ethyl radical formation, via homolytic dissociation of methane [80] and total oxidation of CO to CO$_2$, respectively.

7.1.6 BASICITY OF MIXED AND PROMOTED OXIDES

As the basicity of an oxide is strongly affected by the cation contained by the oxide, doping or promoting a simple metal oxide with another metal is a convenient though hardly predictable method to tune the (mild) basicity of the oxide. Actually, mixed oxides were identified already very early in the OCM history as highly active and selective catalysts for OCM to C_2 conversion [68,81–83]. The role of different sites has also been partially elucidated, leading to a better understanding of competitive adsorption of the species involved [55]. If a promoter modifies the basicity of the support or the main solid phase or influences the stability of surface carbonates [49], it also largely affects the catalytic performance. Hence, strong synergy effects make some catalyst formulations much more effective than the corresponding simple oxides, independently of their basicity. Best combinations reported up to now in literature are Li_2O/MgO, La_2O_3/MgO, $La_2O_3(SrO)/CaO$, and $MnNa_2WO_4/SiO_2$. For the later system, Arndt et al. [61] tried recently to find out the specificities of such a formula.

7.1.7 MAIN TRENDS IN CATALYST FORMULATION

As pointed out earlier, a large number of formulations have been tested, and their performances were reported in the literature [11,67–69,82–85]. The aim of this section is not to analyze all these materials but to highlight the main trends in investigated catalyst formulations, from bare and single oxides to complex mixed systems, as well as in their characteristics.

The supports or main phases used in doped systems are typically metal oxides that are known for their thermal stability in the OCM operating temperature range [74]. Among the most frequently used ones are magnesia, alumina, silica, and lime. All are basic oxides exhibiting no electronic conductivity and moderate ionic oxygen conductivity. Their basicities are ranked according to the following order: $SiO_2 < Al_2O_3 < MgO < CaO$.

As presented by Maitra [49] and by Martin and Mirodatos [39], almost all the metallic elements have been investigated as promoters. The alkali and rare-earth metals have been identified as the best promoters. The resulting materials are solid solutions of mixed oxides, stable at high temperature, favoring the formation of surface lattice defects and oxygen vacancies. Lanthanides also proved to be excellent promoters. However, although promoters such as Li, La, and Ce may have a positive effect on several supports, true synergy effects that further enhance catalytic performances appear only for some combinations, such as Li/MgO, La/CaO, and Ce/ZnO, or for specific multipromoted combinations, such as Ce/Li/MgO and La/Ce/MgO investigated in the works of Bartsch et al. [86,87] and Dedov et al. [88], respectively.

Several authors also reported studies about halide promotion of metal oxides used as catalysts that successfully improve the performances for OCM, mainly Cl [85,89–94] and F [93,94]. However, other studies reported neither positive nor negative effects of chlorine on catalyst performances [85,92,95,96]. Machocki and Jezior [97] explains this discrepancy by differences in experimental conditions and catalytic materials used. This illustrates how many parameters are involved in OCM catalyst performance and that comparison of literature data is not straightforward.

This statement is quite well documented in a recent purely statistical analysis of literature data, performed by Zavyalova et al. [98]. A database comprising 1870 data sets on catalyst compositions was analyzed, and a number of elements and/or binary or ternary combination of elements were identified as present in the most performing OCM systems. As already quoted in the present literature survey, the identified best formulas are mainly based on Mg and La oxides. Alkali (Cs, Na) and alkaline-earth (Sr, Ba) metals used as dopants increase the selectivity of the host oxides, whereas dopants such as Mn, W, and the Cl anion have positive effects on the catalyst activity.

7.1.8 STRATEGY EMPLOYED

As stressed previously and at variance with all the previously quoted statistical-based literature analyses, the original task of the present work consists in a search for relationships between several parameters describing a complex system such as OCM: catalyst properties and kinetics of the reaction.

In order to predict the performances of an OCM catalytic system, the first requirement is to *describe* it by considering relevant phenomena, which might monitor the system performances. Then it can be attempted to establish relationships between the corresponding *descriptors* and the performance of the tested systems. This kind of relationship is typically referred to as *quantitative structure–activity relationship* (QSAR). However, given the particular attention on the catalyst descriptors rather than on the activity descriptors, the term *quantitative structure–descriptor relationship* (QSDR) may be more adequate, because typical nonlinearities related to catalyst activity are accounted for in the microkinetic model. The main steps for establishing this kind of predictive relationships are

- Obtaining experimental data
- Extracting pertinent descriptors from these data
- Establishing correlations between these descriptors and the catalyst properties/synthesis procedure
- Validating correlations by predicting performances from descriptors as determined by the correlations

As the amount of required experimental data increases with the number of descriptors and the complexity of the process, a high-throughput (HT) strategy was chosen for the experimental work. However, although QSAR and QSDR are basically statistical and mathematical relationships, the selection of potential pertinent descriptors essentially depends on the understanding of the reaction kinetics and mechanisms. Such a methodology implies therefore

- Identifying parameters that are characteristic for the system (catalyst + operating conditions + kinetics of the reaction + reactor design)
- Performing experiments, screening the operating window defined by these parameters, and collecting all accessible performance data

- Analyzing the data, identifying relevant parameters, and discarding irrelevant ones to extract the so-called descriptors, when possible, based on a detailed kinetic interpretation
- Employing statistical analysis, establishing quantitative relationships between the so-called descriptors and catalyst structure

Previous works leading to the collection of performance data through HT investigation on many diverse formulations were already published by our team [99,100]. In accordance with the recent conclusions of Zavyalova et al. [98], the best identified catalysts were found to belong to the LaSrCaO system, by combining detailed characterization techniques and focusing on the investigation of various catalyst descriptors. Therefore, the present work will be focused on this catalyst family, in order to demonstrate the efficiency of the methodology presented earlier.

7.2   EXPERIMENTAL

7.2.1   PARALLEL FIXED-BED REACTOR

The fixed-bed parallel reactor system, specifically adapted for testing simultaneously several catalysts at identical operating conditions, was already extensively described in previous work [100].

7.2.2   PERFORMANCE EVALUATION

The following characteristics were considered for performance evaluation.
  Methane and oxygen conversions are calculated as follows:

$$X(CH_4) = 100 \frac{F(CH_{4in}) - F(CH_{4out})}{F(CH_{4in})} \tag{7.9}$$

and

$$X(O_2) = 100 \frac{F(O_{2in}) - F(O_{2out})}{F(O_{2in})} \tag{7.10}$$

with F(i) being the molar inlet or outlet flow rate of component i.
  The carbon atom–based selectivity toward ethane, ethylene, and C_2's is calculated as follows:

$$S(C_2H_6) = 100 \frac{F(C_2H_6)}{F(C_2H_4) + F(C_2H_6) + 0.5F(CO_x) + 0.5F(CH_2O)} \tag{7.11}$$

$$S(C_2H_4) = 100 \frac{F(C_2H_4)}{F(C_2H_4) + F(C_2H_6) + 0.5F(CO_x) + 0.5F(CH_2O)} \tag{7.12}$$

$$S_2 = S(C_2H_6 + C_2H_4) = 100 \frac{F(C_2H_6) + F(C_2H_4)}{F(C_2H_4) + F(C_2H_6) + 0.5F(CO_x) + 0.5F(CH_2O)}$$

$$(7.13)$$

Selectivities of other compounds can be calculated in a similar way.

The combined effect of catalyst activity and selectivity is quantified by the product yield that is calculated as follows:

$$Y(C_2H_6) = \frac{S(C_2H_6) \cdot X(CH_4)}{100} \qquad (7.14)$$

$$Y(C_2H_4) = \frac{S(C_2H_4) \cdot X(CH_4)}{100} \qquad (7.15)$$

$$Y_2 = Y(C_2H_6 + C_2H_4) = \frac{S_2 \cdot X_{CH4}}{100} \qquad (7.16)$$

Finally, the so-called productivity is defined as the mass of desired products obtained per time unit and per catalyst mass:

$$P_2 = F(C_2) \min^{-1} g_{cat}^{-1} \qquad (7.17)$$

The experimental observations are validated by calculating the molar balance over the reactor. Hydrogen and oxygen balances exhibited small deviations related to water condensation and analytical issues. The carbon balance,

$$C_{bal} = \frac{2F(C_2H_4) + 2F(C_2H_6) + F(CO) + F(CO_2) + F(CH_{4out}) + F(CH_2O)}{F(CH_{4in})}$$

$$(7.18)$$

which is primarily used for performance calculations, never deviated more than 4%.

7.2.3 CATALYST SYNTHESIS

The LaSrCaO catalysts were synthesized by coprecipitation. Lanthanum, strontium, and calcium nitrates, purchased from Sigma-Aldrich in a purity exceeding 99.9%, were used as metal precursors. They were mixed in an aqueous solution before addition of oxalic acid. The solution was then evaporated, and the resulting powder was calcined for 6 h in air at 850°C yielding the catalytic material. Before reactor loading, catalysts were crushed and sieved to yield fractions in the particle size of 200–400 μm. Figure 7.2 maps the bulk amount of La and Sr investigated in the synthesized catalyst library.

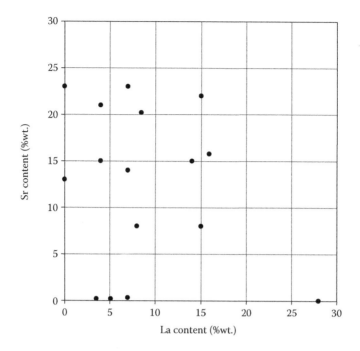

FIGURE 7.2   Mapping of La and Sr content for the investigated LaSrCaO catalysts.

7.2.4   TEXTURAL AND PHYSICOCHEMICAL CHARACTERIZATION TECHNIQUES

Characterization techniques were selected to grant access to structural information concerning either the bulk or the surface of materials. In the current study, the Statistica6© software was used for establishing relationships between catalyst characterization results and performance.

7.2.4.1   Bulk Characterization: Elemental Analysis by ICP

The nominal composition of the prepared catalysts was determined by chemical analysis via inductively coupled plasma and optical emission spectroscopy (ICP-OES) on a Jobin Yvon ACTIVA Spectrometer D ICP-OES instrument. Each measurement was made twice to avoid errors. The measurements were conducted after dissolution of the sample powder in sulfuric and nitric acids and injection of the solution into a plasma torch that atomizes the elements. The ions were then analyzed by a mass spectrometer.

7.2.4.2   Surface Characterization

7.2.4.2.1   BET Surface Area and Pore Size Determination

The BET specific surface area was determined by nitrogen adsorption at 77 K. Isotherm adsorption curves provide information on the pore shape and diameter for pore sizes in the mesoporosity range, that is, exceeding 3 nm. These measurements were carried out on an ASAP 2020 apparatus. Samples were outgassed prior to nitrogen adsorption by heating under secondary vacuum for 4 h at 350°C.

7.2.4.2.2 Surface Composition by XPS

X-ray photoelectron spectroscopy (XPS) was used to determine the surface composition of samples and bond energies of present elements. Analyses were made on an AXIS Ultra DLD spectrometer, from Kratos Analytical. The analysis methodology penetrates about five atomic sublayer deep in the sample.

7.2.4.2.3 CO_2 Temperature-Programmed Desorption

In order to investigate surface basicity, temperature-programmed desorption (TPD) measurements were performed on the most promising catalysts with CO_2 as probe molecule, due to its affinity for basic sites. The sixfold parallel fixed-bed reactor setup, described by Olivier et al. in [99,100], was used for this study.

The testing protocol as shown in Figure 7.3 was used for each catalyst, the setup being loaded with six different materials.

After reaching 850°C, the reactor is cooled down at a rate equal to 10°C min⁻¹. Steps (a) through (c) are performed only for the first catalyst. As the six catalysts are simultaneously subjected to the same treatment, the TPD cycles for the five others begin at step (d) after cooling. The inert used is Ar.

Desorbed CO_2 was quantified and recorded via mass spectrometry during the whole desorption procedure. As TPDs cannot be performed stepwise, the TPDs on six catalysts could not be performed in parallel but sequentially.

In order to get deeper insight in the type of interaction between CO_2 and the catalyst surface during CO_2 TPD experiments, DRIFTS has been used to identify the different carbonate species formed from CO_2 adsorption. An IR spectrometer coupled to a DRIFTS cell designed for high temperature (Harrick) and a gas feeding system was used and allowed recording the IR transmittance of species adsorbed on the surface of the catalyst and in the gas phase. The system allows heating the sample up to 850°C and conducting measurements under the chosen gas flow. Note, however, that the exact temperature of the catalyst bed on the surface, which corresponds

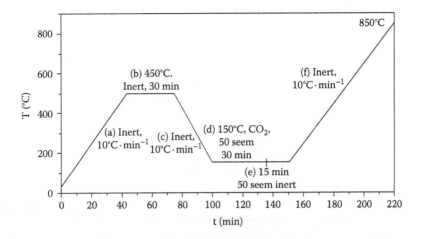

FIGURE 7.3 Protocol used for CO_2 TPDs; catalyst mass is 50 mg.

to the zone explored by the IR beam, may be lower (by almost 100°C at the highest temperatures) than the temperature measured at the bottom of the catalyst bed [101]. The TPD protocol was the same as described earlier, except that He was used as inert gas during DRIFTS experiments. The OMNIC© software was used to acquire and process transmittance spectra.

7.3 RESULTS AND DISCUSSION

7.3.1 PERFORMANCES

Catalyst performances were comparable to the ones reported in our previous work [100], where it was found that CaO with only La or Sr exhibited lower performances than the ternary oxides. Top formulation data obtained in the current study are summarized in Table 7.3.

TABLE 7.3

OCM Performances Obtained on Top LaSrCaO Formulations in Reaction Conditions: 50 sccm, (Ar:CH$_4$:O$_2$) = (48:35:17) during First Hours on Stream, at T$_{oven}$ = 850°C, T$_{oven}$ = 800°C, T$_{oven}$ = 750°C, and T$_{oven}$ = 700°C

La (wt.%)	Sr (wt.%)	X(O$_2$) (%)	X(CH$_4$) (%)	S$_2$ (%)	Y$_2$ (%)	C$_2$H$_4$/C$_2$H$_6$	Y(C$_2$H$_4$) (%)
			T$_{oven}$ = 850°C				
14	15	99	37	41	15	1.3	9
15	8	99	38	42	16	1.5	9
4	15	99	39	49	19	1.7	12
14	22	99	42	57	24	2.1	16
			T$_{oven}$ = 800°C				
14	15	98	39	46	18	1.2	10
15	8	99	38	45	17	1.5	10
4	15	95	37	48	18	1.5	11
14	22	95	40	56	22	1.6	14
			T$_{oven}$ = 750°C				
14	15	95	37	47	17	1.1	9
15	8	98	39	44	17	1.4	10
4	15	87	33	42	14	1.1	7
14	22	62	23	38	9	0.6	3
			T$_{oven}$ = 700°C				
14	15	78	32	41	13	0.8	6
15	8	99	38	42	16	1.2	9
4	15	53	18	14	3	0.3	1
14	22	57	18	20	4	0.3	1

7.3.2 BET Surface Area

For the whole LaSrCaO catalyst family, the BET surface area was found to range between 20 and 50 m² g⁻¹ with average pore size close to 5 nm. No clear relationship between these characteristics and the performances was observed in the studied range of operating conditions, in accordance to other similar studies [102].

7.3.3 CO₂ TPD Combined with In Situ DRIFTS over LaSrCaO Catalysts

7.3.3.1 TPD Curve Analysis

Figure 7.4 gives an example of CO₂ TPD profile analysis including the deconvolution of a well-defined peak at intermediate temperature into two peaks, while a high-temperature badly resolved peak is not further considered. The temperature and amounts of desorbed CO₂ corresponding to a series of LaSr/CaO catalysts are reported in Table 7.4.

The following features can be proposed from these TPD data:

1. Three different CO₂ TPD peaks were observed, revealing at least three different carbonate types, corresponding to various basic strengths.
2. Two peaks corresponding to CO₂ desorbing at intermediate temperature, that is, between 650°C and 750°C, are present in all catalysts. The corresponding amounts of desorbed CO₂ depend strongly on the catalyst composition. However, no clear relationship can be observed in Table 7.4 between the catalyst bulk composition and the temperature or area of these TPD peaks around 700°C.

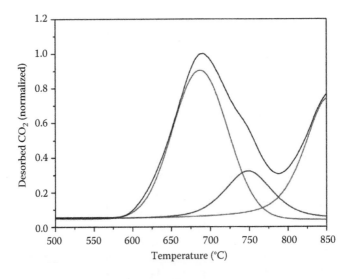

FIGURE 7.4 Decomposition of a typical TPD profile obtained after CO₂ adsorption over LaSr/CaO catalysts. Here, the tested catalyst is 4% La–15% Sr/CaO.

TABLE 7.4

Data Extracted from TPD Profiles as a Function of Catalyst Composition

La Bulk (%wt.)	Sr Bulk (%wt.)	First Peak Max Temp. (T_1) (°C)	First Peak Area (A_1) (CO_2 mol)	Second Peak Max Temp. (T_2) (°C)	Second Peak Area (A_2) (CO_2 mol)	A_1/A_2
15	22	682	1.7×10^{-5}	741	2.9×10^{-5}	0.6
0	23	—	0	723	6.0×10^{-5}	0.0
4	15	687	3.3×10^{-5}	748	1.5×10^{-5}	2.2
14	15	685	4.3×10^{-5}	730	1.4×10^{-5}	3.1
8	8	741	3.1×10^{-4}	779	1.3×10^{-4}	2.4
7	23	—	0	849[a]	0	—
4	21	680	1.5×10^{-5}	716	1.3×10^{-5}	1.2
15	8	726	1.4×10^{-4}	754	1.3×10^{-4}	1.1
7	14	675	4.0×10^{-5}	704	2.4×10^{-5}	1.7
0	13	706	1.1×10^{-4}	718	4.6×10^{-5}	2.4

Temperature and desorbed CO_2 amount corresponding to deconvoluted desorption peaks in the intermediate temperature range.

The non-fully-resolved high-temperature CO_2 desorption peaks are not considered in this table.

[a] This catalyst did not exhibit any CO_2 desorption below 800°C; thus, no value is reported there.

3. Though the high-temperature peaks are not well defined, it was observed that the samples with high Sr contents displayed the largest high-temperature TPD peaks. Thus, a surface enriched in Sr leads to the formation of highly stable carbonates.

7.3.3.2   In Situ DRIFTS under CO_2 TPD Conditions

Figure 7.5 shows IR absorbance spectra obtained upon temperature ramping under inert flow after CO_2 adsorption on four different catalysts selected to illustrate a large span of OCM performances: 14wt.% La, 15wt.% Sr/CaO, 28wt.% La/CaO, 23wt.% Sr/CaO, and bare CaO as reference.

As seen in Figure 7.5, several IR bands were observed between 900 and 1800 cm^{-1}, which can be assigned to the different C–O vibrations of carbonate species, and small bands between 3600 and 3700 cm^{-1}, which can be assigned to the O–H vibration of acid carbonates. All the bands observed in the intermediate wave number range (2000–3000 cm^{-1}) are overtones of the vibration modes identified in the low wave number range, as observed in a typical IR spectrum of calcium carbonate [103].

The various types of carbonates that can be expected upon CO_2 adsorption on basic materials are schematized in Figure 7.6 in accordance with Xu et al. [51], Lavalley [104], and Busca and Lorenzelli [105].

The doublet with relatively small intensity around 1760–1780 and 1170–1180 cm^{-1} can be assigned to $\nu(C = O)$ and $\nu_{asym}(OCO)$ modes, respectively, of bidentate and/ or bridged carbonates, that is, linked to two oxygen anions of an oxide, based on

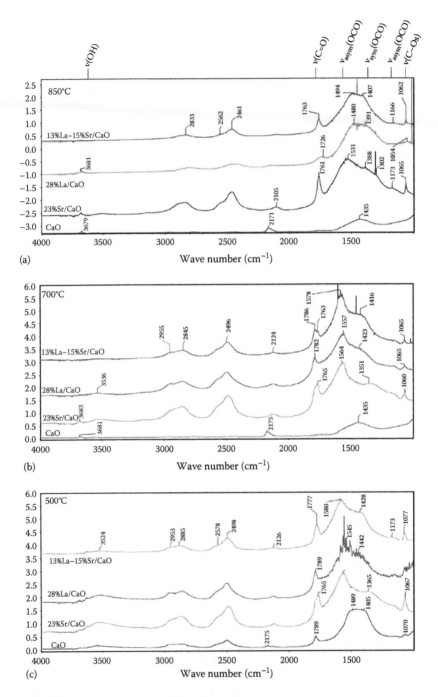

FIGURE 7.5 In situ recorded IR absorbance spectra obtained upon CO₂ TPD for materials with different La and Sr content under inert gas, at 10°C min⁻¹ temperature increase, at (a) 850°C, (b) 700°C, and (c) 500°C.

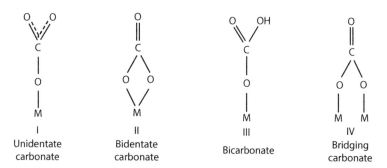

FIGURE 7.6 Carbonate and hydroxycarbonate (or bicarbonate) main structures: (I) unidentate carbonate, (II) bidentate carbonate, (III) bicarbonate, and (IV) bridging carbonate. (From Xu, Y. et al., *Catal. Lett.*, 35, 215, 1995; Lavalley, J.C., *Catal. Today*, 27, 377, 1996; Busca, G. and Lorenzelli, V., *Mater. Chem.*, 7, 89, 1982.)

the large Δv_3 splitting of the two $v(OCO)$ modes, as generally accepted in literature [104,105]. These strongly bounded species are present over the whole temperature range. The intense bands at 1580–1570 and 1410–1350 are assigned to $v_{asym}(OCO)$ and $v_{sym}(OCO)$, respectively, of unidentate, also denoted as *monodentate*, carbonates, that is, linked to a single oxygen anion of the oxide, based on the large Δv_3 splitting of the two $v(OCO)$ modes, while the band at 1065–1060 cm^{-1} is assigned to the $v(C–Os)$ mode. These unidentate carbonates are typical for a CO_2 interaction with strongly basic O^{2-} sites. These more loosely bonded carbonates start to decompose at 700°C, that is, at typical OCM conditions, as expected from a weaker interaction with the surface. They might also partly be transformed into bidentate carbonates [104–106].

The development of bicarbonates at low temperature, that is, from ambient temperature to 500°C, has also been revealed by the presence of $\delta(COH)$ modes around 1300 cm^{-1} or $v_{sym}(OCO)$ mode, together with the band at 3618 cm^{-1} in the position characteristic of the bicarbonate $v(OH)$ mode. This formation of bicarbonates reveals the presence of substantial amounts of hydroxyl groups with weak basic character. Their abundance at low and medium temperature was also revealed unambiguously by XRD measurements [99].

It is worthwhile to stress that these carbonate species detected indirectly by TPD or directly by DRIFTS are essentially bulk species, as revealed by XRD [99], but include surface species as well that are, hence, possibly involved in the OCM catalytic cycle for the most unstable species under reaction conditions (see also Section 7.3.3.3).

7.3.3.3 Comparison of Carbonate Species over LaSrCaO Catalysts

As evident from the TPD profile analysis and IR absorbance spectra obtained during CO_2 TPD, there are several types of basic sites on the various tested oxides, of varying basic strength, all forming carbonates upon CO_2 adsorption. In all temperatures investigated, it can be seen from Figure 7.5 that the performing La-Sr/CaO system displays IR bands that are quite similar to the La/CaO system, while the Sr/CaO material exhibits slightly different wave numbers for the mono- and bidentate carbonates. The IR spectrum for the CaO support is even more different, presenting practically no bidentate species, and much less stable monodentate carbonates. The similarity between La-CaO and La-SrCaO suggests that both systems present close surface composition, which will be assigned to a strong La surface enrichment as evident from XPS measurements (see Section 7.3.4).

Reaching the high-temperature range typical for OCM, that is, between 700°C and 850°C, part of these carbonates, the monodentates in particular, decompose. The amount of carbonates remaining at 850°C is significantly higher for the Sr/CaO system than for the La/CaO or La-Sr/CaO, which confirms that the strontium carbonates are much more stable than the lanthanum ones. On the contrary, it is clear from Figure 7.5 that calcium carbonates are quite unstable.

Similar trends were obtained from XRD patterns obtained before and after OCM on a performing sample such as La 14 wt.% and Sr 15 wt.%/CaO system [99]. Only the $SrCO_3$ phase was still present above 700°C, demonstrating its highest thermal stability as a bulk phase. In contrast, only hydroxides, $La(OH)_3$ and $Ca(OH)_2$, and oxides, La_2O_3 and CaO, were detected by XRD after OCM reaction.

7.3.3.4 Relationship between Carbonate Species and Catalytic Activity

As mentioned previously, carbonates are believed to affect significantly the catalytic activity in OCM [38,50–52]. Figure 7.7 shows the area of monodentate (a) and bidentate (b) peaks from IR spectra on the three CaO-promoted catalysts.

The Sr-promoted CaO catalyst shows carbonates with the highest stability that significantly desorb only at high temperatures (above 800°C). The amount of bidentate species in this material decreases slightly at 850°C, and as a result, the decrease in monodentates in the temperature range from 600°C to 800°C can be attributed to their desorption as no parallel increase in bidentate carbonates is obtained. However, monodentate carbonates are not completely removed from the surface at 850°C with half of them still remaining.

The La-promoted CaO catalyst exhibits carbonates with lower surface stability. Monodentate carbonates disappear in the temperature range of 600°C–800°C, while bidentate ones are present until higher temperatures. In any case, very few carbonate species remain on the surface at temperature around or above 850°C.

The LaSr-promoted CaO catalyst exhibits an intermediate behavior compared to the two aforementioned materials. Monodentate species disappear above 700°C, but this is accompanied by a noticeable increase in the bidentate content at 700°C. Thus, on this catalyst, the transformation of monodentate into bidentate carbonates illustrates the mutual influence of La and Sr on the stability of their carbonates. Bidentate carbonates are more stable on this catalyst than on the La/CaO one; however, at 850°C, a pronounced desorption of these carbonates is observed.

Since the La-doped systems are consistently better performing than the Sr-doped ones, as seen by Olivier et al. [100], it is proposed at this stage of the study that the following equilibrium is essential under reaction conditions:

$$CO_2 + La_2O_3 \leftrightarrow La_2O_2CO_3 \tag{7.19}$$

This equilibrium cannot be achieved on surfaces enriched with Sr due to the too high stability of the strontium carbonates. Moreover, the preferential decomposition of

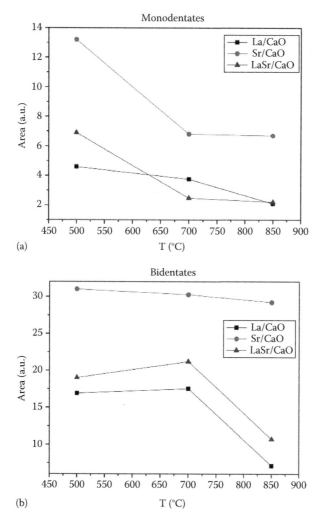

FIGURE 7.7   Area of IR absorbance peaks corresponding to (a) monodentate and (b) bidentate carbonate species on the La/CaO, Sr/CaO, and LaSr/CaO catalysts.

monodentate as compared to bidentate carbonates under OCM conditions strongly suggests that the *active* carbonate species are essentially the most unstable ones, that is, the monodentate carbonates. In turn, the most stable bidentate species can therefore be described as *spectator* species, not directly participating in the reaction or even more *poisoning* species when saturating the surface with catalytically inert species.

7.3.4   XPS Analysis of LaSr/CaO Materials

XPS investigations were performed to identify the precise composition and electronic state of the surface for some selected catalysts (see Table 7.5).

TABLE 7.5

Quantitative Analysis from ICP (Bulk Composition) and XPS (Surface Composition) Obtained for Catalysts at Ambient Temperature, Quenched from 700°C in Air after Initial Calcination

Catalyst (%wt.) (ICP)	Bulk (atomic ratio from [ICP]) La/Sr	Surface (%at. [XPS])				Surface (atomic ratio from [XPS]) La/Sr
		La	O	Ca	Sr	
4% La–21% Sr/CaO	0.11	4.1	42.2	42.2	11.6	0.35
15% La–22% Sr/CaO	0.41	4.6	47.4	35.0	13.0	0.36
8% La–20% Sr/CaO	0.26	7.5	0.0	31.1	7.9	0.95
8% La–8% Sr/CaO	0.64	5.8	0.0	64.2	4.2	1.40
7% La–22% Sr/CaO	0.20	9.9	16.4	52.2	21.6	0.46
23% Sr/CaO	0.00	0.0	14.7	72.5	12.8	0.00

As can be seen, for all cases, a clear surface enrichment in La and depletion in Sr was observed as compared to the nominal bulk composition.

7.3.5 RELATING INTRINSIC CATALYST DESCRIPTORS AND PERFORMANCE

From the results obtained by Olivier et al. [99,100], related to a primary screening of diverse OCM catalysts, poorly predictive relationships were found by considering a number of descriptors essentially related to the bulk properties of the tested materials such as BET surface, optical density (Λ), ionization potential of metallic cations (Ei), formation enthalpies of metal oxides, electronegativity of metallic species, lattice crystal energies (LCEs) of metal oxides, or the difference between promoting cations and support or main phase cation ionic radius (ΔRi). Therefore, it was concluded that intrinsic catalyst descriptors related to the state of the surface rather than of the bulk have to be considered to find relevant and predictive relationships between these descriptors and the OCM performances.

In the present work, this approach was applied to the LaSrCaO catalyst family. Both the bulk and surface catalyst compositions were considered when assessing potential relationships with the catalytic performance to demonstrate differences in significance of the pursued relationships.

Figure 7.8 shows parity diagrams obtained via statistical analysis between the observed C_2 yield and that predicted from the bulk (Figure 7.8a) and catalyst surface (Figure 7.8b) composition, employing a linear regression model. The corresponding linear regression equations drawn as best fit for the predicting models are as follows:

Bulk

$$\langle Y_2 \rangle = 0.54 \text{ La wt.\% (bulk)} - 0.02 \text{ Sr wt.\% (bulk)} \qquad (7.20)$$

Surface

$$\langle Y_2 \rangle = 1.287 \text{ La at.-\% (surf)} - 0.078 \text{ Sr at.-\% (surf)} \qquad (7.21)$$

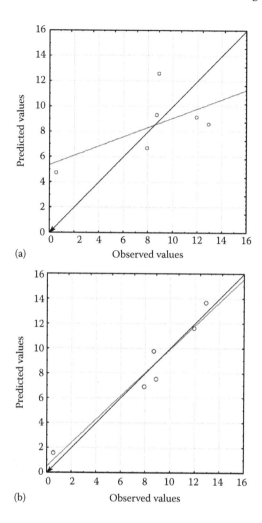

FIGURE 7.8   Parity plot obtained from the QSAR like relationships between C_2 yield and (a) bulk (from ICP measurements) or (b) surface (from XPS measurements) composition.

Whereas the bulk composition remains an irrelevant descriptor as seen from the data scattering in Figure 7.8a, a clear relationship between surface composition, determined per moles units, and OCM performance is observed in Figure 7.8b. It indicates a strongly positive role for La and a slightly negative role for Sr in the range of investigated surface concentrations.

The aforementioned observations can clearly be related to the interaction of CO_2 with the surface of La, Sr, and Ca oxides that was seen to play a significant role via the formation of surface and bulk carbonates.

La was shown to be essentially responsible for catalyst activity in the 650°C–800°C range, which could be partly related to the monodentate carbonate formation/decomposition equilibrium, while an excessive Sr concentration was seen to decrease the catalyst activity at intermediate temperature. At high temperature, catalysts containing both La and Sr in optimized amounts were found to be highly performing, indicating a synergy effect, which will be discussed later on the basis of the molecular requirements for the selective and unselective routes in OCM.

Thus, a clear demonstration of the joint role of surface ions is displayed in the linear relationship, relating the C_2 yield (1) to the La surface concentration, with a clearly positive trend, and (2) to the Sr surface concentration, with a slightly negative trend, indicating that Sr addition above a certain threshold becomes counterproductive, but still necessary for achieving a satisfying parity plot.

7.3.6 DETERMINATION AND OPTIMIZATION OF KINETIC DESCRIPTORS

7.3.6.1 Microkinetic Modeling as a Validation Tool of the Main OCM Mechanistic Features

Initially developed by Couwenberg et al. [107], the present kinetic model is implemented in a 1D heterogeneous reactor model to properly account for the irreducible mass transport limitations that arise due to the high reactivity of gas-phase intermediates [108]. In the work of Sun et al. [17], the use of descriptors was first introduced in the model so as to facilitate knowledge extraction from HT experiments, while, additionally, the description of deep oxidation reactions in the kinetic model was enhanced. Furthermore, in the work of Thybaut et al. [18], the heterogeneous oxidation of ethylene was considered in the network, while the resulting model was utilized to conduct a theoretical study so as to optimize C_2 product yields. Validation of the microkinetic model took place with the experimental data acquired with Li/MgO and Sn/Li/MgO catalysts completed with preliminary data obtained during the screening of the LaSr/CaO catalyst library in the parallel experimental reactor used also in the current work. More recently, Kechagiopoulos et al. [19] further elaborated the model on the reactor as well as on the microkinetic level. Intraparticle concentration profiles for the surface intermediates were accounted for explicitly, additionally to the already implemented for gas-phase molecules and radicals. Moreover, the surface network was considerably expanded by considering H atom abstraction reactions for all gas-phase species and by elaborating on the radical quenching mechanism. The resulting kinetic model comprises a reaction network of 39 gas-phase and 26 catalytic elementary steps that are coupled via the heterogeneous reactor model. Parameter estimation is performed using a combination of Rosenbrock and Levenberg–Marquardt algorithms, while the weighed sum of the squared residuals, between the observed and calculated molar fractions at the reactor outlet, is used as the objective function. The estimated parameters are all of the defined descriptors as described elaborately in the aforementioned publications. Validation of this comprehensive microkinetic model was presented in the recent work of Alexiadis et al. [109]. It was clearly demonstrated, via the good agreement between simulated and experimental results on (Sn/)Li/MgO and Sr/La$_2$O$_3$ catalysts, that the present model describes successfully generic reaction performances based on a single reaction network.

Via sensitivity analysis [17] and comparative analysis of different catalysts [18], four of the catalyst descriptors defined were found to be the most influential in the simulation of the OCM performance, c.q., the C_2 yield, that is, D_1, the reaction enthalpy for the hydrogen abstraction from CH_4; D_2 and D_7, the oxygen and water chemisorption enthalpies; and D_{14}, the density of active sites. These parameters are obviously all directly involved in the key surface steps controlling the rate of the selective pathways: (1) methane activation into methyl radicals; (2) oxygen activation creating the active surface species (O^- or O_2^-) able to generate methyl radicals; (3) water adsorption/desorption determining the fate of the surface OH groups formed during H abstraction from methane and, hence, needed for completing the catalytic cycle; and (4) the density of sites, that is, a descriptor directly related to the state (composition, structure, and texture) of the working surface. This key finding supports the previous assumptions in a quantitative way about the essential role of these surface steps for the selective activation of methane to C_2's.

7.3.6.2 Virtual C_2 Yield Mapping and Relation to Surface Coverage

Two of the aforementioned descriptors, D_1 being the reaction enthalpy for hydrogen abstraction from methane and D_2 the chemisorption enthalpy of oxygen, were identified as the ones primarily impacting the catalyst performance and are, hence, considered as the kinetic descriptors (KDs) to be further investigated.

In our previous work [18], the mapping of simulated C_2 yields as a function of the two above mentioned descriptors, D_1 and D_2, was presented. Similar to the results presented by Su et al. [64], the model predictions indicated that the C_2 yield, with maximum achievable values amounting to approximately 27%, strongly depends on D_1 and D_2. More specifically, two maxima in C_2 yield were identified in that work, the first one corresponding to a moderate value for the oxygen chemisorption enthalpy (~100 kJ mol^{-1}) and a low value for the reaction enthalpy for the hydrogen abstraction from methane (~50 kJ mol^{-1}), whereas the second one was achieved at a much higher value for the oxygen chemisorption enthalpy (~225 kJ mol^{-1}) and a slightly higher value for the reaction enthalpy for methane hydrogen abstraction (~75 kJ mol^{-1}). Given the various enhancements of the model since that work, an updated version of this C_2 yield map versus the D_1 and D_2 descriptors is presented in Figure 7.9. Although the two previously identified maxima are not visible in this figure, there is a clear progression of the C_2 maximum yield from the bottom-left corner of the energy map toward the top-right corner. This C_2 maximum yield trajectory actually overlaps with the original maxima points completely, the latter two regions corresponding to the leftmost and rightmost areas of the former.

The descriptor values, here calculated for one of our best performing LaSr/CaO catalysts, that is, 10 wt.% La–20 wt.% Sr/CaO [100], are reported in Table 7.6. They lead to a C_2 yield of about 26%, as well as for an in silico optimized catalyst leading to a C_2 yield of 27%.

Clearly, these descriptor values for our best experimentally investigated LaSr/CaO catalyst fall in the maximum belonging to lower surface reaction enthalpy values mentioned before, while the values for the optimized catalyst correspond to the C_2 yield maximum with higher surface reaction enthalpy.

The moderate value for the oxygen chemisorption enthalpy and the low value for the reaction enthalpy for hydrogen abstraction from methane would mean that methane activation proceeds relatively easily and, hence, a small surface coverage

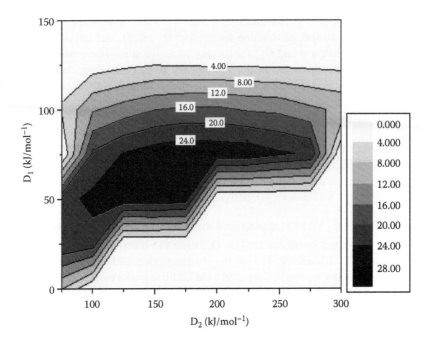

FIGURE 7.9 C_2 yield prediction with respect to the KDs D_1 and D_2, being the reaction enthalpy for hydrogen abstraction from methane and chemisorption enthalpy of oxygen, respectively. (Operating conditions: $T = 1069$ K, $p = 108$ kPa, $CH_4/O_2 = 3.0$, $W/F_{t,0} = 11.0$ kg s mol^{-1}. The simulations were performed with the microkinetic model only varying the values of catalyst descriptors D_1 and D_2 in the range from 0 to 300 kJ mol^{-1} and from 0 to 200 kJ mol^{-1}, respectively.)

TABLE 7.6
Values of D_1 and D_2 Corresponding to the Best Performing LaSr/CaO Catalyst and an *In Silico* Optimized Catalyst

Descriptors	LaSr/CaO	Optimized Catalyst
D_1 (kJ mol^{-1})	59	72.46
D_2 (kJ mol^{-1})	103.2	224.1
C_2 yield (%)	26	27

by oxygen suffices for achieving high methane activation rates. In order to verify this statement, the surface coverage by all the reacting intermediates considered in our OCM mechanism, see also the work by Thybaut et al. [18], as predicted from the microkinetic model were calculated for the D_1 and D_2 values corresponding to the tested LaSr/CaO best catalyst, that is, 50 and 100 kJ mol^{-1}, respectively, as reported in Figure 7.10a. It can be seen that the surface oxygen coverage does not exceed 10%, as anticipated from the previous reasoning.

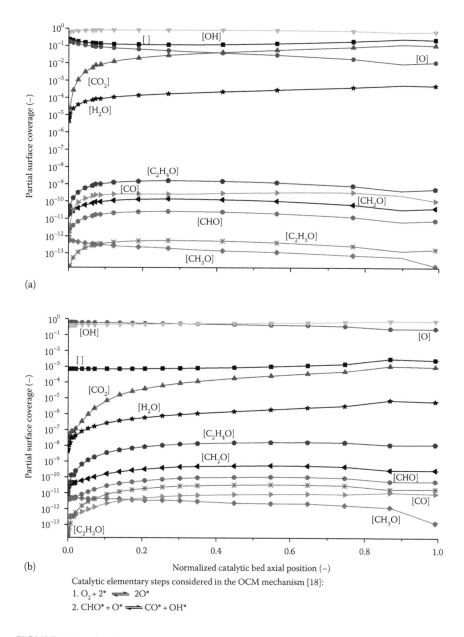

(a)

(b)

Normalized catalytic bed axial position (–)

Catalytic elementary steps considered in the OCM mechanism [18]:

1. $O_2 + 2^* \rightleftharpoons 2O^*$
2. $CHO^* + O^* \rightleftharpoons CO^* + OH^*$

FIGURE 7.10 Surface coverage for the reacting intermediates considered in the OCM mechanism, as predicted by the microkinetic model. (The following list reports all the elementary steps as reported in Table 7.1 from Ref. [18].) (a) $D_1 = 50\ kJ\ mol^{-1}$, $D_2 = 100\ kJ\ mol^{-1}$; (b) $D_1 = 75\ kJ\ mol^{-1}$, $D_2 = 225\ kJ\ mol^{-1}$. (Operating conditions: $T = 1069\ K$, $p = 108\ kPa$, $CH_4/O_2 = 3.0$, $W/F_{1,0} = 11.0\ kg\ s\ mol^{-1}$.) [] ■, [O] ●, [OH] ▼, [CO_2] ▲, [CH_3O] ◆, [CH_2O] ◀, [CO] ▶, [CHO] ⬟, [H_2O] ★, [C_2H_4O] ⬠, [C_2H_3O] ✻.

It can be seen also in Figure 7.10a that in addition to O* species, the most abundant reacting intermediates are OH* adspecies, while all the carbon containing intermediates are in negligible concentration. The latter result is consistent with the very small amount of carbonates remaining at high temperature as observed by DRIFTS in Figures 7.5 and 7.7 and our previous assessment that they are essentially bulk carbonates, mainly linked to Sr cations, not participating to the catalytic cycle, being only *spectators* or poisoning species. In turn, as proposed previously, only unstable carbonates would participate in the catalytic cycle, via anionic basic sites linked to La cations. Note also that the very low concentration of carbon containing intermediates as predicted from the microkinetic model is quite in line with direct measurements of OCM reacting intermediates measured by SSITKA for a close La based catalyst, as reported by Lacombe et al. [27]. Complementary experiments are in progress to identify more precisely the hydroxyl groups under similar reacting conditions as used for the calculated surface occupancy.

If one considers now a virtual increase of the D_1 and D_2 values to 75 and 225 kJ mol^{-1} respectively, it leads to the second maximum in the C_2 yield mapping reported in Table 7.6. For this case, a markedly higher oxygen coverage of approximately 60% is established, as can be seen in Figure 7.10b. It corresponds to a higher oxygen chemisorption enthalpy, which would also promote deep oxidation reactions on the catalyst surface as an excessive number of activated methyl species would be oxidized before desorbing as radicals into the gas phase. At the same time, in case of a higher enthalpy for hydrogen abstraction from methane, a higher oxygen surface coverage would be necessary to allow for a rapid activation of methane to occur. These mechanistic trends would correspond to this second maximum in the C_2 yield mapping.

In principle, more surface oxygen should enhance also the deep oxidation of hydrocarbon species. However, since the surface is practically saturated by O* and OH* species, the fraction of free active sites is significantly reduced (almost two orders of magnitude) and, hence, also the fraction of the surface covered by CO (or methoxy species) is reduced, resulting in a less pronounced deep oxidation. Selective methane activation, occurring via an Eley–Rideal type reaction, hence, not requiring free active sites, would not be affected by these surface coverage effects.

7.3.7 PHYSICAL MEANING AND INTERPRETATION OF INTRINSIC AND KINETIC DESCRIPTORS

The last and probably most challenging objective in the discussion of descriptors is proposing an integrated overview and assigning to all the listed descriptors a general physical meaning and interpretation.

It was already commented that anionic defects or vacancies are required for oxygen activation, with the surface concentration of the former being represented by the active sites density descriptor. As generally agreed, though never unambiguously demonstrated, it is most likely that the anionic vacancies density depends on the surface formulation and on its redox state. Surface basicity should also be taken into account, since it affects directly the competition for surface occupancy between activated oxygen and stabilized carbonates formed on the basic sites.

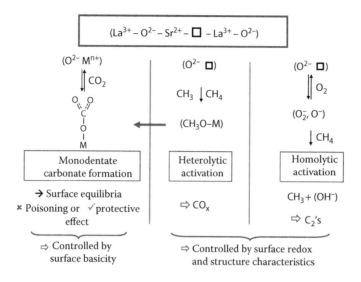

FIGURE 7.11   Global surface pathway competition under OCM conditions.

This kind of competition between defect sites and basic sites density can further be elaborated by reconsidering the status of the working surface under OCM conditions. The surface reaction can be schematized as a competition between two main parallel reaction pathways: the unselective one leading to CO_2 formation and the selective one leading to methyl radicals and further on to C_2 formation. Figure 7.11 tentatively describes this state of the working surface, for the general case of performing mixed oxides such as the LaSr/CaO system selected in the present study.

Generally speaking, the surface reactions are determined by the species involved in the elementary steps and can be classified in a simplified way as follows:

Selective route: The activation of oxygen into active, nucleophilic species such as O^- or O_2^-, requires a lattice vacancy, as seen in the right hand side of Figure 7.11. The following methyl radical formation can derive from the homolytic methane C–H bond cleavage.

Unselective route: Methane can also be activated according to a heterolytic way reacting on a basic pair {vacancy—oxygen reticular anion}, as illustrated in the middle of Figure 7.11. The formed methoxy species are further on oxidized into surface carbonates, which, as indicated on the left side of the figure, are in equilibrium with the gas phase. This required equilibrium would correspond to the less stable adspecies, that is, the unidentate carbonates formed on the strongly basic sites of the mixed oxide overlayers.

Thus, two states of the surface, depending on the activated species, appear to control both the total methane conversion and the C_2 selectivity and, hence, the C_2 yield and productivity. As demonstrated by the direct and predictive QSAR like relationship of the OCM performance to the surface concentrations of La and Sr for the case of the LaSrCaO system, these states globally depend on the surface composition. In more detail, a sharply tuned balance between two types of basic cations, that is, La and Sr, will adjust the surface basicity and the vacancies concentration to the required level, but in a way which cannot be generalized as a unique combination. This surface requirement was further analyzed via a microkinetic analysis.

Thus, the identified optimum C_2 yield was found to correspond to a low oxygen concentration along with a facile methane activation. This property could be obtained from a catalyst with a low number of selective sites/lattice vacancies having a high turnover frequency (TOF). To relate this site efficiency to the above discussed surface composition, it is proposed that (1) the lanthanum oxide provides most of the active sites for the selective and unselective routes, via its property to form surface defects for the selective route and establish a fast formation/decomposition of monodentate carbonates for the unselective route and (2) the addition of strontium will accelerate the redox process of active site regeneration. As a matter of fact, Sr is known not only to accelerate oxygen diffusion through mixed oxides but also to stabilize the carbonates phase. The positive effect of Sr toward the redox process would be dominant at low concentration, as required for the optimized formula, while at higher concentration, Sr will poison the active surface by saturating it with ultra stable carbonates, as observed experimentally.

The second optimum C_2 yield scenario corresponding to a markedly higher surface oxygen coverage, that is, 60% rather than 10%, requires a higher vacancy concentration, but the methane activation on each site should be more demanding, that is to say the active sites should exhibit a lower efficiency. It could be speculated here that less basic materials, avoiding the unselective route, would possess a higher concentration of surface defects and, hence, could meet these surface requirements.

As recently highlighted in the elaborate review by Arndt et al. [61], another complex catalyst, namely, the $Mn-Na_2WO_4/SiO_2$ system, has been proven in a generally reproducible way by many researchers to be highly effective for OCM, leading to CH_4 conversion of 20%–30% and a C_2 selectivity of 70%–80%, while the best performances of the LaSrCaO system correspond a CH_4 conversion of 30%–40% and a C_2 selectivity of 50%–60% [99,100]. Moreover, in the kinetic studies conducted by Pak et al. [26] it was stipulated that the high selectivity of the $Mn-Na_2WO_4/SiO_2$ materials originates from their low reactivity with CH_3 radicals, hence avoiding the unselective production of CO_2 from CH_4. In the same study, unusually high activation energies for CH_4 conversion were calculated that were rationalized by assuming a strongly endothermic dissociative adsorption of O_2. Considering all this information in conjunction with the current modeling results, it appears highly probable that the best performing $Mn-Na_2WO_4/SiO_2$ materials could correspond to this second C_2 yield maximum, in line with preliminary results obtained in our research team. This sensitivity of the main KDs identified in the present study to surface composition and structure obviously opens a new direction for the synthesis of even more advanced materials bearing the potential of still improving OCM performance.

Prior to concluding, it should be stressed that performances obtained during this work were significantly improved based on the reasoning and applied strategy. The initial catalysts tested during the first screening achieved a C_2 yield of less than 19% at a C_2 productivity lower than 1.2×10^{-5} mol s^{-1} g_{cat}^{-1}, while the final materials, obtained after secondary screening, reached a C_2 yield of 24% and C_2 productivity higher than 5×10^{-5} mol s^{-1} g_{cat}^{-1}. The ethylene-to-ethane ratio was also improved from about 1.5 to 5. Although exceeding the 25% yield limitation already mentioned in literature still remains a challenge, continuing the combined strategy of HT experimentation and knowledge extraction via microkinetic analysis can definitely bring this goal within reach.

7.4 CONCLUSIONS

The present study allowed the identification of kinetic and catalyst descriptors for the OCM reaction from experimental data on LaSrCaO catalysts and established relationships between these descriptors and catalyst performances. La positively affects OCM catalyst activity via monodentate carbonate formation/decomposition equilibrium, while an excess of Sr was seen to have a negative effect in the range of investigated surface concentrations. Catalysts containing La and Sr in optimal amounts exhibited a synergy effect leading to enhanced performance, that is, C_2 yields and productivity. Microkinetic modeling validated the essential role of gas-phase oxygen activation in creating the active surface species necessary for the activation of methane to C_2's, while the surface requirements for high OCM performance were identified. An optimal C_2 yield can be achieved either via materials that possess few lattice vacancies and defects but a high turnover frequency or a high concentration of defects and vacancies but a low basicity so as to avoid unselective routes. The current work demonstrated that a combined approach by means of HT strategies and microkinetic modeling applied to the complex OCM process can bring new insights in (1) the design of catalytic materials, (2) the understanding of the key features of the process, and ultimately (3) approaching predicting relations between selected descriptors and the catalyst performance in the OCM reaction.

ACKNOWLEDGMENTS

This work was initiated within the frame of *TOPCOMBI*, an integrated project within the 6th European Framework Programme for Research and Technological Development, belonging to the Thematic Priority *Nanotechnologies and nanosciences, knowledge-based multifunctional materials and new production processes and devices.*

It was continued within the project *OCMOL (Oxidative Coupling of Methane followed by Oligomerization to Liquids).* OCMOL is a large-scale collaborative project supported by the European Commission in the 7th Framework Programme (GA no. 228953). For further information about OCMOL, see: http://www.ocmol. eu or http://www.ocmol.com

REFERENCES

1. Technip brochures, Engineering and technologies—Ethylene production. Technip - Group Communications - September 2013 http://www.technip.com/sites/default/files/technip/publications/attachments/Ethylene_September_2013_Web_0.pdf
2. J. Kjärstad and F. Johnsson, *Energy Policy* 37 (2009), 441–464.
3. A.E. Kontorovich, *Russian Geology and Geophysics* 50 (2009), 237–242.

4. B.V. Vora, T.L. Marker, P.T. Barger, H.R. Nilsen, S. Kvisle, and T. Fuglerud, in: A. Parmaliana et al. (eds.) *Studies in Surface Science and Catalysis*, 107 (1997), 87–98.

5. M.C. Alvarez-Galvan, N. Mota, M. Ojeda, S. Rojas, R.M. Navarro, and J.L.G. Fierro, *Catalysis Today* 171 (2011), 15–23.

6. J.H. Lunsford, *Catalysis Today* 63 (2000), 165–174.

7. T. Ren, M.K. Patel, and K. Blok, *Energy* 33 (2008), 817–833.

8. R.D. Cantrell, A. Ghenciu, K.D. Campbell, D.M.A. Minahan, M.M. Bhasin, A.D. Westwood, and K.A. Nielsen, Catalysts for the oxidative dehydrogenation of hydrocarbons. In: United States Patent and Trademark Office Granted Patent (March 21, 2002), WO0222258 (A2).

9. H.L. Mitchell and R.H. Waghorne, Catalysts for the conversion of relatively low molecular weight hydrocarbons to high molecular weight hydrocarbons and the regeneration of the catalysts, in US Patent (1980), US 4239658 A 19801216.

10. T. Fang and C.-T. Yeh, *Journal of Catalysis* 69 (1981), 227–229.

11. G.E. Keller and M.M. Bhasin, *Journal of Catalysis* 73 (1982), 9–19.

12. F.S. Stone, *Journal of Molecular Catalysis* 59 (1990), 147–163.

13. J.M. Aigler and J.H. Lunsford, *Applied Catalysis* 70 (1991), 29–42.

14. S.J. Conway and J.H. Lunsford, *Journal of Catalysis* 131 (1991), 513–522.

15. J.H. Lunsford, P.G. Hinson, M.P. Rosynek, C.L. Shi, M.T. Xu, and X.M. Yang, *Journal of Catalysis* 147 (1994), 301–310.

16. J.H. Lunsford, *Catalysis Today* 6 (1990), 235–259.

17. J. Sun, J.W. Thybaut, and G.B. Marin, *Catalysis Today* 137 (2008), 90–102.

18. J.W. Thybaut, J. Sun, L. Olivier, A.C. Van Veen, C. Mirodatos, and G.B. Marin, *Catalysis Today* 159 (2011), 29–36.

19. P.N. Kechagiopoulos, J.W. Thybaut, and G.B. Marin, *Industrial and Engineering Chemistry Research* 53 (2014), 1825–1840.

20. G.J. Tjatjopoulos and I.A. Vasalos, *Catalysis Today* 13 (1992), 361–370.

21. E. Miro, J. Santamaria, and E.E. Wolf, *Journal of Catalysis* 124 (1990), 451–464.

22. E.E. Miro, Z. Kalenik, J. Santamaria, and E.E. Wolf, *Catalysis Today* 6 (1990), 511–518.

23. Z. Kalenik and E.E. Wolf, *Catalysis Letters* 9 (1991), 441–449.

24. Z. Kalenik and E.E. Wolf, *Catalysis Today* 13 (1992), 255–264.

25. C. Mirodatos, *Catalysis Today* 9 (1991), 83–95.

26. S. Pak, P. Qiu, and J.H. Lunsford, *Journal of Catalysis* 179 (1998), 222–230.

27. S. Lacombe, C. Geantet, and C. Mirodatos, *Journal of Catalysis* 151 (1994), 439–452.

28. T. Ito, J.X. Wang, C.H. Lin, and J.H. Lunsford, *Journal of the American Chemical Society* 107 (1985), 5062–5068.

29. N.G. Maksimov, G.E. Selyutin, A.G. Anshits, E.V. Kondratenko, and V.G. Roguleva, *Catalysis Today* 42 (1998), 279–281.

30. G.A. Martin, A. Bates, V. Ducarme, and C. Mirodatos, *Applied Catalysis* 47 (1989), 287–297.

31. G.A. Martin, S. Bernal, V. Perrichon, and C. Mirodatos, *Catalysis Today* 13 (1992), 487–494.

32. G.D. Moggridge, J.P.S. Badyal, and R.M. Lambert, *Journal of Catalysis* 132 (1991), 92–99.

33. Y. Moro-oka, *Catalysis Today* 45 (1998), 3–12.

34. J.A. Roos, S.J. Korf, R.H.J. Veehof, J.G. van Ommen, and J.R.H. Ross, *Applied Catalysis* 52 (1989), 131–145.

35. H.B. Zhang, G.D. Lin, H.L. Wan, Y.D. Liu, W.Z. Weng, J.X. Cai, Y.F. Shen, and K.R. Tsai, *Catalysis Letters* 73 (2001), 141–147.

36. L. Lehmann and M. Baerns, *Catalysis Today* 13 (1992), 265–272.

37. J.H. Lunsford, *Angewandte Chemie International Edition in English* 34 (1995), 970–980.

38. S.J. Korf, J.A. Roos, N.A. De Bruijn, J.G. Van Ommen, and J.R.H. Ross, *Applied Catalysis* 58 (1990), 131–146.
39. G. Martin and C. Mirodatos, in: E.E. Wolf (ed.), *Methane Conversion by Oxidative Processes: Fundamental and Engineering Aspects*, Van Nostrand Reinhold Catalysts Series, New York, (1991), pp. 351–381.
40. E. Heracleous and A.A. Lemonidou, *Applied Catalysis A: General* 269 (2004), 123–135.
41. F. Donsì, K.A. Williams, and L.D. Schmidt, *Industrial and Engineering Chemistry Research* 44 (2005), 3453–3470.
42. M.C. Huff and L.D. Schmidt, *AIChE Journal* 42 (1996), 3484–3497.
43. E. Heracleous, A.A. Lemonidou, and J.A. Lercher, *Applied Catalysis A: General* 264 (2004), 73–80.
44. Y. Schuurman, V. Ducarme, T. Chen, W. Li, C. Mirodatos, and G.A. Martin, *Applied Catalysis A: General* 163 (1997), 227–235.
45. E. Heracleous and A.A. Lemonidou, *Journal of Catalysis* 237 (2006), 175–189.
46. R. Burch and R. Swarnakar, *Applied Catalysis* 70 (1991), 129–148.
47. S.A.R. Mulla and V.R. Choudhary, *Journal of Molecular Catalysis A: Chemical* 223 (2004), 259–262.
48. L. Yu, W. Li, V. Ducarme, C. Mirodatos, and G.A. Martin, *Applied Catalysis A: General* 175 (1998), 173–179.
49. A.M. Maitra, *Applied Catalysis A: General* 104 (1993), 11–59.
50. S.J. Korf, J.A. Roos, N.A. de Bruijn, J.G. van Ommen, and J.R.H. Ross, *Catalysis Today* 2 (1988), 535–545.
51. Y. Xu, L. Yu, C. Cai, J. Huang, and X. Guo, *Catalysis Letters* 35 (1995), 215–231.
52. J.L. Dubois and C.J. Cameron, in *Studies in Surface Science and Catalysis*, F.S.L. Guczi and P. Tétényi (eds), 75 (1993), 2245–2248.
53. Z. Stansch, L. Mleczko, and M. Baerns, *Industrial and Engineering Chemistry Research* 36 (1997), 2568–2579.
54. M. Sohrabi, B. Dabir, A. Eskandari, and R.D. Golpasha, *Journal of Chemical Technology and Biotechnology* 67 (1996), 15–20.
55. S. Lacombe, H. Zanthoff, and C. Mirodatos, *Journal of Catalysis* 155 (1995), 106–116.
56. U. Olsbye, G. Desgrandchamps, K.-J. Jens, and S. Kolboe, *Catalysis Today* 13 (1992), 209–218.
57. M. Traykova, N. Davidova, J.-S. Tsaih, and A.H. Weiss, *Applied Catalysis A: General* 169 (1998), 237–247.
58. S.M.K. Shahri and S.M. Alavi, *Journal of Natural Gas Chemistry* 18 (2009), 25–34.
59. M. Daneshpayeh, A. Khodadadi, N. Mostoufi, Y. Mortazavi, R. Sotudeh-Gharebagh, and A. Talebizadeh, *Fuel Processing Technology* 90 (2009), 403–410.
60. Y.T. Chua, A.R. Mohamed, and S. Bhatia, *Applied Catalysis A: General* 343 (2008), 142–148.
61. S. Arndt, T. Otremba, U. Simon, M. Yildiz, H. Schubert, and R. Schomäcker, *Applied Catalysis A: General* 425–426 (2012), 53–61.
62. Z. Fakhroueian, F. Farzaneh, and N. Afrookhteh, *Fuel* 87 (2008), 2512–2516.
63. M.Y. Sinev, Z.T. Fattakhova, V.I. Lomonosov, and Y.A. Gordienko, *Journal of Natural Gas Chemistry* 18 (2009), 273–287.
64. Y.S. Su, J.Y. Ying, and W.H. Green Jr, *Journal of Catalysis* 218 (2003), 321–333.
65. Y. Simon, F. Baronnet, G.M. Côme, and P.M. Marquaire, in: B. Xinhe and X. Yide (eds.), *Studies in Surface Science and Catalysis*, Amsterdam, the Netherlands: Elsevier, 147 (2004), 571–576.
66. S.C. Reyes, E. Iglesia, and C.P. Kelkar, *Chemical Engineering Science* 48 (1993), 2643–2661.
67. V.R. Choudhary and V.H. Rane, *Journal of Catalysis* 130 (1991), 411–422.

68. W. Bytyn and M. Baerns, *Applied Catalysis* 28 (1986), 199–207.
69. J.A.S.P. Carreiro and M. Baerns, *Journal of Catalysis* 117 (1989), 396–403.
70. J.A. Duffy, *Journal of Solid State Chemistry* 62 (1986), 145–157.
71. J.A. Duffy, *Geochimica et Cosmochimica Acta* 57 (1993), 3961–3970.
72. P. Kassner and M. Baerns, *Applied Catalysis A: General* 139 (1996), 107–129.
73. A. Lebouteiller and P. Courtine, *Journal of Solid State Chemistry* 137 (1998), 94–103.
74. A.M. Maitra, *Applied Catalysis A: General* 114 (1994), 65–81.
75. A.M. Maitra, I. Campbell, and R.J. Tyler, *Applied Catalysis A: General* 85 (1992), 27–46.
76. P. Moriceau, B. Taouk, E. Bordes, and P. Courtine, *Catalysis Today* 61 (2000), 197–201.
77. P. Thomasson, O.S. Tyagi, and H. Knozinger, *Applied Catalysis A: General* 181 (1999), 181–188.
78. J.C. Védrine, *Topics in Catalysis* 21 (2002), 97–106.
79. J.-L. Dubois and C.J. Cameron, *Applied Catalysis* 67 (1990), 49–71.
80. D. Dissanayake, J.H. Lunsford, and M.P. Rosynek, *Journal of Catalysis* 146 (1994), 613–615.
81. K. Otsuka, Q. Liu, and A. Morikawa, *Inorganica Chimica Acta* 118 (1986), L23–L24.
82. C.A. Jones, J.J. Leonard, and J.A. Sofranko, *Journal of Catalysis* 103 (1987), 311–319.
83. R. Burch, G.D. Squire, and S.C. Tsang, *Applied Catalysis* 43 (1988), 105–116.
84. S.J. Conway, J.A. Greig, and G.M. Thomas, *Applied Catalysis A: General* 86 (1992), 199–212.
85. J.H. Hong and K.J. Yoon, *Applied Catalysis A: General* 205 (2001), 253–262.
86. S. Bartsch and H. Hofmann, *Catalysis Today* 6 (1990), 527–534.
87. S. Bartsch, J. Falkowski, and H. Hofmann, *Catalysis Today* 4 (1989), 421–431.
88. A.G. Dedov, A.S. Loktev, I.I. Moiseev, A. Aboukais, J.F. Lamonier, and I.N. Filimonov, *Applied Catalysis A: General* 245 (2003), 209–220.
89. Y. Ohtsuka, M. Kuwabara, and A. Tomita, *Applied Catalysis* 47 (1989), 307–315.
90. A. Machocki, A. Denis, T. Borowiecki, and J. Barcicki, *Applied Catalysis* 72 (1991), 283–294.
91. S. Ahmed and J.B. Moffat, *Applied Catalysis* 63 (1990), 129–143.
92. M. Teymouri, C. Petit, L. Hilaire, E. Bagherzadeh, and A. Kiennemann, *Catalysis Today* 21 (1994), 377–385.
93. R. Long, Y. Huang, W. Weng, H. Wan, and K. Tsai, *Catalysis Today* 30 (1996), 59–65.
94. R. Long, J. Luo, M. Chen, and H. Wan, *Applied Catalysis A: General* 159 (1997), 171–185.
95. K. Otsuka, M. Hatano, T. Komatsu, in (Eds) D.M. Bibby, and S. Yurchak, *Studies in Surface Science and Catalysis*, Amsterdam, the Netherlands: Elsevier. 36 (1988), 383–387.
96. A.Z. Khan and E. Ruckenstein, *Journal of Catalysis* 139 (1993), 304–321.
97. A. Machocki and R. Jezior, *Chemical Engineering Journal* 137 (2008), 643–652.
98. U. Zavyalova, M. Holena, R. Schlögl, and M. Baerns, *Chem Cat Chem* 3 (2011), 1935–1947.
99. L. Olivier, PhD thesis, Lyon University, Lyon, France, 512010 (2010).
100. L. Olivier, S. Haag, H. Pennemann, C. Hofmann, C. Mirodatos, and A.C. van Veen, *Catalysis Today* 137 (2008), 80–89.
101. H. Li, M. Rivallan, F. Thibault-Starzyk, A. Travert, and F.C. Meunier, *Physical Chemistry Chemical Physics* 15 (2013), 7321–7327.
102. S. Kuś and M. Taniewski, *Fuel Processing Technology* 76 (2002), 41–49.
103. W. Wu and S.-C. Lu, *Powder Technology* 137 (2003), 41–48.
104. J.C. Lavalley, *Catalysis Today* 27 (1996), 377–401.

105. G. Busca and V. Lorenzelli, *Materials Chemistry* 7 (1982), 89–126.
106. D.A. Constantinou, J.L.G. Fierro, and A.M. Efstathiou, *Applied Catalysis B: Environmental* 90 (2009), 347–359.
107. P.M. Couwenberg, Q. Chen, and G.B. Marin, *Industrial and Engineering Chemistry Research* 35 (1996), 3999–4011.
108. P.M. Couwenberg, Q. Chen, and G.B. Marin, *Industrial and Engineering Chemistry Research* 35 (1996), 415–421.
109. V. Alexiadis, J.W. Thybaut, P.N. Kechagiopoulos, M. Chaar, M. Muhler, A.C. Van Veen, C. Mirodatos, and G.B. Marin, *Applied Catalysis, B: Environmental* 150 (2014), 496–505.

8 New Trends in Catalyst Design for Methane Dehydroaromatization

M. Teresa Portilla, Christiaan H.L. Tempelman,
Cristina Martínez, and Emiel J.M. Hensen

CONTENTS

8.1   INTRODUCTION

Methane, the main component of natural gas, is an interesting source of chemicals and clean liquid fuels and a promising alternative raw material to oil [1–3]. Although conventionally used for heating and electrical power generation, its conversion into higher value products has gained importance along the last decades, and this is due to the following reasons: (1) the composition of natural gas is very little dependent on the source, (2) methane has a high hydrogen-to-carbon (H/C) ratio, (3) the potential reserves of natural gas are larger than the oil reserves, especially if shale gas is considered [4,5]. The extraction of this so-called unconventional natural gas, contained in impermeable shale rock formations, has recently become practically and economically affordable, thanks to the combination of horizontal drilling and hydraulic fracturing [5,6]. Despite the serious environmental concerns that have arisen regarding these extraction techniques [5], it has been fully implemented in some regions, such as the United States, where shale gas extraction has completely transformed the overall energy scenarios, including the energy markets, and has resulted in a strong reduction of natural gas prices as compared to oil. Other countries such as China, Australia, South Africa, or United Kingdom, among others, are in the early stages of their shale gas resource evaluation.

There are different gas to liquid (GTL) technologies available to make use of these large gas reserves by upgrading methane into liquid products with transportation purposes or into higher-value chemicals (see Figure 8.1), and a recent review by Wood et al. [7] gives a very complete picture on this subject. It includes a very interesting analysis of the GTL processes, including not only those that have reached commercial application, but also the commercial plants in development and planning, as well as the market opportunities for the different GTL options and products. At present, all commercially viable processes belong to the group of the so-called indirect routes of methane conversion. Here, methane is transformed in a first step into the more reactive synthesis gas mixture ($CO + H_2$, also named as syngas), via partial oxidation, steam reforming, or autothermic reforming [3,8–10]. Syngas can then be converted into dimethyl ether (DME), light olefins, or gasoline through methanol as an intermediate [1] or into DME in a single stage [2]. Hydrocarbons or higher alcohols can also be produced from the $CO + H_2$ mixture by means of the Fischer–Tropsch synthesis [1,2,11,12]. Hydrocracking of the heavy fraction formed in this way will lead to the production of more valuable fractions such as naphtha, diesel, or lubricants. Among the different GTL strategies, those based on the Fischer–Tropsch synthesis are dominating the large-scale commercial application [7].

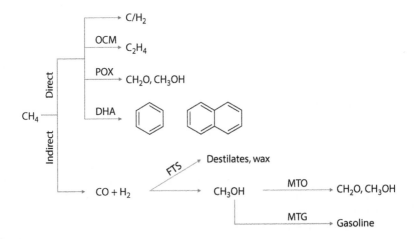

FIGURE 8.1 Main routes from natural gas to value-added products: DHA (dehydroaromatization), POX (partial oxidation), OCM (oxidative coupling of methane), MTG (methanol to gasoline), MTO (methanol to olefins), and FTS (Fischer–Tropsch synthesis).

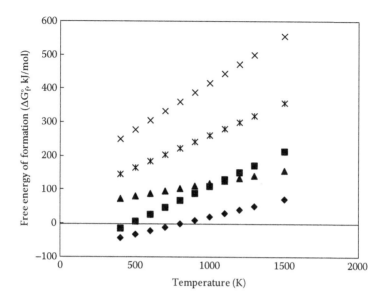

FIGURE 8.2 Gibbs free energy of formation of methane (♦) versus higher hydrocarbons (ethane [■], ethylene [▲], benzene [*], and naphthalene [×]) as a function of temperature.

The other possibility is the direct conversion of methane into heavier hydrocarbons, a challenging task from the thermodynamic and kinetic point of view, as it involves the activation of the highly stable C–H bond to produce CH_x or CH_xO species and their conversion into the desired products in a one-pot single-step process. As shown in Figure 8.2, this direct methane conversion is only favorable at high temperatures, where not only methane but also the products are susceptible to be converted. The higher reactivity of the products, as compared to the starting methane, will lead to low selectivities to the desired compounds if they are not kinetically protected or selectively removed from the reaction mixture [12].

The direct routes for methane conversion can be oxidative or nonoxidative in nature [1,2,11,12]. Regarding the oxidative routes, they take place in the presence of an oxidant, usually oxygen, which will react with excess hydrogen removing it continuously as water, acting in this way as a driving force to overcome thermodynamic limitations. One possible approach is the oxidative coupling of methane (OCM), used for producing mainly ethane and ethene. However, the yields are low and the reaction is highly exothermic, causing serious heat management problems. These two factors have prevented its commercial application up to date [1,12]. The other option, highly interesting, is the direct conversion of methane into methanol or formaldehyde by partial oxidation. However, also in this case, the selectivities obtained decrease significantly at increasing methane conversion due to the higher reactivity of the primary products as compared to methane [1,12]. At present, the direct oxidative routes for conversion of methane still remain a challenge, and their successful applications will depend not only on the design of highly selective catalysts but also on new developments regarding process engineering.

The use of oxygen for the direct conversion of methane is beneficial as it will increase the reaction rate by shifting the thermodynamic equilibrium. However, kinetically, it will limit the yields to the desired primary products that are much more reactive than methane under the experimental conditions used and will therefore be further converted into undesired combustion products. An alternative that has gained much interest in the last decades is the direct aromatization of methane under nonoxidative conditions (MDA [methane dehydroaromatization]). In this case, the selective extraction of H_2, one of the main products, may help surpassing the very unfavorable thermodynamics of this endothermic equilibrium-controlled reaction [13,14], and it would be possible to maximize the selectivity to the desired higher hydrocarbons by means of the adequate catalysts and under optimized process conditions. Moreover, hydrogen demand is expected to grow along the next decade, not only due to its use in traditional processes such as ammonia synthesis, but also due to new applications, such as fuel cells [15].

Considerable progress has been made since the first description of MDA in a downflow fixed-bed reactor by Wang et al. in the early 1990s [16]. They proposed a bifunctional catalyst and, so far, Mo-containing zeolites are still the best catalysts among those described for MDA. Regarding the possible zeolites, the most adequate are those with structures containing pores of dimensions close to the dynamic diameter of benzene, the desired product, such as ZSM-5 [17–24] and MCM-22 [25–30], although other zeolite structures have also been employed for this reaction [31–36]. A bifunctional mechanism is generally accepted for the MDA reaction catalyzed by molybdenum-containing zeolites [37–39]. Methane is first activated on the Mo sites, and the reaction intermediates formed will oligomerize, cyclize, and dehydrogenate on the Brønsted acid sites (BASs). Many efforts have been employed in the last decades directed to the identification of the active species, their formation, and how they are affected by the physicochemical properties of the zeolitic support [21,40–45]. Many publications have also been focused on the understanding of the reaction mechanism [19,46–50].

This chapter will give an overview on the different approaches that have been proposed in the literature from the point of view of catalyst design and activation, which will be completed with the advances obtained during the NEXT-GTL project. In order to fully understand the properties required by an optimum MDA catalyst, the mechanism of the process and the kinetics will also be briefly described.

8.2 MICROPOROUS CATALYSTS FOR MDA REACTION

8.2.1 GENERAL ASPECTS

The first approaches on the use of heterogeneous catalysts for MDA under nonoxidative conditions were already reported in the early 1980s ([37,51] and references therein). However, it was the work based on Mo supported on ZSM-5 zeolite published by Wang et al. in 1993 [16], the one that focused MDA-based research on the use of transition metals (TMs) impregnated on acid zeolites. Zeolites are crystalline aluminosilicates characterized by a well-defined microporous structure with pore sizes in the range of molecular dimensions and a flexible chemical composition. These properties, together with their environmentally benign nature, have turned zeolites into the most widely used materials among the heterogeneous catalysts [52], successfully applied in industrial processes belonging to the fields of refining, petrochemistry, or fine and specialty chemicals. Concerning their application as heterogeneous catalysts, two qualities can be remarked: on the one hand, the shape selectivity effects introduced by their microporous channels [53,54], with dimensions in the range of many reactant molecules, and on the other hand, their outstanding thermal and hydrothermal stability [55]. This is also true when zeolites are applied as catalysts in the MDA process. As we will detail later, by choosing the appropriate zeolite structure, with the adequate pore dimensions, the size of the final aromatic products will be limited by shape selectivity, and this will result in an increased selectivity to the desired benzene, in a lower coke make, and therefore, in a reduced deactivation rate. On the other hand, MDA is a very demanding reaction from the energetic point of view, which takes place at temperatures around 973 K in order to obtain reasonable methane conversions. Thus, thermal stability will be highly convenient. The substitution of Si by Al in tetrahedral framework positions of a zeolite results in a negative charge that has to be compensated by cations located in the zeolite channels or cavities [56]. When the cation in ion exchange positions is a proton, the zeolite presents Brønsted acidity. The amount of BASs will be determined by the chemical composition—Si/Al ratio— of the zeolite, and the higher the Al content, the higher the potential number of BASs. However, the amount of framework Al has a direct effect on the thermal stability of the zeolite framework, and Al-rich zeolites are more susceptible to suffer framework dealumination, generating extra-framework Al (EFAL) species. These EFAL species, when cationic and highly dispersed, may be in charge of compensating sites, reducing in this way the acidity of the zeolite even further [56]. Concerning the crystal size of the zeolite, it may affect its catalytic behavior in several ways. On the one hand, for large crystal size samples, the proportion of external versus internal—micropore— surface will be low, and shape selectivity effects will be enhanced. This will result in increased selectivity to benzene and lower naphthalene make [57]. However, diffusional paths are longer for larger crystals, and this may have a direct influence on the efficient usage—infra-use—of the full crystalline micropore volume and on the deactivation rate, especially by pore blocking. Thus, a compromise between these different factors should give the best performance [58].

Since the early work by Wang et al. [16], many studies have been published on zeolite-supported Mo bifunctional catalysts, trying to improve, not only the already high selectivity to the desired benzene obtained with the original Mo/ZSM-5, but also the catalyst life. Along these two decades, several reviews have been published [37–39,51,59,60], some of them focused on ZSM-5-based catalysts [37,59,60] and others analyzing also alternative zeolite structures [38,39,51]. Among the latter, MCM-22 was seen to be very interesting, with higher activity and lower deactivation rate as compared to ZSM-5 of similar physicochemical properties. The improved catalytic behavior of MCM-22 was attributed to its particular microporous structure. Thus, the channel system and the pore dimensions of the zeolitic support play a decisive role in the overall MDA process.

In the last years, different strategies have been proposed in order to improve, especially, the stability to coking of zeolite-based MDA catalysts, and a deeper overview on this topic will be given in Section 8.4. The medium-pore zeolite ZSM-5 [MFI] [61], with a 2D porous structure and channel dimensions of 5–5.5 Å, close to the dynamic diameter of the benzene molecule, has been the most described zeolite structure for MDA. The zeolite is used in its acid form, and some of its properties, such as Si/Al ratio, Al distribution, and crystallite size, have been seen to have a direct influence on its catalytic behavior [20,23,62,63].

However, acid zeolites are not able to activate methane by themselves, and a second catalytic function is needed. Although the selective activation and dissociation of a first C–H bond without breaking the rest of the bonds is a difficult task, due to the high stability of the methane molecule, theoretical density functional theory (DFT) studies suggested that metal adatoms present on TM catalyst surfaces were able to favor this first dehydrogenation without further decomposition of CH_3 [64,65]. The impregnation of these metal functions on a zeolitic support will have two beneficial consequences: the crystals will be dispersed on the high zeolite surface area, increasing in this way the accessibility of the active sites to the reactants, and the proximity of the TM to the acid sites of the zeolite will enhance the bifunctional nature of the catalyst. Regarding the Mo/ZSM-5 catalysts, it has been proposed that a high density of BASs on the starting zeolite will improve the Mo dispersion by directing the diffusion of the metal into the microporous structure [25,66]. Moreover, the benefits of Al pairs have also been described [63], and the presence of Al pairs is favored by high Al contents. As in other bifunctional zeolite-based catalysts, the effect of the Si/Al ratio cannot be studied individually, because the optimum performance will be obtained by a catalyst with a well-equilibrated acid/metal function [37–39]. Still, the starting BAS density and the metal loading will determine the final amount and strength of the remaining acid sites. Although other metals have been proposed such as Zn, W, Re, Cu, Ni, Fe, Mn, Cr, V, Ga, and even Pt (see [38,39,60] and references therein), molybdenum has been the most reported, due to its high activity and high selectivity for benzene in the MDA process (see Table 8.1).

TABLE 8.1
Comparison MDA Activities of Different TM Supported on HZSM-5 Zeolite

	Reaction Conditions			Selectivity (%)	
Active Metals	T (°C)	Flow (mL·gcat^{-1}·h^{-1})	CH$_4$ Conversion (%)	Benzene	Naphthalene
Mo	730	1500	16.7	60.4	8.1
Zn	700	1500	1.0	69.9	—[b]
W	800	1500	13.3	52.0	—[b]
Re	750	1440	9.3	52.0	0
Co–Ga	700	1500	12.8	66.5	7.2
Fe	750	800[a]	4.1	73.4	16.1
Mn	800	1600	6.9	75.6	11.9
V	750	800[a]	3.2	32.6	6.3
Cr	750	800[a]	1.1	72.0	3.7

Source: Adapted from Ma, S. et al., *J. Energ. Chem.*, 22, 1, 2013.

[a] GHSV/h^{-1}.

[b] Not reported.

8.2.2 Type and Nature of Supported MoO_x Species

The metal incorporation procedure employed for preparing zeolite-based metal-containing catalysts may have a considerable impact on its final catalytic properties [52,67]. Also in the case of Mo/ZSM-5 catalysts for MDA, different methods have been described, and the impregnation by incipient wetness or by slurry impregnation is the usual procedure employed [39], starting from ammonium heptamolybdate (AHM) as the metal precursor. In the as-prepared sample, the bulky molybdate anion [$Mo_7O_{24}^{6-}$] will be located on the external surface of the zeolite. Xu et al. studied the interaction of AHM with the zeolite and followed the location of the different Mo species formed at the consecutive preparation steps by means of several techniques, including FTIR spectroscopy and differential thermal analysis [68]. They observed that AHM decomposes in two steps by calcination at 514 and 619 K forming MoO_3 crystallites, which remain on the outer surface of the support. It is only after calcination at 773 K that these crystallites become finely dispersed on the zeolite surface and diffuse into the zeolite pores. From their study, they conclude that the Mo species clearly interact with the zeolite BASs and that this interaction depends on the Mo loading and calcination temperature. Increasing Mo content and treatment temperature results in a decrease of the BAS density, which may be detrimental from the activity point of view.

Incorporation of Mo by solid-state ion exchange starting from physical mixtures of MoO_3 and H-ZSM-5 has also been described for the preparation of MDA catalysts, and the evolution of the Mo species in Mo/ZSM-5 samples prepared in this way has been thoroughly studied by the group of Iglesia [21,44,45]. As compared to the use of AHM, this procedure avoids the formation of by-products during calcination (N_2, NH_3, H_2O) and allows an easy and accurate measurement of the nature and rate of MoO_x exchange by following the water evolved from condensation reactions of OH groups. Their work confirms the formation of MoO_3 crystallites on the zeolite surface in a temperature range of 623–773 K and the migration of MoO_x species at higher temperatures (773–973 K) into the zeolite channels. There they will react with OH groups to form MoO_2-$(OH)^+$ species that quickly condense to form H_2O and a strong Mo–O–Al bond. The Mo species proposed are (Mo_2O_5)$^{2+}$ dimers interacting with two zeolite exchange sites, as shown in Scheme 8.1, Equation 8.1. The monomeric MoO_2-$(OH)^+$ species can also interact with a second BAS to form a bridging mononuclear [MoO_2^{2-}] species and water as shown in Scheme 8.1, Equation 8.2.

The group of Bao found that when preparing Mo/ZMS-5 by solid ion exchange at 773 K in N_2 atmosphere, Mo was well dispersed on both the external and the internal zeolite surface. Within the zeolite pores, Mo was present as [$Mo_5O_{12}^{6-}$], interacting with the zeolite framework and decreasing in this way the BAS density [69], whereas MoO_3 crystallites were observed on the external surface.

Additional studies have been reported along the last decade, and typically the presence of Mo species inside of the zeolite channels, closely interacting with the BASs, is accepted. The type of Mo species interacting with the zeolitic support has been found to be strongly dependent on the Al content of the zeolite and on the Mo content. A good overview has been given by Ma et al. [39]. We also point out recent work performed using ultrahigh-field solid-state ^{95}Mo NMR spectroscopy [70], which showed good correlation between the rate of aromatics formation and the number of ion-exchanged Mo species.

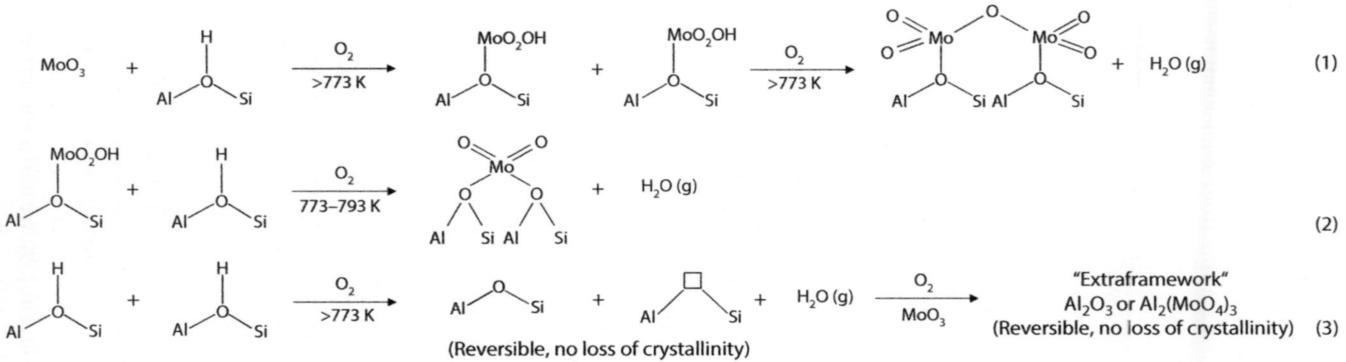

SCHEME 8.1 Interactions between MoO$_x$ species and BASs of HZSM-5. (Adapted from Ma, S. et al., *J. Energ. Chem.*, 22, 1, 2013.)

8.2.3 FORMATION OF THE ACTIVE MO SPECIES BY REDUCTION/ CARBURATION OF MoO$_x$ PRECURSORS

The Mo species in ion exchange positions within the zeolite channels are the precursors of the final carburized molybdenum species, responsible for methane activation. Although the nature of the Mo species active for methane activation under MDA conditions is still under debate, the observation of an induction period during the MDA reaction, which involves methane conversion into CO$_x$, H$_2$O, and H$_2$ but not into hydrocarbons, points to formation of Mo carbides during the first stage of the reaction. In the first studies by the groups of Solymosi [19,71] and Lunsford [85], the formation of Mo$_2$C species was proposed, highly dispersed on the external surface of the zeolite, which were thought to be responsible for the initial activation of methane. Later EXAFS studies by the group of Iglesia suggested that very small Mo$_2$C clusters are formed during the initial reduction and carburization stages, probably by transformation of Mo-oxo dimers [21,43,44]. This leads to the regeneration of one of the two original bridging OH groups, as shown in Scheme 8.2, and thus to the presence of the two catalytic functions required for MDA, MoC$_x$ for C–H bond activation and C$_2$ formation, presumably ethylene, and BASs for the conversion of the C$_2$ intermediate into aromatics.

During reaction, the MoC$_x$ clusters were observed to increase to a size or 0.6 nm, close to that of the ZSM-5 pores, clusters that were estimated to contain around 10 Mo atoms. After prolonged reaction, some of these MoC$_x$ clusters were seen to grow to dimensions larger than the micropore size of ZSM-5 forcing them to migrate to channel intersections, crystalline defects, or to the external surface. However, most of them remained small and stable toward migration and agglomeration. The low vapor pressure and high melting point of Mo carbides, the high dispersion of the MoO$_x$ precursors, and the complexation of MoC$_x$ clusters by oxygens of the zeolite framework were pointed out as possible reasons for the remarkable structural stability of the MoC$_x$/H-ZSM5 materials [43]. Liu et al. also pointed out that Mo species that were strongly interacting with the zeolite were difficult to reduce completely to Mo$_2$C and proposed partial reduction to take place, resulting in MoC$_x$O$_y$ [40–41].

Regarding the reaction mechanism of MDA, it is usually assumed that methane is activated by the Mo carbide species, leading to the first carbon–carbon bonds to be formed in the form of ethylene. Ethylene further oligomerizes on the BASs to final aromatics. The bifunctional nature of the reaction mechanism and the need for exchanged Mo cations and residual BASs are supported by the observation of a maximum in

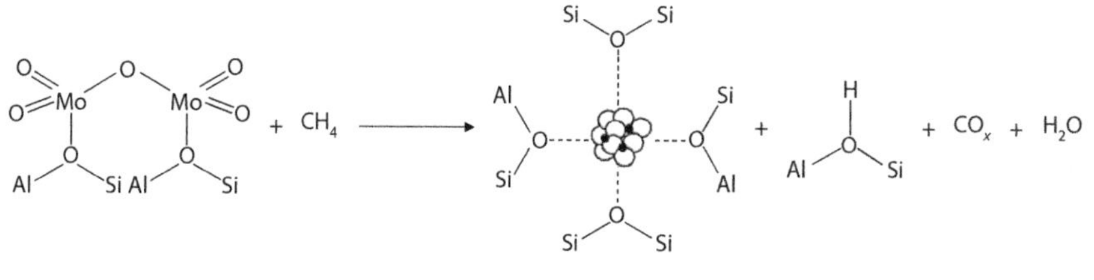

SCHEME 8.2 Reaction of exchanged MoO$_x$/H-ZSM5 with CH$_4$. (Adapted from Ding, W. et al., *J. Phys. Chem. B*, 105, 506, 2001.)

the methane reaction rate for catalysts with intermediate Mo/Al ratios of 0.3–0.5 [21]. However, this view is not unanimously supported, with acetylene being proposed as the alternative reaction intermediate following methane activation [39,72].

8.2.4 ZEOLITE STRUCTURES USED AS ACID SUPPORTS FOR MDA CATALYSTS

Although most studies have employed Mo/ZSM-5 as the catalyst, other zeolite topologies have also been explored as acidic supports for the Mo phase. The early study of Zhang et al. [31] compared various zeolites and related porous materials including ZSM-5, ZSM-8, ZSM-11, Beta, MCM-41, mordenite, X, Y, SAPO-5, SAPO-11, and SAPO-34. It was found that only zeolites with 2D or 3D pore topology and pore sizes near to the kinetic diameter of benzene (H-ZSM-5, H-ZSM-8, H-ZSM-11, beta zeolite) were adequate supports to catalyze the MDA reaction. Large-pore zeolites (mordenite, X, or Y) only produced small amounts of ethylene, while the conversion was very low for low-acidic SAPO-derived catalysts.

The most promising alternative as MDA catalyst to Mo/ZSM-5, Mo/MCM-22, was first described by Shu et al. This catalyst was not only more active but also more stable than Mo/ZSM-5 under similar reaction conditions [73]. Thorough characterization led the authors to conclude that the induction period and the type of active species on Mo/MCM-22 were likely similar to those in Mo/HZSM-5. The higher benzene and lower naphthalene yield were ascribed to the specific pore system of MWW zeolite [61], combined with a proper BAS distribution. The interaction between Mo and the BASs was similar to what had been observed for Mo/HZSM-5 [25,26]. The formation of extra-framework $Al_2(MoO_4)_3$ crystallites and partial destruction of the zeolite lattice were also described for Mo/MCM-22 catalysts. Ha et al. attributed the higher conversion of Mo/MCM-22 to the more open topology of MWW zeolite [72]. The higher accessibility of the MCM-22 structure facilitates the migration of the Mo ions toward ion exchange positions, increasing in this way the dispersion of the Mo-oxide precursor as well as favoring product diffusion.

ITQ-2, a lamellar zeolite obtained from the MCM-22 precursor [74], was also studied as support for Mo-based MDA catalysts [75]. The layers are organized in a *house of card* structure, which confers the final zeolite high external surface area. These sheets have a thickness of 2.5 nm, and they present hexagonal arrays of *cups* (0.7 nm × 0.7 nm) at their surface. The cups are connected by a double 6 MR window to the sheet. These *cups* are the hemicavities related to the 12 MR supercages present in MCM-22. The sheets also contain the 10 MR sinusoidal channel system of 0.40 nm × 0.59 nm size, which has been argued to be responsible for the high benzene selectivity [75]. When compared with MCM-22 zeolite of the same Si/Al ratio, ITQ-2 produced more naphthalene and less benzene. Selective dealumination of the external surface of ITQ-2 significantly improved the benzene selectivity and reduced the formation of naphthalene, indicating that the larger products are preferentially formed on the external surface.

Besides MCM-22 and ITQ-2, other zeolites belonging to the MWW family have also been studied as supports for Mo-containing MDA catalysts. MCM-49 with the same framework topology as MCM-22 but with different Al content and probably different Al distribution [76] was impregnated with 6 wt% Mo by solid-state ion exchange and compared with Mo/ZSM-5. As in the case of Mo/MCM-22, the MCM-49-based catalyst was more stable toward deactivation when compared to Mo/ZSM-5 [32] and more selective to benzene. MCM-36, obtained by swelling and pillaring the layered precursor of MCM-22 [77], was also reported as acidic support for Mo-based MDA catalysts and compared with Mo/MCM-22 [33]. Although initial methane conversion was found to be comparable, the pillared MCM-36-based catalyst was considerably less selective to benzene and deactivated much faster with time on stream. The larger mesoporosity and lower BAS density of Mo/MCM-49 were claimed to explain its lower catalytic performance. ITQ-13, with a tridirectional medium-pore channel system containing 9 MR and 10 MR pores [78], has also been compared with a Mo/ZSM-5 catalyst in a recent publication [34]. Mo incorporation into ITQ-13 results in a larger decrease of the Brønsted acidity than in the case of Mo/ZSM-5. According to the authors, the lower activity of Mo/ITQ-13 is due to the presence of the 9 R channels, which favor the formation of coke. However, no data are given on the amount of coke produced to support these arguments, and the lower acidity could also explain the results. Thus, the potential benefits of using this multipore zeolite as acid support for bifunctional MDA catalyst cannot be ruled out, provided that an adequate preparation procedure is used that will preserve the zeolite acidity in a larger extend. Finally, two new medium-pore zeolites with very complex structures, TNU-9 [79] and IM-5 [80], have also been studied by the same group [35,81]. In both cases, the Mo-containing catalysts based on these two zeolites with unique structure are more active than the reference Mo/ZSM-5, more selective to benzene, and more stable. The optimal Si/Al ratio and Mo loading for TNU-9 and IM-5-based catalysts were 25 and 6 wt%, respectively.

It is clear from all these studies that the activity of Mo/zeolite catalysts does not only depend on the nature of the Mo species and their location, but also the zeolite pore topology, its acidity, and the location of BASs are of pivotal importance to activity, selectivity, and stability in the MDA reaction.

The kinetics of MDA have also been investigated [46,47,49] in order to better understand the reaction mechanism. In the most recent study, obtained within the frame of the NEXT-GTL project, researchers from SINTEF and Ghent University proposed a kinetic model based on elementary steps by fitting it to kinetic data obtained in a fixed-bed reactor under *steady-state* conditions for two different catalysts, Mo/ZSM-5 and Mo/MCM-22 [46]. The model, which relies on kinetic and catalyst descriptors, assumes a bifunctional mechanism where methane is first dimerized into ethene on Mo sites followed by ethene oligomerization into benzene on acid sites, and focuses on the acid-catalyzed steps with the aim of investigating the effect of acid zeolitic support on MDA (see Scheme 8.3). The main differences were found for the physical interaction of the hydrocarbons with the zeolite walls. Thus, the different catalytic behavior of Mo/MCM-22 and Mo/ZSM-5 could be mainly related to topological effects rather than differences in acidity.

As a continuation of this work, and also within the NEXT-GTL collaboration, the kinetic study was further extended to a dynamic 1D, pseudo-homogeneous microkinetic reactor model, which was used to assess the effect of additional zeolite structures via catalyst descriptors [82]. Thus, the results obtained on Mo-supported

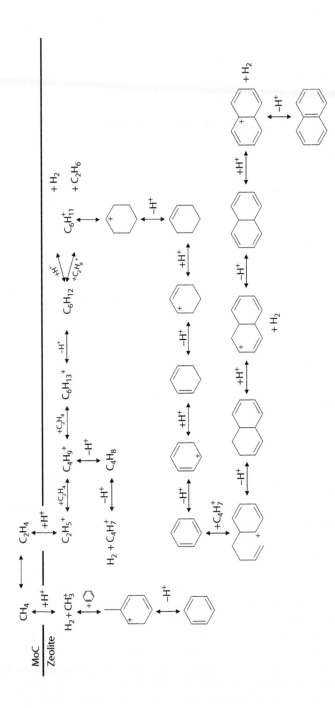

SCHEME 8.3 Proposed reaction mechanism for methane aromatization over Mo/HMCM-22. (Adapted from Wong, K.S. et al., *Micropor. Mesopor. Mater.*, 164, 2012, 302.)

ZSM-5 and MCM-22 were compared with those corresponding to Mo/IM-5 and Mo/TNU-9, and it was found that the chemisorption enthalpy values and acid site deactivation constants were able to assess the different catalytic behavior observed beyond the possible effect of their Brønsted acidity.

8.3   CATALYST DEACTIVATION AND COKE CHARACTERIZATION

Poor catalyst stability is the major hurdle to be overcome in the further development of the MDA reaction into an industrial process. Catalyst deactivation is attributed to the high reaction temperatures required for C–H bond activation, which lead to rapid further dehydrogenation of hydrocarbons and formation and deposition of carbonaceous products [38]. Another problem is sublimation of some Mo-oxide phases [49,83]. Some researchers also attribute deactivation to the transformation of the η-Mo$_3$C$_2$ phase toward less active α-MoC$_{1-x}$ and β-Mo$_2$C phases upon interaction with carbonaceous deposits [84]. Coke formation is widely regarded as the main contributor to catalyst deactivation [50,54,85,86]. BASs located at the external surface are thought to cause extensive carbon deposition [87], with the lack of shape selectivity of the micropores explaining the different product mixture formed from benzene and toluene formed in the micropores. The polynuclear aromatics tend to cover the external surface and block access to the micropores.

Temperature-programmed oxidation and thermographic analysis methods provide quantitative and qualitative information on the bulk carbon composition [85,86,88,89]. The different types of coke are typically distinguished by the variation in oxidation temperature. Usually, two types of carbon species can be identified. An amorphous type of carbon formed in the proximity of Mo can be oxidized at lower temperatures, the alternative interpretation being that it pertains to oligomeric species in the micropores. A polyaromatic type of carbon can only be oxidized at higher temperatures.

8.3.1   A STUDY OF CATALYST DEACTIVATION

Although several aspects of the deactivation of Mo/ZSM-5 catalysts have been investigated, a detailed study of the changes as a function of reaction time has not been undertaken. During the NEXT-GTL project, such an investigation was carried out with the aim to resolve the main contributors to activity loss [90]. Spent catalyst samples were recovered from the reactor after different reaction times, that is, directly after activation and after 5 min, 2, 5, and 10 h on stream in the MDA reaction. This was achieved in consecutive measurements by quenching the reaction in an inert He flow, followed by transfer under exclusion of air and storage in a nitrogen-flushed glove box before further characterization.

Analysis of the textural properties by Ar physisorption (Table 8.2) showed a small decrease of the micropore volume upon activation in CH$_4$/He. During the first 2 h on stream in the reaction mixture, the micropore volume remained nearly constant. After 2 h, maximum formation rate of benzene and naphthalene is observed (see Table 8.3). After this period, the micropore volume gradually decreases. This decrease

TABLE 8.2

Physicochemical Properties of the Parent HZSM-5 Zeolite and the Mo/ZSM-5 Catalysts after Increasing Time on Stream in the MDA Reaction

Sample	Reaction Time (h)	V_{mic} (cm³/g)	Mo/Al (ICP)	I_{FAl} (%)	BAS (%)	C_{Total} (wt%)	C_{MoC} (wt%)	C_{SC} (wt%)	C_{HC} (wt%)
HZSM-5	—	0.13	—	100	100	—	—	—	—
Mo/HZSM-5	—	0.10	1.4	53	98	—	—	—	—
	0	0.07	1.4	37	71	3.7	0.9	2.8	0
	0.083	0.07	1.4	40	65	3.7	0.7	3.0	0
	2	0.07	1.4	34	21	7.8	0	4.0	3.8
	5	0.04	1.4	23	5	11.3	0	4.1	7.2
	10	0.001	1.4	11	0	14.2	0	3.8	10.4

Source: Data taken from Tempelman, C.H.L. and Hensen, E.J.M., Manuscript under preparation.

TABLE 8.3

Catalytic MDA Activity Measurements of Mo/ZSM-5

Sample	Reaction Time (h)	X_{CH_4} (%)	C_6H_6 (mmol/h g_{cat})	$C_{10}H_8$ (mmol/h g_{cat})	$S_{C_6H_6}$ (wt%)	$S_{C_{10}H_8}$ (wt%)	S_{olefin} (wt%)	S_{coke} (wt%)
Mo/	0.083	16.8	0.65	0.07	33	6	2	58
HZSM-5	2	13.5	1.07	0.15	64	14	3	14
	5	11.8	0.80	0.08	55	9	4	25
	10	5.9	0.08	0.07	11	16	13	58

Source: Data taken from Tempelman, C.H.L. and Hensen, E.J.M., Manuscript under preparation.

also leads to a decrease in the benzene formation rate. As it is usually assumed that benzene formation mainly takes place on BASs residing in the micropores, blockage of the micropores explains the decrease in the benzene formation rate. The decrease in naphthalene formation rate is less pronounced, which is consistent with the suggestion that this more bulky product is formed on the BASs on the external zeolite surface [91]. The observation that the naphthalene formation rate also decreases should be due to carbon laydown on the external surface. Even after all micropores have become inaccessible, methane conversion remains substantial so that we conclude that at least part of the molybdenum carbide phase remains accessible at the external zeolite surface.

Characterization of the aluminum speciation was done by ²⁷Al MAS NMR spectroscopy. The spectra show a decrease of the framework aluminum (FAl) peak ($\delta = 55$ ppm) with progressive time on stream in MDA (Figure 8.3). The work of Meinhold and Bibby [92] relates the decrease in FAl peak intensity to increased

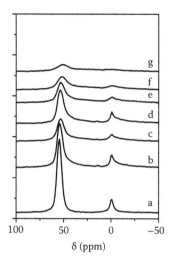

FIGURE 8.3 ^{27}Al MAS NMR spectra of (a) HZSM-5, (b) Mo/ZSM-5, (c) Mo/ZSM-5 (activated), (d) Mo/ZSM-5 (5 min), (e) Mo/ZSM-5 (2 h), (f) Mo/ZSM-5 (5 h), and (g) Mo/ZSM-5 (10 h). (Taken from Tempelman, C.H.L. and Hensen, E.J.M., Manuscript under preparation.)

carbon deposition inside the micropores of ZSM-5. Comparing NMR spectra of Mo/HZSM-5 before and after activation in CH_4/He a decrease is seen in FAl peak intensity (Table 8.2). It suggests some carbon formation in the micropores as molybdenum carbide. Then, the next 2 h on stream, the FAl peak intensity remains constant suggesting that no carbon formation occurs inside the micropores. The observed trends are consistent with the textural trends derived by Ar physisorption on spent samples.

A general problem encountered in the characterization of acidity of coked samples is that applying IR spectroscopy is not possible. In the NEXT-GTL project, H/D exchange in benzene was employed to study the residual acidity of these samples. For this purpose, fresh and spent samples were exposed in a 10-flow parallel reactor to a mixture C_6D_6 and C_6H_6. The H/D exchange is used as a measure of the Brønsted acidity [93] and has, among others, been related to the parent HZSM-5 zeolite (Table 8.2). The corresponding data show a gradual decrease of the acid exchange activity with time on stream. The decrease can be linked to the decreasing accessibility of the micropores. Thus, it is reasonable to conclude that the carbon laydown on the external surface leads to blockage of the micropore entrances.

The coke in the spent catalysts has been further characterized. The importance of carbon deposition at the external surface has been stressed [94]. Following assignments in literature [95], Table 8.2 shows that during the initial stages of reaction, carbon is mainly associated with Mo, directly in the form of Mo carbide (C_{MoC}) and carbon species formed in the proximity of Mo (C_{SC}: soft coke). After longer time on stream, polynuclear aromatics (C_{HC}; hard coke) were formed. It is also seen that the amount of hard coke increased substantially after 2 h time on stream and, together with the deactivation profile, it provides good support for the conclusion that hard coke causes inaccessibility of the micropores and catalyst deactivation. It is also consistent with the significant selectivity change leading to less aromatics and

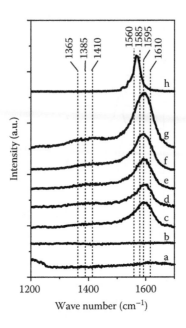

FIGURE 8.4 UV–Raman spectra recorded with a 244 nm laser of (a) HZSM-5, (b) Mo/ZSM-5, (c) Mo/ZSM-5 (activated), (d) Mo/ZSM-5 (5 min), (e) Mo/ZSM-5 (2 h), (f) Mo/ZSM-5 (5 h), (g) Mo/ZSM-5 (10 h), and (h) graphite. (Taken from Tempelman, C.H.L. and Hensen, E.J.M., Manuscript under preparation.)

more ethylene. UV–Raman spectroscopy was used to obtain more information about the coke species formed (Figure 8.4). The work of Li and Stair [96] has been used to assign the bands to various coke species.

After activation of Mo/HZSM-5 in CH_4/He, the Raman spectrum contains a main band at ~1600 cm^{-1} and a broad band at 1385 cm^{-1}, typical for olefinic type of coke. These species are likely located in the micropores. During reaction, the peak at 1600 cm^{-1} shifts toward lower wave numbers in the direction of typical band positions for polyaromatic (1595 cm^{-1}) and graphitic carbon (1585 cm^{-1}). The band at 1385 cm^{-1} shifts toward 1365 cm^{-1}, characteristic for graphite-like species. The analysis of the surface region by x-ray photoelectron spectroscopy (XPS) combined with sputtering with Ar ions revealed that during the MDA reaction, the carbon layer on the external surface grows substantially. Transmission electron micrographs of the spent catalysts (Figure 8.5) confirm the presence of a carbon layer around the zeolite particles including the Mo carbide particles. All these characterization data indicate that intensive hard coke formation, mostly of the polyaromatic/pre-graphitic type, starts after about 2 h and is mainly located at the external surface. This growing coke layer blocks the access to the micropores. This hard coke is formed not only by the BAS on the external surface [94] but also by the Mo carbide phases.

Figure 8.6 shows the relative concentration of various Mo species at the outer surface as determined by XPS. The assignment of Solymosi's paper was followed [97]. With progressing time on stream, these data indicate that the initially highly dispersed Mo_2C agglomerates into larger Mo_2C particles. This growth is consistent with

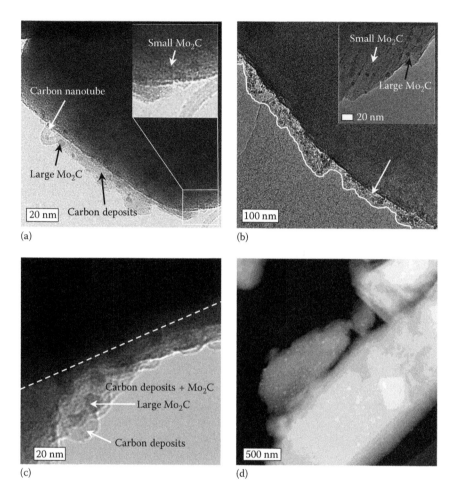

FIGURE 8.5 Transmission electron micrographs of (a) Mo/ZSM-5 (activated), (b) Mo/ZSM-5 (2 h), (c) Mo/ZSM-5 (10 h), and (d) STEM-HAADF micrograph of Mo/ZSM-5 (10 h). (Taken from Tempelman, C.H.L. and Hensen, E.J.M., Manuscript under preparation.)

observations by TEM of spent samples after 2 and 10 h (Figure 8.5). It supports the interpretation of increased rate of carbon formation due to growth of the Mo carbide particles as also suggested by Solymosi et al. [97]. Concomitant with the agglomeration of Mo_2C particles, the Mo/Al and Mo/Si ratios probed by XPS increase with time on stream (Table 8.4), suggesting an increased Mo concentration at the surface. Concentration profiles of the Mo/Al and Mo/Si ratios obtained by sputtering with Ar ions reveal increasing Al concentration deeper into the surface of the spent catalyst. It points to the formation of a carbon layer between the zeolite and the Mo phase, effectively separating them, which is also clearly illustrated by TEM (Figure 8.5c) and STEM-HAADF (Figure 8.5d) images of spent Mo/ZSM-5 after 10 h on stream.

All these results can be summarized in the following model for the evolution of the catalyst with time on stream. Activation of the well-dispersed Mo-oxide

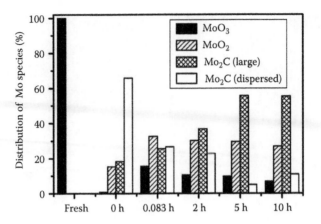

FIGURE 8.6 Molybdenum speciation in the surface region as determined by XPS. (Data taken from Tempelman, C.H.L. and Hensen, E.J.M., Manuscript under preparation.)

TABLE 8.4

XPS Depth Profiling of the Mo/ZSM-5 Catalysts after Increasing Time on Stream in the MDA Reaction

Sample	Reaction Time (h)	Sputtering Time (s)	Si/C (XPS)	Si/C$_{graphite}$ (XPS)	Mo/Al (XPS)	Mo/Si (XPS)
Mo/HZSM-5	0	0	1.14	4.3	1.1	0.045
	0	120	1.59	∞	1.0	0.050
	0	720	2.52	∞	0.9	0.050
	2	0	0.49	3.4	0.8	0.042
	2	120	0.61	3.5	0.9	0.048
	2	720	1.05	5.3	0.8	0.050
	10	0	0.13	0.7	1.7	0.092
	10	120	0.29	1.5	1.2	0.060
	10	720	0.66	4.3	0.9	0.060

Source: Data taken from Tempelman, C.H.L. and Hensen, E.J.M., Manuscript under preparation.

precursor in methane results in a highly dispersed Mo carbide phase, which is predominantly present at the external surface. Besides carbon directly associated with Mo in carbidic or oxycarbidic phases, there is a small amount of coke associated with this Mo phase (soft coke). The loss in micropore volume during the initial stages is limited. It is explained by the observation that only a small part of Mo resides in the micropores. After 2 h on stream, there is a strong increase in the rate of formation of polynuclear aromatics. There is substantial evidence that this hard coke is mainly formed on the external zeolite surface. This hard coke phase forms an increasingly thick layer at the external surface and blocks the entrances of the micropores. Most likely, the formation of polyaromatic species occurs on external BAS. It coincides with the growth of Mo carbide species. This could either mean that larger Mo carbide species produce also coke or, alternatively, that Mo carbide agglomeration regenerates some of the external BASs that were earlier associated with the Mo carbide phase. In any case, the blockage of the micropores results in a decrease of the accessible BAS, a decrease of the benzene selectivity, and a concomitant increase of the ethylene selectivity. All this is consistent with severe deactivation of the acid component compared to the component that activates methane. After 10 h on stream, large Mo carbide particles are formed, which are seen to be separated from the external zeolite surface due to the formation of a thick hard coke layer around the zeolite. This is most likely due to the uncontrolled formation of large carbonaceous deposits on the external surface. Formation of hard coke is the main contributor to catalyst deactivation.

8.3.2 REGENERATION STRATEGIES

Since it appears that coke formation is an intrinsic property of the MDA reaction over Mo-modified zeolites, it is likely that a viable industrial process would need to involve periodic regeneration to remove the carbonaceous deposits. Several regeneration methods have been extensively explored [45,98,99]. It has turned out to be a challenge to control the high exothermicity of the coke oxidation, which will lead to catalyst damage [98]. Sublimation of Mo oxides formed upon regeneration also negatively affects performance of the regenerated catalyst. Typically, oxidation of the coke in air (above 723 K to remove hard coke) leads to decreased activity after consecutive regeneration cycles. In order to prevent catalyst damage, regeneration should thus preferably be carried out at lower temperatures. A promising example is found in the study by Ma et al. [98], who showed the possibility to remove carbon deposits using a small amount of NO (2% NO in air) as promotor. In such mixture, carbon could be fully removed at 623 K, and eight regeneration cycles were reported without activity loss. The regenerability depends also on the degree of coking. Ismagilov et al. [100] showed that it is possible to regain full catalytic activity when regeneration was applied after a period of 6 h on stream for five consecutive regeneration cycles. However, longer periods on stream (15 and 20 h) led to irreversible damage of the catalyst and decrease in activity after each regeneration cycle. This most likely relates to the nature and amount of the different coke species.

8.4 ALTERNATIVE STRATEGIES FOR CATALYST IMPROVEMENT

Besides variation of the pore topology of the zeolitic support, the physical and chemical properties of the catalyst can be adjusted to improve catalyst stability and selectivity. The use of regeneration may also help forward the development of MDA into an industrial process. In this section, several promising routes are discussed, a few of them having been attempted during the NEXT-GTL project. In general, the approaches are focused on minimizing the effect of pore blockage by carbon deposition, by preventing, reducing, or removing the undesired by-product, mainly carbon.

8.4.1 PROMOTERS

In literature, it was shown that the addition of a second metal to the Mo/ZSM-5 catalyst as a copromoter can have beneficial effects on activity, selectivity, and coke resistance. A large part of the periodic system has been screened as a possible additive. A complete overview on the different papers published on the different elements is provided by the review papers by Ma et al. [39] and Majhi et al. [38]. In general, Fe, Co, Ni, Cu, Zn, Ga, Cr, and Ag appear to improve catalytic performance when focusing on catalytic activity, benzene selectivity, and catalyst stability [54]. However, there are large inconsistencies in the open literature due to differences in preparation methods, catalyst activation methods, and reaction conditions. The beneficial effect of Ru is attributed to the reduction of strong acid sites, decreasing coke formation [39]. The use of Fe is argued to play an important role in the in situ coke removal [39].

8.4.2 INTRODUCTION OF MESOPOROSITY IN THE ZEOLITE

For several reactions other than MDA, the introduction of mesoporosity strongly reduces the catalyst deactivation rate. A clear example is the substantially improved stability of a Fe/ZSM-5 catalyst in the benzene to phenol reaction as reported by Koekkoek et al. [101]. Nanostructuring of the zeolite results in better utilization of the micropore space, so that a higher productivity is obtained. Increased mass transport also results in improved activity. A first approach aimed at increasing accessibility in MDA catalysts was presented by Martínez et al. by comparing the catalytic behavior of Mo/MCM-22 with that of Mo/ITQ-2, a catalyst based on the delaminated form of MWW zeolite [75]. Comparison of MCM-22 and ITQ-2-based catalysts with similar Si/Al ratios showed that the latter produced more naphthalene and less benzene. This could be attributed to the larger external surface area of ITQ-2. However, selective dealumination of the external surface of ITQ-2 significantly improved the selectivity to benzene and drastically reduced the formation of naphthalene. These results are consistent with the assertions earlier that the main coking pathway in Mo/ZSM-5 is due to carbon deposition on the external surface.

Common strategies to introduce mesoporosity in zeolite crystals that have been applied to prepare hierarchical MDA zeolite catalysts are silicon extraction by base leaching and carbon black templating. Both methods are appealing because they are relatively straightforward and cheap. They have been applied to the synthesis of mesoporous ZSM-5 and MCM-22 zeolites [102,103]. A higher tolerance toward carbonaceous deposits was shown for the carbon black templated materials resulting in higher catalyst stability and higher aromatics yield. The desilication procedure was found to be more complicated in the sense that various parameters need to be optimized. Su et al. have shown that there is an optimum mesopore volume for the MDA reaction [91]. Too large mesopore volume results in reduced shape selectivity, resulting in more hard coke formation. In the NEXT-GTL project, the effect of mesopore introduction by desilication has been studied in detail [104]. The desilication procedure by Groen et al. [105] was further optimized to obtain high mesopore volume without affecting the crystallinity and acidity of

TABLE 8.5

Catalytic MDA Activity Measurements for Mo/ZSM-5 and Mo/ZSM-5(Meso) With and Without Silylation Treatment

Sample	Time (h)	X_{CH_4} (%)	C_6H_6 (mmol/h gcat)	$C_{10}H_8$ (mmol/h gcat)	S_{arom}[a] (wt%)	$S_{C_6H_6}$ (wt%)	$S_{C_{10}H_8}$ (wt%)	S_{olefin} (wt%)	S_{coke} (wt%)
Mo/ZSM-5	0.5	14.7	0.86	0.05	55	48	5	3	42
Mo/ZSM-5(Si)	0.5	14.4	1.35	0.03	78	72	3	3	19
Mo/ZSM-5(meso)	0.5	9.6	0.66	0.04	64	55	6	3	32
Mo/ZSM-5(meso, Si)	0.5	14.6	0.92	0.04	60	55	4	2	38
Mo/ZSM-5	2	9.0	0.55	0.05	61	49	8	7	32
Mo/ZSM-5(Si)	2	10.3	1.04	0.03	87	78	4	6	6
Mo/ZSM-5(meso)	2	5.7	0.32	0.04	55	43	9	8	30
Mo/ZSM-5(meso,Si)	2	9.5	0.64	0.03	66	58	5	6	27
Mo/ZSM-5	6	2.9	0.11	0.05	56	30	24	27	17
Mo/ZSM-5(Si)	6	4.7	0.35	0.03	70	56	8	24	6
Mo/ZSM-5(meso)	6	3.4	0.11	0.03	38	24	13	15	37
Mo/ZSM-5(meso,Si)	6	5.9	0.27	0.04	55	42	8	15	29
Mo/ZSM-5	10	1.2	0.02	0.05	54	9	43	37	9
Mo/ZSM-5(Si)	10	1.9	0.06	0.03	46	23	19	39	15
Mo/ZSM-5(meso)	10	2.5	0.05	0	16	15	0	19	22
Mo/ZSM-5(meso,Si)	10	3.4	0.13	0.04	48	32	13	26	23

Source: Data taken from Tempelman, C.H.L. et al., Manuscript under preparation.
[a] Total aromatics.

the zeolite. Table 8.5 shows that such hierarchical catalyst (Mo/ZSM-5(meso)) exhibited improved resistance against coking, although it has to be admitted that the methane conversion rate and aromatics selectivity were lower compared to a microporous Mo/ZSM-5 zeolite. It was found that enhanced mesoporosity led to higher rate of coke formation and this appears to tie in with the role of the external surface in hard coke formation.

The lower aromatics selectivity is partially attributed to the slightly lower bulk BAS concentration (Table 8.6) of the starting hierarchical ZSM-5. The hierarchical zeolite is more defective as evidenced by increased intensity of the silanol band (3745 cm^{-1}) and decreased intensity of the BAS band (3612 cm^{-1}) (Figure 8.7). After the modification of the zeolite with Mo, the bulk BAS concentration of the Mo/ZSM-5(meso) catalyst was found to be lower compared to Mo/ZSM-5.

Another artifact that became apparent during intensive characterization of microporous and mesoporous Mo/ZSM-5 was that the much higher external/mesopore surface area in the latter resulted in a much higher Mo-oxide dispersion, bringing the active phase closer to the micropores with the adverse consequence that more $Al_2(MoO_4)_3$ was formed in the mesoporous zeolite-derived catalyst. Formation of $Al_2(MoO_4)_3$ is evidenced by NMR peaks at 14 and −11 ppm, in the ^{27}Al-MAS-NMR spectrum (Figure 8.8). $Al_2(MoO_4)_3$ formation implies that Al is

TABLE 8.6

Physicochemical Properties of Mo/ZSM-5 and Mo/ZSM-5(Meso) Catalysts

Sample	V_{mic} (cm³/g)	V_{meso} (cm³/g)	I_{Si}/I_{Al} (XPS)	BAS$_{(bulk)}$ (µmol/ gcat)	BAS$_{(external)}$ (µmol/ gcat)	C_{Total}[a] (wt%)	C_{SC}[b] (wt%)	C_{HC-Ext}[b] (wt%)	C_{HC-Int}[b] (wt%)
HZSM-5	0.13	0.03	22	679	61	—	—	—	—
HZSM-5(meso)	0.14	0.31	15	544	66	—	—	—	—
Mo/ZSM-5	0.07	0.07	49	533	51	0.13	0.022	0.108	0
Mo/ZSM-5(Si)	0.04	0.02	58	314	21	0.10	0.024	0.076	0
Mo/ZSM-5(meso)	0.07	0.23	20	471	71	0.19	0.095	0.082	0.013
Mo/ZSM-5(meso, Si)	0.13	0.34	30	356	69	0.21	0.210	0	0

Source: Data taken from Tempelman, C.H.L. et al., Manuscript under preparation.

[a] Total coke.

[b] SC, soft coke; HC-Ext, hard coke on external surface; HC-Int, hard coke on internal surface.

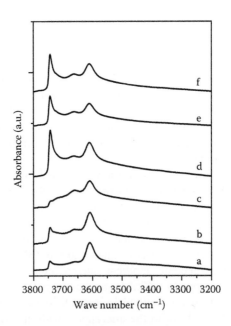

FIGURE 8.7 FTIR spectra of (a) ZSM-5, (b) Mo/ZSM-5, (c) Mo/ZSM-5(silylated), (d) ZSM-5(meso), (e) Mo/ZSM-5(meso), and (f) Mo/ZSM-5(meso, silylated). (Data taken from Tempelman, C.H.L. et al., Manuscript under preparation.)

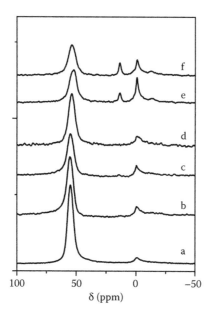

FIGURE 8.8 ^{27}Al MAS NMR spectra of (a) ZSM-5, (b) Mo/ZSM-5, (c) Mo/ZSM-5(silylated), (d) ZSM-5(meso), (e) Mo/ZSM-5(meso), and (f) Mo/ZSM-5(meso,silylated). (Data taken from Tempelman, C.H.L. et al., Manuscript under preparation.)

extracted from the framework, reducing the BAS concentration. It has also been suggested that $Al_2(MoO_4)_3$ cannot be activated to form the active phase for methane activation [106].

The decrease in aromatics selectivity is compensated by the increased selectivity toward coke for the mesoporous catalyst. Characterization of the spent samples by TGA (Table 8.6) revealed similar hard coke (C_{HC}) content for conventional and hierarchical zeolite catalysts. When taking into account the lower activity of Mo/ZSM-5(meso), hard coke formation is slightly higher. This is in line with the slightly increased external acidity as determined by collidine adsorption experiments (see Table 8.6). However, the main increase and constituent of the carbon deposits at the surface of Mo/ZSM-5(meso) is soft coke (C_{SC}), which is about four times higher compared to Mo/ZSM-5. Soft coke is mainly formed in the proximity of Mo. The increase of C_{SC} should be attributed to the higher dispersion of the Mo phase in the mesopores. Furthermore, the MoO_3 particles deposited in the mesopores are more sensitive for agglomeration due to the siliceous nature of the mesopore surface [107,108]. Solymosi et al. [97] have shown that large MoO_3 crystallites carburize to Mo_2C more easily than small Mo-oxide particles. These authors have also contended that larger Mo_2C particles favor formation of carbonaceous deposits and, accordingly, they are prone to fast deactivation. Typically, MoO_x species interact with the siliceous surface of the external or mesopore surface. The BASs facilitate better immobilization of the Mo species, stabilizing smaller particles and improving interaction between the support and the metal oxide [107,108] in this way facilitating spreading of Mo over the surface. It has also been shown that zeolite desilication results in a relatively low BAS density on the mesopore surface [109]. This appears to provide an explanation for the more facile sintering of the initially dispersed Mo phase in mesoporous Mo/ZSM-5.

The previous findings are in good agreement with findings of Su et al. [110] that too large mesopore/external surface area is disadvantageous for MDA. Desilication reduces the acidity of the catalyst resulting in lower benzene selectivity. Interaction of the Mo-oxo species with the FAl results in $Al_2(MoO_4)_3$. This process reduces the zeolite acidity and results in loss of Mo into a stable phase. The decrease of BAS in the mesopores upon silicon extraction assists in the agglomeration of MoO_3 into large particles, and large MoO_3 particles possess a high selectivity toward coke and are prone to fast deactivation.

8.4.3 EXTERNAL SURFACE DEACTIVATION

As discussed in Section 8.3.1, deactivation of the catalyst is mainly attributed to polyaromatic carbon formation at the external surface BAS, covering the micropore entrances of the zeolite. We explored deactivation of external BAS by silylation, which is a common strategy for ZSM-5 zeolite [111–113]. We employed the method of Zheng et al. [70] using tetraethyl orthosilicate (TEOS) as the silicon source. Several approaches were compared, and it turned out to be most favorable to first load the Mo in the form of ammonium heptamolybdate by wet impregnation and calcination followed by silylation. This approach is different from the usual one involving the loading Mo on the silylated ZSM-5 zeolite [111]. Characterization of the silylated Mo/HZSM-5 revealed that deactivation of the external BAS is not only due to silica deposition. Formation of a Mo-oxo layer over the external surface also contributes to deactivation of the external acid sites. The formation of such a layer covering the zeolite crystal was suggested by the complete disappearance of the silanol groups (3745 cm^{-1}) in the FTIR spectrum (Figure 8.7). It suggests reaction of Mo-oxo species with surface silanols. The spreading of Mo-oxo species eventually forming a Mo-oxo surface layer leads also to Mo diffusion into the micropores. Inside the micropores, Mo can react with the BAS, evidenced by the strong decrease of the FAl signal in the ^{27}Al MAS NMR spectrum of silylated Mo/HZSM-5 (Figure 8.8). Pyridine adsorption measurements show a decrease in the amount of BAS (Table 8.6). The Mo migration into the micropore space is also seen in the decreased micropore volume upon silylation. The improved Mo-oxide spreading originates from external surface BAS deactivation. These BASs form strong Mo–O–Al bonds that immobilize the Mo phase and limit surface spreading. Elimination of some BAS leads to an increased mobility of Mo-oxo species over the surface. The Mo/ZSM-5 catalysts exhibited substantially improved catalytic properties after silylation (Table 8.5). The coke selectivity decreased, consistent with the lower coke content as determined by TGA (Table 8.6). Especially, the amount of polynuclear aromatics/hard coke is lowered upon silylation, which should steam from deactivation of the external BAS. The benzene selectivity is higher after silylation because of the lower rate of heavy hydrocarbon formation.

8.5 SUMMARY AND OUTLOOK

Despite substantial effort dedicated during the last two decades to understand methane dehydrogenation to aromatics under nonoxidative conditions over Mo-promoted zeolites, controlling and tailoring this reaction remains a considerable challenge, especially from the industrial application perspective. Although more selective to the desired product than typical oxidative direct routes, methane conversion is thermodynamically limited, and the best catalysts reported so far, Mo on acid zeolitic supports, suffer from fast deactivation on stream. Much progress has been made regarding the understanding of the reaction mechanism and the active species participating in the different steps of the process, but there remain many uncertainties that also hamper guidance in further catalyst improvements. The increased insight into the reaction mechanism and active species has led to the preparation of improved catalysts, more selective to the desired benzene product, and more stable toward deactivation. Although Mo-impregnated ZSM-5 and MCM-22 are reported to be the best performing catalysts so far, further improvements should be possible by using new zeolite structures with adequate pore dimensions, with proper BAS density and distribution. An alternative approach is optimization of hierarchically organized zeolites, combining high accessibility, with sufficiently preserved acidity and shape selectivity for the correct dispersion of the TM and the control of the oxide crystallite size, as well as improved selectivity for the formation of benzene. In each case, the Mo loading will have to be fine-tuned, as its optimum content and dispersion/location will be highly affected by the physicochemical properties of the zeolite host. Although several strategies have been explored to decrease the adverse effect of carbon laydown, complete suppression of this unselective reaction will likely not be achieved. Thus, regeneration protocols need to be optimized, and there is sufficient room for novel process layout strategies with the aim to optimize reaction–regeneration cycles for the MDA reaction.

ACKNOWLEDGMENTS

The research leading to these results has received funding from the European Union Seventh Framework Program FP7/2007–2013 via the NEXT-GTL project under grant agreement no. 229183. M. T. Portilla and C. Martínez would also like to thank the Spanish Government-MINECO through Consolider Ingenio 2010-Multicat, MAT2012–31657, and the *Severo Ochoa Programme* for the financial support.

REFERENCES

1. J.H. Lunsford, *Catalysis Today*, 63 (2000) 165–174.
2. E.F. Sousa-Aguiar, L.G. Appel, C. Mota, *Catalysis Today*, 101 (2005) 3–7.
3. A. Caballero, P.J. Pérez, *Chemical Society Reviews*, 42 (2013) 8809–8820.
4. J.N. Armor, *Journal of Energy Chemistry*, 22 (2013) 21–26.
5. Q. Wang, X. Chen, A.N. Jha, H. Rogers, *Renewable and Sustainable Energy Reviews*, 30 (2014) 1–28.

6. Annual Energy Outlook 2013 (AEO2013), U.S. Energy Information Administration (EIA), U.S. Energy Information Administration, Office of Integrated and International Energy Analysis, U.S. Department of Energy, Washington DC, http://www.eia.gov/forecasts/steo/report/natgas.cfm, 2013. Accessed January 28, 2014.

7. D.A. Wood, C. Nwaoha, B.F. Towler, *Journal of Natural Gas Science and Engineering*, 9 (2012) 196–208.

8. K. Aasberg-Petersen, I. Dybkjær, C.V. Ovesen, N.C. Schjødt, J. Sehested, S.G. Thomsen, *Journal of Natural Gas Science and Engineering*, 3 (2011) 423–459.

9. R.C. Baliban, J.A. Elia, C.A. Floudas, *AIChE Journal*, 59 (2013) 505–531.

10. C.B. Roberts, N.O. Elbashir, *Fuel Processing Technology*, 83 (2003) 1–9.

11. A. Martínez, G. Prieto, A. García-Trenco et al., Advanced catalysts based on micro- and mesoporous molecular sieves for the conversion of natural gas to fuels and chemicals. In *Zeolites and Catalysis: Synthesis, Reactions and Applications*, J. Cejka, A. Corma, S. Zones(eds.), Wiley-VCH Verlag GmbH & Co. KGaA. 2010, pp. 649–685.

12. M. Gharibi, F.T. Zangeneh, F. Yaripour, S. Sahebdelfar, *Applied Catalysis A: General*, 443–444 (2012) 8–26.

13. F. Larachi, H. Oudghiri-Hassani, M.C. Iliuta, B.P.A. Grandjean, P.H. McBreen, *Catalysis Letters*, 84 (2002) 183–192.

14. O. Rival, B.P.A. Grandjean, C. Guy, A. Sayari, F. Larachi, *Industrial and Engineering Chemistry Research*, 40 (2001) 2212–2219.

15. R.M. Navarro, M.A. Peña, J.L.G. Fierro, *Chemical Reviews*, 107 (2007) 3952–3991.

16. L. Wang, L. Tao, M. Xie, G. Xu, J. Huang, Y. Xu, *Catalysis Letters*, 21 (1993) 35–41.

17. Y. Xu, S. Liu, X. Guo, L. Wang, M. Xie, *Catalysis Letters*, 30 (1995) 135–149.

18. W. Liu, Y. Xu, S.T. Wong, L. Wang, J. Qiu, N. Yang, *Journal of Molecular Catalysis A: Chemical*, 120 (1997) 257–265.

19. F. Solymosi, J. Cserényi, A. Szöke, T. Bánsági, A. Oszkó, *Journal of Catalysis*, 165 (1997) 150–161.

20. X. Bao, W. Zhang, D. Ma, X. Han, X. Liu, X. Guo, X. Wang, *Journal of Catalysis*, 188 (1999) 393–402.

21. R.W. Borry Iii, Y.H. Kim, A. Huffsmith, J.A. Reimer, E. Iglesia, *Journal of Physical Chemistry B*, 103 (1999) 5787–5796.

22. A. Sarioğlan, O.T. SavaŞçi, A. Erdem-Şenatalar, A. Tuel, G. Sapaly, Y. Ben Taârit, *Journal of Catalysis*, 246 (2007) 35–39.

23. A. Martínez, E. Peris, A. Vidal-Moya, Modulation of zeolite acidity by post-synthesis treatments in Mo/HZSM-5 catalysts for methane dehydroaromatization. In *Zeolites and Related Materials: Trends, Targets and Challenges*, A. Gedeon, P. Massiani, F. Babonneau, (eds.), Amsterdam: Elsevier B.V., 2008, pp. 1075–1080.

24. J.P. Tessonnier, B. Louis, S. Rigolet, M.J. Ledoux, C. Pham-Huu, *Applied Catalysis A: General*, 336 (2008) 79–88.

25. D. Ma, Y. Shu, X. Han, X. Liu, Y. Xu, X. Bao, *Journal of Physical Chemistry B*, 105 (2001) 1786–1793.

26. Z. Sobalík, Z. Tvarůžková, B. Wichterlová, V. Fíla, Š. Špatenka, *Applied Catalysis A: General*, 253 (2003) 271–282.

27. J. Bai, S. Liu, S. Xie, L. Xu, L. Lin, *Reaction Kinetics and Catalysis Letters*, 82 (2004) 279–286.

28. L. Liu, D. Ma, H. Chen, H. Zheng, M. Cheng, Y. Xu, X. Bao, *Catalysis Letters*, 108 (2006) 25–30.

29. A. Smiešková, P. Hudec, N. Kumar, T. Salmi, D.Y. Murzin, V. Jorík, *Applied Catalysis A: General*, 377 (2010) 83–91.

30. X. Yin, N. Chu, J. Yang, J. Wang, Z. Li, *Catalysis Communications*, 43 (2014) 218–222.

31. C.L. Zhang, S. Li, Y. Yuan, W.X. Zhang, T.H. Wu, L.W. Lin, *Catalysis Letters*, 56 (1998) 207–213.

32. D.Y. Wang, Q.B. Kan, N. Xu, P. Wu, T.H. Wu, *Catalysis Today*, 93–95 (2004) 75–80.
33. P. Wu, Q. Kan, D. Wang, H. Xing, M. Jia, T. Wu, *Catalysis Communications*, 6 (2005) 449–454.
34. C. Xu, J. Guan, S. Wu, M. Jia, T. Wu, Q. Kan, *Reaction Kinetics, Mechanisms and Catalysis*, 99 (2010) 193–199.
35. H. Liu, S. Yang, S. Wu, F. Shang, X. Yu, C. Xu, J. Guan, Q. Kan, *Energy*, 36 (2011) 1582–1589.
36. H. Liu, J. Hu, Z. Li, S. Wu, L. Liu, J. Guan, Q. Kan, *Kinetics and Catalysis*, 54 (2013) 443–450.
37. Z.R. Ismagilov, E.V. Matus, L.T. Tsikoza, *Energy & Environmental Science*, 1 (2008) 526–541.
38. S. Majhi, P. Mohanty, H. Wang, K.K. Pant, *Journal of Energy Chemistry*, 22 (2013) 543–554.
39. S. Ma, X. Guo, L. Zhao, S. Scott, X. Bao, *Journal of Energy Chemistry*, 22 (2013) 1–20.
40. H. Liu, W. Shen, X. Bao, Y. Xu, *Journal of Molecular Catalysis A: Chemical*, 244 (2006) 229–236.
41. H. Liu, X. Bao, Y. Xu, *Journal of Catalysis*, 239 (2006) 441–450.
42. H.S. Lacheen, E. Iglesia, *Physical Chemistry Chemical Physics*, 7 (2005) 538–547.
43. W. Ding, S. Li, G.D. Meitzner, E. Iglesia, *Journal of Physical Chemistry B*, 105 (2001) 506–513.
44. W. Li, G.D. Meitzner, R.W. Borry III, E. Iglesia, *Journal of Catalysis*, 191 (2000) 373–383.
45. Y.H. Kim, R.W. Borry III, E. Iglesia, *Microporous and Mesoporous Materials*, 35–36 (2000) 495–509.
46. K.S. Wong, J.W. Thybaut, E. Tangstad, M.W. Stöcker, G.B. Marin, *Microporous and Mesoporous Materials*, 164 (2012) 302–312.
47. B. Yao, J. Chen, D. Liu, D. Fang, *Journal of Natural Gas Chemistry*, 17 (2008) 64–68.
48. A. Bhan, S.H. Hsu, G. Blau, J.M. Caruthers, V. Venkatasubramanian, W.N. Delgass, *Journal of Catalysis*, 235 (2005) 35–51.
49. M.C. Iliuta, I. Iliuta, B.P.A. Grandjean, F. Larachi, *Industrial and Engineering Chemistry Research*, 42 (2003) 3203–3209.
50. S. Liu, L. Wang, R. Ohnishi, M. Ichikawa, *Kinetics and Catalysis*, 41 (2000) 132–144.
51. Y. Xu, L. Lin, *Applied Catalysis A: General*, 188 (1999) 53–67.
52. C. Martínez, A. Corma, *Coordination Chemistry Reviews*, 255 (2011) 1558–1580.
53. P.B. Weisz, V.J. Frilette, *Journal of Physical Chemistry*, 64(3) (1960) 382.
54. J.T.F. Degnan, *Journal of Catalysis*, 216 (2003) 15, 32–46.
55. A. Corma, *Chemical Reviews* (Washington, DC), 97 (1997) 2373–2420.
56. C. Martínez, A. Corma, Zeolites. In *Comprehensive Inorganic Chemistry II*, J. Reedijk, K. Poeppelmeier (eds.), Oxford, U.K.: Elsevier, 2013, pp. 103–131.
57. Y. Cui, Y. Xu, J. Lu, Y. Suzuki, Z.-G. Zhang, *Applied Catalysis A: General*, 393 (2011) 348–358.
58. J. Perez-Ramirez, C.H. Christensen, K. Egeblad, C.H. Christensen, J.C. Groen, *Chemical Society Reviews*, 37 (2008) 2530–2542.
59. Z.R. Ismagilov, E.V. Matus, M.A. Kerzhentsev, L.T. Tsikoza, I.Z. Ismagilov, K.D. Dosumov, A.G. Mustafin, *Petroleum Chemistry*, 51 (2011) 174–186.
60. W. Wei, *Journal of Natural Gas Chemistry*, 9 (2000) 76–86+88.
61. L.Y. Chen, L.W. Lin, Z.S. Xu, X.S. Li, T. Zhang, *Journal of Catalysis*, 157 (1995) 190–200.
62. C. Xu, H. Liu, M. Jia, J. Guan, S. Wu, T. Wu, Q. Kan, *Applied Surface Science*, 257 (2011) 2448–2454.
63. J. Dědeček, Z. Sobalík, B. Wichterlová, *Catalysis Reviews—Science and Engineering*, 54 (2012) 135–223.
64. G. Fratesi, S. De Gironcoli, *Journal of Chemical Physics*, 125 (2006). 044701 1–7.

65. A. Kokalj, N. Bonini, S. De Gironcoli, C. Sbraccia, G. Fratesi, S. Baroni, *Journal of the American Chemical Society*, 128 (2006) 12448–12454.

66. D. Ma, X. Han, D. Zhou, Z. Yan, R. Fu, Y. Xu, X. Bao, H. Hu, S.C.F. Au-Yeung, *Chemistry—A European Journal*, 8 (2002) 4557–4561.

67. C.H. Bartholomew, R.J. Farrauto, *Fundamentals of Industrial Catalytic Processes*, 2nd edn. Hoboken, New Jersey: John Wiley & Sons, Inc., 2006.

68. Y. Xu, Y. Shu, S. Liu, J. Huang, X. Guo, *Catalysis Letters*, 35 (1995) 233–243.

69. B. Li, S. Li, N. Li, H. Chen, W. Zhang, X. Bao, B. Lin, *Microporous and Mesoporous Materials*, 88 (2006) 244–253.

70. S. Zheng, H.R. Heydenrych, A. Jentys, J.A. Lercher, Influence of Surface Modification on the Acid Site Distribution of HZSM-5, *J. Phys. Chem. B*, 106 (2002) 9552–9558.

71. F. Solymosi, L. Bugyi, A. Oszkó, *Catalysis Letters*, 57 (1999) 103–107.

72. V.T.T. Ha, L.V. Tiep, P. Meriaudeau, C. Naccache, *Journal of Molecular Catalysis A: Chemical*, 181 (2002) 283–290.

73. Y. Shu, D. Ma, L. Xu, Y. Xu, X. Bao, *Catalysis Letters*, 70 (2000) 67–73.

74. A. Corma, V. Fornes, S.B. Pergher, T. Maesen, J.G. Buglass, *Nature* (London), 396 (1998) 353–356.

75. A. Martínez, E. Peris, G. Sastre, *Catalysis Today*, 107–108 (2005) 676–684.

76. S.L. Lawton, A.S. Fung, G.J. Kennedy, L.B. Alemany, C.D. Chang, G.H. Hatzikos, D.N. Lissy et al., *Journal of Physical Chemistry*, 100 (1996) 3788–3798.

77. C.T. Kresge, W.J. Roth, K.G. Simmons, J.C. Vartuli, US5229341 (1993).

78. R. Castaneda, A. Corma, V. Fornes, J. Martinez-Triguero, S. Valencia, *Journal of Catalysis*, 238 (2006) 79–87.

79. S.B. Hong, H.-K. Min, C.-H. Shin, P.A. Cox, S.J. Warrender, P.A. Wright, *Journal of the American Chemical Society*, 129 (2007) 10870–10885.

80. C. Baerlocher, F. Gramm, L. Massüger, L.B. McCusker, Z. He, S. Hovmöller, X. Zou, *Science*, 315 (2007) 1113–1116.

81. H. Liu, S. Wu, Y. Guo, F. Shang, X. Yu, Y. Ma, C. Xu, J. Guan, Q. Kan, *Fuel*, 90 (2011) 1515–1521.

82. C. Martínez, M.T. Portilla, K.S. Wong, J.W. Thybaut, E. Tangstad, M. Stöcker, G.B. Marin, Non-oxidative aromatization of methane: Catalyst descriptors obtained by a dynamic microkinetic reactor model for assessment of zeolite framework effects, in: *17th International Zeolite Conference*, Moscow, Russia, 2013. Book of Abstracts, 110.

83. D. Ren, X. Wang, G. Li, X. Cheng, H. Long, L. Chen, *Journal of Natural Gas Chemistry*, 19 (2010) 646–652.

84. M. Nagai, T. Nishibayashi, S. Omi, *Applied Catalysis A*, 253 (2003) 101–112.

85. B. Weckhuysen, M. Rosynek, J. Lunsford, *Catalysis Letters*, 52 (1998) 31–36.

86. D. Ma, D. Wang, L. Su, Y. Shu, Y. Xu, X. Bao, *Journal of Catalysis*, 208 (2002) 260–269.

87. C. Descorme, P. Gelin, C. Lecuyer, A. Primet, *Applied Catalysis B*, 13 (1997) 185–195.

88. H. Liu, L. Su, H. Wang, W. Shen, X. Bao, Y. Xu, *Applied Catalysis A: General*, 236 (2002) 263–280.

89. K. Honda, X. Chen, Z.-G. Zhang, *Catalysis Communications*, 5 (2004) 557–561.

90. C.H.L. Tempelman, E.J.M. Hensen, A model study of the deactivation of Mo/HZSM-5 in methane dehydroaromatization, Manuscript under preparation.

91. L. Su, L. Liu, J. Zhuang, H. Wang, Y. Li, W. Shen, Y. Xu, X. Bao, *Catalysis Letters*, 91 (2003) 155–167.

92. R.H. Meinhold, D.M. Bibby, *Zeolites*, 10 (1990) 146–150.

93. E.J.M. Hensen, D.G. Poduval, D.A.J.M. Ligthart, J.A.R.V. Veen, M.S. Rigutto, *Journal of Physical Chemistry C*, 114 (2010) 8363–8374.

94. T. Behrsing, H. Jaeger, J.V. Sanders, *Applied Catalysis*, 54 (1989) 289–302.

95. L. Su, L. Liu, J. Zhuang, H. Wang, Y. Li, W. Shen, Y. Xu, X. Bao, *Catalysis Letters*, 91 (2003) 155–156.

96. C. Li, P. Stair, *Catalysis Today*, 33 (1997) 353–360.
97. F. Solymosi, A. Szöke, J. Cserényi, Conversion of methane to benzene over Mo_2C and Mo_2C/ZSM-5 catalysts, *Catalysis Letters*, 39 (1996) 157–162.
98. H. Ma, R. Kojima, R. Ohnishi, M. Ichikawa, *Applied Catalysis A: General*, 275 (2004) 183–187.
99. R.W. Borry III, E.C. Lu, Y.H. Kim, E. Iglesia. Non-oxidative catalytic conversion of methane with continuous hydrogen removal. In *Studies in Surface Science and Catalysis*, A. Parmaliana, D. Sanfilippo, F. Frusteri, A. Vaccari, F. Arena (eds.), Elsevier B.V., 1998, Vol. 119, pp. 403–410.
100. Z.R. Ismagilov, L.T. Tsikoza, E.V. Matus, G.S. Litvak, I.Z. Ismagilov, O.B. Sukhova, *Eurasian Chemico-Technological Journal*, 7 (2005) 115–121.
101. A.J.J. Koekkoek, C.H.L. Tempelman, V. Degirmenci, M. Guo, Z. Feng, C. Li, E.J.M. Hensen, *Catalysis Today*, 168 (2011) 96–111.
102. A. Martinez, E. Peris, M. Derewinski, A. Burkat-Dulak, *Catalysis Today*, 169 (2011) 75–84.
103. N.B. Chu, J.Q. Wang, Y. Zhang, J.H. Yang, J.M. Lu, D.H. Yin, *Chemistry of Materials*, 22 (2010) 2757–2763.
104. C.H.L. Tempelman, V.O.D. Rodrigues, E.H.R.V. Eck, P.C.C.M. Magusing, E.J.M. Hensen, Desilication and silylation of Mo/HZSM-5 for methane dehydroaromatization, Manuscript under preparation.
105. J.C. Groen, W. Zhu, S. Brouwer, S.J. Huynink, F. Kapteijn, J.A. Moulijn, J. Pérez-Ramírez, *Journal of the American Chemical Society*, 129 (2007) 355–360.
106. W. Zhang, D. Ma, X. Han, X. Liu, X. Bao, X. Guo, X. Wang, *Journal of Catalysis*, 188 (1999) 393–402.
107. J. Leyrer, D. Mey, H. Knözinger, *Journal of Catalysis*, 124 (1990) 349–356.
108. J. Leyrer, R. Margraf, E. Taglauer, H. Knözinger, *Surface Science*, 201 (1988) 603–623.
109. M.H.F. Kox, E. Stavitski, J.C. Groen, J. Pérez-Ramírez, F. Kapteijn, B.M. Weckhuysen, *Chemistry—A European Journal*, 14 (2008) 1718–1725.
110. L. Su, L. Liu, J. Zhuang, H. Wang, Y. Li, W. Shen, Y. Xu, X. Bao, *Catalysis Letters*, 91 (2003) 155–168.
111. W. Ding, G.D. Meitzner, E. Iglesia, *Journal of Catalysis*, 206 (2002) 14–22.
112. H. Liu, Y. Li, W. Shen, X. Bao, Y. Xu, *Catalysis Today*, 93–95 (2004) 67–73.
113. S. Kikuchi, R. Kojima, H. Maa, J. Bai, M. Ichikawa, *Journal of Catalysis*, 242 (2006) 349–356.

9 Structured Catalysts on Metal Supports for Light Alkane Conversion

Leonid M. Kustov

CONTENTS

9.1 INTRODUCTION

Continuous progress in catalytic technology requires development of new catalysts and catalyst supports with improved gas-dynamic characteristics, heat and mass transfer, etc. Catalysts supported on metal carriers such as monoliths, metal foams, and wire grids are promising systems for various catalytic processes. These catalysts possess high mechanical strength, resistance to thermal shocks, and high thermal conductivity and provide conditions for perfect heat and mass transfer. Also, they are characterized by a low pressure drop. Catalyst supports based on metal wire gauzes and sintered metal fibers are especially perspective systems because of higher accessible surface area and gas transfer compared to ceramic or metallic monoliths, foils, and foams. An important advantage of structured catalysts of this type is their low resistance to the flow of liquids and gases passing through a reactor filled with the structured material, which is significantly lower compared to that of powdered or granulated catalysts.

Gauze catalysts based on wires of noble metals (Pt, Ru, Ag) are used in production of nitric acid, hydrocyanic acid, and aldehydes [1]. Such monometallic wires are extremely expensive and therefore have limited application. Silver gauzes used for the conversion of methanol and ethanol to aldehydes operate at 630°C–700°C. Catalytic wire gauzes were used also for ammonia oxidation. Presently, diverse gauzes with a deposited catalytic phase received significant attention as reactor fillers used for oxidation of other compounds. Excellent reviews of the preceding state-of-the-art gauze catalysts [2,3] have been published in the 1990s.

Structured materials produced from inexpensive metals such as Ni, Al, Fe, stainless steel, and alloys like Inconel, Fecralloy, and Ni alloys are now commercially available from diverse vendors and promise a wide range of potential applications, including catalytic processes. Fibers and gauzes are flexible materials and provide unlimited variety of shapes, knitting styles, and forms, which is not the case of conventional powdered or granulated catalysts. Fabrics or mats produced from woven metal wool or sintered metal fibers may be packed or constructed for any catalytic application and built in any type of reactor geometry.

The distinctive feature of structured metal catalysts is their high geometrical surface area, safer operation, and easy scale-up. Also, they are the carrier of choice in the case of highly exothermal reactions. When such reactions as oxidation or hydrogenation occur on granulated catalysts, hot spots are formed, which are the source of decreasing selectivity, coke formation, and deactivation. In the case of metal carriers, the ideal heat transfer mitigates the problem of hot spots, and thus higher selectivities, better productivities, and longer life times can be expected from structured metal catalysts. Actually, this is not the case of honeycomb-type materials and oxide fibers (like silica), which, due to the insulating nature of a carrier, do not conduct heat and the unexpected overheating and side reactions may occur on these catalysts.

The efficient flow resistance of metal-structured catalysts was studied and modeled in Ref. [4]. The flow and transfer properties of stacked gauzes have been sufficiently examined. The theoretical mathematical model of the flow passing through wire gauzes based on Ergun's approach is put forward and tested. It is noteworthy that the model does not include any empirical constants. The model prediction was found to be in good agreement with the available experimental results. The model provides a tool for the classification of wire gauzes for the application as reactor fillers.

The study of mass transfer occurring on wire gauzes is presented in Ref. [5], whereas the flow resistance problem is explicitly addressed in Ref. [4,6].

The specific area of metallic fibers per se, however, is very small, and this limits their use by catalysis of very fast processes. To achieve a high surface area, porous oxides or their precursors can be used for coating the metal carrier with a layer of a high surface area. Thus, the use of the metal wire gauzes as catalysts requires to solve the problem of formation of a uniform secondary support layer with a high surface area and porosity onto the metal wire. The crucial problem of the design and application of structured catalysts based on metal carriers is the deposition of appropriate thin, uniform, and catalytically effective films of either secondary oxides or active phases with a good adhesion to the metal surface on the metallic wires. For this reason, several advanced deposition techniques have been designed, and new catalysts characterized with a very high activity are now becoming available for the wide use. Interesting examples of these developments are the gauzes developed by Ahlstroem-Silversand and Odenbrand [7].

Among the catalytic reactions studied using structured catalysts based on metal carriers (foils, gauzes, foams, wool, sintered metal fibers), of special interest are the highly exothermal reactions such as hydrogenation and partial or complete oxidation. Quite a number of examples of applications of structured metal catalysts for pollution control systems have been described in the literature and patents, but this area is outside the scope of this review. Also, it is not the intention of the author to discuss diverse hydrogenation processes, although outstanding properties of such materials, especially sintered metal fibers, have been already demonstrated [8]. This review will be focused exclusively on the conversion of light alkanes by their partial or complete oxidation.

9.2 PREPARATION OF STRUCTURED CATALYSTS BASED ON METAL CARRIERS

9.2.1 METAL CARRIERS

Most frequently used forms of metal carriers include metal foams, foils, wool, gauzes (meshes), and sintered metal fibers. Foams have a number of limitations because of the presence of dead volume (pores with dead ends). Foils are good candidates but their disadvantage is the possible formation of flow channels. Catalytic packing arranged from metal wires (d = 10–100 mkm) presents an open regular structure and acceptable mechanical properties. As a result of their high thermal conductivity, hot spot formation is suppressed in exothermic reactions. The creation of a porous oxide layer on a metal surface is necessary for active phase deposition.

Sintered metal fiber filters (MFFs) consisting of thin metal filaments (d = 2–30 mkm) [9] combine the advantages of metal wires and gauzes with the properties of fibrous materials. Sintered metal fibers are mechanically and chemically stable. They are now commercially available in the form of panels or mats of different thickness and are characterized by uniformly highly porous 3D macrostructures with porosity ranging between 70% and 90%. High thermal conductivity of the metal fibers provides a radial heat transfer coefficient in a catalytic bed about two times higher compared to conventional packed beds [10], which translates into nearly isothermal conditions in a reactor.

A fiber or gauze acts as a micron-scale mixer, which allows one to avoid the formation of flow channels. One of the examples of sintered metal fibers is the product Bekipor (made from Inconel) in the form of a mat with a thickness of 0.5 mm (Bekaert Fibre Technology Company, Belgium). It consists of metal fibers with a diameter of 6–12 mkm and is characterized with a porosity of 81% and wetness capacity of ~40 wt%.

9.2.2 Supporting a Secondary Oxide Coating

In order to bring the metal support the porosity and enhanced surface area, they are coated with a secondary oxide layer. This can be done by different methods, like immersion/precipitation of the oxide precursor solution, electroplating or electrophoretic deposition (EPD), chemical or physical vapor deposition, thermal spray deposition, and laser-induced deposition, or by other more complicated methods requiring the use of high-energy sources (plasma, microwave treatments). In any case, the deposition is followed by adequate oxidative treatment and calcination to convert the precursor into the porous oxide state and destroy the precursor.

The authors of Ref. [11] investigated the catalysts prepared from wire meshes (stainless steel) coated by thermal spray deposition. This method was used before for the production of Raney nickel and iron catalysts for hydrocarbon conversion [12]. An oxide layer with a good adhesion to the metal substrate was produced by this method. The porosity and the specific surface area of the oxide layer may be increased by adding a composite alumina–polymer powder to the thermally sprayed matrix by treating it with different sols or by precipitation before introducing a catalytically active phase.

One of the most promising methods for producing very thin catalytic structures is cold plasma deposition from metal–organic precursors. This method (plasma-enhanced metal–organic chemical vapor deposition) has already been used to prepare a number of thin-film materials for various applications. Well-adhered coatings can be obtained by plasma spraying (Figure 9.1).

The sol–gel solvent evaporation method can also be used for supporting secondary oxide coating onto metal carriers [13,14], for instance, silica, alumina, and mesoporous SBA-15 films on sintered metal fibers (Figure 9.2).

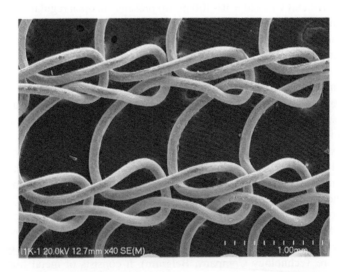

FIGURE 9.1 SEM image (40×) of the wire gauze deposited with nanocrystalline CO_3O_4 films by plasma deposition.

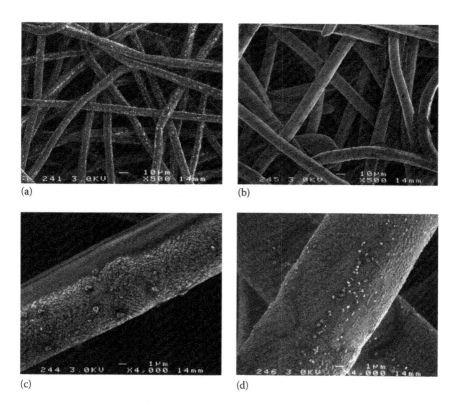

FIGURE 9.2 SEM images of noncoated and coated sintered MFFs: (a) initial MFF; (b,d) microporous SiO_2-coated MFF; (c) mesoporous SBA-15-coated MFF. (From Ogawa, M., *J. Am. Chem. Soc.*, 116, 7941, 1994. With permission.)

The wash-coating method is commonly used for the preparation of oxide layers on ceramic or metallic monoliths. However, this method cannot be applied for deposition of coatings on a metal gauze with a wire thickness as low as 50 mkm because of the poor adhesion and nonuniformity of the resulting coating [15].

EPD is an efficient and energy-saving method of preparation of new types of gauze or metal fiber catalysts, including those prepared from nonnoble metals (stainless steel, Al-containing alloys of high thermal stability like Fecralloy) [16,17]. Typically, wools are 10–100 µm thick. Each of the filaments is coated with a thin film (1–20 µm) of a secondary support oxide layer (α-Al_2O_3, SiO_2, TiO_2) with a controllable surface area and porosity. Catalytically active phases such as metals (for instance, platinum) or metal oxides (Mn_2O_3, Na_2WO_4, PbO_x, BiO_x, SnO_x, CeO_x, LnO_x where Ln = La, Sm, Yb, etc.) can be supported on oxide films. The procedure of coating a metal wire with a supported oxide film is the most important stage in the preparation of oxide catalysts supported on metal gauzes. In the case of supports based on conducting materials such as metal gauzes, the method of electrochemical deposition of metals can be efficiently used for modification of a metal gauze surface and for supporting a precursor of a secondary support oxide layer. EPD method can be chosen as the most attractive method for deposition of oxide layers on metal gauzes and fibers in order to

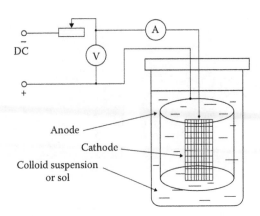

FIGURE 9.3 Coating by EPD.

prepare new catalyst carriers. It is based on the deposition of particles of a colloidal suspension on the surface of an electrode under the influence of an electric field. The EPD technique is an inexpensive and simple method characterized by fast coating formation. The scheme of EPD is shown in Figure 9.3. Oxide coatings serve also to enhance high-temperature resistance, diffusion barriers, and corrosion resistance of a metal fiber for following usage in metal matrix composites. EPD has been used as an alternative to wash coating to produce more uniform thickness of the supported material on the metal foil or gauze surface. A number of catalyst supports based on stainless steel gauze coated with 1–10 μm oxide layers of Al_2O_3, ZrO_2, SiO_2, mixed oxides, and coatings containing H-ZSM-5 zeolite microcrystallites were prepared by EPD method. The samples obtained by the EPD method were compared with the catalysts prepared by other available methods, such as plasma spray deposition and wash coating in terms of the preparation efficiency and quality of the coating. The properties of the obtained coatings were studied by SEM, XPS, and DRIFTS. It was shown that alumina coatings obtained by EPD (α- or γ-Al_2O_3) have a controllable specific surface area (from 10 to 450 m^2/g), porosity, and acidic properties and can be used as catalyst supports.

The mechanisms and kinetics of EPD and its applications for preparation of various ceramic materials were considered in the review [18]. Earlier, the EPD technique has been used for the deposition of alumina coating on a metal surface for the preparation of reinforcement ceramic composites [19–21].

9.2.3 Preparation of Catalysts Containing Different Phases

9.2.3.1 Alumina Coating

In order to obtain the optimum conditions of the deposition by EPD, the influence of applied voltage on the thickness and quality of the coating was studied. The voltage was varied from 1 to 50 V. The quality of the obtained coatings was controlled by SEM and by measuring the Al/Fe atomic ratio calculated from the Al 2p and Fe 2p line intensities in XPS spectra. The increase in the Al/Fe ratio with increasing voltage indicates that the fraction of the uncoated surface characterized by the

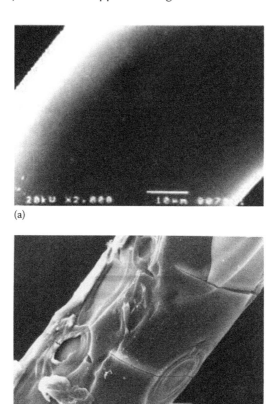

FIGURE 9.4   SEM pictures of Al_2O_3 coatings: (a) aqueous solution, 5 V; (b) aqueous solution, 15 V.

concentration of detectable iron decreases and the quality of the coating is improved. The SEM pictures of the alumina layer on the stainless steel gauze at different voltages are presented in Figure 9.4. The metal surface is perfectly coated with a uniform oxide layer composed of fine particles (about 100 nm) without formation of large agglomerates. The intensity of the Fe 2p and Cr 2p lines in XPS spectra of gauze samples coated with alumina appreciably decreases, whereas the high intensity of Al 2s and Al 2p lines appeared. The use of the voltage higher than 10 V leads to the formation of a thick layer of the oxide film composed of large agglomerates that can be easily peeled off after drying. The decrease in the coating quality at the higher voltage is probably the result of both the intensification of formation of hydrogen bubbles and the high rate of deposition resulting in an increase in thickness. To prevent the crack formation, slow drying at a low temperature and at high humidity was performed. The quality of the coating was shown to be improved when larger particles of alumina (about 1 µm) were introduced into the sol. To decrease the hydrogen evolution, the initial sol was diluted by ethanol in the volume ratio of 1:3. The high Al/Fe atomic ratio for the samples coated in the water–ethanol solution (Figure 9.5)

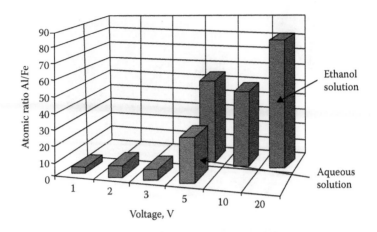

FIGURE 9.5 XPS data for the Al_2O_3-coated stainless steel gauzes.

indicates that the alumina layer is more uniform than that obtained in the aqueous solution. Noteworthy, the structure of the Al_2O_3/gauze interface also appears to be a function of the solution composition. Thus, XPS analysis of the sample prepared in the aqueous solution testifies that iron on the Al_2O_3/gauze interface exists as Fe^{3+}. In the sample obtained in the water–alcohol solution, a mixture of Fe^{2+} and Fe^{3+} ions was found. This effect may be accounted for by the better quality of the Al_2O_3 layer in the second sample, which effectively protects the gauze surface against oxidation under calcination. In turn, the presence of different forms of iron on the Al_2O_3/gauze interface can also result in the better adhesion of the Al_2O_3 layer, thus improving the quality of the coating. At the initial deposition stage, the thickness of the coating increases proportionally to the deposition time. At longer deposition times, the rate of the deposition decreases due to the increase in the electrical resistance of the coating. It should be pointed out that the specific surface area of the gauze coated with alumina determined by BET was rather high and increased with an increase in the thickness of the deposited layer. For the sample with a coating thickness of 5 pm, the surface area after calcination at 550°C was equal to 40 m²/g. The specific surface area per gram of the deposited alumina was about 450 m²/g, that is, two times higher than that of the bulk alumina sample prepared from the alkoxide sol.

Thus, the crucial factors determining the quality of coating the metal carrier with a secondary oxide layer are the following:

- Voltage
- Sol composition
- Deposition time
- Drying conditions

The mechanism of the oxide layer formation on the surface of a metal wire gauze can be presented as follows. The initial alumina sol is composed of nanoparticles of boehmite AlO(OH), positively charged in an acid media. Under the influence of an electric field, the boehmite particles migrate toward the cathode and coagulate on its surface, forming a dense coating. The weight of the particles deposited on the unit of the electrode surface is a function of the quantity of electricity passed through the sol, particle concentration, and deposition time. At a low voltage (1–2 V), the rate of deposition was low so that 10–20 min was required for the complete coating of the metal surface. Voltages higher than 15 V led to the formation of a thick layer of the oxide film composed of large agglomerates that can be easily exfoliated and cracked after drying. Additionally, at a higher voltage, formation of hydrogen bubbles was intensified and the quality of the coating was affected. Hence, the optimum voltage for the deposition of the alumina film should be lower than 10 V.

The properties of the surface acid sites of alumina deposited on the stainless steel gauze were studied by DRIFT spectroscopy. The spectrum in the region of stretching vibrations of OH groups revealed strong absorption at 3470 cm^{-1} with shoulders at 3770, 3715, and 3660 cm^{-1} assigned to hydrogen-bonded OH groups and terminal OH groups and to two- and three-folded bridging OH groups of alumina, respectively. The higher intensity of the band at 3470 cm^{-1} in comparison with conventional bulk alumina could be associated with the enhancement of the H-bonding in the more developed fine porous structure of alumina obtained by EPD. The acid sites of alumina were studied by adsorption of CD_3CN and CO as probe molecules. The difference spectra of adsorbed acetonitrile-d$_3$ revealed the bands 2260 and 2315 cm^{-1} assigned to vibrations of nitrile groups of CD_3CN physically adsorbed on the alumina surface and adsorbed on the Lewis acid sites, respectively.

9.2.3.2 Zirconia Coatings

The main trends found for alumina coatings apply in the case of zirconia and other oxides deposited onto metal gauzes and foils. The SEM picture of the zirconium hydroxide coating obtained under the optimized conditions after drying at room temperature is shown in Figure 9.6. The surface of the stainless steel wire is completely coated with a thin and uniform zirconium hydroxide layer. However, during calcination, the coating is extremely prone to cracking and peeling off due to the

FIGURE 9.6 SEM picture of ZrO$_2$/stainless steel coating.

shrinkage as a result of dehydration and hydroxide-to-oxide phase transformation. The deposition of ZrO_2 finely dispersed powder together with the sol was also performed to obtain the zirconia-based coating resistant to cracking. For preparation of the coating based on ZrO_2, the suspension of zirconia particles in the alumina sol diluted with ethanol was also used.

9.2.3.3 Silica Coatings

The deposition of SiO_2 layers can be performed on the anode surface because the silica sol is stable in alkali solutions and silica particles bear a significant negative charge when neutral and basic conditions are used. EPD was performed in the presence of sodium dodecyl sulfate (SDS) as the surfactant to prevent the anodic dissolution of stainless steel. The weight ratio of SDS to silica was about 1–15. The SEM picture of the obtained sample of stainless steel gauze coated with a SiO_2 layer of the thickness about 2 mkm is presented in Figure 9.7. Furthermore, the coating from the suspension of a SiO_2 powder in the alumina sol was also effectively prepared.

9.2.3.4 Zeolite Coatings

The deposition of zeolites, in particular, the ZSM-5 zeolite layer, can be performed using a suspension of zeolite fine particles in the alumina sol. The obtained coating is composed of large zeolite particles with a size of about 1–5 mkm bound together and with the metal surface by alumina nanoparticles (Figure 9.8). The prepared catalyst was tested in the reaction of partial oxidation of benzene to phenol with N_2O. Before the catalytic tests, the zeolite catalyst was calcined at 900°C in an N_2 flow, and the stainless steel gauze was found to be not destroyed and zeolite remained fixed on the metal surface. When the reaction was carried out with the gauze-supported zeolite, the heavy condensation by-products were not detected. This fact provides an evidence for the more effective heat and mass transfer in the gauze-based catalyst.

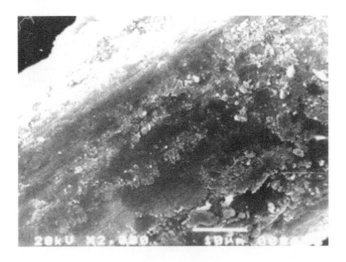

FIGURE 9.7   SEM picture of SiO_2/stainless steel coating.

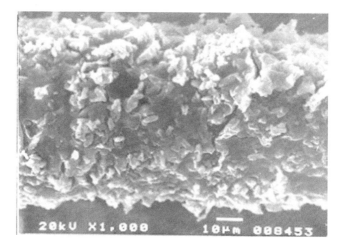

FIGURE 9.8   SEM picture of HZSM-5 zeolite on stainless steel coating.

9.3   CATALYSIS BY STRUCTURED CATALYSTS BASED ON METAL CARRIERS

9.3.1   COMPLETE OXIDATION OF LIGHT ALKANES

The reaction studied in most detail on structured metal catalysts, including gauzes and fibers, is the complete oxidation of CO and hydrocarbons, like the early example of methane combustion on Pd/gauze catalysts [7]. Extensive modeling studies have been also carried out [22]. The reactor chosen for modeling was based on the Ce_3O_4 catalyst on gauze doped with CeO_2 and PdO to increase the activity in the catalytic combustion of CH_4. The reaction kinetics was determined by CSTR catalytic combustion tests. A 1D plug-flow model of structured catalytic reactor was used in modeling. The results demonstrated that the wire gauzes can be regarded as a serious alternative to the ceramic monoliths in the catalytic combustion processes.

There is quite a good number of papers on the use of metal wire grids [23–25] that turned out to be suitable for the design of structured combustion reactors. Metal nickel and copper grids are shown to be suitable supports for structured combustion catalysts [25]. The specific surface area was increased by creating the porous outer layer with the structure of Raney metal on the wire surface. Metal (Co, Cu, Mn, Cr) oxides deposited on the support were tested in propane oxidation. Cobalt oxide demonstrated the highest activity. Reductive pretreatment resulted in a more active catalyst than the activation in oxygen. This effect was explained by partial reduction of Co_3O_4 to metallic Co detected by XPS.

Novel efficient structured combustion catalysts based on sintered MFFs were developed with increased specific surface areas [26]. For this purpose, the metal fibers were coated by porous oxides of SiO_2, Al_2O_3, porous glass, and mesoporous SBA-15 silica. The composite materials have a uniform open macrostructure of MFF and were used as supports for deposition of catalytically active phases (Pd, Pt, and Co_3O_4) tested in CH_4 and C_3H_8 complete oxidation. Co_3O_4 supported on MFF (6.8% Co_3O_4/MFF) demonstrated the highest activity in the propane oxidation. The best performance in methane oxidation was found for a 0.5% Pd/SBA-15/MFF catalyst. The average Pd or Pt particle size on this type of catalysts was ~2.0 nm. The metal nanoparticles encapsulated in the mesopores were stable toward sintering. Pd supported on MFF coated by the microporous SiO_2 was Al_2O_3 films, prepared by sol–gel technique, and deactivated due to metal sintering. The enhanced catalytic activity was found in the adiabatic catalytic reactor for propane oxidation due to a synergy of the 0.5% Pd/SBA-15/MFF and the 0.5% Pt/SBA-15/MFF catalytic layers assembled to form a gradient catalytic bed. The order of activities in propane oxidation at 300°C was as follows: 6.8% Co_3O_4/MFF > 0.5% Pt/SBA-15/MFF > 0.5% Pd/SBA-15/MFF > 0.5% Pd/PG/MFF > 0.5% Pd/SiO_2/MFF > 0.5% Pd/Al_2O_3/MFF > 1.4% CO_3O_4/SBA-15/MFF.

The authors of Ref. [27] reported on the design of structured reactor for methane combustion built from stacked catalytic knitted wire gauzes. Cobalt oxide was deposited on the gauze surface using the plasma-enhanced metal–organic chemical vapor deposition method. The spectral analysis of the catalyst surface revealed the formation of a cobalt oxide spinel with crystallites of about 5 nm. The kinetic studies of combustion showed that the reaction follows first-order kinetics.

Bimetallic Pd–Pt/Al_2O_3 catalysts deposited by dip coating on Fecralloy fibers were studied in the total combustion of methane [28]. The dispersion of Pd and Pt was shown to be a crucial factor for the catalytic performance. The bimetallic Pd–Pt catalyst has a better catalytic activity compared to the Pd catalyst because of the sintering of the latter. This was explained by the cooperation between PtO and PdO.

Nowadays, catalytic systems based on Fecralloy fibers and gauzes gained an increasing interest and are extensively used in catalytic combustion of methane [29,30]. The reasons behind this interest are as follows:

1. High catalytic activity at low combustion temperatures
2. Good resistance to deactivation
3. High thermal stability
4. Very low pressure drop
5. More compact design of the combustion chamber and the heat exchanger
6. Good resistance to flashback phenomena
7. Capability of being shaped to fit nearly any geometry
8. Good resistance to thermal and mechanical shocks
9. Lower air excess levels, allowing higher thermal efficiency

The systems prepared by EPD using stainless steel gauzes and FeCrAl foils modified with a secondary alumina layer with further supported Pt (1 wt%) or $CuO–MnO_2$ oxide composition demonstrated about twice higher activity in methane and butane oxidation as compared to the bulk catalysts [31,32].

9.3.2 PARTIAL OXIDATION OF METHANE INTO SYNTHESIS GAS

This process attracts nowadays a growing attention as a possible method of syngas production from strained and natural gas with downstream Fischer–Tropsch or methanol synthesis. The general energetic scheme (Figure 9.9) demonstrates that

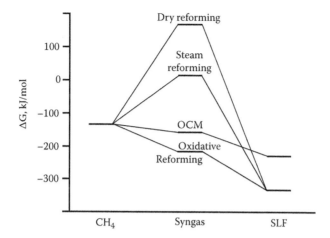

FIGURE 9.9 Free energy diagram of methane conversion via different routes.

unlike steam and dry (CO_2 assisted) reforming, oxidative reforming is very advantageous in terms of the energy. OCM stands for oxidative coupling of methane, which is also advantageous, but, unfortunately, to the best of our knowledge, there are no publications so far devoted to OCM on structured metal catalysts prepared from gauzes and metal fibers.

Thus, the partial oxidation of methane to syngas was carried out on the Pt gauze [33] at 680°C–1000°C and contact times of 0.02–0.2 ms, pressure, 130–240 kPa; methane/oxygen ratio, 2–5; and helium dilution, 40%–80%. Under these conditions, the methane conversion was pretty low (2%–12%), as well as the oxygen conversion, 9%–46%. The reaction is strongly limited by heat and mass transport. The highest selectivity to CO was ~50%.

Davis et al. [34] examined Pt, Pt–10%Rh, and Ni gauzes in the partial oxidation of methane. The relative activities of these catalysts follow the order Pt > Pt–10%Rh > Ni. Nickel gauze was found to be rather inert in an excess of oxygen. OH radicals were assumed to be formed in the gas phase. The authors proposed that the heterogeneous and homogeneous reactions are spatially well separated and chemically decoupled: rather than observing a supporting effect of the catalytic chemistry on homogeneous combustion, the substantial conversion of methane on the surface leads to the suppression of downstream combustion. In contrast to that study, a strong interplay between heterogeneous and homogeneous reactions was found for rhodium foams [35]. The rhodium additive reduces metal carryover in the course of a high-temperature process.

Partial oxidation of methane to syngas on Pt–10%Rh was also studied by Hofstad et al. [36] at an atmospheric pressure and a temperature of 200°C–1057°C, contact times ranging from 0.15 to 5 ms, and the ratio $CH_4/O_2/Ar = 2{:}1{:}10$. The products were CO_x, H_2O, H_2, and traces of C_2 hydrocarbons. This gauze material provides a higher conversion of methane and a higher selectivity to syngas than the Pt gauze catalyst [37]. At a contact time of 0.21 ms, a methane conversion of 28.2% and an oxygen conversion of 74% are achieved; the highest selectivity to CO was 95.6% at a gauze temperature of 1000°C. Under these conditions, the water–gas shift reaction and steam reforming do not occur at temperatures below 1050°C at a contact time of 0.21 ms. At 960°C, CO and H_2 are formed without H_2O as a primary product on the Pt–10%Rh catalyst.

Partial oxidation of methane into syngas was also studied in [38–40], including the novel Rh-based structured catalysts prepared by the electrosynthesis of Rh/Mg/Al hydrotalcite (HT) precursors on a FeCrAl foam. The electrosynthesis conditions (potential and pH of the plating solution) exert a substantial effect on the surface morphology and chemical composition of the samples as well as on the catalytic activity. By increasing the cathodic potential from −1.2 to −1.3 V, but keeping the synthesis time constant (1000 s), the required conditions to obtain HT precursors were achieved faster. The catalytic performance depended on the synthesis conditions, the best values being achieved by the catalyst obtained from the HT precursor prepared at −1.3 V for 1000 s [40].

Bulk metal-structured catalysts, including noble metal gauzes or sponges [37,41,42] and nickel foams [43,44], have been applied in the partial oxidation of methane, although they usually show low surface areas. On the other hand, the use of structured catalysts based on Fecralloy and Nicrofer metallic monoliths [45–47] as well as a Fecralloy metallic foam [48] resulted in a significant reduction of the amount of the precious metals, without any drop of the catalytic activity.

Some modeling studies have been also performed for the high-temperature catalytic partial oxidation of methane over a platinum gauze using 3D numerical simulations of the flow field, heat transport, and microkinetics including gas-phase and surface reaction mechanisms [49,50]. The theoretical data are consistent with the experiments over Pt gauzes. The conversions of CH_4 and O_2 increased with an increased contact time and were constant in the temperature range of 1000–1200 K. The selectivity to CO linearly increases with temperature. H_2 was found only above 1200 K. The contribution of heterogeneous steps in the overall process is significant, but gas-phase reactions become important at certain temperatures, pressures, and residence times. Simulations predicted significant formation of ethane and ethylene via methane oxidative coupling with increasing pressure and contact time.

de Smet and other authors [33,51] considered geometric effects in the course of methane partial oxidation in steady-state experiments in a continuous flow reactor with a single platinum gauze. Modeling of a single monolith channel coated with platinum incorporating both gas-phase and surface mechanisms was performed as well [52].

The catalytic partial oxidation of methane with oxygen was studied at atmospheric pressure in a continuous-flow reactor containing a single Pt metal gauze [33]. Heat-transport limitations are taken into account by measuring the catalyst temperature. The conversions of methane and oxygen were determined by transport phenomena. On the contrary, the CO selectivity was affected mostly by the kinetics of the reactions. The reactor consisting of two rows of parallel flat plates was validated by comparing the results to 3D FLUENT simulations of heat and mass transfer for the simple surface reaction on the gauze catalyst.

A series of papers by the group of A. Holmen is devoted to partial oxidation of alkanes under very short contact time conditions. Catalytic partial oxidation of methane has been studied over Pt, Pt/Rh, Pt/Ir, and Pd gauze catalysts at contact times in

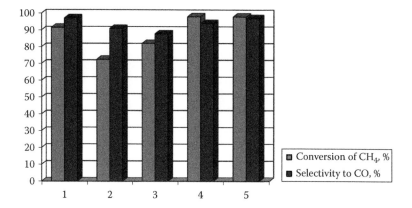

FIGURE 9.10 Conversion/selectivity numbers at the Y axes should be introduced as well as the legend to the columns: 1-1%Pd/gauze, 2-1%Pt/gauze, 3-1%Rh/gauze, 4-etched FeCrAl foil, 5-1%Pd/sintered metal fiber (FeCrAl).

the range 0.00021–0.00033 s [53]. The tests were run at 1 bar and 700°C–1100°C using a single gauze. The feed consisted of CH_4 and O_2 with different ratios between them. High selectivities to CO were observed at high temperatures. On the contrary, the selectivity to H_2 was quite low (<30%) with incomplete conversion of O_2. Doping the Pt gauze with Rh improves the selectivity to synthesis gas. The effect of the CH_4/O_2 ratio depends on the composition of the gauze. The selectivity to synthesis gas decreased with increasing content of O_2 on the Pt gauze, unlike the Pt/Rh gauze.

The performance of the best catalytic systems based on metal carriers in the partial oxidation of methane into synthesis gas onto stainless steel gauzes with 1% loading of the noble metal, etched FeCrAl foil, and sintered FeCrAl fiber (1% Pd) at T = 950°C, GHSV = 10,000 h^{-1}, and O_2/CH_4 = 0.5 is shown in Figure 9.10. The ratio H_2/CO for these catalysts ranges between 2.1 and 2.6, while the conversion and selectivity reach 98%. It is noteworthy that the activity of the etched FeCrAl materials containing no noble metals is comparable with those of the noble metal-containing catalysts. This was explained by the enrichment of the surface in chromium and the formation of coordinatively unsaturated chromium species at the FeCrAl surface responsible for the methane activation.

9.3.3 DRY METHANE REFORMING

This reaction is considered as an alternative to steam reforming and partial oxidation of methane to produce syngas, although the energetics of the process is very unfavorable and the product ratio H_2/CO is not quite good for both Fischer–Tropsch and methanol syntheses:

$$CH_4 + CO_2 \rightarrow 2H_2 + 2CO \quad \Delta H_{298} = 247 \text{ kJ/mol}$$

A honeycomb-type nickel-based catalyst for this reaction was prepared by sol–gel and electroless plating methods on a stainless steel (Figure 9.11) [54]. The catalytically active phase was a 7 mkm porous alumina layer and nickel particles 70–150 nm in diameter.

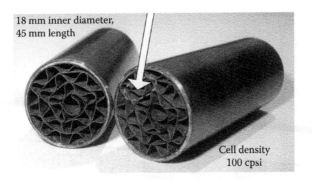

FIGURE 9.11 A honeycomb Ni/alumina/stainless steel catalyst for dry methane reforming. (From Fukuhara, C. et al., *Appl. Catal. A Gen.*, 468, 18, 2013. With permission.)

This nickel honeycomb catalyst exhibited a high activity in methane dry reforming, indicating that the catalyst degradation was insignificant under severe CO_2/CH_4 conditions. The amount of coke formed on the nickel catalyst was much smaller compared to the commercial catalyst. The deposited coke on the nickel catalyst formed whiskers.

A 8 mm copper microfibrous entrapped Ni/Al_2O_3 composite catalyst was proposed for the CO_2 reforming of methane process [55]. Computational fluid dynamics was applied to demonstrate the enhancement of the heat transfer for the microfibrous structured catalyst. The average bed temperature was 1039 K, which is 75 K higher compared to packed bed granulated catalyst Ni/Al_2O_3. Therefore, the resistance of the catalyst bed was significantly improved together with the observed conversion enhancement. It is noteworthy that at 1073 K, a fourfold reduction of the carbon deposition rate was found for the composite catalyst, whereas the CH_4 conversion increased from 84% to 89% compared to the bulk catalyst at the gas hourly space velocity of 20,000 h^{-1}. The proposed microfibrous entrapment technology produces the open structure with a large void volume (71.3 vol%), thus providing enhanced mass transfer and high permeability (a low pressure drop). Lower coking rate was also noticed (about 20% decrease compared to packed-bed Ni/alumina).

9.3.4 STEAM METHANE REFORMING

Steam methane reforming is a mature process and a number of commercial catalysts and plants are running. Nevertheless, the quest for new catalysts and catalytic reactor technologies that can decrease the energy burdens of this highly endothermic process is going on. Catalytically active surface of nickel plates for steam methane reforming (Figure 9.12) was enhanced in Ref. [56] by consecutive oxidation–reduction pretreatments and deposition of Pt and Al_2O_3. The effect of the calcination time, temperature, order, and number of coatings on the catalytic activity was revealed. The most active and stable phases were produced by depositing layers of Pt, Al_2O_3, and then again Pt after calcination for 12 h in two cycles of oxidation–reduction treatments of the nickel plate at 700°C. The stability was tested for 70 h at 500°C–650°C. Deactivation due to carbon deposition was insignificant. Ultrasonic cleaning of the nickel surface before depositions helped to improve the resulting activity.

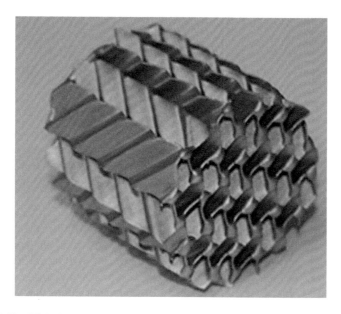

FIGURE 9.12 Nickel plates for steam methane reforming. (From Obradovic, A. et al., *Int. J. Hydr. Energ.*, 38, 1419, 2013. With permission.)

Methane steam reforming was also studied on Ru supported onto a metalloceramic composite [57]. The core–shell structured Al_2O_3–Al composite with an Al metal core and a high-surface-area Al_2O_3 overlayer was prepared by hydrothermal oxidation of aluminum metal powder. A hierarchical secondary structure with macrosize pores is formed during such a treatment. This core–shell composite is characterized by the enhanced heat conductivity and a high surface area suitable for the dispersion of Ru. The Ru/Al_2O_3–Al catalyst exhibits a very stable and significantly higher CH_4 conversion via steam reforming compared to the conventional Ru/Al_2O_3 catalyst.

The hydrothermal synthesis of such metalloceramics was first proposed by Tikhov et al. [58]. But their application in catalysis is limited [59,60].

9.3.5 CONVERSION OF ALKANES TO OTHER PRODUCTS

Hydrogen production by methane catalytic cracking on nickel gauzes under periodic reactor operation was studied by Monnerat et al. [61].The authors proposed the catalytic cracking of methane over the nickel gauze at 450°C–550°C to generate CO-free hydrogen. The catalyst is however prompt to coke formation. That is why the reactor operated periodically with 4 min reaction/4 min regeneration periods.

The oxidative dehydrogenation of C_2H_6 has been studied at very short contact times of 5×10^{-5} s using Pt or Pt/Rh gauze catalysts [62]. The reaction was run at 600°C–1050°C and 1 bar pressure. The C_2H_6/O_2 ratio was ranged within 1.3–2.5 with Ar and H_2O serving as diluents. The C_2H_4 selectivity is higher in the empty reactor or with monoliths. Similar selectivities were found for the gauze at conversions above 55%–60%. The gauze causes fast heating of the gas mixture due to the increased contribution of the combustion, while the oxidative dehydrogenation reaction takes place predominantly in the gas phase. The product distribution was slightly influenced by the composition of the gauze. The synthesis gas formed as a side product with the H_2/CO ratio of 1.5 at severe conditions when the formation of C_2H_4 takes place.

Similarly, short-contact-time oxidative dehydrogenation of propane was explored on a Pt/10%Rh gauze catalyst and a catalyst consisting of VMgO composition deposited on a monolithic substrate at 500°C–950°C and propane/oxygen ratios between 1.1 and 2.3 at 1 atm [63]. The role of homogeneous and heterogeneous reactions was unraveled. Ethene, methane, and synthesis gas (with a H_2/CO ratio close to 1.0) were the products at high C_3H_8 conversions. Propene appears at lower conversions, especially on the VMgO catalyst. Poor selectivity to ethene and high selectivity to carbon oxides were found for the VMgO monolith at low substrate conversions (<20%). Mostly gas-phase reactions occur at high temperatures and high conversions. Traces of ethyne, propyne, and benzene are present in some runs. The O_2 conversion and the selectivity to CO_x were higher over both catalysts compared to the empty reactor, particularly at lower conversions. The catalysts however ignite the process and facilitate the production of heat for the gas-phase formation of olefins. The flow rate effect for the VMgO system provides evidence for the heterogeneous reactions controlled by transport limitations.

Butane oxidation on diverse platinum wires revealed the local temperature fluctuations (flickering) determined by an infrared detector [64]. The amplitude and frequency of the temperature fluctuations increase with decreasing wire diameter. A similarity exists between flickering and turbulent velocity fluctuations of the reacting stream.

9.4   CONCLUSIONS

It is seen from the preceding brief overview that information on the use of metal gauzes and fibers in methane and other alkane activations is rather scarce. For oxidation processes, the catalyst plays a dual role by (1) catalyzing the main reaction or certain stages of the overall process and (2) igniting (catalyzing) the gas-phase reactions. In most cases, catalysts lower the light-off temperature and make it possible to carry out the reaction in an autothermal regime. Complications of the processes are coke formation and metal carryover.

The uniform oxide coatings on wire gauzes or sintered metal fibers can be obtained by different methods, with EPD and plasma spraying being the most efficient. For alumina coatings, the use of alcohol-diluted sol is preferable to aqueous sol.

Novel nanostructured and nanoengineered catalysts based on sintered metal fibers and gauzes may change the modern face of catalytic reactor technologies, not necessarily in respect to the processes discussed in this short review. Moreover, hydrogenation and DeNOx processes may be more suitable for the application of metal-structured catalysts. Of particular interest seems to be the opportunity of designing gradient reactions with different packings and assemblies of structured metal catalysts based on gauzes and fibers with the composition/concentration gradient of active components in different directions.

ACKNOWLEDGMENT

The financial support from Russian Science Foundation (grant No. 14-33-00001) is acknowledged.

REFERENCES

1. C.N. Satterfield, *Heterogeneous Catalysis in Industrial Practice*, 2nd edn., McGraw-Hill, New York, 1991.
2. B.T. Horner, Knitted Gauze for Ammonia Oxidation, *Platinum Met. Rev.* 35 (1991) 58.
3. A. Cybulski, J.A. Moulijn, Monoliths in heterogeneous catalysis, *Catal. Rev. Sci. Eng.* 36 (1994) 179.
4. A. Kołodziej, J. Łojewska, Experimental and modelling study on flow resistance of wire gauzes, *Chem. Eng. Process.* 48 (2009) 816–822.
5. A. Kołodziej, J. Łojewska, Mass transfer for woven and knitted wire gauze substrates: Experimental and modelling, *Catal. Today* 147 (2009) S120–S124.
6. A. Kołodziej, J. Łojewska, Flow resistance of wire gauzes, *AIChE J.* 55 (2009) 264–267.
7. A.F. Ahlstroem-Silversand, C.U.I. Odenbrand, Modelling catalytic combustion of carbon monoxide and hydrocarbons over catalytically active wire meshes, *Chem. Eng. J.* 73 (1999) 205–216.
8. A. Renken, L. Kiwi-Minsker, Microstructured catalytic reactors, *Adv. Catal.* 53 (2010) 47–122.
9. R. De Bruyne, *Multipass Test Performance of Metal Fibre Filter Media*, World Filtration Congress III, Downingtown, PA, 1982, p. 400.
10. D.R. Cahela, B.J. Tatarchuk, Permeability of sintered microfibrous composites for heterogeneous catalysis and other chemical processing opportunities, *Catal. Today* 69 (2001) 33.
11. A.F Ahlstrom-Silversand, C.U.I. Odenbrand, Thermally sprayed wire-mesh monolith catalysts and activity studies, *Appl. Catal. A.* 53 (1997) 177.
12. A. Kapoor, S.K. Goyal, N.N. Bakhasni, Fischer—Tropsch studies on a plasma-sprayed iron catalyst in a tube-wall reactor, *Can. J. Chem. Eng.* 64 (5) (1986) 792.
13. M. Ogawa, K. Kuroda, J. Mori, Aluminum-containing mesoporous silica films as nano-vessel for organic photochemical reactions, *Langmuir* 18 (2002) 744.
14. M. Ogawa, Formation of novel oriented transparent films of layered silica-surfactant nanocomposites, *J. Am. Chem. Soc.* 116 (1994) 7941.
15. P.G. Menon, M.F.M. Zwinkels, E.M. Johansson, S.G. Jaras, *Kinet. Katal.* 39 (1998) 670.
16. M.P. Vorobieva, A.A. Greish, A.V. Ivanov, L.M. Kustov, Preparation of catalyst carriers on the basis of alumina supported on metallic gauzes, *Appl. Catal. A: Gen.* 199 (2000) 257–261.
17. M.P. Vorob'eva, A.A. Greish, A.Y. Stakheev, N.S. Telegina, A.A. Tyrlov, E.S. Obolonkova, L.M. Kustov, New catalyst supports and catalytic systems on the basis of metallic gauzes coated with II–IV group element oxides, *Stud. Surf. Sci. Catal.* 130B (2000) 1127–1132.
18. P. Sarkar, P.S. Nicholson, Electrophoretic Deposition (EPD): Mechanisms, kinetics, and applications to ceramics, *J. Am. Ceram. Soc.* 79 (1996) 1987.
19. D.E. Clark, W.J. Dalzell, D.C. Folz, *Ceram. Eng. Sci. Proc.* 9 (1998) 1111.
20. L. Shaw, R. Abbaschian, Al2O3 coatings as diffusion barriers deposited from particulate-containing sol-gel solutions, *J. Am. Ceram. Soc.* 78 (1995) 3376.
21. A.R. Boccaccini, P.A. Trusty, Electrophoretic deposition infiltration of metallic fabrics with a boehmite sol for the preparation of ductile-phase-toughened ceramic composites, *J. Mater. Sci.* 33 (1998) 933.

22. P.J. Jodłowski, J. Kryca, M. Iwaniszyn, R. Jedrzejczyk, J. Thomas, A. Kołodziej, J. Łojewska, Methane combustion modelling of wire gauze reactor coated with Co_3O_4–CeO_2, Co_3O_4–PdO catalysts, *Catal. Today* 216 (2013) 276–282.

23. W. Gürtler, J. Ackermann, G. Emig, Neuartige Oxidationskatalysatoren auf der Basis von Drahtgestricken für die katalytische Nachverbrennung, *Chem. Ing. Technol.* 68 (1996) 1438.

24. W. Gürtler, A metal catalyst for the oxidation of hydrocarbon(s), Patent No. 19611395C1, Germany (1997).

25. I. Yuranov, N. Dunand, L. Kiwi-Minsker, A. Renken, Metal grids with high-porous surface as structured catalysts: Preparation, characterization and activity in propane total oxidation, *Appl. Catal. B: Environ.* 36 (2002) 183.

26. I. Yuranov, L. Kiwi-Minsker, A. Renken, Structured combustion catalysts based on sintered metal fibre filters, *Appl. Catal. B: Environ.* 43 (2003) 217–227.

27. A. Kołodziej, J. Łojewska, J. Tyczkowski, P. Jodłowski, W. Redzynia, M. Iwaniszyn, S. Zapotoczny, P. Kusrtrowski, Coupled engineering and chemical approach to the design of a catalytic structured reactor for combustion of VOCs: Cobalt oxide catalyst on knitted wire gauzes, *Chem. Eng. J.* 200–202 (2012) 329–337.

28. A. Maione, F. Andre, P. Ruiz, Structured bimetallic Pd-Pt/γ-Al2O3 catalysts on FeCrAlloy fibers for total combustion of methane, *Appl. Catal. B: Environ.* 75 (2007) 59–70.

29. S.R. Vaillant, A.S. Gastec, Catalytic combustion in a domestic natural gas burner, *Catal. Today* 47 (1999) 415–420.

30. G. Saracco, I. Cerri, V. Specchia, R. Accornero, Catalytic pre-mixed fibre burners, *Chem. Eng. Sci.* 54 (1999) 3599–3608.

31. M.P. Vorob'eva, New catalysts and catalyst supports based on metal gauzes, Thesis of Candidate of Science Dissertation, Institute of Organic Chemistry, Moscow, Russia, 2000.

32. K.I. Slovetskaya, M.P. Vorob'eva, A.A. Greish, L.M. Kustov, Deep oxidation of methane on granulated and monolith copper-manganese oxide catalysts, *Bull. Russ. Acad. Sci. Ser. Khim.* 9 (2001) 1512–1515.

33. C.R.H. de Smet, M.H.J.M. de Croon, R.J. Berger, G.B. Marin, J.C. Schouten, An experimental reactor to study the intrinsic kinetics of catalytic partial oxidation of methane in the presence of heat-transport limitations, *Appl. Catal. A Gen.* 187 (1990) 33.

34. M.B. Davis, M.D. Pawson, G.Veser, L.D. Schmidt, Methane oxidation over noble metal gauzes: An LIF study, *Combust. Flame* 123 (2000) 159.

35. C.T. Goralski, R.P. O'Connor, L.D. Schmidt, Modeling homogeneous and heterogeneous chemistry in the production of syngas from methane, *Chem. Eng. Sci.* 55 (2000) 1357.

36. K.H. Hofstad, T. Sperle, O.A. Rokstad, A. Holmen, Partial oxidation of methane to synthesis gas over a Pt/10% Rh gauze, *Catal. Lett.* 45 (1997) 97.

37. K.H. Hofstad, O.A. Rokstad, A. Holmen, Partial oxidation of methane over platinum metal gauze, *Catal. Lett.* 36 (1996) 25.

38. D.A. Goetsch, P. Witt, L.D. Schmidt, *Prepr. Am. Chem. Soc. Div. Petrol. Chem.* 41 (1996) 150.

39. R.P. O'Connor, L.D. Schmidt, Oxygenates and olefins from catalytic partial oxidation of methane in monolyth chemical reactors, *Stud. Surf. Sci. Catal.* 133 (2001) 289.

40. F. Basile, P. Benito, G. Fornasari, M. Monti, E. Scavetta, D. Tonelli, A. Vaccari, Novel Rh-based structured catalysts for the catalytic partial oxidation of methane, *Catal. Today* 157 (2010) 183–190.

41. K. Heitnes, S. Lindberg, O.A. Rokstad, A. Holmen, Catalytic partial oxidation of methane to synthesis gas, *Catal. Today* 24 (1995) 211.

42. D.A. Hickman, L.D. Schmidt, Synthesis gas formation by direct oxidation of methane over Pt monoliths, *J. Catal.* 138 (1992) 267.
43. Y.P. Tulenin, M.Y. Sinev, V.V. Savkin, V.N. Korchak, Dynamic behaviour of Ni-containing catalysts during partial oxidation of methane to synthesis gas, *Catal. Today* 91–92 (2004) 155.
44. L.J.I. Coleman, E. Croiset, W. Epling, M. Fowler, R.R. Hudgins, Evaluation of foam nickel for the catalytic partial oxidation of methane, *Catal. Lett.* 128 (2009) 144.
45. B.C. Enger, J. Walmskey, E. Bjorgun, R. Lodeng, P. Pfeifer, K. Schubert, A. Holmen, H.J. Venvik, Performance and SEM characterization of Rh umpregnated microchannel reactors in the catalytic partial oxidation of methane and propane, *Chem. Eng. J.* 144 (2008) 489.
46. H. Jung, W.L. Yoon, H. Lee, J.S. Park, J.S. Shin, H. Lab, J.D. Leehave, Fast start-up reactor for partial oxidation of methane with electrically-heated metallic monolith catalyst, *J. Power Sourc.* 124 (2003) 76.
47. J.-H. Ryu, K.-Y. Lee, H.-J. Kim, J.-I. Yang, H. Jung, Promotion of palladium-based catalysts on metal monolith for partial oxidation of methane to syngas, *Appl. Catal. B: Environ.* 80 (2008) 306.
48. A. Shamsi, J.J. Spivey, Partial oxidation of methane on NiO-MgO catalysts supported on metal foams, *Ind. Eng. Chem. Res.* 44 (2005) 7298.
49. R. Quiceno, O. Deutschmann, J. Warnatz, J. Perez-Ramirez, Rational modeling of the CPO of methane over platinum gauze: Elementary gas-phase and surface mechanisms coupled with flow simulations, *Catal. Today* 119 (2007) 311–316.
50. R. Quiceno, J. Perez-Ramirez, J. Warnatz, O. Deutschmann, Modeling the high-temperature catalytic partial oxidation of methane over platinum gauze: Detailed gas-phase and surface chemistries coupled with 3D flow field simulations, *Appl. Catal. A: Gen.* 303 (2006) 166–176.
51. M.F. Reyniers, C.R. De Smet, P. Govind, G. Marin, Catalytic partial oxidation. Part I. Catalytic processes to convert methane: Partial or total oxidation, *CATTECH 6* (2002) 140.
52. O. Deutschmann, L.D. Schmidt, Modeling the partial oxidation of methane in a short contact time reactor, *AIChE J.* 44 (1998) 2465.
53. M. Fathi, K. Heitnes Hofstad, T. Sperle, O.A. Rokstad, A. Holmen, Partial oxidation of methane to synthesis gas at very short contact times, *Catal. Today* 42(3) (1998) 205–209.
54. C. Fukuhara, R. Hyodo, K. Yamamoto, K. Masuda, R. Watanabe, A novel nickel-based catalyst for methane dry reforming: A metal honeycomb-type catalyst prepared by sol–gel method and electroless plating, *Appl. Catal. A: Gen.* 468 (2013) 18–25.
55. W. Chen, W. Sheng, F. Cao, Y. Lu, Microfibrous entrapment of Ni/Al$_2$O$_3$ for dry reforming of methane: Heat/mass transfer enhancement towards carbon resistance and conversion promotion, *Int. J. Hydr. Energy* 37 (2012) 18021–18030.
56. A. Obradovic, B. Likozar, J. Levec, Catalytic surface development of novel nickel plate catalyst with combined thermally annealed platinum and alumina coatings for steam methane reforming, *Int. J. Hydr. Energy* 38 (2013) 1419–1429.
57. H.C. Lee, Y. Potapova, D. Lee, A core-shell structured, metal–ceramic composite-supported Ru catalyst for methane steam reforming, *J. Power Sources* 216 (2012) 256–260.
58. S.F. Tikhov, A.N. Salanov, Y.V. Palesskaya, V.A. Sadykov, G.N. Kustova, G.S. Litvak, N.A. Rudina, V.A. Zaikovskii, S.V. Tsybulya, Mechanism of formation of porous Al$_2$O$_3$ composites in hydrothermal conditions, *React. Kinet. Catal. Lett.* 64 (1998) 301–308.
59. S.F. Tikhov, V.E. Romanenkov, V.A. Sadykov, V.N. Parmon, A.I. Rat'ko, Physicochemical principles of the synthesis of porous composite materials through the hydrothermal oxidation of aluminum powder, *Kinet. Catal.* 46 (2005) 641–659.

60. N. Burgos, M. Paulis, M.M. Antxustegi, M. Montes, Deep oxidation of VOC mixtures with platinum supported on Al$_2$O$_3$/Al monoliths, *Appl. Catal. B: Environ.* 38 (2002) 251–258.

61. B. Monnerat, L. Kiwi-Minsker, A. Renken, Hydrogen production by catalytic cracking of methane over nickel gauze under periodic reactor operation, *Chem. Eng. Sci.* 56(2) (2001) 633–639.

62. R. Lødeng, O.A. Lindvag, S. Kvisle, H. Reier-Nielsen, A. Holmen, Short contact time oxidative dehydrogenation of C2 and C3 alkanes over noble metal catalysts, *Stud. Surf. Sci. Catal.* 119 (1998) 641–646.

63. M. Fathi, R. Lødeng, E.S. Nilsen, B. Silberova, A. Holmen, Short contact time oxidative dehydrogenation of propane, *Catal. Today* 64(1–2) (2001) 113–120.

64. W.M. Edwards, F.L. Worley Jr., D. Luss, Temperature fluctuations (flickering) of catalytic wires and gauzes—II experimental study of butane oxidation on platinum wires, *Chem. Eng. Sci.* 28(7) (1973) 1479–1491.

10 Methane Dry Reforming on Nanocomposite Catalysts
Design, Kinetics, and Mechanism

V.A. Sadykov, L.N. Bobrova, N.V. Mezentseva,
S.N. Pavlova, Yu. E. Fedorova, A.S. Bobin,
Z. Yu. Vostrikov, T.S. Glazneva, M. Yu. Smirnova,
N.N. Sazonova, V.A. Rogov, A. Ishchenko,
Guy B. Marin, Joris W. Thybaut, V.V. Galvita,
Y. Schuurman, and C. Mirodatos

CONTENTS

10.1 STATE OF THE ART

10.1.1 GENERAL PROBLEMS

Electricity and hydrogen are now considered as the dominant energy carriers in modern green chemical and process engineering. Nowadays, almost all of the hydrogen is produced commercially from natural gas (NG) via steam reforming to synthesis gas, also denoted as *syngas*, that is, a mixture comprising H_2 and CO. Synthesis gas production via alternative routes to traditional methane steam reforming has recently attracted considerable attention due to both environmental and commercial reasons [1,2]. Transformation of NG using carbon dioxide and reforming of oxygenates derived from fast pyrolysis of biomass to synthesis gas are the most promising processes [3–14]. Dry reforming transforms cheap and undesirable greenhouse gases such as methane and carbon dioxide and is particularly important in the case of biogas or gas fields containing a significant amount of both compounds.

Methane dry reforming (MDR) by carbon dioxide is a highly endothermic reaction and requires higher temperatures being equally favored by lower pressures:

$$CH_4 + CO_2 \leftrightarrow 2CO + 2H_2; \quad \Delta H_{298} = 247.3\,\text{kJ/mol}; \quad \Delta G^\circ = 61{,}770 - 67.32T \quad (10.1)$$

The synthesis gas produced in this reaction having a H_2/CO ratio close to unity is a favorable feedstock for further chemical processes such as the Fischer–Tropsch synthesis as well as oxygenates production followed by liquid fuel synthesis [3–8]. Water formation cannot be avoided, the reverse water–gas shift reaction, as a side reaction, being its main formation route:

$$CO_2 + H_2 \rightleftharpoons CO + H_2O; \quad \Delta H^{\circ}_{298} = 41\,kJ/mol; \quad \Delta G^{\circ} = -8545 + 7.84T \quad (10.2)$$

The generalized reaction sequence corresponds to the following overall reaction stoichiometry [15]:

$$CH_4 + 2CO_2 \rightleftharpoons H_2 + H_2O + 3CO \quad (10.3)$$

The following catalytic aspects will be presented and discussed in what follows with respect to MDR: catalytically active metals, the catalyst supports, catalyst stability toward carbon deposition, reaction mechanism, and kinetic details.

10.1.2 Survey on the Catalytic Aspects of Methane Dry Reforming

The active catalysts for steam/dry reforming of methane and oxygenates are usually based on supported noble metals (Pt, Pd, Rh, Ru) and transition metals (Ni, Co, Cu, Cr) [16–36]. Fast deactivation due to coking and metal sintering on traditional supports such as alumina and aluminosilicates, especially in the case of Ni-containing catalysts, is a major obstacle for the industrial implementation of these processes [16–27].

10.1.2.1 Coking Effect

Generally, the coking effect is intimately related to the complex reaction mechanism [37]. The key reaction for carbon deposition is methane cracking [38]:

$$CH_4 \rightarrow C + 2H_2; \quad \Delta H^{\circ}_{298} = 74.85\,kJ/mol; \quad \Delta G^{\circ} = 21,960 - 26.45T \quad (10.4)$$

Upon carbon formation in MDR, the H_2/CO ratio is enhanced. At $CO_2/CH_4 > 1$, an endothermic reaction such as methane decomposition (10.4) has only a minor effect on carbon deposition as CH_4 is the limiting reactant. At temperatures between 550°C and 700°C and under stoichiometric MDR, carbon deposition may also occur via CO disproportionation, that is, the Boudouard reaction:

$$2CO \rightarrow C + CO_2; \quad \Delta H_{298} = -172\,kJ/mol; \quad \Delta G^{\circ} = 39,810 - 40.87T \quad (10.5)$$

while above 820°C, the reverse water–gas shift (10.2) and the Boudouard reaction (10.5) are not favored by thermodynamics. Thereby, carbon deposition is thermodynamically possible for a CO_2/CH_4 reforming feed ratio of 1:1 at temperature up to 870°C at 1 atm. At a given pressure, the temperature limit increases as the CO_2/CH_4 feed ratio decreases [39]. Topor et al. [40] suggested the use of carbon dioxide excess in the dry reforming reaction. Thus, using a fuel mixture of 30% methane and 70% carbon dioxide, little carbon fouling and optimum performance of a microtubular zirconia cell with lanthanum strontium manganite cathode and nickel/zirconia/ceria anode were observed compared to solid oxide fuel cells (SOFC) operating on pure methane [41]. Nevertheless, in order to maintain a high selectivity toward syngas, CO_2/CH_4 ratios as low as 1:1 are preferred in many applications, for example, for further synthesis of various chemicals and internal reforming SOFCs.

It is known that carbon formation, even in conditions under which it is thermodynamically favorable, can be kinetically impeded by selecting the appropriate catalyst. Thereby, the design and elaboration of suitable catalyst with high activity, selectivity, and redox stability for MDR is a key issue for the industrial application of NG dry reforming. In the attempts to reduce catalyst deactivation and coke deposition, many different concepts and approaches have been tried and tested [4–13].

10.1.2.2 Nanocomposite Materials as Catalysts for MDR Reaction

To date, much attention has been paid to fluorite-like (Ln–Ce–Zr–O) and perovskite-like oxides as supports or additives to traditional supports due to their high surface/lattice oxygen mobility, excellent redox properties, and good thermal stability [42–86].

Generally, the structure of complex perovskite-like precursors, referred to as ABO_3 or A_2BO_4, can contain a large variety of different atoms with A-site atoms being usually an alkali-earth or rare-earth element and B being a transition metal, Pt group metals, etc. The transition metal cation B is in the octahedral surrounding of oxygen anions, and the cation A is surrounded by 12 oxygen ions. The structural and thermal stability as well as activity of complex perovskite-like precursors can be improved through substitution of various cations on A and/or B sites. Although the idealized perovskite structure does not contain oxygen vacancies, the structure is extremely versatile. Generation of oxygen vacancies in the structure through, for example, doping with aliovalent cations significantly affects the electronic energy levels and ionic mobility. The relative ease of the redox process facilitates methane activation and enhances mobility and reactivity of oxygen species required for methane transformation into syngas [19–36,43,45–76].

As a general rule, the catalytic reactions benefit from highly dispersed active metal particles stabilized by supports [51]. Excellent catalytic activity and coking stability in MDR can be achieved via controlling the surface/bulk oxygen mobility and reactivity in complex oxide supports, for example, by changing the type and content of A (La, Pr, Sm) and B (Fe, Cr, Mn) cations in perovskites and/or by adding fluorite-like oxides (Ln–Ce–Zr–O) possessing a high lattice mobility as well as by optimizing the size and composition of metal alloy nanoparticles. Under high-temperature reducing conditions in the presence of H_2, CH_4, CO, etc., restructuring of the perovskite-type oxides occurs. Some metals located at B sites (Ni, Co, Cu, Ru, etc.) are reduced to the metallic state, while others (Mn, Cr, Fe) decrease their oxidation degree. The active metals dispersed inside the bulk migrate to the surface forming metallic nanoparticles (Ni, Ru, Co, their alloys, etc.), which are homogeneously dispersed. The metal–support interaction is getting stronger than that obtainable by the usual supporting methods. Formation of highly dispersed metal nanoparticles surrounded by LnO_x patches epitaxially bound with disordered Ln–(Fe, Mn, Cr)–O perovskite particles provides higher active surface area and limits metal sintering [46–48,64,65].

For Ni-containing perovskite-type oxides with general formula $A_{1-x}A_x'Ni_{1-y}B_y'$ (A, A'—La, Ca, Sr, Ce, Pr; B'—Fe, Co, Ru; with x, y being the substitution degree) used as precursors of MDR catalysts, both the high dispersion of metals depending on the nature of A, A' and/or B' cations and participation of mobile oxygen of oxide matrix improve catalytic activity and, most of all, stability toward coking [42–72]. Thus, a $La_{0.9}Pr_{0.1}NiO_3$ catalyst exhibits exceptionally high sintering- and coking-resistance properties in MDR, which are due to the higher Ni dispersion in the reduced catalyst as well as due to redox properties of praseodymium oxide [65]. For $Ln_{1-x}Ca_xRu_{0.8}Ni_{0.2}O_3$ (Ln = La, Sm, Nd) catalysts, a partial substitution of Ni with Co [56–59], Fe [60–64], and/or Ru [43,45,65–70] leading to formation of Ni–Me alloys under reducing MDR reaction conditions also improves catalyst activity and stability.

Ceria–zirconia fluorite-like solid solutions with fluorite-like structure also demonstrated their high efficiency as supports for precious metals, transition metals, and their combinations in design of catalysts for environment protection and transformation of fuels into syngas. Key factors in providing a high performance and stability of these catalysts are phase stability in contact with reaction feeds with a broad variation of oxygen fugacity, strong metal–support interaction, and high mobility/reactivity of surface and near-surface oxygen forms.

10.2 CATALYST DESIGN

10.2.1 STRUCTURAL AND TEXTURAL FEATURES OF NANOCOMPOSITE CATALYSTS

Complex oxides with perovskite-like and fluorite-like structures were prepared via the Pechini route (citric acid–ethylene glycol polyester precursors). Perovskite–fluorite nanocomposites including nanocomposites consisting of perovskite $La_{0.8}Pr_{0.2}Mn_{0.2}Cr_{0.8}O_3$ in combination with NiO and YSZ ($Y_{0.08}Zr_{0.92}O_{2-\delta}$) were prepared via one-pot Pechini route. Promoting metals, such as Pt, Ru, Ni, and their combinations, were supported via incipient wetness impregnation with respective solutions of inorganic salts [33,42,64,65,78,81–86]. Structural and surface properties of these materials as well as their reactivity and oxygen mobility are described in detail in earlier publications. In what follows, a brief overview will be given for some selected systems as required for illustration and understanding of their catalytic properties.

10.2.1.1 Pt-Supported $Pr_{0.3}Ce_{0.35}Zr_{0.35}O_{2-x}$ Catalysts

X-ray diffraction (XRD) patterns for these catalysts correspond to the typical fluorite structure [79,86]. For dispersed $Pr_{0.3}Ce_{0.35}Zr_{0.35}O_{2-x}$ oxides calcined at 900°C, the specific surface area amounts to 29 m²/g. The specific surface area decreases with increasing Pt content in promoted samples and varies in the range of 14.6–13.5 m²/g. Apparently, during preparation, a part of supported Pt is encapsulated within the support due to their surface activation by acidic H_2PtCl_6 solution followed by high-temperature calcination [79]. In addition, for Pt-promoted samples with a Pt loading between 1.6 and 4.9 wt.%, reflections at $2\theta \sim 39°$, $47°$, $68°$, $83°$ corresponding to metal Pt particles with typical sizes of ~40–60 nm were observed, and their intensity increased with increasing Pt content. As revealed by analysis of integral intensity of these reflections using a Pt/corundum sample as a standard, only a small (<10%) part of Pt is present as metallic particles detected by XRD, which is a typical feature of Pt-supported doped ceria (ceria–zirconia) oxides [42,79]. This conclusion is supported by TEM data (Figures 10.1 and 10.2). It was revealed (Figure 10.1) that in as-prepared samples, only small (typical sizes up to several nanometers) Pt clusters strongly interacting with support and possibly covered by support oxidic

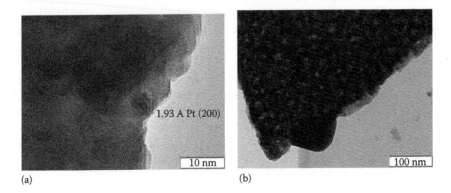

(a)   (b)

FIGURE 10.1 Typical morphology of Pt particles in oxidized Pt/PrCeZrO samples. (a) Disordered Pt cluster on the surface of 1.6% Pt/PrCeZrO sample; (b) big Pt particle at the edge of support platelet in 4.9% Pt/PrCeZrO sample.

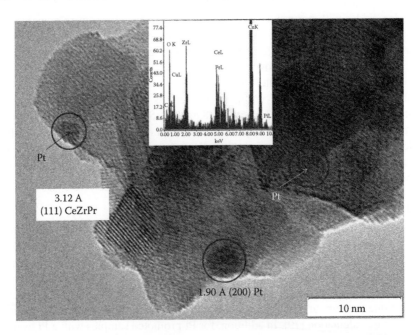

FIGURE 10.2 Typical morphology of Pt clusters on the surface of 1.6% Pt/PrCeZrO sample treated in the reaction at high temperatures and respective EDX spectra.

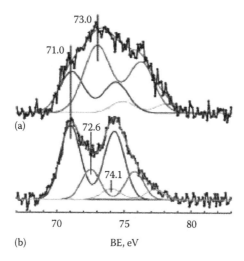

FIGURE 10.3 (a) XPS spectra for 1.6% Pt/PrCeZrO sample after oxidizing pretreatment and (b) after high-temperature treatment in the reaction feed 7% CH_4 + 7% CO_2 in He.

fragments exist. During reaction (7% CH_4 + 7% for 5 h at 800°C), the number of Pt clusters per the support surface unit and their size appear to be somewhat increased, without any pronounced sintering (Figure 10.2). In general, XPS results (Figure 10.3) support the XRD and TEM data.

For oxidized samples, three different Pt states (BE 71–73, and 75 eV) corresponding to species in 0, 2^+, and 4^+ states have been observed (Figure 10.3). For the sample with the lowest Pt loading, metallic Pt was not detected at all. The ratio of surface concentrations of Pt/Pr + Ce + Zr increases from ~0.2% to ~0.45% and ~1% with increasing Pt loading from 0.5 to 1.6 and 4.9 wt.%, respectively. Such a low surface concentration of Pt suggests that Pt cations are incorporated into the surface/subsurface layers of fluorite-like oxides. A strong interaction between Pt and the support, that is, decoration of Pt clusters by support oxidic species, takes place [79]. Pretreatment of a sample with 1.6% Pt loading in the reaction feed at high temperature results in reduction of Pt^{2+} to metallic Pt (Figure 10.3), while the Pt/Pr + Ce + Zr ratio only slightly decreases from ~0.45% to ~0.4%. High-temperature reduction of this sample by H_2 further decreases the Pt/Pr + Ce + Zr ratio to ~0.2%.

10.2.1.2 Ni–Ru-Supported Fluorite-Like and Perovskite-Like Catalysts

According to TEM data (Figure 10.4), PrSmCeZrO oxide support particles consist of stacked nanodomains. RuO_x species incorporated into the surface of these domains are revealed by EDX, while NiO layers were detected as regions with 2.44 Å spacing corresponding to (111) planes of NiO (PDF 47-1049).

After reaction, the fluorite-like oxide support morphology was not changed (Figure 10.5). Along with a lattice spacing of 2.41 Å corresponding to NiO nanoparticles, which could develop due to Ni oxidation by air after sample discharge from the reactor, a digital diffraction pattern revealed reflection corresponding to 1.87 Å spacing of Ni_xRu_y intermetallide (PDF #65-4309).

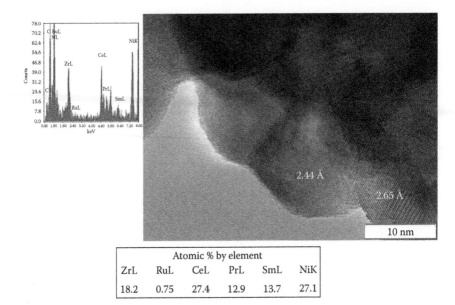

Atomic % by element					
ZrL	RuL	CeL	PrL	SmL	NiK
18.2	0.75	27.4	12.9	13.7	27.1

FIGURE 10.4 Typical morphology of Ni–Ru/PrSmCeZrO sample calcined under air at 800°C with respective EDX data.

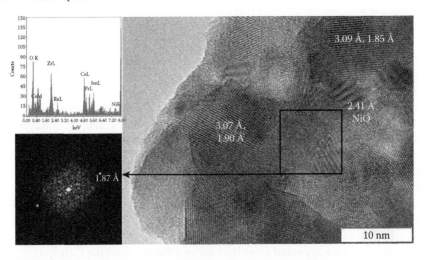

FIGURE 10.5 Typical morphology and EDX data of Ni–Ru/PrSmCeZrO sample after reaction of CH_4 dry reforming at 850°C. Atomic composition of selected area estimated from EDX corresponds to 32.5% Zr, 32.02% Ce, 13.86% Pr, 14.77% Sm, 3.81% Ru, and 3.04% Ni.

In the case of Ni + Ru-containing perovskites, after contact with concentrated CH_4 + CO_2 feed, Ni–Ru alloy nanoparticles strongly interacting with perovskite matrix are revealed as well (Figure 10.6).

As-prepared $Ru/La_{0.8}Pr_{0.2}Mn_{0.2}Cr_{0.8}O_3$ + 10% NiO + 10% YSZ nanocomposites are composed of perovskite-like and fluorite-like phases with Ni and Ru cations dissolved in their particles/surface layers [42,86]. Addition of fluorite-like oxides such

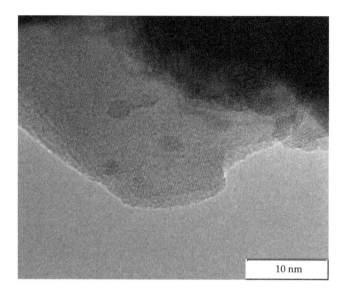

FIGURE 10.6 TEM image of $PrFe_{0.65}Ru_{0.05}Ni_{0.3}O_3$ catalyst particles after reaction.

as yttria-stabilized zirconia (YSZ) helps to disorder the perovskite structure and prevents sintering due to generation of interfaces between perovskite and fluorite domains, with the specific surface area remaining up to 20 m/g after calcination at 900°C. Similar to the case of perovskite- and fluorite-based catalysts, after contact with reaction feed, nanoparticles of Ru + Ni alloy in a strong epitaxy with support were observed as well.

10.2.2 CATALYST PRETREATMENT EFFECTS

Pretreatment is often required for activation of reforming catalysts via reducing supported oxides to the metal phase [43,45,60–67,77]. Nanocomposite catalysts can exhibit significant differences in redox potential, depending on the nature of the support, type of structural promoters, temperature, time, and characteristics of the atmosphere. The lowest reduction temperature and the highest degree of reducibility are typically considered as desirable properties of the promoted fluorite-like/perovskite-like oxide catalysts. Both textural and structural factors govern the redox behavior of these mixed oxides.

10.2.2.1 Pretreatment of Fluorite-Like Oxide Catalysts

High oxygen vacancy concentration and oxygen mobility are well-known properties of fluorite-like oxides. Subjecting complex fluorite-like oxides, such as Pt-supported $Pr_{0.3}Ce_{0.35}Zr_{0.35}O_{2-x}$ catalysts, to severe reducing treatment in H_2 at 800°C for 0.5 h strongly deactivates the catalysts. Conversion of CH_4 and CO_2 in MDR was decreased from ~40%–50% to ~10% at 700°C and a residence time of 15 ms. The TEM data (Figure 10.7) demonstrate that quite large Pt particles emerged on the support surface.

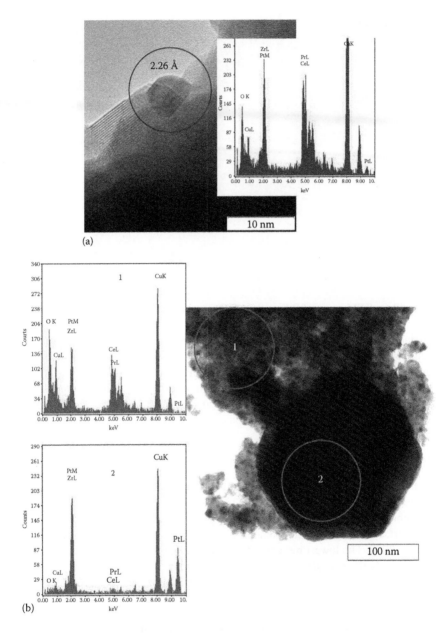

FIGURE 10.7 Typical morphology of Pt particles (a) on the surface of 1.6% Pt/PrCeZrO sample after reduction by H_2 at 800°C and (b) respective EDX spectra. (1) Small ordered Pt platelet, (2) big Pt particle. *(Continued)*

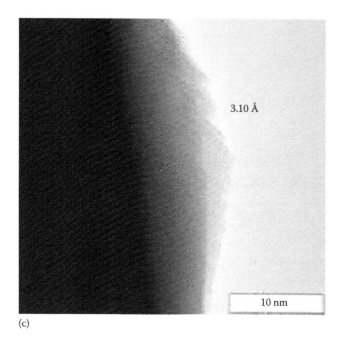

3.10 Å

10 nm

(c)

FIGURE 10.7 (CONTINUED) Typical morphology of Pt particles (c) high-resolution image of the edge of big Pt particle.

According to XPS data, reduction of 1.6% $Pt/Ce_xZr_{1-x}O_2$ sample by H_2 at 800°C strongly decreases the Pt dispersion, expressed as Pt/Pr + Ce + Zr ratio, from ~0.45% in the as-prepared sample to ~0.2% [78], while after contact with the reaction feed 7% CH_4 + 7% CO_2 in He at 800°C, it remains close to the initial value (~0.4%). Irregularly shaped large Pt particles are observed after reduction by hydrogen at high temperatures. The EDX spectrum from the edge of such a particle demonstrates that it is essentially free from any traces of Pr, Sm, or Zr cations [78]. Hence, the deactivation of this catalyst after high-temperature reduction could not be assigned to any effects of strong metal–support interactions such as formation of surface PtCe alloys or ceria overlayers [79]. Rather, this can be ascribed to ordering the structure of Pt particles and decreasing their surface coverage by oxidic fragments. After reduction, the degree of Pt coordinative unsaturation is apparently decreased, resulting in a lower reactivity to C–H bond rupture [80]. Pt/PrCeZrO catalyst was found to be in the most active and selective state after oxidizing pretreatment in O_2 at 900°C for 1 h [78].

10.2.2.2 Pretreatment of the Perovskite-Oxide Catalysts

The redox properties of the promoted perovskite-oxide catalysts were studied [65] using $LnFeNi(Ru)O_3$ perovskite as a promising precursor of active and stable-to-coking catalysts of MDR. Typical temperature-programmed reduction profiles of $LaFe_{0.6}Ru_{0.1}Ni_{0.3}O_{3-\delta}$ and $LaFe_{0.7}Ni_{0.3}O_{3-\delta}$ samples after repeated reduction–reoxidation experiments are shown in Figure 10.8. For each TPR run, samples were

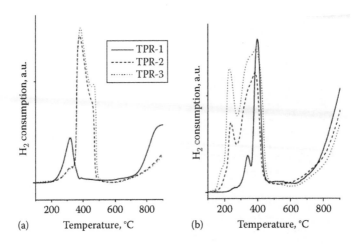

FIGURE 10.8 Repeated H_2 TPR after reoxidation for (a) $LaFe_{0.7}Ni_{0.3}O_{3-\delta}$ and (b) $LaFe_{0.6}Ru_{0.1}Ni_{0.3}O_{3-\delta}$.

previously oxidized in an oxygen flow for 1 h at 500°C. Then samples were heated from 25°C to 900°C with a heating rate of 10°C/min in flowing H_2 reducing environment, that is, 10% H_2 in Ar.

The total area under the TPR spectrum corresponds to the amount of hydrogen consumed and, hence, relates to the reduction of cations in perovskite structure, which can go up to the metallic state for Ru and Ni cations. Repeated reduction–reoxidation experiments revealed a reactivity evolution of perovskites that was reflecting the variation in their real structure/microstructure. In the case of perovskites without Ru reoxidized after the first TPR run, the first peak at ~315°C remains only as a shoulder of a new intense asymmetric peak with the main maximum at ~420°C. Since the amount of hydrogen consumed in these low-temperature peaks corresponds to complete reduction of all Ni in perovskite to the Ni^0 state, this clearly indicates the increase in sample reactivity after reoxidation. Nearly identical H_2 TPR spectra for the second and the third runs demonstrate the reproducibility of the sample's reactivity after reoxidation. Since reduction of both NiO and Ni-containing perovskites to the metallic phase is a topochemical process including generation of Ni^0 nuclei followed by their growth, the observed variation in perovskites reactivity after redox cycles can be explained by microheterogeneity of reoxidized perovskites caused by preferential location of Ni cations in the surface layer and within domain boundaries of defect La ferrite without regeneration of the bulk structure of $LaFe_{0.7}Ni_{0.3}O_{3-\delta}$ perovskite.

For the $LaFe_{0.6}Ni_{0.3}Ru_{0.1}O_{3-\delta}$ sample, TPR spectra also change between consecutive runs, revealing a reactivity increase (Figure 10.8b). In the third run, practically all Ni and Ru cations in the sample are reduced to the metal state in the low-temperature range. Since the main peak is positioned at ~400°C, which is the same as for the Ru-free sample, it clearly corresponds to the topochemical reduction of NiO_x species. Peaks at 180°C–230°C appeared due to reduction of RuO_x species formed as a result of ruthenium segregation from the perovskite structure during the first reduction

[43,68], while the peak at 340°C–390°C can be assigned to the reduction of the mixed Ni–Ru oxide species [87]. The oxidation/reduction cycles in the perovskite catalysts cause metals to migrate into and out of the perovskite matrix and, hence, to maintain their catalytic activity by regenerating the metal nanoparticles and preventing metal particle growth.

10.2.3 ACTIVATION AT REACTIVE CONDITIONS

10.2.3.1 Perovskites

In typical experiments, fresh (as-prepared) LnFeNi(Ru)O$_3$ perovskite catalysts were exposed to the feed stream comprising 1% CH$_4$ + 1% CO$_2$ in He, without any reducing pretreatment, and were heated from 25°C to 800°C with a heating ramp 5°/min. Reactant and product concentration variations are monitored as a function of temperature (see Figures 10.9 and 10.10). For LnFeNi$_{0.3}$ perovskites, the reaction starts at ~600°C. CH$_4$ and CO$_2$ consumption was immediately followed by CO formation, while hydrogen evolution required higher temperatures. Such a sequence of product

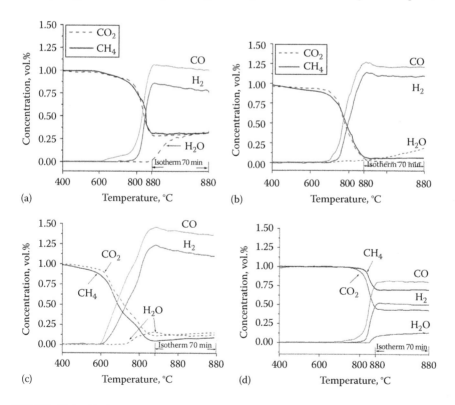

FIGURE 10.9 Typical curves of reagents and product concentration in the temperature-programmed reaction of CH$_4$ dry reforming for LnFNi$_{0.3}$ catalyst pretreated in O$_2$. Contact time 0.005 s, feed 1% CH$_4$ + 1% CO$_2$ in He; (a) PrFe$_{0.7}$Ni$_{0.3}$O$_3$, (b) LaFe$_{0.7}$Ni$_{0.3}$O$_3$, (c) La$_{0.8}$Sr$_{0.2}$Fe$_{0.7}$Ni$_{0.3}$O$_3$, and (d) La$_{0.9}$Ce$_{0.1}$Fe$_{0.7}$Ni$_{0.3}$O$_3$.

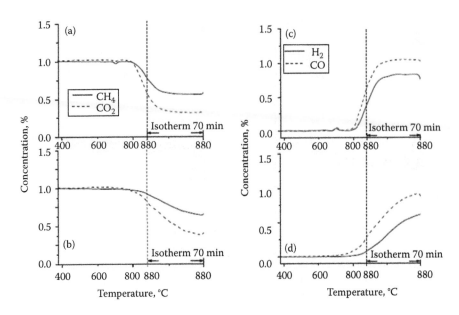

FIGURE 10.10 Typical curves of reagents (a and b) and product (c and d) concentration in the temperature-programmed reaction over $PrFe_{0.65}Ru_{0.05}Ni_{0.3}O_{3-\delta}$ (a and c) and $LaFe_{0.65}Ru_{0.05}Ni_{0.3}O_{3-\delta}$ (b and d).

formation indicates that the reverse water–gas shift reaction occurs simultaneously with the reforming reactions at temperatures below 700°C, giving rise to larger CO_2 conversions (Figure 10.9) and higher selectivity to CO, which, in turn, increases the CO/H_2 molar ratio. The highest light-off temperature, that is, about 700°C, along with the lowest activity, in terms of H_2 and CO steady-state concentrations, is observed for the $La_{0.9}Ce_{0.1}Fe_{0.7}Ni_{0.3}O_3$ sample, which suggests strong stabilization of Ni cations in the perovskite structure by Ce doping.

For as-prepared Ru-containing perovskites, product concentration profiles during temperature-programmed reaction of MDR presented in Figure 10.10 are, in general, similar to those for Ru-free samples. A partial substitution of Fe by Ru results in increased light-off temperatures up to 680°C–800°C, which is not expected due to well-known easy reducibility of Ru oxides. According to XRD data [65], this phenomenon can be explained by praseodymium or lanthanum oxychloride phase formation, the content of which increases with the Ru concentration, caused by the use of $Ru(OH)Cl_3$ as a Ru source. Apparently this phase hampers methane activation by blocking surface sites. Such an effect of oxychloride is more pronounced for La-containing samples due to more basic nature of La cation. Even in more concentrated reaction feeds, the activation of Ru-containing perovskites prepared using $Ru(OH)Cl_3$ salt proceeds rather slowly (Figure 10.11). Nevertheless, the final activity that could be achieved correlates with Ru content, as could be expected. Hence, from the practical point of view, chlorine-free salts are to be used for preparation of perovskite-like precursors, or high-temperature pretreatment in H_2 is required.

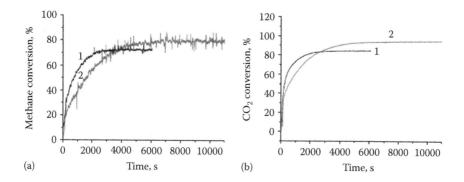

FIGURE 10.11   Time dependence of (a) CH_4 and (b) CO_2 conversion in MDR over oxidized $PrFe_{0.7-x}Ru_xNi_{0.3}O_{3-\delta}$ ($1 - x = 0.05$, $2 - x = 0.1$) at 850°C, feed 10% CH_4 + 10% CO_2, He balance, contact time 0.015 s.

10.2.3.2   Fluorites

In contrast to catalysts based on perovskite-like precursors (see aforementioned discussion), Ru-promoted fluorite-like catalysts ($Ru(OH)Cl_3$ salt was used for impregnation as well) are easily activated in diluted reaction feed at moderate (~400°C) temperatures providing a high activity level (Figure 10.12). Hence, in this case, HCl is apparently easily removed from the catalyst surface mainly at the catalyst preparation stage (calcination under air at 900°C) and in the reaction mixture due to a higher stability of doped ceria–zirconia complex oxides to chlorination.

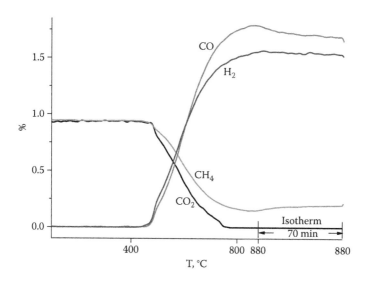

FIGURE 10.12   Typical curves of reagents and product concentration in the temperature-programmed reaction of CH_4 dry reforming for 1% Ru/80Sm + 10 NiO + 10 YSZ catalyst pretreated in O_2. Contact time 0.005 s, feed 1% CH_4 + 1% CO_2 in He.

10.2.4 Steady-State Catalytic Activity

10.2.4.1 Perovskites

To analyze the effect of the perovskite chemical composition and the pretreatment on the catalytic activity, the specific effective first-order rate coefficients of methane transformation were estimated using a plug-flow reactor model. The MDR specific rate coefficients at 800°C for all catalysts activated in the reaction mixture and in the flow of H_2 at 800°C are shown in Figure 10.13. These data show that, as in the case of Ru-free samples, the highest performance has been demonstrated by Pr-based catalysts. Moreover, the catalyst activity increases with the Ru concentration independently of the activation mode. However, the activity of the catalysts activated by the reaction mixture is slightly inferior due to the presence of PrOCl (see aforementioned discussion). The hydrogen pretreatment leads to segregation of active metal alloy particles and partial decomposition of PrOCl blocking active centers that increases catalyst activity. The specific rate coefficients for samples activated under reaction conditions are close due to comparatively easy decomposition of PrOCl. In contrast to Pr-containing samples, for $LaFe_{1-x}Ru_xNi_{0.3}O_{3-\delta}$ catalysts, the activity order strongly depends on the activation mode due to a higher stability of LaOCl.

For perovskite-based catalysts without Ru (Figure 10.14), activity levels very close to that of Pr–Fe–Ni–O can be achieved by using Sm instead of La or even by substitution 10% of La by Pr. A high CH_4 conversion is achieved for the latter catalyst even at high NG concentrations, though longer contact times are required (see Figure 10.15). This could be important for practical application when using Cl-free salts of Pt or Ru for promoting Ni-containing catalysts.

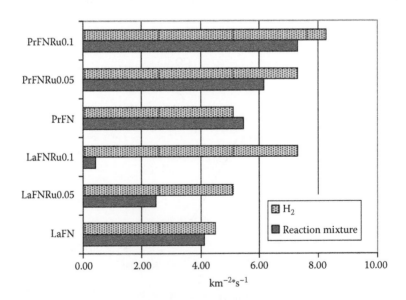

FIGURE 10.13 Specific rate constant of CH_4 dry reforming over perovskites reduced at 800°C in H_2 or activated in the reaction mixture. Feed 10% CH_4 + 10% CO_2 in He, 800°C, contact time 0.01 s.

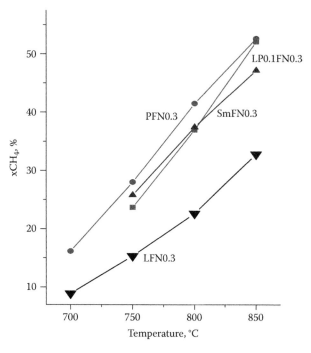

FIGURE 10.14 Temperature dependence of methane conversion in CH_4 DR over $LnFe_{0.7}Ni_{0.3}O_{3-\delta}$ (Ln = La, LaPr, Pr, Sm) activated in the reaction mixture. Feed 10% CH_4 + 10% CO_2 in He, contact time 0.01 s.

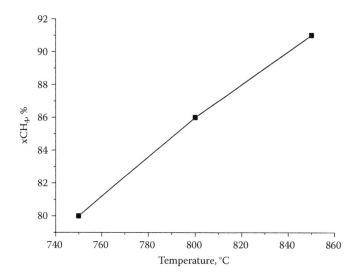

FIGURE 10.15 Temperature dependence of methane conversion in CH_4 DR on $La_{0.9}Pr_{0.1}Fe_{0.7}Ni_{0.3}$ catalyst. Feed composition 47.5% NG + 47.5% CO_2 + 5% N_2, contact time 0.16 s.

10.2.4.2 Fluorites

As follows from Figure 10.16, for a SmPrCeZr fluorite-like complex oxide support, the specific catalytic activity of supported Ru catalysts is higher than that with Pt. The activity obtained with supported Ni is much lower—specific rate coefficient only amounts up to ca. 0.25 s^{-1} m^{-2} at 800°C, which is explained by the catalyst coking (as could be concluded from O_2 TPO data not shown here for brevity). This agrees with the much higher activity of the $LaNiO_3$-supported sample, when small Ni clusters generated under reaction conditions are stabilized and prevented from coking by decoration with La hydroxycarbonate species. Combination of Ru and Ni provides the nonadditive increase of activity due to formation of Ru–Ni alloy nanoparticles in reaction conditions. At the same time, combination of Ru with $LaNiO_3$ gives substantially lower activity, perhaps due to excessive decoration of metal alloy nanoparticles by La hydroxycarbonate species in reaction media. Indeed, the accessible metal surface for RuLaNi/SmPrCeZr sample estimated from CO pulse chemisorption results at room temperature, that is, ~0.05 m^2/g, is an order of magnitude lower than that obtained for the Ru + Ni/SmPrCeZr sample, that is, 0.5 m^2/g. Hence, the specific rate coefficient related to the unit of accessible metal surface is two to three times higher for RuLaNi/SmPrCeZr sample due to the positive effect of developed metal–oxide interface.

The effect of the fluorite-like support chemical composition is also quite clear: without Sm as codopant with Pr, the performance is lower (Figure 10.17). This suggests that disordering of fluorite-like ceria–zirconia solid solution by doping with two cations helps to increase the oxygen mobility and disorder Ru or Ni + Ru

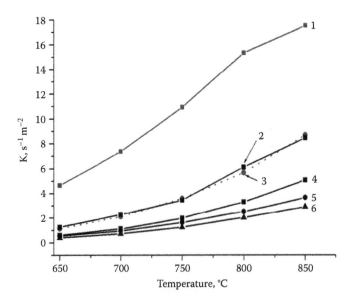

FIGURE 10.16 Temperature dependence of specific first-order rate constants for samples with $Sm_{0.15}Pr_{0.15}Ce_{0.35}Zr_{0.35}O_2$ oxide support (SmPr): (1) 6.6% Ni + 1.4% Ru/SmPr, (2) 1.6% Ru/SmPr, (3) 8% $LaNiO_3$ + 1.4% Ru/SmPr, (4) 8% $LaNiO_3$/SmPr, (5) 8% $LaNiO_3$ + 1.4% Pt/SmPr, (6) 1.6% Pt/SmPr. Feed 10% CH_4 + 10% CO_2 in He, contact time 0.01 s.

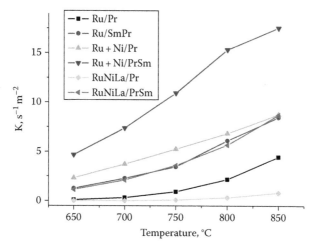

FIGURE 10.17 Temperature dependence of specific first-order rate constants for samples with Ru/Ru–Ni supported on fluorite-like oxides $Sm_{0.15}Pr_{0.15}Ce_{0.35}Zr_{0.35}O_2$ (PrSm) or $Pr_{0.3}Ce_{0.35}Zr_{0.35}O_2$ (Pr). Feed 10% CH_4 + 10% CO_2 in He, contact time 0.01 s.

nanoparticles as well, which favor more efficient activation of C–H bond in CH_4. When La is added by supporting a perovskite-like precursor on fluorite-like oxide, the activity is decreased as well. This can be explained either by excessive decoration of metal nanoparticles by La hydroxycarbonate or by hampering oxygen mobility in fluorite-like supports due to La cation incorporation into the surface layer or domain boundaries [33,82,84,86].

An optimized catalyst based on Pr + Sm-doped ceria–zirconia demonstrates a high activity and performance stability in dry reforming of real NG in concentrated feeds (see Figures 10.18 and 10.19).

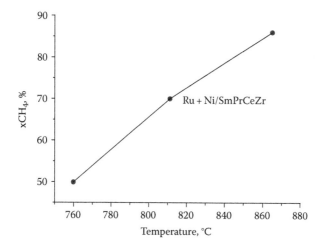

FIGURE 10.18 Temperature dependence of CH_4 conversion for Ru + Ni/SmPrCeZrO catalyst in concentrated feed 50% CH_4 + 50% CO_2, contact time 0.12 s.

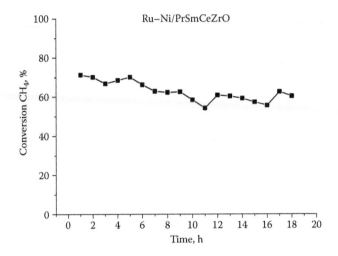

FIGURE 10.19 Variation of CH_4 conversion with time on-stream at 810°C for Ru + Ni/ SmPrCeZrO catalyst in concentrated feed 50% CH_4 + 50% CO_2, contact time 0.12 s (equilibrium conversion 85%).

10.2.4.3   Catalysts Based on Perovskite + Fluorite Nanocomposites

Typical results obtained for $PrFe_{0.7}Ni_{0.3}O_3–Ce_{0.9}Gd_{0.1}O_2$ (PFN–GDC) nanocomposites are shown in Figure 10.20. As follows from these results, a limited, that is, up to 5 wt.%, addition of GDC allows to enhance the catalyst activity, while at higher GDC content, the activity decreases again due to dilution or by modification of the surface layer of disordered Pr ferrite by Gd and Ce affecting surface oxygen mobility.

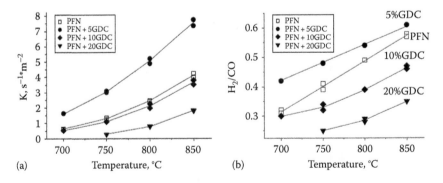

FIGURE 10.20 Temperature dependence of the (a) specific rate constant and (b) H_2/CO ratio for $PrFe_{0.7}Ni_{0.3}O_{3–\delta}$ and composites containing 5%–20% GDC prepared via one-pot method. Samples were activated in the reaction mixture at 850°C. Feed 10% CH_4 + 10% CO_2 in He, contact time 0.015 s.

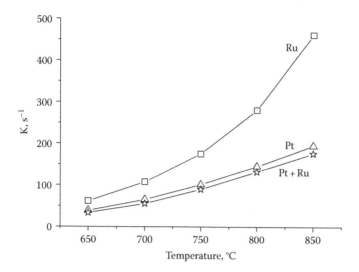

FIGURE 10.21 Effect of precious metals (loading 1 wt.%) supported on nanocomposite 80% $Sm_{0.3}Ce_{0.35}Zr_{0.35}O$ + 10 NiO + 10 YSZ on activity in CH_4 DR. Feed 7% CH_4 + 7% CO_2 in He, contact time 15 ms.

For nanocomposite 80% $Sm_{0.3}Ce_{0.35}Zr_{0.35}O$ + 10 NiO + 10 YSZ prepared via one-pot Pechini route [86], promotion by Ru allowed to provide a high catalytic activity as well (Figure 10.21). Addition of YSZ to doped ceria–zirconia helps to improve sintering resistance of these catalysts [86].

Similarly, high activity was demonstrated for nanocomposite Ru/Ni/LaMnCrPr/YSZ catalyst as well in dry reforming of pure methane (Figure 10.22) as well as real NG (Figure 10.23).

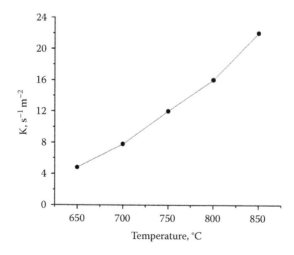

FIGURE 10.22 Temperature dependence of specific rate constant for Ru/Ni/LaMnCrPr/YSZ. Feed 10% CH_4 + 10% CO_2 in He, contact time 15 ms.

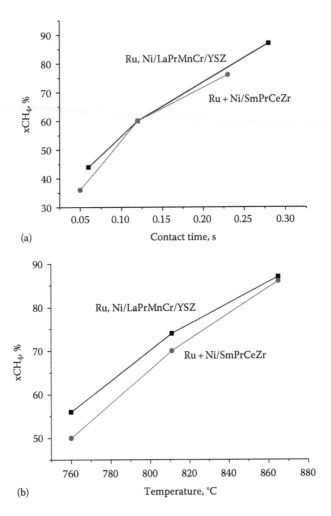

FIGURE 10.23   (a) Methane conversion at 800°C versus contact time in CH$_4$ DR on best supported Ru + Ni catalysts. Feed composition 47.5% NG + 47.5% CO$_2$ + 5% N$_2$. (b) Temperature dependence of methane conversion in CH$_4$ DR on best supported Ru + Ni catalysts at contact time 0.2 s. Feed composition 47.5% NG + 47.5% CO$_2$ + 5% N$_2$, contact time 0.12 s.

Note that for NG reforming, very close and high performance was demonstrated for catalysts consisting of Ru + Ni nanoparticles on fluorite-like SmPrCeZr oxide support and LaMnCrPr/YSZ nanocomposite (Figure 10.23).

For the latter catalyst, the NG dry reforming performance was good and stable in feeds with CO$_2$ excess as well (see Figures 10.24 and 10.25).

For optimized catalysts based on Ru + Ni nanoparticles on perovskite-like or fluorite-like complex oxides with high oxygen mobility and reactivity, both TEM studies and temperature-programmed oxidation after reactions did not reveal any carbon deposition (see Figure 10.26).

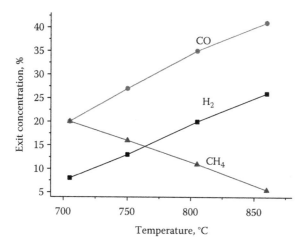

FIGURE 10.24 Temperature dependence of exit concentrations in the process of NG dry reforming on 2.8% Ru/LaPrMnCr/10% YSZ/10 NiO at contact time 0.09 s, feed 53% CO_2 + 38% NG + N_2.

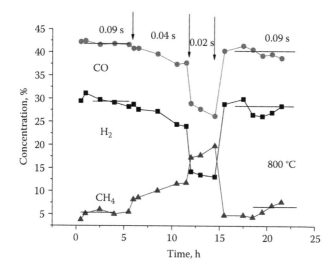

FIGURE 10.25 Variation of exit concentrations with time on-stream in the process of NG dry reforming on fresh 2.8% Ru/LaPrMnCr/10% YSZ/10 NiO catalyst heated to 800°C in the stream of N_2 followed by switch to 53% CO_2 + 38% NG + N_2 stream.

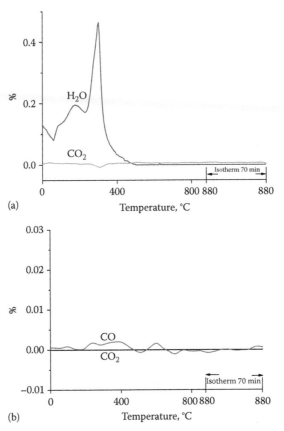

(a)

(b)

FIGURE 10.26 O_2 TPO spectra for $PrFe_{0.6}Ru_{0.1}Ni_{0.3}$ (a) and $Ru + Ni/SmPrCeZrO$ (b) catalyst after testing in NG DR. 1% O_2 in He, temperature ramp 5°/min.

10.3 MECHANISTIC ASPECTS OF METHANE DRY REFORMING OVER NANOCOMPOSITE CATALYSTS

10.3.1 General Schemes

The main factor controlling the catalytic activity during MDR is believed to be the reactant activation. The dissociative adsorption and activation of both CH_4 and CO_2 are structure sensitive, that is, they depend on the structure of nanocatalysts based on fluorite-like/perovskite-like oxides, as far as they depend on electronic as well as on geometric factors [3,88]. The existence of the strong metal–support interaction between active metal particles and support in catalysts drastically suppresses chemisorption of both H_2 and CO facilitating synthesis gas production [89].

It is generally accepted that methane is dissociatively adsorbed on the metal species to form CH_x fragments, whereas CO_2 activation may depend on the type of support used (see Figure 10.27).

If CO_2 is adsorbed on the support and activated at the interface between the metal particle and support (path I), rapid oxidation of CH_x on metal surface occurs.

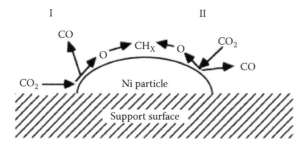

FIGURE 10.27 Scheme of the MDR reaction. (From Tomishige, K. et al., *Catal. Today*, 45, 35, 1998.)

Alternatively, CO_2 can be activated on the metal surface (path II). It was suggested that path I was more effective for the inhibition of carbon formation than path II. For nanocatalysts, oxygen vacancies formed on the support during prereduction can act as active sites for dissociative adsorption of CO_2. For the irreducible oxide supports, CO_2 dissociation is supposed to be promoted by the H (ads) originating from the CH_4 dissociation, which can be assisted by oxygen atoms on the support. Also, the reactive intermediates in the reaction mechanism are mostly support-related species [91–95]. Thus, Bradford and Vannice [93] have suggested for nickel-supported catalysts that CO_2 participates in the reaction mechanism through the reverse water–gas shift to produce surface OH groups. The surface OH groups react with adsorbed CH_x intermediates being formed through CH_4 decomposition, yielding a formate-type intermediate, CH_xO. CH_xO decomposition leads to the principal reforming products, that is, H_2 and CO. Decomposition of both CH_4 and CH_xO is a slow kinetic step. According to observations by O'Connor et al. [94] over both Pt/Al_2O_3 and Pt/ZrO_2 catalysts, methane decomposition takes place over platinum. The main difference between the two catalysts concerns the carbon dioxide dissociation. XPS and DRIFTS data obtained for a Rh/La_2O_3 catalyst [92] indicate that the only visible surface species for the used catalyst are lanthanum oxycarbonate and mainly $Rh°$, whereas both types Ia oxycarbonate and II-$La_2O_2CO_3$, as well as both linear and bridge-bonded CO adsorbed on metallic rhodium, are present during MDR.

The metal–support interface may also contain active sites for subsequent CHO formation and decomposition. Hence, the support may affect the catalyst activity by altering the stability of any intermediate species at the metal–support interface. To gain insights into the specificity of MDR over Me-supported fluorite-like doped ceria–zirconia catalysts, a combination of transient kinetic methods including TAP and SSITKA with pulse microcalorimetry and spectral studies, mass spectrometry such as in situ Fourier Transform Infrared Spectroscopy (FTIRS) techniques has been used [81]. One of the most significant findings of this study is that both CH_4 and CO_2 dissociate independently of each other on metal and support sites, respectively. However, metal alloy nanoparticles (Pt, Ru, Ni + Ru) are not only involved in CH_4 activation, but may mediate CO_2 activation also. Thus, in case of Ru + Ni/PrSmCeZr catalyst, for CO_2 molecules adsorbed at the metal–support interface, the C–O bond rupture is aided by the Ni + Ru surface species.

10.3.2 TAP STUDIES

Generally, the Ni + Ru-supported catalysts are most active for the complete oxidation, c.q., combustion, of methane in their oxidized state. The catalysts pretreated by reduction with hydrogen are effective for dry reforming [84–86]. The Temporal Analysis of Products (TAP) studies were performed with the reduced catalyst [81]. Successive pulses of CO_2 and CH_4 are efficiently transformed into CO and CO + H_2, respectively. Furthermore, the amount of CO generated during a CO_2 pulse is similar to that produced during a CH_4 pulse, indicating that no carbon is left on the catalyst surface. In the pump–probe experiments with heavy isotope oxygen (^{18}O)-labeled carbon dioxide, formation of $C^{16}O$ was observed only when the reduced Ru + Ni/PrSmCeZr catalyst was exposed to pulse of $C^{18}O_2$. Since the only possible source of ^{16}O in this experiment was the support, it was suggested that $C^{18}O_2$ was able to exchange oxygen with the surface very fast.

According to Sadykov et al. [82,84,104] and Galdikas et al. [96], a complex exchange mechanism is involved with two oxygen atoms participating in the hetero-exchange. For $Ce_xZr_{(1-x)}O_2$ solid solutions [96], the diffusion coefficient was found to depend on Ce percentage and was the highest, that is, ~1.6 × 10^{-18} m²/s at 850°C, for $Ce_{0.15}Z_{0.85}O_2$. Oxygen diffusion coefficients along interfaces, that is, surface as well as domain boundaries, for both catalysts, that is, Pt/PrCeZr and LaNiPt/PrSmCeZr, were found to be much higher (see Table 10.1). Application of $C^{18}O_2$ SSITKA allowed to estimate the oxygen mobility in catalysts under steady-state MDR conditions when the catalysts are in the reduced state. The oxygen diffusion coefficients along interfaces remained high, thus demonstrating that, indeed, the surface diffusion can provide the required fast transfer of reactive oxygen species from support sites to metal–support interface under reaction conditions.

TABLE 10.1
Oxygen Diffusion Coefficients in Pt/PrCeZr and LaNiPt/PrSmCeZr Catalysts as Measured by $^{18}O_2$ and $C^{18}O_2$ SSITKA

Sample	D_{eff},[a] s⁻¹	D_{bulk},[b] ×10^{-18} m²/s	$D_{interfaces}$,[c] ×10^{-16} m²/s
Pt/PrCeZr			
$^{18}O_2$ [31]	0.04	4	>33
$C^{18}O_2$	0.003	—	>2
LaNiPt/PrSmCeZr			
$^{18}O_2$ [31]	>0.03	3	>25
$C^{18}O_2$	0.008	—	>5

Source: Bobin, A.S. et al., *Top. Catal.*, 56, 958, 2013.

[a] D_{eff}—effective average oxygen diffusion coefficient for catalyst estimated by solving the system of differential equations describing SSITKA data.

[b] D_{bulk}—oxygen diffusion coefficient within oxide domains estimated by using D_{eff} and domain size.

[c] $D_{interfaces}$—oxygen diffusion coefficient along domain boundaries estimated with a regard for the relative amount of oxygen located within domain boundaries [81,104].

During subsequent CH_4 pulses, only $C^{16}O$ was produced, indicating that oxygen was able to be supplied from the support during 0.6 s time lag between CO_2 and CH_4 pulses. After a prolonged use of $C^{18}O_2$ pulses without any CH_4 present, the response spectra for $C^{18}O$ appeared. Eight thousand pulses were required for this, indicating that the support oxygen was partially replenished during the previous CO_2 pulse series. Therefore, either $C^{18}O_2$ dissociation takes place near the metal/support interface or it occurs on oxygen vacancies on the support, and the oxygen isotope is in equilibrium with the surface oxygen at the metal/support interface. This may be related to the fact that a mobile oxygen pool is present in the fluorite-like doped ceria–zirconia.

In the subsequent series of $^{13}CO_2$ and $^{12}CH_4$ pulses upon pump–probe experiments with heavy isotope (^{13}C)-labeled carbon dioxide (see Figures 10.28 and 10.29), only ^{13}CO was observed during the first pulse, while a $^{12}CH_4$ pulse resulted in ^{12}CO production. The findings do not display any exchange of carbon atoms between the reactants, which excludes the existence of any common intermediate.

Isotopic labeling studies have shown that the rate of the catalyst reoxidation by CO_2 as an oxygen supplier greatly exceeds that of its reduction by methane. CH_4 dissociation is the rate-limiting step. Hence, the rate coefficients for the elementary steps involved in CH_4 and CO_2 dissociation were estimated at ~1–10 and >10^2 s^{-1}, respectively. Moreover, the removal of carbon remaining on the metal sites occurs during CO_2 pulses due to the reactivity of the oxygen species and the mobility of the surface carbon species on the catalyst. Rapid oxygen redistribution between adsorption sites located on metal and oxide surface occurs during catalysis. Being strongly adsorbed on basic supports, carbon dioxide can be activated through formation of the surface carbonate species.

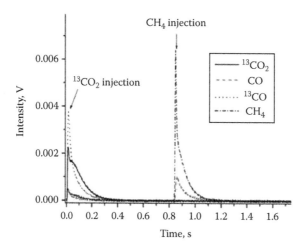

FIGURE 10.28 $^{13}CO_2$, ^{12}CO, ^{13}CO, and CH_4 responses corresponding to $^{13}CO_2$ and $^{12}CH_4$ pump–probe experiment over reduced Ru + Ni/PrSmCeZr catalyst at 750°C. Injection times were 0 s for $^{13}CO_2$ and 0.8 s for CH_4.

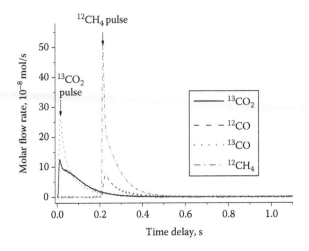

FIGURE 10.29 $^{13}CO_2$, ^{12}CO, ^{13}CO, and CH_4 outlet molar flow rates corresponding to $^{13}CO_2$/CH_4 alternating pulse experiment over reduced Ru + Ni/PrSmCeZr catalyst at 1023 K. Time lag for CH_4 was 0.2 s.

10.3.3 Pulse Microcalorimetry Studies

In general, the changes in CH_4 conversion and CO/CO_2 selectivity with the CH_4 pulse number and, hence, the reduction degree of the catalyst sample (see Figure 10.30), reasonably well agree with the trends observed for the TAP data (see previous discussion). The CO formation that already occurs during the first CH_4 pulse admitted onto the oxidized sample surface supports the hypothesis about a primary route of syngas formation via a CH_4 pyrolysis partial oxidation route. The observation of rather high degrees of CH_4 conversion after the removal of about one monolayer of oxygen from the sample again underlines a high rate of oxygen diffusion from the bulk of oxide particle to the surface, in order to compensate the used surface

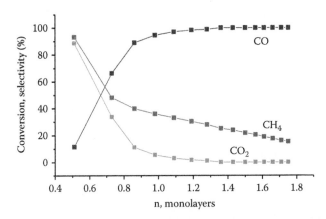

FIGURE 10.30 Dependence of CH_4 conversion, CO, and CO_2 selectivity on the degree of Pt/PrCeZrO sample reduction (n) by pulses of 7% CH_4 in He at 600°C.

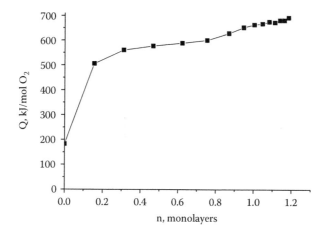

FIGURE 10.31 Enthalpy of oxygen adsorption (Q) versus reduction degree (n) estimated from the heats of 1.4% $Pt/Pr_{0.3}Ce_{0.35}Zr_{0.35}O_2$ sample reduction by CO pulses at 600°C.

oxygen during the preceding pulses. The average heat of oxygen adsorption on a partially reduced surface, that is, ~600 kJ/mol O_2 (see Figure 10.31), is close to values corresponding to bonding strength of bridging (M_2O) oxygen forms located at Ce cations [81].

The CH_4 interaction enthalpy with the catalyst (see Figure 10.32), endothermic process, increases with the reduction degree corresponding to the enthalpy of its transformation into deep and partial oxidation products with a due regard for syngas selectivity and variation of the average oxygen bonding strength. Some decline of the heat of reduction at a reduction degree exceeding 1.5 monolayers can be explained by the increasing contribution of CH_4 cracking in agreement with CH_4 TPR results [81]. A practically linear, that is, rather weak, dependence of the CH_4 conversion on the

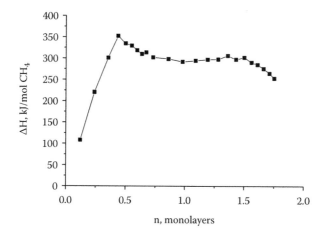

FIGURE 10.32 Enthalpy of CH_4 interaction with 1.4% $Pt/Pr_{0.3}Ce_{0.35}Zr_{0.35}O_2$ sample as a function of reduction degree (n) at 600°C.

reduction degree agrees with this weak variation of the heat of CH_4 transformation. Since the surface diffusion rate, that is, reverse oxygen spillover from the support to Pt, is high, this provides some coverage of Pt by adsorbed oxygen, hence favoring CH_4 activation.

For all investigated catalysts in steady-state MDR, reactant conversions in mixed pulses and in pulses containing only pure components were practically identical (see Figures 10.33 and 10.34). Moreover, the product selectivity was the same, that is, methane producing $CO + H_2$, CO_2, and CO, in several pulses as long as less than ~30% of oxygen is removed/replenished. In agreement with the isotope transient studies and TAP, this underlines the independent activation of the two reactants on different active sites with rapid oxygen migration between them.

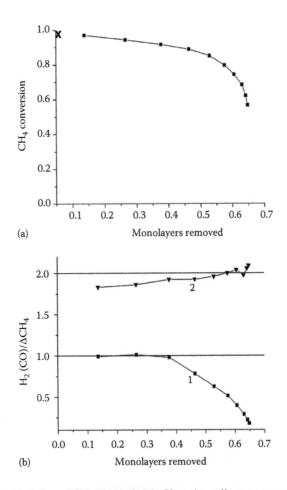

(a)

(b)

FIGURE 10.33   Variation of CH_4 conversion (a, X on the ordinate axes marks CH_4 conversion in the mixed $CH_4 + CO_2$ pulse) and product selectivity (b, 1—$CO/\Delta CH_4$; 2—$H_2/\Delta CH_4$) in the course of steady-state Ru + Ni/PrSmCeZr catalyst reduction by pulses of 7% CH_4 in He at 700°C.

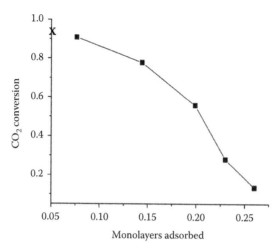

FIGURE 10.34   Variation of CO_2 conversion in the course of steady-state Ru + Ni/PrSmCeZr catalyst oxidation by pulses of 7% CO_2 in He at 700°C; x on the ordinate axes marks CO_2 conversion in the mixed CH_4 + CO_2 pulse.

TABLE 10.2

Characteristics of Bonding Strength of Reactive Bridging Oxygen Forms for Catalysts in the Steady State by Reduction of CH_4 or CO Pulses and Reoxidation by CO_2 or O_2 Pulses at 700°C

Catalyst Composition	Heat of Oxygen Desorption, kJ/mol O_2			
	CO_2[a]	CH_4[a]	CO[a]	O_2[a]
Pt/PrCeZr	640	650	640	640
LaNi/PrSmCeZr	660	630	650	560
Ru + Ni/PrSmCeZr	630	670	635	550
Ru + Ni/PrSmCeZr/YSZ	640	740	645	550

[a] Estimated by using pulses of respective component.

The measured heat values correspond to removal/replenishing of the bridging surface/interface oxygen forms with a desorption heat ~600–650 kJ/mol O_2 (Table 10.2).

Note that CO_2 conversion rapidly declines with the pulse number as the catalyst is reoxidized (see Figure 10.34). In agreement with calorimetric data demonstrating that the heat of surface reoxidation by CO_2 is constant, such kinetics reasonably fitted by a first-order equation suggests that the surface sites are occupied following a uniform adsorption energy. The rate coefficients for CO_2 consumption are very close for Pt and $LaNiO_3$-supported samples, that is, ~10^2 s^{-1}, while being an order of magnitude higher for Ni + Ru-supported sample, that is, ~10^3 s^{-1}. In agreement with SSITKA results (see aforementioned discussion), this suggests that Ni–Ru alloy nanoparticles participate in CO_2 activation, perhaps, by favoring the C–O bond rupture in CO_2 molecules adsorbed at metal–support interface.

10.3.4 SSITKA STUDIES

Figure 10.35 illustrates typical SSITKA responses corresponding to feed switches from ^{12}C normal isotope composition to that containing labeled $^{13}CH_4$. As can be seen, the transients are fast indicating that the steady-state surface coverage by carbon-containing species is quite low, that is, not exceeding 10% of a monolayer, calculated on the basis of the metallic surface. After switching the feed stream from $^{12}CH_4 + ^{12}CO_2 + He$ to $^{12}CH_4 + ^{13}CO_2 + He + 1\% Ar$ for the Ru + Ni/PrSmCeZr catalyst at 650°C, the fractions of ^{13}C in CO and CO_2 in the effluent increased without any delay relative to the Ar tracer concentration, and at each moment, the total number of

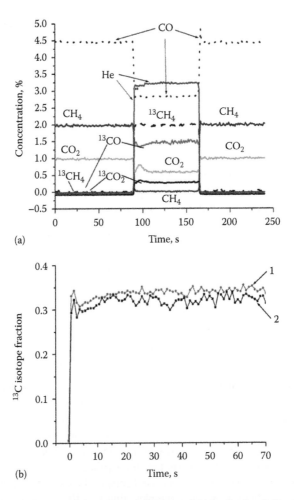

(a)

(b)

FIGURE 10.35 (a) Steady-state isotopic transients after the feed switches $CH_4 + CO_2 + He \rightarrow ^{13}CH_4 + CO_2 + He \rightarrow CH_4 + CO_2 + He$ for Pt/PrCeZr sample at 830°C, 15 ms contact time, and inlet concentration of CH_4 and CO_2 4%. (b) Time dependence of ^{13}C isotope fraction in CO (1) and CO_2 (2) after switching $CH_4 + CO_2 + He \rightarrow ^{13}CH_4 + CO_2 + He$ for Pt/PrCeZr sample at 830°C, 15 ms contact time, and inlet concentration of CH_4 and CO_2 4%.

[13]C atoms in CO and CO_2 was equal to that in the inlet [13]CO_2, so there was no carbon isotope accumulation on the surface, in addition to the amount of carbon adsorbed on the catalyst surface under steady-state conditions. This demonstrates that the concentration of C-containing intermediates, that is, carbonates and carbides, on the steady-state surface remains small but stable. This also suggests that the fraction of ^{13}C in CO should be equal to half of the sum of ^{13}C fractions in CO_2 and CH_4:

$$\alpha_{CO} = \frac{\alpha_{CO_2} + \alpha_{CH_4}}{2}$$

In the present experiments, the ^{13}C fractions in the reactants were equal to $\alpha_{CO_2} = 0.7$, $\alpha_{CH_4} = 0$, respectively. In this case, the estimated ^{13}C fraction in CO should be equal to 0.35, while the experimental value was found equal to 0.4. At the same time, the ^{13}C fraction of CO_2 in the effluent is much lower than that in the inlet feed being equal to 0.46. This means that in the course of reaction, some exchange of carbon atoms between CO and CO_2 proceeds, while there is no transfer of ^{13}C into CH_4. This proves that the interaction of the catalyst with CH_4 and CO_2 occurs independently, the first one being irreversible and the second one reversible.

The simplest mechanism corresponding to this statement can be presented as follows:

1. $[ZO] + CH_4 \rightarrow [Z] + CO + 2H_2$
2. $[Z] + CO_2 \leftrightarrow [ZO] + CO$

Here, methane irreversibly interacts with oxidized sites leading to CO and H_2 formation. Most likely metallic sites are involved in these steps. The methane transformation is followed by a reversible reoxidation of the reduced sites by carbon dioxide.

The total CO formation rate according to this scheme is equal to

$$w_{\sum CO} = w_1 + w_2 - w_{-2}$$

In the case of labeled CH_4, the redistribution of ^{13}C among all C-containing feed compounds is observed. The kinetic parameters characterizing the ^{13}C exchange rate and specific rates of CH_4 and CO_2 consumption are shown in Table 10.3. The rate coefficients of CO_2 transformation as estimated by SSITKA exceed those of CH_4 consumption by ca. an order of magnitude. As already mentioned, these results are in line with the resistance to coking of the studied catalysts.

Hence, under steady-state conditions, the MDR rate is limited by methane interaction with, c.q., activation by, the catalyst. Note that both w_1, that is, rate of CH_4 transformation, and w_2, that is, rate of catalyst reoxidation by CO_2, at a lower (550°C) temperature are apparently higher for catalysts containing Ru + Ni as compared with those for Pt/PrCeZr at a higher (735°C) temperature. Since the specific activity of Pt/PrSmCeZr is even lower than that of Pt/PrCeZr (see Table 10.3), this difference could not be explained by the effect of support composition. Hence, this result suggests that Ru + Ni alloy nanoparticles are not only involved in CH_4 activation but also help to activate carbon dioxide as well. This confirms the key role of interface sites for this bifunctional reaction mechanism.

TABLE 10.3

Elementary Step Rate Assessment in MDR Using SSITKA

T, °C	α_{CO_2}	α_{CO}	w_2/w_1	w_1 mkmol/(g min)	w_2 mkmol/(g min)	w_{-2} mkmol/(g min)
Pt/PrCeZr						
735	0.12	0.14	42	0.6	25	24
785	0.21	0.23	34	1.0	34	33
830	0.32	0.34	30	1.6	48	46
Ru + Ni/PrSmCeZr						
550	0.52	0.46	7.7	4.2	32	28
600	0.49	0.42	6.0	6.2	36	30
650	0.46	0.39	5.5	7.6	42	34
Ru + Ni/PrSmCeZr/YSZ						
550	0.49	0.46	11.5	1.5	17	15
600	0.44	0.42	24	2.0	47	45
650	0.40	0.39	43	2.5	106	103

10.3.5 FTIRS IN SITU STUDIES

The question whether the carbonate phase participates in the catalytic cycle by providing oxygen that reacts with the carbon issued from methane activated on the metal sites is still open. To elucidate whether the main route of reductive transformation of CO_2 at reaction conditions proceeds via carbonates as intermediates, the reactivity of carbonate species on La-containing catalysts was studied, because carbonates bound to La cations were expected to be very stable. Interaction of CO_2 with the catalysts could yield a variety of carbonate-like species due to the surface basicity.

Results of in situ FTIRS study are shown in Figure 10.36. In this study, thin pellets, that is, 2.8–3.6 cm² in area and 60–90 mg in weight, of the $LaNi/PrSmCeZrO_2$ catalyst in the IR cell adjusted to experimental temperature (600°C and 650°C) were firstly exposed to CO_2. After fluorescence steady-state spectra were obtained, carbon dioxide was removed from the gas phase by freezing a CO_2 vessel with liquid nitrogen, and again, the IR spectrum was registered. Then CH_4 was injected, and the IR steady-state spectra were registered.

No matter which is the active metal, lanthanum oxycarbonates are formed by the interaction of La_2O_3 with CO_2. The lanthanum oxycarbonate bands are observed in the range 1300–1500 cm⁻¹. As follows from Figure 10.36, bridged surface carbonates, bands at 1385 and 1460 cm⁻¹, are visualized, which are consistent with a lanthanum carbonate structure. Though their intensity strongly declines after removing gas-phase CO_2, they are still retained on the surface. The existence of a pool of carbonate species, mostly dioxomonocarbonate $La_2O_2CO_3$, at 700°C–800°C under a partial pressure of CO_2 that corresponds to the reforming conditions was demonstrated by Slagtern et al. [97]. The type of carbonate species seems to be different for the various supports. Fast equilibrium was found to be achieved between the carbonates and the carbon dioxide in the gas phase, but essential irreversibility was

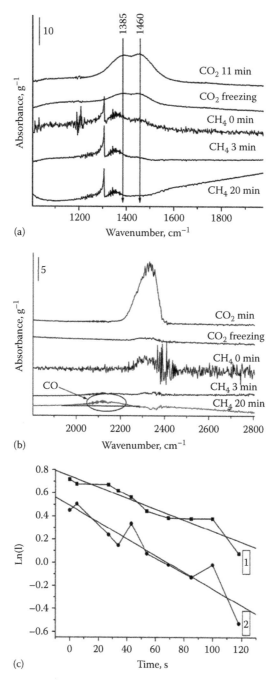

FIGURE 10.36 FTIR spectra of LaNi/PrSmCeZrO$_2$ sample in the range of surface carbonates (a) and CO and CO$_2$ gas-phase molecules (b); vibrations under contact with CO$_2$ followed by admission of CH$_4$ at 600°C; variation of the intensity of carbonate bands with time under contact with CH$_4$ in coordinates of the first-order rate equation (c): (1) band 1385 cm^{-1}, (2) 1460 cm^{-1}

TABLE 10.4

Rate Constants of the Interaction of the Surface Carbonates with Gas-Phase CH_4

Sample	Temperature, °C	Constant, s^{-1}
Ru/PrSmCeZr	600	0.003
LaNi/PrSmCeZr	600	0.007
LaNi/PrSmCeZr	650	0.009

observed under steady-state reaction conditions [97]. The lanthanum carbonate species readily react with CH_4 with the formation of gas-phase CO, since respective bands disappear (see Figure 10.36a). Indeed, as it was studied by Krylov et al. [98], surface carbonate species may react with carbon formed via methane decomposition. Again, admission of CH_4 into the cell causes the formation of new carbonates (Figure 10.36b). Linearization of the intensity variation in coordinates of the first-order equation revealed their identical slopes for both bands, which confirms their assignment to the same bridging carbonate (Figure 10.36c).

The rate coefficients of the first-order reaction of gas-phase CH_4 with the surface carbonates over the nanocomposite fluorite-like doped ceria–zirconia catalysts were found to be very small compared to that of the rate-limiting step of CH_4 dissociation. Therefore, the surface carbonate may assist with continuous supply of activated oxygen to the metal sites, preserving them from deactivation. However, they can also be considered as spectator species only in the catalytic cycle (Table 10.4).

10.3.6 SUMMARY OF MECHANISM

The investigation of the Me/PrSmCeZr catalysts has revealed the most important features of nanocomposite catalysts for MDR:

- Reducibility of the support
- Oxygen mobility on the surface and in the bulk of the specific catalyst
- Reoxidation of the support by CO_2

The mechanistic scheme of syngas formation during MDR on Me/PrSmCeZr catalyst and sequence of elementary steps over $Ni–Ru–Sm_{0.15}Pr_{0.15}Ce_{0.35}Zr_{0.35}O_2$ catalyst can be proposed as it is presented in Figure 10.37.

The mechanistic approach suggests that during reduction, methane is activated and decomposed to carbon and hydrogen on the metal sites. Hence, methane dissociation on Ru may result in the formation of different hydrocarbonaceous species, namely, methylidyne (CH), vinylidene (CCH_2), and ethylidyne (CCH_3) species, which, at temperatures exceeding 700 K, may transform into the graphitic phase [99]. Hydrogen species might be present on the surface either in the form of H_s or CH_{xs}. When the metal particle is preoxidized, H_2O is formed as a result of the reduction of the metal sites. The cationic metal particles cause labilization of lattice and surface oxygen. The lattice oxygen from oxide support is consumed for CO or H_2O formation, and oxygen vacancies are generated again. Similar to

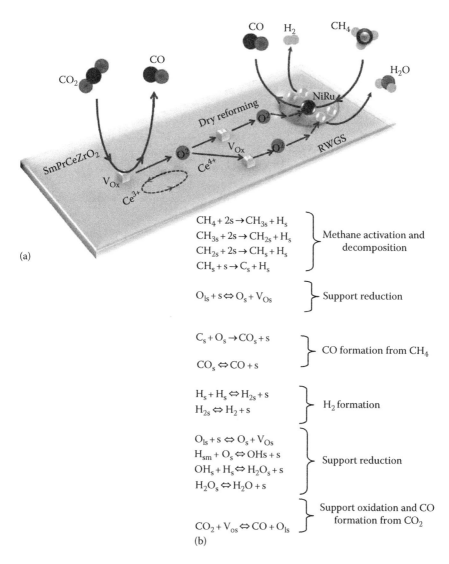

(a)

$$CH_4 + 2s \rightarrow CH_{3s} + H_s$$
$$CH_{3s} + 2s \rightarrow CH_{2s} + H_s$$
$$CH_{2s} + 2s \rightarrow CH_s + H_s$$
$$CH_s + s \rightarrow C_s + H_s$$

Methane activation and decomposition

$$O_{ls} + s \Leftrightarrow O_s + V_{Os}$$

Support reduction

$$C_s + O_s \rightarrow CO_s + s$$
$$CO_s \Leftrightarrow CO + s$$

CO formation from CH_4

$$H_s + H_s \Leftrightarrow H_{2s} + s$$
$$H_{2s} \Leftrightarrow H_2 + s$$

H_2 formation

$$O_{ls} + s \Leftrightarrow O_s + V_{Os}$$
$$H_{sm} + O_s \Leftrightarrow OHs + s$$
$$OH_s + H_s \Leftrightarrow H_2O_s + s$$
$$H_2O_s \Leftrightarrow H_2O + s$$

Support reduction

$$CO_2 + V_{Os} \Leftrightarrow CO + O_{ls}$$

Support oxidation and CO formation from CO_2

(b)

FIGURE 10.37 (a) Simplified scheme of methane dry reforming and (b) sequence of elementary steps over Ni–Ru–Sm$_{0.15}$Pr$_{0.15}$Ce$_{0.35}$Zr$_{0.35}$O$_2$ catalyst ($\rightarrow$ – irreversible step; $\Leftrightarrow$ – reversible steps; s – surface sites; O$_{ls}$ – lattice oxygen atoms; V$_{Os}$ – lattice oxygen vacancy).

the literature findings [90,100], it was suggested that CO produced from methane originates from the removal of carbon that is deposited on the metal particle during CH_4 decomposition. The carbon originating from methane partially reduces the oxide support near the perimeter of the metal particle, thus creating oxygen vacancies in the support. The oxygen formed during CO_2 dissociation replenishes of the oxygen in the lattice, thus providing a redox mechanism. The long-term activity of the composite catalyst would depend upon the balance between the ability for CO_2 dissociation, oxygen exchange, and removal of carbon from the metal surface.

10.4 KINETIC ASSESSMENT OF DRY REFORMING OF METHANE ON THE NANOCOMPOSITE CATALYSTS ON THE BASE OF FLUORITE-LIKE COMPLEX OXIDES

The complex MDR reaction mechanism is often simplified to a mechanistic picture with a unifying kinetic treatment involving CH_4–CO_2, CH_4–H_2O, and CH_4 decomposition, as well as the water–gas shift reaction. CH_4 reactions are limited by C–H bond activation and unaffected by the concentration of coreactants or by the presence of the reaction product [101]. CH_4–CO_2 reaction rates can simply be described by a first-order dependence of CH_4 and a zero-order dependence in CO_2:

$$r_f = k \cdot P_{CH_4} \tag{10.6}$$

Equation 10.6 is consistent with CH_4 activation on metal surfaces as the sole kinetically relevant elementary step and with fast steps involving recombinative hydrogen desorption to form H_2 and reactions of CO_2 with CH_4-derived chemisorbed species to form CO.

10.4.1 METHANE DRY REFORMING OVER Ni/CeO$_2$–ZrO$_2$ CATALYSTS

10.4.1.1 Kinetic Modeling with Power-Law Kinetic Model

Experimental data for kinetic modeling were taken from the literature [102]. The data were collected at atmospheric pressure, at 873, 923, and 973 K, in a plug-flow reactor on 50 mg of Ni/CeO$_2$–ZrO$_2$ catalyst. The data were collected after 3 h time on-stream to ensure stable catalyst performance at the considered operating conditions.

A generalized power-law model (10.7), in which CH_4 and CO_2 partial reaction orders could deviate from 1 and 0, was used to describe the MDR rate over Ni/CeO$_2$–ZrO$_2$ catalysts:

$$r = k \cdot P_{CH_4}^n \cdot P_{CO_2}^m \tag{10.7}$$

Measured reaction rates were corrected for the approach to equilibrium using thermodynamic data and prevalent pressure of reactants and products to give forward rates for CH_4–CO_2 reaction:

$$\eta = \frac{P_{CO}^2 P_{H_2}^2}{P_{CH_4} P_{CO_2}} \frac{1}{K_{eq}} \tag{10.8}$$

$$r = k_0 e^{-E/RT} \cdot P_{CH_4}^n \cdot P_{CO_2}^m \cdot (1 - \eta) \tag{10.9}$$

The power-law model is able to predict the reaction rate satisfactorily with an F value amounting to 205. Simplicity is the major advantage of the power-law model.

TABLE 10.5
Kinetic Parameter Values Obtained from Model Regression to Experimental Data

Parameter	Estimate	Unit
k_o	4.95×10^8	$\mu mol_{CH_4}/(s\,Pa\,kg_{cat})$
E	135.56	kJ/mol

Despite the anticipated generality of the reaction orders, a first-order dependence on CH_4 and zero-order dependence on CO_2 were found to best describe the experimental data. Once the reverse reaction rate is accounted for through Equation 10.9, this expression is capable of describing MDR rates at all temperatures. However, the power-law model might not be meaningful for predicting net rates of carbon monoxide and carbon dioxide formation.

The catalytic performance of catalysts in the carbon dioxide reaction shows that the reverse water–gas shift reaction was rather fast and affects the H_2/CO ratio in the syngas production. RWGS is the most common side reaction in CO_2 reforming of CH_4. It is important to take its occurrence into account. Activated CO_2 can also react with the surface hydrogen species that originated from the decomposition of CH_4 and was assumed at equilibrium with the gaseous H_2. The occurrence of RWGS results in the H_2 consumption and CO formation, which is confirmed by CO/H_2 ratios exceeding unity.

After including the RWGS reaction and assuming it to be at thermodynamic equilibrium, the power-law model is able to reproduce the reaction rate slightly better with an F value for the global significance of the regression amounting to 381. Kinetic parameter values obtained from the model regression are shown in Table 10.5.

10.4.1.2 Stepwise Kinetic Modeling

In the literature, numerous examples of Langmuir–Hinshelwood type of kinetic models are used to simulate MDR kinetics [3,94,103]. In the Langmuir–Hinshelwood kinetic model (10.10), the reversible adsorption and dissociation of CH_4 was assumed to be the initial step. Methane is supposed to adsorb and dissociate on metal ensembles, leading to H_2 formation as well as to C species. It was supposed that the CH_4 adsorption was at equilibrium while CH_4 dissociation was a slow, c.q., rate-determining, step:

$$r_{CH_4} = \frac{k_1 P_{CH_4}}{\left(1 + K_{CH_4} P_{CH_4} + \sqrt{K_{H_2} P_{H_2}} + K_{CO} P_{CO}\right)^2} \left(1 - \frac{P_{CO_2}^2 P_{H_2}^2}{P_{CH_4} P_{CO_2}} \frac{1}{K_{eq}}\right) \quad (10.10)$$

A comparison of the experimental CH_4 and CO_2 consumption rates and CO and H_2 production over the Ni/CeO$_2$–ZrO catalyst with those calculated by the Langmuir–Hinshelwood type of kinetic model (10.10) showed that such model extensions could not improve the modeling results.

10.4.2 TRANSIENT MODELING OF MDR OVER Pt/PrSmCeZrO COMPLEX OXIDE COMPOSITE

A more detailed analysis of the MDR mechanism over nanocomposite catalysts shows that both metal and support sites could be involved. A schematic representation of the near-surface layer in the Pt/PrSmCeZrO complex oxide composite catalyst is shown in Figure 10.38.

According to the kinetic scheme (10.11) proposed for the Pt/PrSmCeZrO complex oxide composite, methane transformation occurs on the active cationic Pt centers, that is, [PtO] and [PtCO$_3$], whereas carbon dioxide interacts with [PtO] and oxygen vacancies [V$_s$] of the support. Hence, the MDR reaction mechanism can be assumed to comprise the following steps to simulate CO, H$_2$, and H$_2$O formation, accounting for the main reaction routes (10.12) [78]:

$$
\begin{aligned}
&1. \quad CO_2 + [PtO] \leftrightarrow [PtCO_3] \\
&2. \quad CH_4 + [PtCO_3] \rightarrow 2CO + 2H_2 + [PtO] \\
&3. \quad CH_4 + [PtO] \rightarrow CO + 2H_2 + [Pt] \\
&4. \quad [Pt] + [O_s] \leftrightarrow [PtO] + [V_s] \\
&5. \quad CO_2 + [V_s] \rightarrow CO + [O_s] \\
&6. \quad H_2 + [PtO] \rightarrow H_2O + [Pt]
\end{aligned}
\qquad (10.11)
$$

where
[PtO] and [Pt] denote the oxidized and vacant Pt centers
[PtCO$_3$] is the carbonate complex
[O$_s$] and [V$_s$] are the oxidized and vacant sites inside the lattice layer

So the catalyst provides the following routes for MDR to occur:

$$
\begin{aligned}
&1. \quad CH_4 + CO_2 \leftrightarrow 2CO + H_2 \\
&2. \quad CO_2 + H_2 \leftrightarrow CO + H_2O
\end{aligned}
\qquad (10.12)
$$

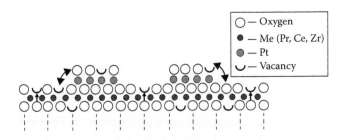

○ — Oxygen	
● — Me (Pr, Ce, Zr)	
◉ — Pt	
∪ — Vacancy	

FIGURE 10.38 Scheme of the near-surface layer of Pt/PrCeZrO catalyst.

According to the mass conservation law, the rates of catalytic steps can be calculated as follows:

$$r_1 = k_1 c_{CO_2} \theta_1, \quad r_{-1} = k_{-1} \theta_2$$
$$r_2 = k_2 c_{CH_4} \theta_2$$
$$r_3 = k_3 c_{CH_4} \theta_1$$
$$r_4 = k_4 (1 - \theta_1 - \theta_2) \theta_3, \quad r_{-4} = k_{-4} \theta_1 (1 - \theta_3) \tag{10.13}$$
$$r_5 = k_5 c_{CO_2} (1 - \theta_3)$$
$$r_6 = k_6 c_{H_2} \theta_1$$

where

r_i is the rate of the ith catalytic step ($i = 1, 2, ..., 6$)

c_{CH_4}, c_{CO_2}, and c_{H_2} are the CH_4, CO_2, and H_2 concentrations in the gas mixture

θ_1, θ_2, and θ_3 are the relative surface concentrations of oxidized Pt centers [PtO], carbonate complexes [PtCO$_3$], and oxidized lattice layer sites [O$_s$], respectively

k_i is the rate coefficient of the ith catalytic step ($i = 1, 2, ..., 6$)

It is assumed that the total number of active Pt centers is equal to

$$[ZO] + [ZCO_3] + [Z] = \alpha N_\theta \tag{10.14}$$

and the number of active lattice layer sites available for spillover is equal to

$$[ZO^*] + [Z^*] = \beta N_\theta \tag{10.15}$$

where

N_θ is the total number of active centers on the catalyst surface

α is the relative surface concentration of Pt centers

β is the ratio of the number of active lattice layer sites under the surface to the total number of active surface centers

Since Pt^{4+} cations are easily transformed into Pt^{2+} even after catalyst purging by He at high temperatures after pretreatment in O_2, and reoxidation of Pt^{2+} cations to Pt^{4+} state by oxygen atoms migrating from support is apparently impossible under reaction conditions, Pt^{4+} cationic species revealed by XPS data were not considered separately in the suggested reaction scheme.

The following equations are included in the mathematical model [78]:

- The first-order partial differential equations of the mass balance of the reagents and products in the gas phase, that is, CO_2, CH_4, CO, and H_2, which reflect both the convective term due to the gas flow along the catalyst fragment length and the term related to the chemical transformations of the reactive mixture components on the catalyst surface at every cross section of the channel.
- The second-order parabolic-type equation at each axial position describing the unsteady oxygen transport from the near-surface layers of the catalyst lattice toward the active Pt centers on the catalyst surface where the catalytic reaction proceeds (the diffusion mechanism is assumed).

- Two ordinary differential equations describing the time behavior of the concentrations of oxidized Pt centers and carbonate complexes on the catalyst surface at each axial cross-section point; the dynamics depends on the local rates of catalytic transformations at the surface and of oxygen diffusion through the catalyst lattice.

Thus, the developed mathematical model is formulated as an initial-boundary value problem for a system of differential equations of different type. The proposed algorithm is based on a second-order finite-difference approximation with respect to two spatial coordinates, that is, the length of the catalyst fragment, the distance between a point inside the catalyst lattice and the catalyst surface, and time. At each time step, the method of alternating directions is used to construct the numerical solution of the discrete approximation equations.

A number of computational transient runs have been performed with the process parameters corresponding to the experimental response curves obtained with the catalyst pretreated in O_2 stream: gas feed rate $u = 10.6$ and 18 L/h, temperature $T = 750°C$, inlet concentrations of both CH_4 and CO_2 are 7%, and contact time $\tau = 4.7$ and 8 ms. It has been assumed in these computations that the total number of active centers on the catalyst surface is equal to the maximum monolayer coverage $N_\theta = 1.28 \times 10^{15}$ at O/cm^2. Before reaction mixture feeding, at time $t = 0$, all active centers are oxidized, that is, $\theta_1 = \theta_2 = 1$.

For verification of the proposed kinetic scheme (10.11), the computational runs of transient regimes have been performed on the basis of mathematical model with the simple redox kinetic scheme that includes only three steps, that is, steps 3–5. The carbonate intermediates formation and consumption rates as well as the RWGS rate are assumed to equal zero. Modeling has revealed that this scheme does not describe the measured transients. Hence, a route should exist providing an efficient CO_2 transformation on the oxidized surface. In the computations accounting for the overall kinetic scheme (10.11), the initial approximation for model parameters has been estimated from the steady-state experimental data and the data of XPS, FTIRS of adsorbed CO, and oxygen isotope exchange [82,104–107].

Simulation results revealed that in the investigated range of operating conditions, the relaxation time for achieving steady-state surface coverages for species involved in catalytic cycle could not exceed a few seconds, which agrees with the simple analytic estimations following Temkin's approach [108]. Analyzing the transient curves up to 300 s, it has been convincingly demonstrated that

- The character and shape of response curves are mainly defined by carbonate formation and consumption rates, the rates of *spillover*, and the subsequent interaction of CO_2 with vacant centers on the catalyst surface
- The transients are determined by parameters characterizing the catalyst structure and the lattice oxygen mobility, such as the catalyst-specific surface S_b, the quantity of active centers on the catalyst surface N_θ, the fractions of the surface Pt centers α and the lattice layer sites β, and the rate and characteristic length of the oxygen bulk diffusion

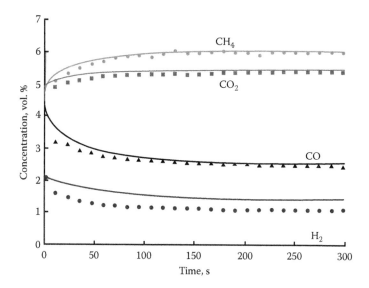

FIGURE 10.39 Comparison of the experimental (points) and computed (lines) time dependences of concentrations in transient experiments. 750°C, gas feed rate 18 L/h, feed 7%
CH_4 + 7% CO_2 in He, contact time 4.7 ms. The values of the model parameters: $\tau = 4.7$ ms,
$k_1 = 200$ s^{-1}, $k_{-1} = 0.6$ s^{-1}, $k_2 = 600$ s^{-1}, $k_3 = 60$ s^{-1}, $k_4 = 300$ s^{-1}, $k_{-4} = 5$ s^{-1}, $k_5 = 55$ s^{-1}, $k_6 = 1100$ s^{-1}, $\alpha = 0.04$, $\beta = 0.2$, $S_b = 20 \times 10^4$ cm^{-1}, $D = 2.5 \times 10^{-13}$ cm^2/s, $H = 5 \times 10^{-6}$ cm.

Figure 10.39 demonstrates that the suggested kinetic scheme reflects the main pecu-
liarities of the catalyst behavior rather well. The experimental values of gas concentrations and the modeling data are in a good agreement.

Figure 10.40 shows the relative surface concentrations of Pt centers occupied by
oxygen and carbonates. After feeding the reaction mixture, the number of oxidized

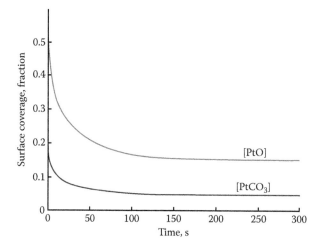

FIGURE 10.40 Time dependence of the surface coverage fraction with intermediates [PtO]
and [PtCO$_3$].

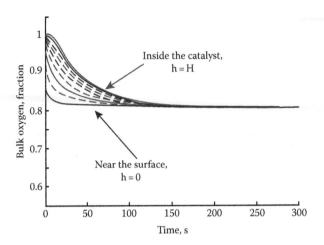

FIGURE 10.41   Time dependence of the bulk oxygen fraction.

Pt sites quickly decreases due to high CH_4 and CO_2 adsorption rates, and hence, the surface site coverage by carbonate species has a sharp peak during a few seconds. Due to the interaction with the oxygen surface species, intermediate concentrations of such as those of [PtO] and [PtCO$_3$] slowly decrease toward the steady-state values (see Figure 10.41). The bulk oxygen concentration follows a similar trend, reaching a steady-state level within ~300 s (see also Figure 10.41). Hence, it could be concluded that the transients caused by the catalyst reduction in the reaction feed occur within ~300 s. Transients on a much longer timescale, that is, up to 3000 s, were also experimentally observed and could be caused by slower processes such as Pt aggregation, generally leading to a decreased Pt–support interface/interaction and ordering of Pt clusters. Temperature-programmed oxidation by oxygen after reaction of this sample as well as TEM studies revealed that no carbon had been deposited.

In the absence of gas-phase oxygen, PtO sites can be regenerated only by the reactive oxygen species migrating from the support (step 4). According to (10.11), direct reoxidation of Pt sites by CO_2 is disabled. Apart from pure kinetic aspect, it is possible that carbonates can be stabilized only by Pt cations having reducible cations such as Ce or Pr as neighbors. Decoration of Pt cluster surfaces by Pr or Ce oxidic species could also stabilize such carbonates coordinated to Pt cations. The routes for dry reforming and RWGS corresponding to steps 3–6 also involve PtO sites. However, these steps could not reliably be differentiated, and hence, they are included in the Pt active site balance [ZO] + [ZCO$_3$] + [Z] = αN_0, in Equation 10.14.

It was revealed that for the first seconds of contact with the reaction mixture, the CH_4 consumption rate over the oxidized catalyst was considerably affected by the transient behavior of the reaction mechanism (10.11): the ratio of the rates of steps 2 and 3 was about ~3. The decrease of the oxidation degree of the surface resulted in lowering the fraction of Pt centers occupied by carbonates, and the ratio of step rates 2 and 3 decreased to 1.6.

10.4.3   MDR over 1.2% Pt/80% $Pr_{0.15}Sm_{0.15}Ce_{0.35}Zr_{0.35}O_x$ + 10% NiO + 10% YSZ Catalyst

In this investigation, a 1.2% Pt/80% $Pr_{0.15}Sm_{0.15}Ce_{0.35}Zr_{0.35}O_x$ + 10% NiO + 10% YSZ catalytic material (Pt + Ni/PrSmCeZrO/YSZ) [86] was tested at 650°C–850°C and 1 atm total pressure. Diluted by an inert gas (N_2), methane and CO_2 were used in the kinetic experiments. The CH_4 and CO_2 concentrations were changed over a wide range.

The Pt + Ni/PrSmCeZrO/YSZ catalyst was benchmarked against a complex Pt-promoted fluorite-like doped ceria–zirconia oxide sample of similar atomic fractions of the corresponding elements, that is, 1.4% Pt/80% $Pr_{0.15}Sm_{0.15}Ce_{0.35}Zr_{0.35}O_x$, in MDR. The Ni–$Y_2O_3$-stabilized zirconia-containing catalyst has a higher syngas production compared to the Pt/PrSmCeZrO sample.

At steady state, the Pt + Ni/PrSmCeZrO/YSZ material (first catalyst) exhibits considerably higher methane conversions at all the temperatures as compared to the Pt/PrSmCeZrO sample (second catalyst). The greater H_2/CO ratio in the synthesis gas is also detected in case of the composite with doped NiO and YSZ, while only a small difference is observed in the CO_2 conversions on the two catalysts (see Figure 10.42).

10.4.3.1   Analysis of Kinetic Effects

10.4.3.1.1   Effect of Residence Time

Figure 10.43 represents the catalytic activity of the Pt + Ni/PrSmCeZrO/YSZ catalyst with respect to the residence time over the sample. As clearly shown in Figure 10.43, increasing the residence time increases the reactant conversion, hydrogen and CO yields, as well as the H_2/CO molar ratio in the produced syngas. It can be seen that under operating conditions with the selected reaction temperature of 850°C and

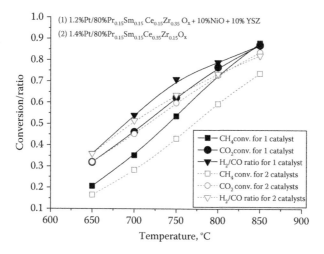

FIGURE 10.42   Effect of the temperature on the CH_4 and CO_2 conversions, the H_2/CO ratio for the compared catalysts: (1) 1.2% Pt/80% $Pr_{0.15}Sm_{0.15}Ce_{0.35}Zr_{0.35}O_x$ + 10% NiO + 10% YSZ (density of 1.45 g/cm²); (2) 1.4% Pt/80% $Pr_{0.15}Sm_{0.15}Ce_{0.35}Zr_{0.35}O_x$ (density of 1.94 g/cm²).

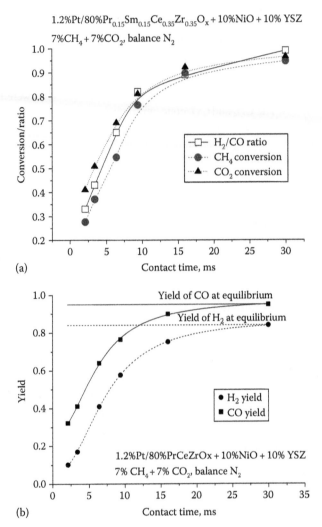

FIGURE 10.43 Effect of contact time on conversion and H_2/CO ratio (a) and yield of the products (H_2 and CO) at 850°C (b).

low residence time, that is, ≤30 ms, the yield of the main products is far from equilibrium. Highest reactant conversion and equilibrium product yields were achieved at 30 ms residence time (see Figure 10.43).

10.4.3.1.2   *Effects of Variation of the CO_2 and CH_4 Feeding Concentrations*

The effect of CO_2 and CH_4 concentration on the ratio of the components in the product gas at the temperatures of 650°C and 850°C was investigated by their variation between 3.5 and 17.5 vol.%, while keeping the other one constant at about 7 vol.% (nitrogen—balance). The results are shown in Figure 10.44.

As shown in Figure 10.44a, the H_2/CO ratio in the product gas increases when the CH_4 concentration, that is, the source of hydrogen atoms, increases in the feed.

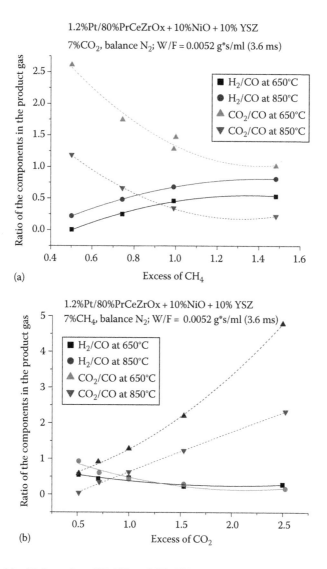

FIGURE 10.44   Molar ratios of H_2/CO and CO_2/CO in the product gas depending on excess of (a) CH_4 and (b) CO_2 over stoichiometric values in the feed.

For the specified temperatures, they seem to change with the CH_4 excess at the similar rate. Conversely, the effect of increasing CO_2/CH_4 ratio from 0.5 to 2.5 in the feed (see Figure 10.44b) is negative for synthesis gas selectivity: the H_2/CO ratio slightly decreases with the CO_2 excess in the feed. Whenever the temperature is higher, the H_2/CO ratio decreases more intensively. At 850°C, the H_2/CO ratio at $CO_2/CH_4 \geq 1.5$ becomes even lower when compared to the corresponding experimental results for 650°C. The decrease in the H_2/CO ratio in syngas is presumably attributed to the side reverse water–gas shift reaction in which H_2 produced reacts with CO_2 to form water and CO.

10.4.3.1.3 Effect of the Reaction Temperature

The temperature effect was investigated at atmospheric pressure in the range from 650°C to 850°C and the reactant gas composition with 7% CH_4, 7% CO_2, and nitrogen as balance. Both CH_4 and CO_2 reaction rates were found to increase monotonically with the temperature. Higher temperatures were found to result in higher H_2/CO and lower CO_2/CO ratios. Hence, more synthesis gas is produced at higher temperatures, which was more pronounced at higher residence times. A syngas concentration of 22.7% (v/v) with the H_2/CO ratio of 0.975 was obtained at 750°C (77% H_2 yield and 90% CO yield), 23.7% (v/v) with 0.982 at 800°C (83.5% H_2 yield and 93.7% CO yield), and 23.7% (v/v) with 0.972 at 850°C at a residence time of 30 ms ($F = 5.64$ L/h, $W = 68$ mg). Simultaneously, the corresponding values of CO_2/CO ratios were 0.04, 0.03, and 0.02.

10.4.3.2 Steady-State Kinetic Modeling

There is no generally accepted kinetic model available for CO_2 reforming of CH_4. A Langmuir–Hinshelwood type of model has been most often used in the kinetic studies of CO_2 reforming of CH_4 [3,95,109–117].

For the Pt + Ni/PrSmCeZrO/YSZ composite catalyst, a kind of Pt–Ni interactions certainly exists, which is responsible for variations in the dissociative adsorption capacity of methane and carbon dioxide. It would be interesting to analyze the kinetic scheme (10.11) to deduce a kinetic equation for the rate of CH_4 transformation, when the reaction occurs under steady-state conditions. For the composite catalyst, the $[O_s]$ intermediate concentration on the catalyst surface is rather high and does not vary much. The adsorption microcalorimetry and TAP data show that the surface coverage by carbonates is rather low and that they are reversibly adsorbed on the surface sites. These results allow to take some assumptions and to linearize the reaction network. In this case, the variable θ_3 is assumed to be practically constant, and the impact rates r_4 and r_{-4} are minor, such that θ_3 could be expressed as function $f(CO_2)$ at steady conditions.

To obtain the first approximation for the kinetic equation under steady-state conditions, the following simplifications could be used for the rates of catalytic steps also: (1) $[PtCO_3]$ concentration on the surface is significantly lower than the surface concentration of oxidized Pt centers $[PtO]$, (2) the rate of step 4 in the forward direction r_4 is significantly lower than in the reverse direction r_{-4}, and (3) $[PtO]$ concentration varies negligibly in step 6.

Under steady-state conditions, the time derivatives of θ_1 and θ_2 are equal to zero, and the concentration of intermediates is determined by reagent concentrations in the gas phase. For the assumptions given earlier, θ_1 value is defined by steps 3 and 4, and θ_2 value is expressed from the rates of steps 1 and 2. Preliminary simplified forms of kinetic equations for the rates of CH_4 transformation look as follows:

$$r_2 = \frac{k_1 k_2 k_4 c_{CH_4} c_{CO_2}^2}{\left(k_{-1} + k_2 c_{CH_4}\right)\left(k_3 c_{CH_4} + k_4 c_{CO_2}\right)} = \frac{k_1 k_2 k_4 c_{CH_4} c_{CO_2}}{\left(k_{-1} + k_2 c_{CH_4}\right)\left(k_4 + k_3 (c_{CH_4}/c_{CO_2})\right)}$$

$$r_3 = \frac{k_3 k_4 c_{CH_4} c_{CO_2}}{\left(k_3 c_{CH_4} + k_4 c_{CO_2}\right)}$$

(10.16)

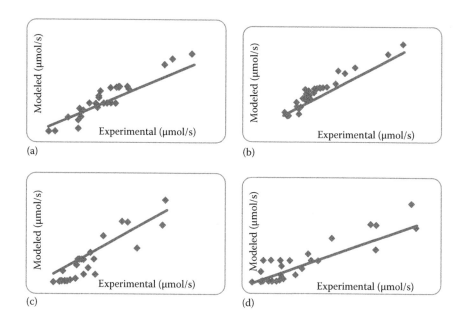

FIGURE 10.45 Parity diagrams of CO_2 reforming of CH_4: (a) methane, (b) carbon dioxide, (c) carbon monoxide, and (d) hydrogen.

A regression analysis of the previously discussed equations against a data set consisting of 36 measurements has revealed that the dry reforming rates are practically independent of the CO_2 partial pressure in the feed (see Figure 10.45). And again, a first-order rate equation in CH_4 and zero order in CO_2 results in the best agreement between model simulations and experimental data with an F value for the global significance of the regression amounting to 112 (see parity diagrams in Figure 10.45).

It is an indication that most of the methane is consumed via reaction 3 from the aforementioned reported scheme (10.11), since the corresponding reaction rate equation is reduced to a proportional relationship with the methane partial pressure on the condition that $k_4 \gg k_3$. Activation energy of 114 kJ/mol is in agreement with typically reported values for dry reforming [3,93,94,103] (see Table 10.6).

TABLE 10.6
Kinetic Parameter Values Obtained from Model Regression to Experimental Data (T_{ave} = 950 K)

Parameter	Estimate	Unit
$k_{T_{ave}}$	2.80×10^2	$\mu mol_{CH_4}/(s\,Pa\,kg_{cat})$
E	114	kJ/mol

10.5 STRUCTURED CATALYSTS

Since for any practical application under investigation, structured catalysts are well known to present several advantages [118–124], nanocomposite active components were successfully deposited on ceramic honeycombs, microchannel cermets, or metallic substrates (see Table 10.7) and tested in realistic feeds at lab-scale and pilot-scale levels using specifically designed reactors and installations allowing to broadly tune the operational parameters.

At some excess of CO_2 in dry reforming of NG containing up to 5% C_2–C_4 alkanes and some admixture of sulfur compounds, stable and comparable performance was achieved for structured catalyst 1, that is, active component based on Ni + Ru/LaPrMnCr–YSZ nanocomposite, and 7, that is, active component based on Ni + Ru/fluorite-like oxide support, on microchannel substrates (see Table 10.8 and Figures 10.46 and 10.47).

TABLE 10.7
Characteristics of Substrates

No.	Chemical Composition, wt.%	Density, g/cm³	Pore Size/ Porosity	Heat Conductivity, W/(m·K)	Shape and Size, mm
1	Fechraloy microchannel plate/5 µm corundum sublayer	7.3	n/a		20/50/1.5, one side flat, one side with 15 channels of 0.4 mm depth
2	Fechraloy gauze with 5 µm corundum sublayer	n/a	n/a		Squares 35 × 35, woven of wires of 0.2 mm diameter with 0.2 mm spacing
3	Ni–Al alloy, Al 10%	2.5	60 ppi/65.6%	9.8	35 × 35 × 1
4	SiC 40%; Al_2O_3–SiO_2 rest	0.5	30 ppi/75.5%	3.5	38 × 38 × 5
5	α-Al_2O_3 45%, Al_2O_3–SiO_2 rest	0.6	30 ppi/78.6%	1.7	38 × 38 × 5
6	Ni–Al alloy, Al 10%	0.36	30 ppi/95.0%	5.5	38 × 38 × 10

TABLE 10.8
Basic Types of Structured Catalysts

	Sample Description	
No.	Type of Substrate	Active Component Composition and Loading, wt.%
1	Five stacked plates No. 1,	4% ($La_{0.8}Pr_{0.2}Mn_{0.2}Cr_{0.8}O_3$ + 10% NiO + 10% YSZ) +2% Ru
2	Six plates No. 3 + 6 gauzes No. 2	4% ($La_{0.8}Pr_{0.2}Mn_{0.2}Cr_{0.8}O_3$ + 10% NiO + 10% YSZ) +2% Ru
3	One plate No. 6	10% (10 wt.% $LaNi_{0.95}Ru_{0.05}O_3$/Mg-doped alumina)
4	Two plates No. 4	10% (10 wt.% $LaNi_{0.95}Ru_{0.05}O_3$/Mg-doped alumina)
5	Two plates No. 5	10% (10 wt.% $LaNi_{0.95}Ru_{0.05}O_3$/Mg-doped alumina)
6	Two plates No. 4	4% ($La_{0.8}Pr_{0.2}Mn_{0.2}Cr_{0.8}O_3$ + 10% NiO + 10% YSZ) + 2% Ru
7	Five stacked plates No. 1	4% SmPrCeZrO + 1% NiO + 1% Ru

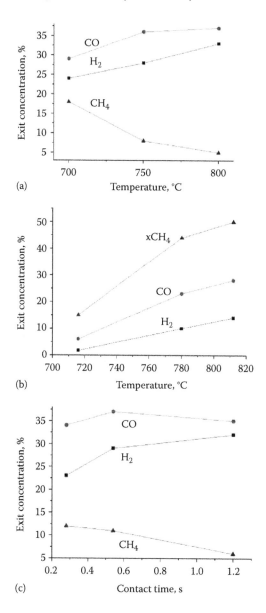

FIGURE 10.46 Characteristics of NG dry reforming process for the package of microchannel plates (catalyst 1, Table 10.8) in feed 50% CO_2 + 40% NG +N_2. Contact times at (a) 1.2 s and (b) 0.1 s, and (c) temperature of 800°C.

The values of main process parameters for the structured catalysts tested in the NG reforming are shown in Table 10.9.

For a stack of gauzes and high-density Ni–Al foam plates (catalyst 2, Table 10.8), the NG dry reforming performance is quite close to that of catalyst 1. Apparently, these types of structured substrates also provide quite efficient heat and mass transfer,

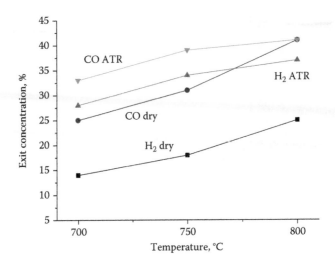

FIGURE 10.47 Characteristics of NG dry or ATR reforming process for the package of microchannel plates (catalyst 7, Table 10.8) in feed 47% CO_2 + 44% NG + N_2 (dry) or 41% CO_2 + 44% NG + 9% O_2 (ATR). Contact time at 0.5 s (dry) or 0.29 s (ATR).

TABLE 10.9

Main Parameters of NG Reforming on Structured Catalysts

Type of Structured Catalyst	T, °C	Contact Time, s	Feed, mol.%	Concentration of Products, vol.% (Exit/Equilibrium)			
				H_2	CO	CO_2	CH_4
1	800	1.2	50% CO_2 + 40% NG + N_2	33/43	35/47	12/3	5/3
2	800	0.9	50% CO_2 + 40% NG + N_2	18	33	16	10
2	800	0.5	45% NG + 45% CO_2 + 8% O_2 + N_2	18	25		
2	800	0.3	30 CO_2 + 31 NG + 3O_2 + 37 H_2O	21	7	26	20
2	900	0.3	The same	40	27	17	8
3	850	1.9	56% CO_2 + 44% NG + N_2	14	30		
4	850	1.9	56% CO_2 + 44% NG + N_2	22	40		
5	850	1.9	56% CO_2 + 44% NG + N_2	12	26		
6	800	3.8	50% CO_2 + 40% NG + N_2	16	31	25	14

---

which agrees with our results on steam reforming of NG for catalysts on these substrates [85,123,124]. Addition of oxygen to the feed improves syngas yield at shorter contact times (see Figure 10.47 and Table 10.9), while addition of water increases hydrogen yield (see Table 10.9).

For both types of active components supported on the low-density foams, a reasonable performance is achieved only at much longer contact times (see Table 10.9), catalysts 3–6, which apparently can be explained by lower values of the active surface per the unit of volume. For the same active component 2, the lowest syngas yield is observed for a pure ceramic foam as substrate (structured catalyst 5). Hence, in the

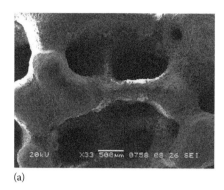

(a)   (b)

FIGURE 10.48 Typical SEM images of supported LaNiRuO/alumina layers on alumina/Al–Si–O foam substrate after testing in natural gas dry reforming in concentrated feeds. (a) general view, (b) a porous layer of supported active component on the walls of foam cells.

case of foam substrates with low density, the heat transfer from the reactor wall into the catalyst package could affect the catalyst performance in strongly endothermic reaction of NG dry reforming.

To assess the long-term stability of the catalyst's performance with respect to coking, the catalysts were contacted with reaction feed at selected experimental conditions after a standard pretreatment in O_2 at 700°C and kept for 6–8 h/day followed by reactor purging with Ar stream and cooling to the room temperature. The testing was resumed the next day by heating the catalyst in an Ar stream to the operating temperature followed by switching to the reaction feed stream. The long-term test of the structured catalysts has confirmed that these nanocomposite active components can also retain their high activity and coking/sintering stability when supported on much bigger substrates (microchannel/foam plates, gauzes). Apparently, the development of a given catalytic process has always been closely related to optimization of both specific active component and structured substrates to provide reliable and stable performance.

SEM data for catalysts on foam substrates discharged after reaction (Figure 10.48) demonstrate absence of coke deposits/fibers and strong adherence of supported layers to substrates without cracks and spallation.

10.6 CONCLUSION

The most efficient approach to the design of efficient nanocomposite catalysts for NG reforming into syngas that are stable to coking consists of (1) the application of complex perovskite precursors containing Ni, Ru, and transition metal cations with broadly varying redox properties as catalyst precursors and (2) using complex fluorite-like doped ceria–zirconia oxides as supports for Ni + Ru nanoparticles. Fast transfer of oxygen species from the oxide sites to the metal–oxide interface where they transform activated CH_xO_y molecules into syngas ensures the operation of a bifunctional redox mechanism for NG reforming. High catalytic activity and coking stability is provided by tuning the oxygen mobility and reactivity in complex oxide supports (via changing their chemical composition) as well as by optimizing the size and composition of metal alloy nanoparticles and nature of starting salts. Strong metal–oxide–support interaction provides also stability of metal nanoparticles to sintering in real operation conditions.

The main features of the MDR kinetics investigated for such systems revealed its specificity caused by strong metal–support interaction, especially in unsteady-state conditions or in feeds with excess of CO_2.

Best compositions of nanocomposite active components supported as porous layers on heat-conducting substrates demonstrated a high and stable performance in dry reforming of real NG in concentrated feeds.

ACKNOWLEDGMENTS

The authors gratefully acknowledge support from FP7 OCMOL Project, Integration Project 8 of SB RAS-NAN Belarus, Project 50 of Presidium RAS, Russian Foundation of Basic Research (RFBR) projects RFBR-CNRS 09-03-93112 and 12-03-93115, Federal Program *Scientific and Educational Cadres of Russia*.

REFERENCES

1. Graves, C., Ebbesen, S.D., Mogensen, M., and Lackner, K.S., Sustainable hydrocarbon fuels by recycling CO_2 and H_2O with renewable or nuclear energy, *Renewable and Sustainable Energy Reviews*, 7, 1–23, 2011.
2. Vlachos, D.G. and Caratzoulas, S., The roles of catalysis and reaction engineering in overcoming the energy and the environment crisis, *Chemical Engineering Science*, 65, 18–29, 2010.
3. Bradford, M.C.J. and Vannice, M.A., CO_2 reforming of CH_4, *Catalysis Reviews*, 41(1), 1–42, 1999.
4. Spivey, J.J. and Inui, T., Reforming of CH_4 by CO_2, O_2 and/or H_2O, in: James, J. S. (ed.), *Catalysis*, Vol. 16, The Royal Society of Chemistry, London, U.K., 2002, pp. 133–151.
5. Hu, Y.H. and Ruckenstein, E., Catalytic conversion of methane to synthesis gas by partial oxidation and CO_2 reforming, *Advances in Catalysis*, 48, 297–345, 2004.
6. York, A.P.E., Xiao, T., Green, M.L.H., and Claridge, J., Methane oxyforming for synthesis gas production, *Catalysis Reviews*, 49, 511–560, 2007.
7. Ma, J., Sun, N., Zhang, X., Zhao, N., Xiao, F., Wei, W., and Sun, Y., A short review of catalysis for CO_2 conversion, *Catalysis Today*, 148, 221–231, 2009.
8. Adesina, A.A., The role of CO_2 in hydrocarbon reforming catalysis: Friend or foe? *Current Opinion in Chemical Engineering*, 1, 272–280, 2012.
9. Demirbas, A., Progress and recent trends in biofuels, *Progress in Energy and Combustion Science*, 33, 1–18, 2007.
10. Sarkar, S. and Kumar, A., Large-scale biohydrogen production from bio-oil, *Bioresource Technology*, 101, 7350–7361, 2010.
11. Gupta, K.K., Rehman, A., and Sarviya, R.M., Bio-fuels for the gas turbine: A review, *Renewable and Sustainable Energy Reviews*, 14, 2946–2955, 2010.
12. Serrano-Ruiz, J.C. and Dumesic, J.A., Catalytic routes for the conversion of biomass into liquid hydrocarbon transportation fuels, *Energy & Environmental Science*, 4, 83–99, 2011.
13. Yung, M.M., Jablonski, W.S., and Magrini-Bair, K.A., Review of catalytic conditioning of biomass-derived syngas, *Energy & Fuels*, 23, 1874–1887, 2009.

14. Kırtay, E., Recent advances in production of hydrogen from biomass, *Energy Conversion and Management*, 52, 1778–1789, 2011.

15. Bradford, M.C.J. and Vannice, M.A., CO_2 reforming of CH_4 over supported Pt catalysts, *Journal of Catalysis*, 173, 157–171, 1998.

16. Gonzalez-Delacruz, V.M., Ternero, F., Pereniguez, R., Caballero, A., and Holgado, J.P., Study of nanostructured Ni/CeO_2 catalysts prepared by combustion synthesis in dry reforming of methane, *Applied Catalysis A: General*, 384, 1–9, 2010.

17. Daza, C.E., Kiennemann, A., Moreno, S., and Molina, R., Dry reforming of methane using Ni–Ce catalysts supported on a modified mineral clay, *Applied Catalysis A: General*, 364, 65–74, 2009.

18. Juan-Juan, J., Roman-Martınez, M.C., and Illan-Gomez, M.J., Effect of potassium content in the activity of K-promoted Ni/Al_2O_3 catalysts for the dry reforming of methane, *Applied Catalysis A: General*, 301, 9–15, 2006.

19. Effendi, A., Hellgardt, K., Zhang, Z.-G., and Yoshida, T., Characterisation of carbon deposits on Ni/SiO_2 in the reforming of CH_4–CO_2 using fixed- and fluidized-bed reactors, *Catalysis Communications*, 4, 203–207, 2003.

20. Yang, R., Xing, C., Lv, C., Shi, L., and Tsubaki, N., Promotional effect of La_2O_3 and CeO_2 on Ni/γ-Al_2O_3 catalysts for CO_2 reforming of CH_4, *Applied Catalysis A: General*, 385, 92–100, 2010.

21. Bellido, J.D.A. and Assaf, E.M., Effect of the Y_2O_3–ZrO_2 support composition on nickel catalyst evaluated in dry reforming of methane, *Applied Catalysis A: General*, 352, 179–187, 2009.

22. Biswas, P. and Kunzru, D., Oxidative steam reforming of ethanol over Ni/CeO_2–ZrO_2 catalyst, *Chemical Engineering Journal*, 136, 41–49, 2008.

23. Haga, F., Nakajima, T., Miya, H., and Mishima, S., Catalytic properties of supported cobalt catalysts for steam reforming of ethanol, *Catalysis Letters*, 48, 223–227, 1997.

24. Rodrigues, C.P., da Silva, V.T., and Schmal, M., Partial oxidation of ethanol on Cu/Alumina/cordierite monolith, *Catalysis Communications*, 10, 1697–1701, 2009.

25. Resini, C., Delgado, M.C.H., Presto, S., Alemany, L.J., Riani, P., Marazza, R., Ramis, G., and Busca, G., Yttria-stabilized zirconia (YSZ) supported Ni–Co alloys (precursor of SOFC anodes) as catalysts for the steam reforming of ethanol, *International Journal of Hydrogen Energy*, 33, 3728–3735, 2008.

26. Velu, S., Satoh, N., Gopinath, C.S., and Suzuki, K., Oxidative reforming of bio-ethanol over CuNiZnAl mixed oxide catalysts for hydrogen production, *Catalysis Letters*, 82(1), 145–152, 2002.

27. Yaseneva, P., Pavlova, S., Sadykov, V., Alikina, G., Lukashevich, A., Rogov, V., Belochapkine, S., and Ross., J., Hydrogen production by steam reforming of methanol over Cu–$CeZrYO_x$-based catalysts, *Catalysis Today*, 137, 23–28, 2008.

28. Souza, M.M.V.M., Aranda, D.A.G., and Schmal, M., Reforming of methane with carbon dioxide over $Pt/ZrO_2/Al_2O_3$ catalysts, *Journal of Catalysis*, 204, 498–511, 2001.

29. Safariamin, M., Tidahy, L.H., Abi-Aad, E., Siffert, S., and Aboukais, A., Dry reforming of methane in the presence of ruthenium-based catalysts, *Comptes Rendus Chimie*, 12, 748–753, 2009.

30. Carrara, C., Roa, A., Cornaglia, L., Lombardo, E.A., Mateos-Pedrero, C., and Ruiz, P., Hydrogen production in membrane reactors using Rh catalysts on binary supports, *Catalysis Today*, 133–135, 344–350, 2008.

31. Chen, J., Yao, C., Zhao, Y., and Jia, P., Synthesis gas production from dry reforming of methane over $Ce_{0.75}Zr_{0.25}O_2$-supported Ru catalysts, *International Journal of Hydrogen Energy*, 35, 1630–1642, 2010.

32. Domine, M.E., Iojoiu, E.E., Davidian, T., Guilhaume, N., and Mirodatos, C., Hydrogen production from biomass-derived oil over monolithic Pt- and Rh-based catalysts using steam reforming and sequential cracking processes, *Catalysis Today*, 133–135, 565–573, 2008.

33. Sadykov, V., Mezentseva, N., Alikina, G., Bunina, R., Rogov, V., Krieger, T., Belochapkine, S., and Ross, J., Composite catalytic materials for steam reforming of methane and oxygenates: Combinatorial synthesis, characterization and performance, *Catalysis Today*, 145, 127–137, 2009.

34. Busca, G., Montanari, T., Resini, C., Ramis, G., and Costantino, U., Hydrogen from alcohols: IR and flow reactor studies, *Catalysis Today*, 143, 2–8, 2009.

35. Takanabe, K., Aika, K., Seshan, K., and Lefferts, L., Sustainable hydrogen from bio-oil—Steam reforming of acetic acid as a model oxygenate, *Journal of Catalysis*, 227, 101–108, 2004.

36. Vaidya, P.D. and Rodrigues, A.E, Insight into steam reforming of ethanol to produce hydrogen for fuel cells, *Chemical Engineering Journal*, 117, 39–49, 2006.

37. Kroll, V.C.H., Swaan, H.M., and Mirodatos, C., Methane reforming reaction with carbon dioxide over Ni/SiO$_2$ catalyst-I. Deactivation studies, *Journal of Catalysis*, 161(1), 409–422, 1996.

38. Khoshtinat Nikoo, M. and Amin, N.A.S., Thermodynamic analysis of carbon dioxide reforming of methane in view of solid carbon formation, *Fuel Processing Technology*, 92, 678–691, 2011.

39. Wang, S., Lu, G.Q. (Max), and Millar, G.J., Carbon dioxide reforming of methane to produce synthesis gas over metal-supported catalysts: State of the art, *Energy & Fuels*, 10, 896–904, 1996.

40. Topor, L., Bejan, L., Ivana, E., and Georgescu, N., Formarea negrului de fum in reactia de conversie a metanului cu CO, *Revista de Chimie (Bucharest)*, 30, 539–541, 1979.

41. Kendall, K., Finnerty, C.M., Saunders, G., and Chung, J.T., Effects of dilution on methane entering an SOFC anode, *Journal of Power Sources*, 106, 323–327, 2002.

42. Sadykov, V.A., Pavlova, S.N., Kharlamova, T.S., Muzykantov, V.S., Uvarov, N.F., Okhlupin, Yu.S., Ishchenko, A.V. et al., Perovskites and their nanocomposites with fluorite-like oxides as materials for solid oxide fuel cells cathodes and oxygen-conducting membranes: Mobility and reactivity of the surface/bulk oxygen as a key factor of their performance, in: Borowski, M. (ed.), *Perovskites: Structure, Properties and Uses*, Nova Science Publishers, Hauppauge, NY, 2010, pp. 67–178.

43. Goldwasser, M.R., Rivas, M.E., Pietri, E., Pérez-Zurita, M.J., Cubeiro, M.L., Grivobal-Constant, A., and Leclercq, G., Perovskites as catalysts precursors: Synthesis and characterization, *Journal of Molecular Catalysis A: Chemical*, 228, 325–331, 2005.

44. Tanaka, H. and Misono, M., Advances in designing perovskite catalysts, *Current Opinion in Solid State and Materials Science*, 5, 381–387, 2001.

45. Goldwasser, M.R., Rivas, M.E., Pietri, E., Pérez-Zurita, M.J., Cubeiro, M.L., Gingembre, L., Leclercq, L., and Leclercq, G., Perovskites as catalysts precursors: CO$_2$ reforming of CH$_4$ on Ln$_{1-x}$Ca$_x$Ru$_{0.8}$Ni$_{0.2}$O$_3$ (Ln = La, Sm, Nd), *Applied Catalysis A: General*, 255, 45–57, 2003.

46. Hayakawa, T., Suzuki, Sh., Nakamura, J., Uchijima, T., Hamakawa, S., Suzuki, K., Shishido, T., and Takehira, K., CO$_2$ reforming of CH$_4$ over Ni/perovskite catalysts prepared by solid phase crystallization method, *Applied Catalysis A: General*, 183(2), 273–285, 1999.

47. Mawdsley, J.R. and Krause, T.R., Rare earth-first-row transition metal perovskites as catalysts for the autothermal reforming of hydrocarbon fuels to generate hydrogen, *Applied Catalysis A: General*, 334, 311–320, 2008.

48. Batiot-Dupeyrat, C., Gallego, G.A.S., Mondragon, F., Barrault, J., and Tatibouët, J.-M., CO$_2$ reforming of methane over LaNiO$_3$ as precursor material, *Catalysis Today*, 107–108, 474–480, 2005.

49. Pereniguez, R., Gonzalez-DelaCruz, V.M., Holgado, J.P., and Caballero, A., Synthesis and characterization of a LaNiO$_3$ perovskite as precursor for methane reforming reactions catalysts, *Applied Catalysis B: Environmental*, 93, 346–353, 2010.

50. Moradi, G.R., Rahmanzade, M., and Sharifni, S., Kinetic investigation of CO_2 reforming of CH_4 over La–Ni based perovskite, *Chemical Engineering Journal*, 162, 787–791, 2010.

51. Nam, J.W., Chae, H., Lee, S.H., Jung, H., and Lee, K.-Y., Methane dry reforming over well-dispersed Ni catalyst prepared from perovskite-type mixed oxides, *Studies in Surface Science and Catalysis*, 119, 843–848, 1998.

52. Valderrama, G., Goldwasser, M.R., Urbina de Navarro, C., Tatibouet, J.M., Barrault, J., Batiot-Dupeyrat, C., and Martınez, F., Dry reforming of methane over Ni perovskite type oxides, *Catalysis Today*, 107–108, 785–791, 2005.

53. Rynkowski, J., Samulkiewicz, P., Ladavos, A.K., and Pomonis, P.J., Catalytic performance of reduced $La_{2-x}Sr_xNiO_4$ perovskite-like oxides for CO_2 reforming of CH_4, *Applied Catalysis A: General*, 263, 1–9, 2004.

54. Chettapongsaphan, C., Charojrochkul, S., Assabumrungrat, S., and Laosiripojana, N., Catalytic H_2O and CO_2 reforming of CH_4 over perovskite-based $La_{0.8}Sr_{0.2}Cr_{0.9}Ni_{0.1}O_3$: Effects of pre-treatment and co-reactant/CH_4 on its reforming characteristics, *Applied Catalysis A: General*, 386(1–2), 194–200, 2010.

55. Valderrama, G., Kiennemann, A., and Goldwasser, M.R., La–Sr–Ni–Co–O based perovskite-type solid solutions as catalyst precursors in the CO_2 reforming of methane, *Journal of Power Sources*, 195, 1765–1771, 2010.

56. Mota N., Álvarez-Galván, M.C., Al-Zahrani, S.M., Navarro, R.M., and Fierro, J.L.G., Diesel fuel reforming over catalysts derived from $LaCo_{1-x}Ru_xO_3$ perovskites with high Ru loading, *International Journal of Hydrogen Energy*, 37, 7056–7066, 2012.

57. Gallego, G.S., Batiot-Dupeyrat, C., Barrault, J., Florez, E., and Mondragon, F., Dry reforming of methane over $LaNi_{1-y}ByO_{3\pm\delta}$ (B = Mg, Co) perovskites used as catalyst precursor, *Applied Catalysis A: General*, 334, 251–258, 2008.

58. Valderrama, G., Kienneman, A., and Goldwasser, M.R., Dry reforming of CH_4 over solid solutions of $LaNi_{1-x}Co_xO_3$, *Catalysis Today*, 133, 142–148, 2008.

59. Goldwasser, M.R., Rivas, M.E., Lugo, M.L., Pietri, E., Pe´rez-Zurita, J., Cubeiro, M.L., Griboval-Constant, A., and Leclercq, G., Dry reforming of methane over Ni perovskite type oxides, *Catalysis Today*, 107–108, 106–113, 2005.

60. Provendier, H., Petit, C., Estournes C., and Kiennemann, A., Dry reforming of methane. Interest of La–Ni–Fe solid solutions compared to $LaNiO_3$ and $LaFeO_3$, *Studies in Surface Science and Catalysis*, 119, 741–746, 1998.

61. Provendier, H., Petit, C., Estournes, C., and Kiennemann, A., Steam reforming of methane on $LaNi_xFe_{1-x}O_3$ ($0 \leq x \leq 1$) perovskites. Reactivity and characterisation after test, *Comptes Rendus de l'Académie des Sciences—Series IIC—Chemistry*, 4, 57–66, 2001.

62. Provendier, H., Petit, C., Estournes, C., Libs, S., and Kiennemann, A., Stabilisation of active nickel catalysts in partial oxidation of methane to synthesis gas by iron addition, *Applied Catalysis A: General*, 180, 163–173, 1999.

63. Falcon, H., Baranda, J., Campos-Martin, J.M., Pena, M.A., and Fierro, J.L.G., Structural features and activity for CO oxidation of $LaFe_xNi_{1-x}O_{3+\delta}$ catalysts, *Studies in Surface Science and Catalysis*, 130, 2195–2200, 2000.

64. Kapokova, L., Pavlova, S., Bunina, R., Alikina, G., Krieger, T., Ishchenko, A., Rogov, V., and Sadykov, V., Dry reforming of methane over $LnFe_{0.7}Ni_{0.3}O_{3-\delta}$ perovskites: Influence of Ln nature, *Catalysis Today*, 164, 227–233, 2011.

65. Pavlova, S., Kapokova, L., Bunina, R., Alikina, G., Sazonova, N., Krieger, T., Ishchenko, A. et al., Syngas production by CO_2 reforming of methane using $LnFeNi(Ru)O_3$ perovskites as precursors of robust catalysts, *Catalysis Science & Technology*, 2, 2099–2108, 2012.

66. Gallego, G.S., Marín, J.G., Batiot-Dupeyrat, C., Barrault, J., and Mondragon, F., Influence of Pr and Ce in dry methane reforming catalysts produced from $La_{1-x}A_xNiO_{3-\delta}$ perovskites, *Applied Catalysis A: General*, 369, 97–103, 2009.

67. Rivas, M.E., Fierro, J.L.G., Goldwasser, M.R., Pietri, E., Pérez-Zurita, M.J., Griboval-Constant, A., and Lecler, G., Structural features and performance of $LaNi_{1-x}Rh_xO_3$ system for the dry reforming of methane, *Applied Catalysis A: General*, 344, 10–19, 2008.

68. Araujo, G.C., Lima, S.M., Assaf, J.M., Pena, M.A., Fierro, J.L.G., and Rangel, M.D.C., Catalytic evaluation of perovskite-type oxide $LaNi_{1-x}Ru_xO_3$ in methane dry reforming, *Catalysis Today*, 133–135, 129–135, 2008.

69. Mota, N., Alvarez-Galvána, M.C., Navarro, R.M., Al-Zahrani, S.M., Goguet, A., Daly, H., Zhang,W., Trunschke, A., Schlögl, R., and Fierro, J.L.G., Insights on the role of Ru substitution in the properties of $LaCoO_3$-based oxides as catalysts precursors for the oxidative reforming of diesel fuel, *Applied Catalysis B: Environmental*, 113–114, 271–280, 2012.

70. Labhsetwar, N.K., Watanabe, A., Mitsuhashi, T., and Haneda, H., Thermally stable ruthenium-based catalyst for methane combustion, *Journal of Molecular Catalysis A: Chemical*, 223, 217–223, 2004.

71. Chen, H., Yu, H., Peng, F., Yang, G., Wang, H., Yang, J., and Tang, Y., Autothermal reforming of ethanol for hydrogen production over perovskite $LaNiO_3$, *Chemical Engineering Journal*, 160, 333–339, 2010.

72. Liu, J.Y., Lee, C.C., Wang, C.H., Yeh, C.T., and Wang, C.B., Application of nickel-lanthanum composite oxide on the steam reforming of ethanol to produce hydrogen, *International Journal of Hydrogen Energy*, 35, 4069–4075, 2010.

73. de Lima, S.M., da Silva, A.M., da Costa, L.O.O., Assaf, J.M., Jacobs, G., Davis, B.H., Mattos, L.V., and Noronha, F.B., Evaluation of the performance of Ni/La_2O_3 catalyst prepared from $LaNiO_3$ perovskite-type oxides for the production of hydrogen through steam reforming and oxidative steam reforming of ethanol, *Applied Catalysis A: General*, 377, 181–190, 2010.

74. Chen, S.Q., Wang, H., and Liu, Y., Perovskite La–St–Fe–O (St = Ca, Sr) supported nickel catalysts for steam reforming of ethanol: The effect of the A site substitution, *International Journal of Hydrogen Energy*, 34, 7995–8005, 2009.

75. Chen, S.Q., Li, Y.D., Liu, Y., and Bai, X., Regenerable and durable catalyst for hydrogen production from ethanol steam reforming, *International Journal of Hydrogen Energy*, 36, 5849–5856, 2011.

76. de Lima, S.M., da Silva, A.M., da Costa, L.O.O., Assaf, J.M., Mattos, L.V., Sarkari, R., Venugopale, A., and Noronha, F.B., Hydrogen production through oxidative steam reforming of ethanol over Ni-based catalysts derived from $La_{1-x}Ce_xNiO_3$ perovskite-type oxides, *Applied Catalysis B: Environmental*, 121–122, 1–9, 2012.

77. Juan-Juan, J., Roman-Martinez, M.C., and Illan-Gomez, M. J., Nickel catalyst activation in the carbon dioxide reforming of methane: Effect of pretreatments, *Applied Catalysis A: General*, 355, 27–32, 2009.

78. Sadykov, V.A., Gubanova, E.L., Sazonova, N.N., Pokrovskaya, S.A., Chumakova, N.A., Mezentseva, N.V. et al., Dry reforming of methane over Pt/PrCeZrO catalyst: Kinetic and mechanistic features by transient studies and their modeling, *Catalysis Today*, 171, 140–149, 2011.

79. Bernal, S., Calvino, J.J., Gatica, J.M., Cartes, C.L., and Pintado, J.M., Chemical and nano-structural aspects of the preparation and characterisation of ceria and ceria-based mixed oxide-supported metal catalysts, in: Trovarelli, T. (ed.), *Catalysis by Ceria and Related Materials*, Imperial College Press, London, U.K., 2002, pp. 85–168.

80. Wei, J. and Iglesia, E., Isotopic and kinetic assessment of the mechanism of reactions of CH_4 with CO_2 or H_2O to form synthesis gas and carbon on nickel catalysts, *Journal of Catalysis*, 224, 370–383, 2004.

81. Bobin, A.S., Sadykov, V.A., Rogov, V.A., Mezentseva, N.V., Alikina, G.M., Sadovskaya, E.M., Glazneva, T.S. et al., Mechanism of CH_4 dry reforming on nanocrystalline doped ceria–zirconia with supported Pt, Ru, Ni, and Ni–Ru, *Topics in Catalysis*, 56, 958–968, 2013.

82. Sadykov, V.A., Mezentseva, N.V., Alikina, G.M., Lukashevich, A.I., Borchert, Yu.V., Kuznetsova, T.G., Ivanov, V.P. et al., Pt-supported nanocrystalline ceria–zirconia doped with La, Pr or Gd: Factors controlling syngas generation in partial oxidation/autothermal reforming of methane or oxygenates, *Solid State Phenomena*, 128, 239–248, 2007.

83. Sadykov, V., Mezentseva, N., Alikina, G., Bunina, R., Pelipenko V., Lukashevich, A., Tikhov, S. et al., Nanocomposite catalysts for internal steam reforming of methane and biofuels in solid oxide fuel cells: Design and performance, *Catalysis Today*, 146, 132–140, 2009.

84. Sadykov, V., Mezentseva, N., Muzykantov, V., Efremov, D., Gubanova, E., Sazonova, N., Bobin, A. et al., Real structure—Oxygen mobility relationship in nanocrystalline doped ceria-zirconia fluorite-like solid solutions promoted by Pt, *Materials Research Society Symposium Proceedings*, 1122, 36–41, 2009.

85. Sadykov, V., Sobyanin, V., Mezentseva, N., Alikina, G., Vostrikov, Z., Fedorova, Y., Pelipenko, V. et al., Transformation of CH_4 and liquid fuels into syngas on monolithic catalysts, *Fuel*, 89, 1230–1240, 2010.

86. Sadykov, V., Mezentseva, N., Alikina, G., Bunina, R., Pelipenko, V., Lukashevich, A., Vostrikov, Z. et al., Nanocomposite catalysts for steam reforming of methane and biofuels: Design and performance, in: *Nanocomposite Materials, Theory and Applications*, INTECH, Vienna, Austria, 2011, pp. 909–946.

87. Crisafulli, C., Scirè, S., Minicò, S., and Solarino, L., Ni–Ru bimetallic catalysts for the CO_2 reforming of methane, *Applied Catalysis A: General*, 225(1–2), 1–9, 2002.

88. Wei, J. and Iglesia, E., Structural requirements and reaction pathways in methane activation and chemical conversion catalyzed by rhodium, *Journal of Catalysis*, 225, 116–127, 2004.

89. Tauster, S.J., Strong metal–support interactions, *Accounts of Chemical Research*, 20, 389–394, 1987.

90. Tomishige, K., Yamazaki, O., Chen, Y., Yokoyama, K., Li, X., and Fujimoto, K., Development of ultra-stable Ni catalysts for CO_2 reforming of methane, *Catalysis Today*, 45, 35–39, 1998.

91. Erdöhelyi, A., Cserényi, J., Papp, E., and Solymosi, F., Catalytic reaction of methane with carbon dioxide over supported palladium, *Applied Catalysis A: General*, 108, 205–219, 1994.

92. Múnera, J.F., Irusta, S., Cornaglia, L.M., Lombardo, E.A., Cesar, D.V., and Schmal, M., Kinetics and reaction pathway of the CO_2 reforming of methane on Rh supported on lanthanum-based solid, *Journal of Catalysis*, 245, 25–34, 2007.

93. Bradford, M.C.J. and Vannice, M.A., Catalytic reforming of methane with carbon dioxide over nickel catalysts. II. Reaction kinetics, *Applied Catalysis A: General*, 142, 97–122, 1996.

94. O'Connor, A.M., Schuurman, Y., Ross, J.R.H., and Mirodatos, C., Transient studies of carbon dioxide reforming of methane over Pt/ZrO_2 and Pt/Al_2O_3, *Catalysis Today*, 115, 191–198, 2006.

95. Tsipouriari, V.A. and Verykios, X.E., Kinetic study of the catalytic reforming of methane with carbon dioxide to synthesis gas over Ni/La_2O_3 catalyst, *Catalysis Today*, 64, 83–90, 2001.

96. Galdikas, A., Bion, N., Duprez, D., Virbickas, V., and Maželis, D., Modeling of diffusion process in the isotopic oxygen exchange experiments of $Ce_xZr_{1-x}O_2$ catalysts, Materials science = Medžiagotyra, *Kaunas University of Technology, Academy of Sciences of Lithuania, Kaunas*, 19(1), 83–88, 2013.

97. Slagtern, Å., Schuurman, Y., Leclercq, C., Verykios, X., and Mirodatos, C., Specific features concerning the mechanism of methane reforming by carbon dioxide over Ni/La_2O_3 catalyst, *Journal of Catalysis*, 172, 118–126, 1997.

98. Krylov, O.V., Mamedov, A.K., and Mirzabekova, S.R., Interaction of carbon dioxide with methane on oxide catalysts, *Catalysis Today*, 42, 211–215, 1998.

 99. Choudhary, T.V., Aksoylu, E., and Goodman, D.W., Non-oxidative activation of methane, *Catalysis Reviews: Science and Engineering*, 45, 151–203, 2003.
100. Stagg-Williams, S.M., Noronha, F.B., Fendley, G., and Resasco, D.E., CO$_2$ reforming of CH$_4$ over Pt/ZrO$_2$ catalysts promoted with La and Ce oxides, *Journal of Catalysis*, 194, 240–249, 2000.
101. Wei, J. and Iglesia, E., Reaction pathway and site requirements for the activation and chemical conversion of methane on Ru-based catalysts, *Journal of Physical Chemistry B*, 108, 7253–7262, 2004.
102. Akpan, E., Sun, Y., Kumar, P., Ibrahim, H., Aboudheir, A., and Idem, R., Kinetics, experimental and reactor modeling studies of the carbon dioxide reforming of methane (CDRM) over a new Ni/CeO$_2$-ZrO$_2$ catalysts in a packed bed tubular reactor, *Chemical Engineering Science*, 62, 4012–4024, 2007.
103. Maestri, M., Vlachos, D.G., Beretta, A., Groppi, G., and Tronconi, E., Steam and dry reforming of methane on Rh: Microkinetic analysis and hierarchy of kinetic models, *Journal of Catalysis*, 259, 211–222, 2008.
104. Sadykov, V., Muzykantov, V., Bobin, A., Mezentseva, N., Alikina, G., Sazonova, N., Sadovskaya, E., Gubanova, L., Lukashevich, A., and Mirodatos, C., Oxygen mobility of Pt-promoted doped CeO$_2$-ZrO$_2$ solid solutions: Characterization and effect on catalytic performance in syngas generation by fuels oxidation/reforming, *Catalysis Today*, 157, 55–60, 2010.
105. Sadykov, V., Kuznetsova, T., Frolova-Borchert, Yu., Alikina, G., Lukashevich, A., Rogov, V., Muzykantov, V. et al., Fuel-rich methane combustion: Role of the Pt dispersion and oxygen mobility in a fluorite-like complex oxide support, *Catalysis Today*, 117, 475–483, 2006.
106. Yang, W., Ma, Y., Tang, J., and Yang, X., "Green synthesis" of monodisperse Pt nanoparticles and their catalytic properties, *Colloids and Surfaces A: Physicochemical and Engineering Aspects*, 302, 628–633, 2007.
107. Sadykov, V., Mezentseva, N., Alikina, G., Lukashevich, A., Muzykantov, V., Kuznetsova, T., Batuev, L. et al., Nanocrystalline doped ceria–zirconia fluorite-like solid solutions promoted by Pt: Structure, surface properties and catalytic performance in syngas generation, *Materials Research Society Symposium Proceedings*, 988, QQ06-04.1-6, 2007.
108. Temkin, M.I., Relaxation of two-stage catalytic reaction rate, *Kinetika i Kataliz*, 17, 1095–1099, 1976 (in Russian).
109. Sazonova, N.N., Sadykov, V.A., Bobin, A.S., Pokrovskaya, S.A., Gubanova, E.L., and Mirodatos, C., Dry reforming of methane over fluorite-like mixed oxides promoted by Pt, *Reaction Kinetics and Catalysis Letters*, 98, 35–41, 2009.
110. Boudart, M. and Djega-Mariadassou, G., *The Kinetics of Heterogeneous Catalytic Reactions*, Princeton University Press, Princeton, NJ, 1984.
111. Wang, H.Y. and Ruckenstein, E., Carbon dioxide reforming of methane to synthesis gas over supported rhodium catalysts: The effect of support, *Applied Catalysis A: General*, 204, 143–152, 2000.
112. Wei, J. and Iglesia, E., Structural and mechanistic requirements for methane activation and chemical conversion on supported iridium clusters, *Angewandte Chemie International Edition*, 43, 3685–3688, 2004.
113. Yamaguchi, A. and Iglesia E., Catalytic activation and reforming of methane on supported palladium clusters, *Journal of Catalysis*, 274, 52–63, 2010.
114. Zhang, J., Wang, H., and Dalai, A.K., Kinetic studies of carbon dioxide reforming of methane over Ni–Co/Al–Mg–O bimetallic catalyst, *Industrial Engineering and Chemical Research*, 48, 677–684, 2009.
115. Zhang, Z. and Verykios, X.E., Mechanistic aspects of carbon dioxide reforming of methane to synthesis gas over Ni catalysts, *Catalysis Today*, 21, 589–595, 1994.

116. Olsbye, U., Wurzel, T., and Mleczko, L., Kinetic and reaction engineering studies of dry reforming of methane over a Ni/La/Al$_2$O$_3$ catalyst, *Industrial Engineering and Chemical Research*, 36, 5180–5188, 1997.

117. Nandini, A., Pant, K.K., and Dhingra, S.C., Kinetic study of the catalytic carbon dioxide reforming of methane to synthesis gas over Ni–K/CeO$_2$–Al$_2$O$_3$, *Applied Catalysis A: General*, 308, 119–127, 2006.

118. Sadykov, V., Bobrova, L., Pavlova, S., Simagina, V., Makarshin, L., Parmon, V., Ross, J.R.H., Van Veen, A.C., Mirodatos, C., *Syngas Generation from Hydrocarbons and Oxygenates with Structured Catalysts*, Series: Energy Science, Engineering and Technology, Sadykov, V., Parmon,V., Ross, J.R.H., Mirodatos, C. (eds.), Nova Science Publishers, Inc., New York, 2012, 140pp.

119. Bobrova, L., Zolotarskii, I., Sadykov, V., Pavlova, S., Snegurenko, S., Tikhov, S., Korotkich, V., Kuznetsova, T., Sobyanin, V., and Parmon V., Syngas formation by selective catalytic oxidation of liquid hydrocarbons in a short contact time adiabatic reactor, *Chemical Engineering Journal*, 107, 171–179, 2005.

120. Bobrova, L., Korotkich, V., Sadykov, V., and Parmon, V., Syngas formation from gasoline in adiabatic reactor: Thermodynamic approach and experimental observations, *Chemical Engineering Journal*, 134, 145–152, 2007.

121. Bobrova, L., Vernikovskaya, N., and Sadykov, V., Conversion of hydrocarbon fuels to syngas in a short contact time catalytic reactor, *Catalysis Today*, 144, 185–200, 2009.

122. Vernikovskaya, N., Bobrova, L., Pinaeva, L., Sadykov, V., Zolotarskii, I., Sobyanin, V., Buyakou, I., Kalinin, V., and Zhdanok, S., Transient behavior of the methane partial oxidation in a short contact time reactor: Modeling on the base of catalyst detailed chemistry, *Chemical Engineering Journal*, 134, 180–189, 2007.

123. Smorygo, O., Sadykov, V., Mikutski, V., Marukovich, A., Ilyushchanka, A., Yarkovich, A., Mezentseva, N. et al., Porous substrates for intermediate temperature SOFCs and in-cell reforming catalysts, *Catalysis for Sustainable Energy*, 1, 90–99, 2012.

124. Smorygo, O., Mikutski, V., Marukovich, A., Vialiuha, Y., Ilyushchanka, A., Mezentseva, N., Alikina, G. et al., Structured catalyst supports and catalysts for the methane indirect internal steam reforming in the intermediate temperature SOFC, *International Journal of Hydrogen Energy*, 34, 9505–9514, 2009.

11 Development of Bifunctional Ni-Based Catalysts for the Heterogeneous Oligomerization of Ethylene to Liquids

A. Martínez, M.A. Arribas, and S. Moussa

CONTENTS

11.1 INTRODUCTION AND SCOPE OF THE CHAPTER

Despite environmental concerns that promote the increasing utilization of alternative energy resources such as renewables, most of the world's energetic demand for the next decades will still rely on the use of fossil fuels. Among fossil fuels, natural gas is a more attractive primary energy source as compared to oil and coal owing to its lower concentration of contaminants (S, N) and higher hydrogen-to-carbon ratio, related to its main constituent methane, resulting in reduced harmful emissions. Natural gas demands are, hence, forecasted to grow nearly 70% worldwide between 2002 and 2025 at an average annual consumption rate of 2.3%, surpassing that projected for oil and coal [1]. Increasing natural gas consumption not only would help in alleviating the impact of fossil fuels on climate change by reducing greenhouse gas (GHG) emissions but also in increasing the energy security of supply for non-oil-producing countries, which is one of the major goals of the EU energy policy. Based on the data provided by BP, the total world proven natural gas reserves at the end of 2012 exceeded 187 trillion cubic meters, which could provide more than 55 years of production [2]. These values might be even greater if new technologies/processes become available in the near future so as to make the extraction and use of unconventional natural gas reserves such as deep and tight natural gas, shale gas, coal bed methane, and methane hydrates economically viable. More than 30% of the world's natural gas is classified as *stranded*, meaning that it cannot be used locally or it is too far away from markets to be economically transported using traditional technologies such as pipelines or condensation to LNG. Gas to liquid (GTL) processes, in which methane is indirectly converted to synthetic fuels or chemicals via *syngas*, represent a cost-effective option for monetizing large remote natural gas reservoirs. However, the huge capital investments and operational costs associated to conventional GTL complexes become unaffordable when dealing with small stranded gas fields. In such a case, the development of alternative cost-effective technologies becomes ineluctable.

In this context, the oxidative coupling of methane followed by oligomerization to liquids (OCMOL) project was conceived in order to provide a technically and economically feasible chemical route particularly suited to the exploitation of small gas reservoirs [3]. The very basic process scheme proposed in the OCMOL project comprises the direct conversion of methane to ethylene via OCM and the subsequent oligomerization of ethylene to transportable liquids. To maximize the energetic efficiency of the process while minimizing CO_2 emissions, the heat released in the highly exothermic OCM reaction is used to drive the endothermic reforming of methane (RM) with CO_2 recycled from the OCM product stream to produce *syngas* that is then converted to oxygenates (MeOH/DME) and finally to liquid fuels in two consecutive catalytic steps, as schematically shown in Figure 11.1.

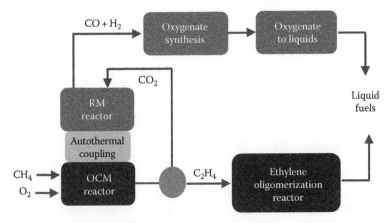

FIGURE 11.1 Catalytic steps involved in the fully integrated methane-to-liquid route as proposed in the OCMOL project.

Separation steps, which are vital to the success of the project, have been intentionally omitted in the scheme for the sake of simplicity.

A key catalytic step in the unconventional GTL integrated approach proposed in OCMOL (see Figure 11.1) is the oligomerization of the OCM main product, ethylene, to liquid fuels. At this point, it is worthy to mention that the state-of-the-art technologies in commercial ethylene oligomerization rely on the use of homogeneous catalysts based on transition metal, that is, Ni, Co, Cr, and Fe complexes [4]. Apart from not being the most ideal from the environmental viewpoint, the homogeneous processes suffer from several drawbacks such as the issue of catalyst–product separation, product contamination, and catalyst reusability. It is, hence, essential for the OCMOL project to develop alternative eco-friendly heterogeneous catalysts enabling the efficient oligomerization of ethylene to liquids at mild conditions. To this aim, an exhaustive screening study was undertaken at the beginning of the project in order to identify the most suitable catalysts. The study involved the preparation and testing of a library of materials comprising more than 100 solid catalysts displaying distinct chemical, structural, and textural characteristics. The screening revealed that catalysts comprised of Ni loaded on acidic porous aluminosilicates displayed the highest potential in terms of activity, selectivity to liquid oligomers, and stability with time under continuous feeding of pure ethylene. More specifically, the catalysts selected for further optimization along the project comprised Ni loaded on the following three acidic porous aluminosilicates: nanocrystalline zeolite beta (Ni-beta), amorphous silica–alumina (Ni–SiO_2–Al_2O_3), and mesostructured Al-MCM-41. The two main targets pursued during the catalyst optimization, as contemplated in the initial OCMOL objectives, were a productivity to liquid oligomers of at least 5 mmol/(kg_{cat}·s) and a lifetime of at least 60 h on stream. Thus, the present chapter gathers the main advances achieved within the OCMOL project regarding the development of Ni-based catalysts for the heterogeneous oligomerization of ethylene to liquids. In the last part of the chapter, the tolerance of Ni-containing catalysts to the main feed impurities accompanying ethylene in the gaseous stream leaving the OCM reactor (CH_4, CO, CO_2) is also addressed.

11.2 GENERAL ASPECTS OF THE OLIGOMERIZATION OF ETHYLENE ON BIFUNCTIONAL Ni-BASED CATALYSTS

Oligomerization of propylene and higher olefins is well known to readily occur on purely acidic catalysts, such as H-zeolites or supported phosphoric acid. In the case of ethylene, however, the energetically unfavorable primary carbocation that would form upon protonation of the double bond makes the monofunctional acid-catalyzed route very unlikely at moderate temperatures. In fact, it has been shown that ethylene oligomerization hardly proceeds at temperatures below 300°C on purely acidic catalysts [5–10]. This issue has been circumvented by preparing bifunctional catalysts comprising Ni ions loaded on a variety of acidic porous materials. This is not surprising at all taking into account that organonickel compounds are among the most efficient homogenous catalysts applied in industrial ethylene oligomerization processes [11]. Whereas, as mentioned earlier, monofunctional acid catalysts are practically inactive toward ethylene at mild temperatures, dimerization into butenes is the predominant reaction when using Ni loaded on nonacidic or weakly acidic supports, such as SiO_2 [12,13], zeolite NaY [14,15], and the pure silica form of the ordered mesoporous MCM-41 material [16]. Therefore, it appears that both Ni species and Brønsted acid sites (H^+) need to be present in the catalyst in order to effectively promote the formation of liquid oligomers, that is, the targeted products in OCMOL.

The possible reaction paths that may occur during the heterogeneous oligomerization of ethylene on bifunctional Ni–H^+ catalysts are schematically illustrated in Figure 11.2. Although some authors have suggested that both nickel and acid sites are required for the dimerization/oligomerization of ethylene [17], it is generally assumed that *isolated* nickel cations in the bifunctional catalysts are responsible for the initial activation and oligomerization of ethylene at low temperatures [5,7–11,14,15,18–22], presumably through a coordination–insertion mechanism analogous to that proposed for homogeneous organonickel catalysts. The Ni-mediated oligomerization pathway is usually referred to as *true oligomerization* and typically produces a mixture of even-numbered carbon linear α-C_{4+} olefins displaying a Schulz–Flory-type distribution ($C_4 > C_6 > C_8 >$ etc.). Then the C_{4+} olefins formed on the Ni sites undergo further (co-)oligomerization reactions on the Brønsted-type acid sites, via carbocation intermediates, of the acidic support to produce a mixture of oligomers with an increased average chain length. The acid-catalyzed route is commonly referred to as *hetero-oligomerization*. Besides (co-)oligomerization, other acid-catalyzed reactions such as double bond and skeletal isomerizations and cracking of heavier oligomers may also occur to different extents depending on reaction conditions and support acidity.

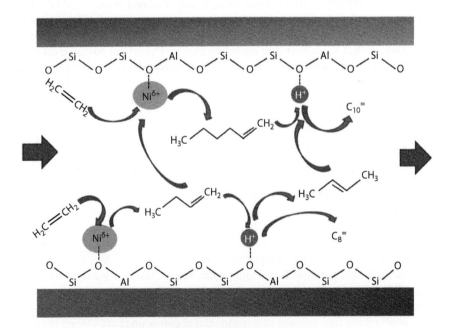

FIGURE 11.2 Schematic representation of the heterogeneous oligomerization of ethylene on bifunctional Ni-containing catalysts. (Adapted from Lallemand, M. et al., *Appl. Catal. A*, 338, 37, 2008.)

Cracking reactions account for the formation of branched olefins with an odd number of carbon atoms and are, obviously, favored at increasing temperatures.

From the point of view of catalyst design, the most relevant parameters affecting the activity, selectivity, c.q., oligomer distribution, and stability of bifunctional Ni-based catalysts are the active Ni species concentration, the density and strength of Brønsted acid sites, and, particularly, the porous structure of the acidic carrier. Among the different acidic supports investigated in earlier studies, microporous crystalline materials, such as zeolites, and mesoporous amorphous solids, such as silica–alumina and Al-MCM-41, have attracted the most interest. This fact concurs with the results obtained in the initial catalyst screening study performed in the OCMOL project, where three types of catalysts based on Ni loaded on beta zeolite, amorphous silica–alumina, and Al-MCM-41 displayed the best performance and were selected for further optimization. The most relevant features of these three types of oligomerization catalysts according to the previous literature are discussed hereinafter.

11.2.1 Ni-Zeolite Catalysts

The large availability of topologies with pores of molecular dimensions and the possibility to tune, a priori, the nature, density, and strength of the active sites make zeolites very appealing materials for many different catalytic applications [23]. As stated earlier, *isolated* Ni cations are generally believed to be the active sites for the activation of ethylene and its oligomerization into mainly linear α-olefins with a typical Schulz–Flory distribution. Such *isolated* nickel ions can be easily generated in zeolites by replacing cations compensating the negative charge associated to four-fold coordinated Al^{3+} species in the zeolite framework, for example, H^+, NH_4^+, and Na^+, with Ni ions through ion exchange procedures, followed by appropriate activation treatments. Due to their relatively high ion exchange capacity and large pores, Ni^{2+}-exchanged in Al-rich faujasite-type zeolites, such as X and Y, were studied for ethylene oligomerization [10,14,15,24]. For instance, Ni–NaY catalysts showed two regions of high activity: a low temperature region at 100°C–150°C where a typical ASF-type product distribution was obtained and a high temperature region (>300°C) characterized by a nonnegligible contribution of acid-catalyzed side reactions such as isomerization, oligomerization, and cracking, to the product slate [14]. Increasing the Ni loading increased the catalytic activity and the selectivity to diesel-range oligomers when working in the low-temperature region [14].

A major disadvantage of Ni-zeolite catalysts is their rapid deactivation during the oligomerization reaction [9,10,14,15]. Such a deactivation mainly occurs by the formation of bulky oligomers and polymeric compounds that remain strongly adsorbed in the micropores and whose formation becomes favored by a high density and strength of the acid sites. Indeed, dealumination of Y zeolite was seen to improve the oligomerization activity of Ni–Y catalysts, thanks to an increased pore accessibility and mild acidity as compared to the parent zeolite [10]. In this line, Ni-MCM-36 catalysts containing mesopores in the interlayer space of the pillared MCM-36 zeolite displayed better activity and stability than the structurally related purely microporous Ni-MCM-22 zeolite [9].

11.2.2   Ni–Silica–Alumina Catalysts

Ni^{2+}-containing silica–alumina catalysts have been extensively investigated for ethylene oligomerization. Of particular relevance are the studies performed by Nicolaides and coworkers [6,7,18,25–29]. These authors found that the activity and selectivity of Ni–silica–alumina catalysts are strongly dependent on the nickel concentration [27], the acidity of the support [26], and reaction conditions [25]. An increase in the Ni concentration increased the ethylene oligomerization activity and produced a shift in the molecular weight distribution of the formed oligomers toward lighter products [27]. However, a more efficient utilization of the Ni species resulting in a higher intrinsic activity, that is, the activity per Ni site or turnover frequency, was found at lower Ni concentrations in ion-exchanged catalysts. The authors ascribed this effect to a selective exchange of Ni^{2+} on the most acidic sites of the silica–alumina carrier at low Ni loadings [27]. In fact, the activity and selectivity to heavier oligomers of Ni–silica–alumina were observed to increase with increasing acid strength of the carrier [7,26].

The ethylene conversion on Ni–silica–alumina catalysts showed two regions of high activity as a function of the reaction temperature: one showing a volcano-type dependence at low temperatures with a maximum activity at 120°C, at a total pressure amounting to 35 bar and mass hourly space velocity (MHSV) of 2 $g_{feed}/(g_{cat}$ h), and the other at temperatures of ca. 300°C [6,7]. The exact reason for such activity-temperature dependence is not fully understood, though according to the authors, it could be ascribed to a different nature of the catalytically active species and/or reaction mechanism predominating in the two temperature regions [7]. These assumptions were supported by the different product selectivity obtained in the low- and high-temperature regions. Thus, at low temperatures, the reaction resulted almost exclusively in C_4–C_{20} oligomers with an even number of carbon atoms, while a significant amount of oligomers with an odd number of carbon atoms were produced at high temperatures. This trend is indicative of a change in the oligomerization mode from predominantly *true oligomerization* to predominantly *hetero-oligomerization* [28]. Furthermore, a stable catalytic performance without apparent signs of deactivation was observed during a period of 22 days on stream when working at high pressures, that is, 35 bar, and at the reaction temperature of 120°C (MHSV = 2 $g_{feed}/[g_{cat} \cdot h]$) [6]. Conversely, deactivation of the Ni–silica–alumina catalysts was evidenced in the high-temperature regime, being particularly rapid above 300°C [7]. On the other hand, increasing the pressure and, more specifically, lowering the MSHV increased the average molecular weight and decreased the linearity of the oligomers.

11.2.3   Ni–Al-MCM-41 Catalysts

MCM-41 materials display very high surface areas (>700 m^2/g) and are characterized by a hexagonal arrangement of non-interconnected uniformly sized cylindrical pores in the mesopore range (2–10 nm). The incorporation of Al in tetrahedral coordination in the amorphous walls of the material during the synthesis and subsequent removal of the organic matter by calcination allows the generation of Brønsted acid sites of a lower strength than those present in zeolites and confers the material with a certain ion exchange capacity. These properties make meso-porous Al-MCM-41 very appealing materials to be used as supports for preparing bifunctional Ni-based catalysts. In fact, Ni–Al-MCM-41 catalysts were reported to display high activity for the oligomerization of ethylene at mild reaction conditions [5,19,20,22,30]. Typically, the preparation of Ni–Al-MCM-41 catalysts in most previous reports involved the ionic exchange of H^+ in the calcined material by Ni^{2+} species followed by a thermal activation treatment. As for other Ni-based materials, the oligomerization activity of Ni–Al-MCM-41 becomes affected by the concentration of Ni and the Si/Al ratio and, hence, by the balance between Ni and acid sites. To this respect, by preparing ion-exchanged Ni–Al-MCM-41 catalysts with Ni loadings of 0.3–0.6 wt% and variable Si/Al ratios in the 10–80 range, Hulea and Fajula [5] observed a lower deactivation for the less acidic sample (Si/Al = 80) during the oligomerization of ethylene in a batch reactor. The main reaction products over Ni–Al-MCM-41 at 150°C were C_4–C_{12} alkenes. Reaction temperatures above 150°C promoted the formation of heavier oligomers but accelerated catalyst deactivation [5]. Operation in a continuous mode significantly improved the stability of Ni–Al-MCM-41 catalysts. For instance, a stable ethylene conversion of 95% was observed during 7 days for a catalyst with Si/Al = 30 and containing 0.5 wt% Ni, introduced by ionic exchange, operating in a continuous stirred tank reactor (CSTR) at T = 30°C, P = 30 bar, and WHSV = 2.1 h^{-1} [19]. Under these conditions, the carbon number distribution of the oligomers followed the typical Anderson–Schulz–Flory statistics with predominance of C_4 and C_6 alkenes.

The effect of Ni loading and pore size of the Al-MCM-41 carrier was addressed in a recent work by Lacarriere et al. [20]. In their study, these authors prepared samples with Ni loadings in the range of 1.4–7.5 wt% by either ionic exchange, for Ni contents of up to 2 wt%, or impregnation, for higher loadings, starting from an Al-MCM-41 sample with a Si/Al ratio of 9 and pore diameter of 3.5 nm. The oligomerization activity, in the batch reactor used, was found to increase with increasing the Ni content up to 2 wt%, reached a plateau, and then declined for Ni contents above 5.5 wt% due to a partial pore blockage by bulk NiO particles formed at such high metal loadings. This behavior suggested that the catalytic sites for the oligomerization of ethylene are *isolated* nickel cations in exchange positions, which, according to H_2-TPR measurements, were the only nickel species present in Ni–Al-MCM-41 catalysts at low loadings (<2 wt%). Bulk NiO particles, which were detected at high loadings, were presumed to be inactive for the reaction. Interestingly, at equivalent Ni loadings, the oligomerization activity of Ni–Al-MCM-41 was seen to increase with increasing the pore diameter of the Al-MCM-41 matrix from 3.5 to 10 nm [20], likely due to an enhanced mass transfer of the formed oligomers through the mesopores. Under the reaction condition used, the Ni–Al-MCM-41 catalysts were highly selective to C_4–C_{10} olefins with an ASF-type distribution, with butenes (40%–50%) and hexenes (30%–35%) being the most abundant products. Introduction of an acidic H–Al-MCM-41 sample to the reactor together with the Ni–Al-MCM-41 catalyst increased both the productivity and the average chain length of the oligomers by promoting the reaction of the oligomers initially produced on Ni–Al-MCM-41 with the acid sites of the H-MCM-41 co-catalyst [20].

11.3   CONTRIBUTION OF OCMOL TO THE DEVELOPMENT OF Ni-BASED CATALYSTS

11.3.1   Ni-Beta Catalysts

As mentioned earlier, conventional microporous Ni-zeolites are not suitable catalysts for ethylene oligomerization owing to a rapid deactivation caused by the buildup of bulky oligomers that remain trapped in the zeolite pores and cavities. Conversely, an improved stability was attained by using zeolites such as MCM-36 having a meso-pore system facilitating the diffusion of the oligomers [9]. The approach followed within OCMOL to design an active and stable Ni-zeolite catalyst was based on the use of a nanocrystalline large-pore zeolite beta displaying a relatively high and somehow organized mesoporosity associated to the intercrystalline voids. The zeo-lite employed for preparing the Ni-beta catalysts was a commercial H-beta sample (Si/Al = 12, Zeolyst International) with an average crystallite size of ca. 25 nm, as shown in Figure 11.3 in a representative image obtained by transmission electron microscopy (TEM) and the corresponding particle size distribution analysis.

Starting from this commercial H-beta zeolite, two series of Ni-beta catalysts with Ni loadings in the range of 1–10 wt% were prepared by either ionic exchange (series EX) or incipient wetness impregnation (series IM) using aqueous $Ni(NO_3)_2$ solu-tions, followed by drying and calcination in flowing air at 550°C for 3 h. The maxi-mum amount of Ni that could be incorporated after multiple ion exchange steps was 2.5 wt%. The main physicochemical properties of the parent nanocrystalline H-beta and the Ni-beta samples are shown in Table 11.1. No peaks related to bulk-like NiO nanoparticles were detected by x-ray diffraction (XRD) for the series prepared by ionic exchange. For the impregnated series, diffractions corresponding to bulk-like NiO crystallites were only observed at Ni loadings ≥3.8 wt%. The mean size of the NiO crystallites, estimated from the respective XRD by applying the Scherrer equa-tion, increased from 7 to 16 nm with increasing the Ni loading from 5 to 10 wt%.

These catalysts were evaluated for the oligomerization of pure ethylene in a bench-scale fixed bed reactor purposely designed and constructed for the project.

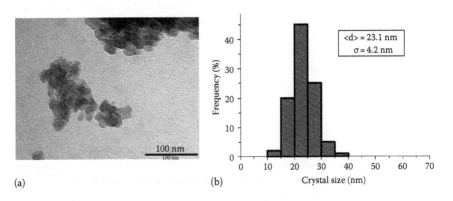

(a)   (b)

FIGURE 11.3   (a) Representative TEM image and (b) corresponding histogram for the nanocrystalline H-beta zeolite.

TABLE 11.1

Main Characteristics of the Parent Nanocrystalline H-Beta and Ni-Beta Catalysts

Sample	Ni Loading (wt%)	BET Area (m²/g)	Micropore Volume (cm³/g)	Brønsted Acidity[a] ($\mu mol_{py}/g_{cat}$)
H-beta	—	608	0.190	160
Ni–B-IE-1	1.0	602	0.187	118
Ni–B-IE-2	1.7	601	0.186	103
Ni–B-IE-3	2.0	610	0.189	93
Ni–B-IE-4	2.5	600	0.186	70
Ni–B-IM-1	1.1	581	0.182	120
Ni–B-IM-2	2.7	581	0.184	65
Ni–B-IM-3	3.8	552	0.175	58
Ni–B-IM-4	5.0	551	0.176	48
Ni–B-IM-5	5.8	539	0.172	42
Ni–B-IM-6	10.0	504	0.165	45

[a] Density of Brønsted acid sites as determined by FTIR of adsorbed pyridine at a desorption temperature of 250°C.

First, the catalytic performance of Ni-beta catalysts was compared at the standard reaction conditions of 120°C, 35 bar total pressure—comprising 26 bar ethylene partial pressure and Ar as balance gas—and WHSV of 2.1 h⁻¹. Prior to starting the reaction, the catalysts were pretreated in situ in flowing dried nitrogen at a temperature of 300°C for 16 h. No apparent deactivation was evidenced for any of the Ni-beta catalysts under the applied reaction conditions in runs lasting ca. 8–9 h regardless of the Ni loading and method of Ni incorporation, that is, ion exchange or impregnation, as shown in Figure 11.4 for representative samples.

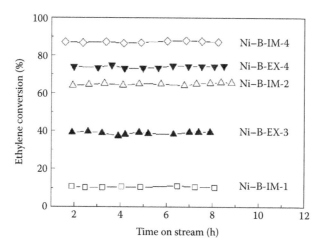

FIGURE 11.4 Ethylene conversion as a function of TOS for Ni-beta catalysts. Reaction conditions: T = 120°C, P_{tot} = 35 bar, P_{ethyl} = 26 bar, and WHSV = 2.1 h⁻¹.

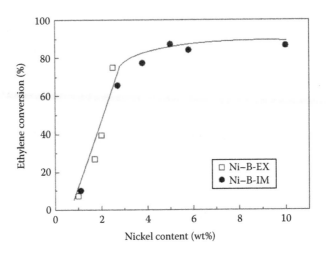

FIGURE 11.5   Ethylene conversion as a function of Ni loading for Ni-beta catalysts. Reaction conditions: $T = 120°C$, $P_{tot} = 35$ bar, $P_{ethyl} = 26$ bar, and WHSV = 2.1 h^{-1}.

The high stability of Ni-beta catalysts contrasts with the fast deactivation previously reported for Ni–Y [10] and Ni-MCM-22 [9] zeolites, which can be attributed to an enhanced mass transport of the heavy oligomers favored by the very small crystallites and the associated interparticle mesoporosity of the nanocrystalline zeolite beta. As observed in Figure 11.5, the ethylene conversion was strongly dependent on the Ni loading. For Ni contents of up to ca. 2.5 wt%, the conversion increased linearly with the concentration of Ni for both the ion-exchanged and impregnated samples and then leveled off until a constant value of around 85%–90% was attained at Ni loadings of 5 wt% and above. It is worth noting that, in the low Ni loading range (≤2.5 wt%), the activity of Ni-beta was equivalent irrespective of whether the nickel is introduced by ionic exchange or impregnation.

The activity trend shown in Figure 11.5 indicates a more efficient utilization of the nickel species in the low Ni loading range, concurring with previous observations for other bifunctional Ni-based catalysts [6,7,20]. The apparent higher intrinsic activity per mass of nickel observed at low Ni loadings goes in favor of the general belief that *isolated* Ni cations in exchange positions are the active sites for the activation and *true oligomerization* of ethylene. The nearly constant activity found at Ni loadings ≥5 wt% can be explained by the formation of inactive bulk-like NiO particles, as detected by XRD. In our Ni-beta catalysts, a nearly linear relationship was found between the activity for ethylene conversion and the amount of Ni^{2+} in exchange positions (see Figure 11.6). The latter values were estimated from the disappearance of Brønsted acid sites, as measured from FTIR-pyridine experiments, from the bare H-beta support upon Ni loading assuming that one Ni^{2+} replaced two H^+. Although a more thorough discussion about the controversial issue of the nature and oxidation state of the active nickel sites is beyond the scope of this chapter, the aforementioned correlation suggests that divalent Ni cations are the active sites, in contrast to the most extended opinion in favor of low-valent Ni^+ cations [17,30–33]. In fact, a stronger support for the assignment of active sites in

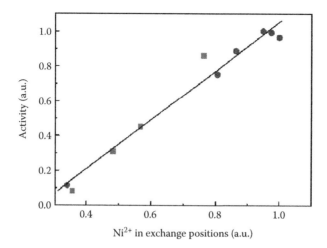

FIGURE 11.6 Correlation between the activity for ethylene conversion and the concentration of *isolated* Ni^{2+} cations occupying exchange positions in Ni-beta catalysts. Reaction conditions: $T = 120°C$, $P_{tot} = 35$ bar, $P_{ethyl} = 26$ bar, and WHSV = 2.1 h^{-1}.

Ni-beta catalysts to Ni^{2+} species was obtained in a thorough characterization study performed in the frame of OCMOL employing low-temperature CO adsorption and CO-ethylene co-adsorption FTIR experiments [34]. In this study, Ni^+ cations were not detected in the freshly activated catalysts, but, instead, they formed by reduction of Ni^{2+} with ethylene at the reaction temperature and were assigned to the role of mere spectators during the oligomerization reaction [34].

The distribution of the oligomers formed on Ni-beta catalysts varied with conversion and, hence, with the Ni loading. As shown in Figure 11.7a, the selectivity to liquid oligomers (C_{5+}) increased with increasing ethylene conversion as more butenes, the primary products formed on the Ni sites (*true oligomerization*), become available for further reaction on the zeolite acid sites (*hetero-oligomerization*). Under the investigated operating conditions, selectivities to C_{5+} oligomers of around 60%–65% were achieved at conversions ≥80%. The carbon number distribution for the liquid fraction (C_{5+}) obtained on the most active Ni-beta sample impregnated with 5 wt% Ni (Ni–B-IM-5) is presented in Figure 11.7b as an example.

As observed, the liquid oligomers did not obey the statistical ASF distribution that could be expected for the *true oligomerization* pathway on the nickel sites. It indicates a nonnegligible contribution of *hetero-oligomerization* occurring on the zeolitic Brønsted acid sites. For this particular catalyst and at the investigated conditions, gasoline-range oligomers ($C_5–C_{12}$) were the predominant liquid products, representing about 80 wt% of the total liquids produced. It is worth noting that oligomers with an odd number of carbon atoms were formed in relatively low amounts, which is indicative of a moderate cracking activity of the zeolite at the reaction temperature of 120°C.

On the other hand, the liquid oligomers with more than six carbon atoms contained a significant amount of branched products. Since cracking reactions did not occur to a great extent, branched oligomers should mostly originate from linear olefins through

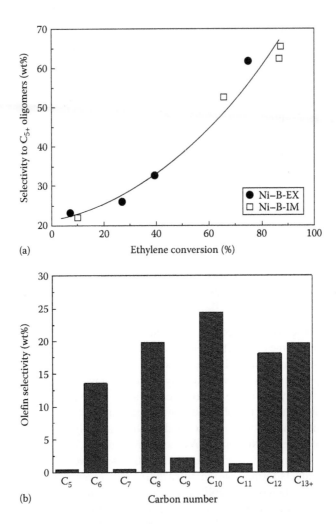

(a)

(b)

FIGURE 11.7 (a) Selectivity to liquid oligomers as a function of ethylene conversion for Ni-beta catalysts and (b) distribution of liquid oligomers for Ni–B–IM-4 catalyst (5 wt% Ni, impregnation). Reaction conditions: T = 120°C, P_{tot} = 35 bar, P_{ethyl} = 26 bar, and WHSV = 2.1 h^{-1}.

acid-catalyzed skeletal isomerization reactions. In the case of Ni–B–IM-4 catalyst, for instance, 89% and 98% of the C_8 and C_{10} oligomers were branched with a relatively high concentration of multibranched isomers, for example, ca. 70% of the C_8 fraction. Hence, the C_8–C_{12} fraction obtained on bifunctional Ni-beta catalysts could be suitably used, after a simple hydrogenation step, as a high-octane synthetic gasoline component.

As mentioned in the introduction, the OCMOL process targeted the development of an efficient heterogeneous ethylene oligomerization catalyst featuring a minimum productivity to liquid oligomers of 5 mmol/(kg$_{cat}$·s) and a lifetime of, at least, 60 h on stream in a continuous operation mode. For the prepared Ni-beta catalysts, the productivity to liquids at the standard reaction conditions (T = 120°C, P_{tot} = 35 bar,

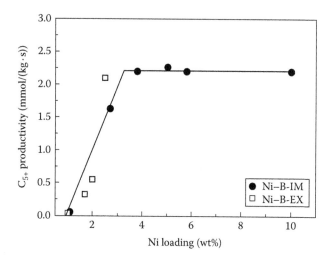

FIGURE 11.8 Influence of Ni loading on the productivity to liquid oligomers for Ni-beta catalysts. Reaction conditions: $T = 120°C$, $P_{tot} = 35$ bar, $P_{ethyl} = 26$ bar, and WHSV = 2.1 h^{-1}.

$P_{ethyl} = 26$ bar, WHSV = 2.1 h^{-1}) followed a trend with Ni loading similar to that found for conversion, that is, it increased up to a Ni content of 3.8 wt% and then remained nearly constant reaching a maximum value of 2.2 mmol/(kg$_{cat}$·s) at higher metal loadings (Figure 11.8). This value represents only about 45% of the project's target.

In an attempt to improve the productivity to liquids, additional optimization experiments were carried out with the most active Ni–B-IM-4 sample on which 5 wt% Ni was impregnated. The operating conditions were varied in a relatively wide range in runs lasting ca. 8 h: $T = 50°C–230°C$, $P_{total} = 8–35$ bar, $P_{ethyl} = 7–26$ bar, and WHSV = 1.4–13.8 h^{-1}. This study revealed a maximum initial, that is, after 2 h on stream, productivity to liquids of 4.5 mmol/(kg$_{cat}$·s) (90% of the target) at $T = 200°C$, $P_{total} = 35$ bar, $P_{ethyl} = 26$ bar, and WHSV = 10.8 h^{-1}. However, under these conditions, the catalyst experienced a gradual loss of activity during the initial reaction stages and then reached a pseudo-steady productivity value of ca. 3 mmol/(kg$_{cat}$·s) after 6 h on stream. On the other hand, the maximum productivity that could be attained on this catalyst without any evidence of deactivation during the ca. 8 h runs was 2.6 mmol/(kg$_{cat}$·s) at $T = 230°C$, $P_{total} = 35$ bar, $P_{ethyl} = 26$ bar, and WHSV = 2.1 h^{-1}. This productivity to liquids was still far from the target, and consequently, further efforts in the project were concentrated on the optimization of Ni–SiO$_2$–Al$_2$O$_3$ and Ni–Al-MCM-41 catalysts, as it will be discussed in the following sections.

11.3.2 Ni–SiO$_2$–Al$_2$O$_3$ Catalysts

A series of Ni–SiO$_2$–Al$_2$O$_3$ catalysts were prepared by impregnation of several commercial silica–aluminas differing in chemical composition and texture at two levels of Ni loading, namely, 2 and 6 wt% (nominal). The nomenclature and main physicochemical properties of SiO$_2$–Al$_2$O$_3$ supports and Ni–SiO$_2$–Al$_2$O$_3$ catalysts are listed in Table 11.2. Siral-30 (supplied as silica–alumina hydrate, with alumina being in

TABLE 11.2

Properties of Commercial Silica–Alumina Supports and Ni–SiO$_2$–Al$_2$O$_3$ Catalysts

Support	Catalyst	Chemical Composition[a]		Textural Properties[b]		Brønsted Acidity[c]
		Si/Al Ratio	Ni (wt%)	BET (m^2/g)	TPV (cm^3/g)	(μmol$_{py}$/g$_{cat}$)
Siral-30	—	0.33	—	427	0.657	20
Siralox-30	—	0.33	—	192	0.622	22
ASA-25	—	2.40	—	365	0.698	54
ASA-13	—	5.00	—	488	0.680	64
Siral-30	Ni(2.2)-Sir30	0.33	2.2	424	0.839	16
	Ni(6.5)-Sir30	0.32	6.5	399	0.799	14
Siralox-30	Ni(2.1)-Slox30	0.33	2.1	219	0.697	21
	Ni(5.5)-Slox30	0.33	5.5	180	0.596	16
ASA-25	Ni(2.9)-ASA25	2.50	2.9	266	0.628	32
	Ni(6.4)-ASA25	2.50	6.4	256	0.482	23
ASA-13	Ni(1.7)-ASA13	5.40	1.7	461	0.730	46
	Ni(6.2)-ASA13	5.30	6.2	421	0.673	32

[a] Chemical composition determined by chemical analysis (ICP-OES).

[b] From N$_2$ adsorption ($-196°$C). TPV is the total pore volume obtained with the BJH formalism applied to the adsorption branch of the isotherm.

[c] Density of Brønsted acid sites measured by FTIR of adsorbed pyridine at a desorption temperature of 150°C.

the form of boehmite) and Siralox-30 were obtained from Sasol materials, while ASA-25 and ASA-13 having lower Al contents and higher BET surface areas were supplied by Crossfield. The total pore volume did not differ much for the different silica–alumina carriers (0.6–0.7 cm^3/g) with most of the pores falling in the meso-pore range. Moreover, the ASA supports and their corresponding catalysts contained a higher density of Brønsted acid sites, as measured by FTIR-pyridine.

The incorporation of Ni generally provoked a modest decrease in the BET area of the silica–alumina supports as well as in the Brønsted acidity, which becomes more notorious at greater Ni contents. The decrease in the Brønsted acid site density was, at a constant Ni loading, higher for the catalysts based on the silica–aluminas supplied by Crossfield (ASA) having higher Si/Al ratios. No diffractions related to bulk NiO were evidenced by XRD for the catalysts loaded with 2 wt% Ni. By contrast, characteristic reflections for the NiO phase were detected at higher Ni loadings (6 wt%) for the catalysts based on Silarox-30 and ASA supports but were absent for that prepared from Siral-30. This indicates that a better dispersion of the Ni species can be achieved when the alumina phase in the carrier at the impregnation stage is boehmite (AlOOH or hydrated alumina), as in Siral-30, instead of the oxidic Al$_2$O$_3$ form. This fact is probably due to a stronger interaction of the Ni precursor with the hydrated alumina surface containing a higher concentration of surface –OH groups acting as *anchoring* points for cationic Ni species and inhibiting the formation of relatively large NiO clusters during the calcination step.

Preliminary experiments performed at different reaction temperatures in the 50°C–220°C range revealed a maximum productivity to liquid oligomers at 120°C

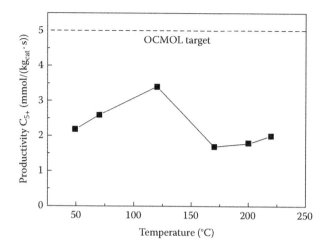

FIGURE 11.9 Productivity to liquid oligomers as a function of reaction temperature for the Ni(2.1)-Slox30 catalyst. Conditions: P_{total} = 35 bar, P_{ethyl} = 26 bar, and WHSV = 2.1 h^{-1}.

(P_{total} = 35 bar, P_{ethyl} = 26 bar, and WHSV = 2.1 h^{-1}). The productivity-temperature dependence is exemplified for the Ni(2.1)-Slox30 catalyst in Figure 11.9. Therefore, the catalytic performance of the prepared Ni–SiO$_2$–Al$_2$O$_3$ catalysts was compared at the optimum reaction temperature of 120°C.

Under these conditions, all the catalysts displayed a stable activity in experiments lasting 8–12 h. In general, a higher productivity to liquids was obtained for catalysts loaded with 6 wt% Ni. Among the different Ni–SiO$_2$–Al$_2$O$_3$ catalysts prepared, those based on Siralox-30 were, by far, the most productive to liquid oligomers, with sustained values of around 3.5 mmol/(kg$_{cat}$·s) at 120°C and WHSV of 2.1 h^{-1} during 8–12 h on stream. The reasons for the higher oligomerization activity of the Siralox-30 catalysts remain, at present, unclear. Indeed, these catalysts were characterized by relatively low specific surface areas and Brønsted acidities as compared to the other silica–aluminas (Table 11.2). Increasing WHSV from 2.1 to 5.1 h^{-1} increased the productivity of Ni(2.1)-Slox30 catalyst to 5.6 mmol/(kg$_{cat}$·s) and hence to values above the OCMOL target. Therefore, an extended run was carried out at the aforementioned conditions in order to assess the stability of the catalyst with time on stream (TOS). The results of this extended run are shown in Figure 11.10. As seen there, productivities to liquids above the target value were obtained during the first 35 h on stream. At higher reaction times, the productivity slightly declined below the targeted value of 5 mmol/(kg$_{cat}$·s) and reached ca. 4.5 mmol/(kg$_{cat}$·s) after 60 h. Nevertheless, an average productivity slightly above the target, 5.2 mmol/(kg$_{cat}$·s), was obtained during the whole extended run.

11.3.3 Ni–Al-MCM-41 Catalysts

For this study, Ni–Al-MCM-41 catalysts with different Ni loading (1–10 wt%) and atomic Si/Al ratio (14–43) were prepared by incipient wetness impregnation with nickel nitrate solutions in ethanol. For the support with Si/Al = 14, an additional

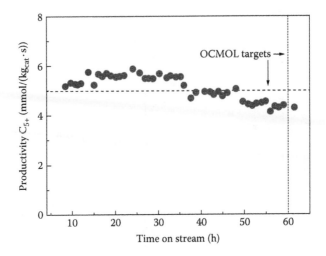

FIGURE 11.10   Extended run showing the evolution of the productivity to liquid oligomers with time for Ni(2.1)-Slox30 catalyst at the reaction conditions of T = 120°C, P_{total} = 35 bar, P_{ethyl} = 26 bar, and WHSV = 5.1 h^{-1}.

catalyst was also prepared by ion exchange, that is, sample Ni(1.3)-M41(14)-ex. After impregnation or ion exchange, the samples were dried and finally calcined in air at 500°C for 3 h. The nomenclature and main physicochemical properties of the Al-MCM-41 carriers and corresponding Ni–Al-MCM-41 catalysts are summarized in Table 11.3.

TABLE 11.3

Nomenclature and Physicochemical Properties of Al-MCM-41 Supports and Ni–Al-MCM-41 Catalysts

Sample	Atomic Si/Al Ratio	Ni Loading (wt%)	BET Area (m²/g)	APD^a (nm)	Brønsted Acidity^b (μmol$_{py}$/g$_{cat}$)
M41(14)	14	—	923	3.4	64
M41(22)	22	—	968	3.4	52
M41(43)	43	—	1017	3.3	39
Ni(1.3)-M41(14)-ex	14	1.3	904	3.3	43
Ni(1.3)-M41(14)-im	14	1.3	901	3.2	51
Ni(2.8)-M41(14)-im	14	2.8	880	3.2	34
Ni(5.2)-M41(14)-im	14	5.2	829	3.2	36
Ni(9.8)-M41(14)-im	14	9.8	781	3.2	32
Ni(5.4)-M41(22)-im	22	5.4	861	3.4	23
Ni(5.0)-M41(43)-im	43	5.0	908	3.2	23

^{a} APD: Average pore diameter obtained by applying the BJH formalism to the adsorption branch of the N$_2$ adsorption isotherms measured at −196°C.

^{b} Amount of Brønsted acid sites determined by FTIR of adsorbed pyridine at a desorption temperature of 150°C.

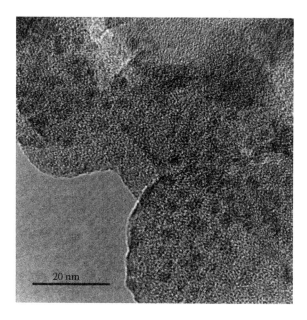

FIGURE 11.11   Representative TEM image showing the presence of uniform NiO nanoparticles in the Ni(5.2)-M41(14)-im catalyst.

All the materials displayed very high specific surface areas (>800 m^2/g) and average pore diameters of 3.3 ± 0.1 nm. Low-angle XRD and TEM measurements indicated that in all catalysts, the ordered mesoporous structure was well preserved after incorporation of nickel and calcination. Moreover, no diffraction peaks of bulk NiO were detected by XRD in the high angle range even for the highly loaded catalysts, indicating a very high dispersion of the Ni species in these materials. In fact, very small NiO nanoparticles located inside the mesopores with a uniform size of about 3 nm were observed by TEM, as illustrated in Figure 11.11 for the Ni(5.2)-M41(14)-im sample. It can also be seen in Table 11.3 that loading of Ni species reduced the density of Brønsted acid sites initially present in the Al-MCM-41 support. The reduction in Brønsted acidity can be mostly related to a partial exchange of protons (H$^+$) in Al-MCM-41 by Ni^{2+} species, though a certain blockage of acid sites at high loadings by NiO nanoparticles encapsulated in the mesopores is not discarded.

The Ni–Al-MCM-41 catalysts were evaluated for the continuous oligomerization of ethylene in a fixed bed reactor. No deactivation was observed for any of the catalysts in runs lasting ca. 8 h under the investigated reaction conditions. First, the influence of the Ni incorporation method and Ni loading was investigated for the series of catalysts prepared from the Al-MCM-41 carrier with Si/Al = 14. The ethylene conversion and the productivity to liquid oligomers obtained with these catalysts at T = 120°C, P$_{tot}$ = 35 bar, P$_{ethyl}$ = 26 bar, and WHSV = 2.1 h^{-1} are presented in Figure 11.12 as a function of Ni content.

As observed in Figure 11.12a, at equivalent Ni loading (1.3 wt%), the catalyst prepared by impregnation was significantly more active than that obtained

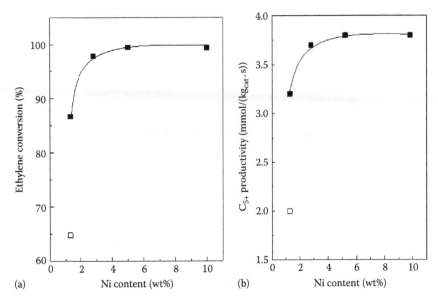

FIGURE 11.12 (a) Effect of Ni loading on ethylene conversion and (b) productivity to liquid oligomers for Ni–Al-MCM-41 catalysts (Si/Al = 14) prepared by either ionic exchange (open symbol) or impregnation (closed symbol). Reaction conditions: T = 120°C, P_{tot} = 35 bar, P_{ethyl} = 26 bar, and WHSV = 2.1 h⁻¹.

by ion exchange, despite the latter containing a higher proportion of Ni^{2+} species in exchange positions, as inferred from the larger decrease in Brønsted acidity (see Table 11.3) as well as from H_2-TPR measurements (not shown). This result suggests, in contrast to the general belief, that ion-exchanged Ni cations are not the only active species in Ni–Al-MCM-41 catalysts, and calls for a more thorough characterization study in order to further elucidate the nature of the active nickel sites in these materials. For the impregnated series, nearly 100% ethylene conversion was obtained under the studied conditions for Ni loadings of 5 wt% and above. The fact that a maximum catalytic activity was obtained for catalysts with Ni contents well above the maximum theoretical exchange capacity of the Al-MCM-41 carrier further questions the assignment of cationic Ni in exchange positions as the only active nickel species in these catalysts. The selectivity to C_{5+} oligomers increased with Ni loading up to 5 wt%, reaching a value as high as ca. 90% at the studied conditions. Most of the C_{5+} oligomers fell within the gasoline range (C_5–C_{10}) irrespective of Ni loading. The degree of branching for the most abundant C_8 and C_{10} gasoline-range oligomers was very high (>0.9) and hardly changed with the Ni content. As seen in Figure 11.12b, the productivity to liquid oligomers followed a similar trend to that of conversion as a function of Ni loading. In that case, a maximum productivity to liquids of 3.8 mmol/(kg_{cat}· s) was obtained at Ni loadings of 5–10 wt%. On the other hand, for catalysts containing 5 wt% Ni prepared by impregnation, the Si/Al ratio of the starting Al-MCM-41 material had a minor effect on the catalytic performance of the Ni–Al-MCM-41 catalysts, with productivities to liquid oligomers of 3.8–3.9 mmol/(kg_{cat}· s).

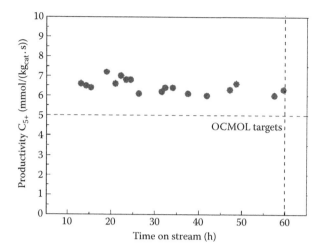

FIGURE 11.13 Evolution of the productivity to liquid oligomers with time for the Ni(5.2)-M41(14)-im catalyst at T = 120°C, P_{total} = 35 bar, P_{ethyl} = 26 bar, and WHSV = 5.1 h^{-1}.

At the investigated operating conditions, the maximum productivity to liquids for the most active Ni–Al-MCM-41 catalysts discussed earlier, that is, 3.8–3.9 mmol/(kg$_{cat}$·s), did not reach the targeted value of 5 mmol/(kg$_{cat}$·s). Therefore, additional experiments were performed in order to assess the impact of reaction temperature and WHSV on productivity for the most active Ni(5.2)-M41(14)-im catalyst. From this study, a maximum productivity to liquids of ca. 6.5–7 mmol/(kg$_{cat}$·s) could be obtained at a reaction temperature of 120°C and WHSV of 4 h^{-1} during 8 h on stream. This productivity was about 30%–40% higher than the targeted value of 5 mmol/(kg$_{cat}$·s). More importantly, an average productivity above the target could be sustained for a reaction period of at least 60 h, as ascertained in an additional extended run (Figure 11.13), thus also accomplishing the catalyst lifetime target initially established in the project for the development of an efficient ethylene oligomerization catalyst.

11.4 TOLERANCE OF Ni-BASED CATALYSTS TO FEED IMPURITIES

An assessment of the effect on the oligomerization performance of the main impurities in the ethylene stream from the OCM reactor, even after separation, was a critical issue for the overall OCMOL process integration. The tolerance of the Ni-based catalysts to potential poisons in the OCM effluent will largely determine the requirements and the particular technologies to be applied in the previous separation steps before the ethylene-rich stream enters the oligomerization reactor. Among the impurities present in the OCM effluent, CO and CO_2 were identified as the most critical for the performance of Ni-based oligomerization catalysts. In a preliminary study, it was found that methane, a by-product of the OCM reaction, had no appreciable effect, at least in concentrations of 12–37 vol%, on the catalytic performance of Ni-based catalysts, acting as a simple diluent in the oligomerization reactor.

The study of the effect of CO and CO_2 in the ethylene feed was performed using an amorphous Ni–SiO_2–Al_2O_3 catalyst prepared by impregnation with a Ni content of 2.1 wt%. This catalyst did not experience any appreciable deactivation in experiments lasting about 8 h using pure ethylene feed at the reaction conditions employed in the poisoning study.

First, the impact of CO was addressed according to the following experimental protocol. Initially, a CO-free ethylene/Ar/N_2 gas mixture was fed to the reactor, Ar being used as internal standard for GC analyses and N_2 as balance gas to maintain the ethylene partial pressure constant at the desired value. When a steady-state behavior was reached at a certain reaction temperature, the CO-free feed mixture was replaced by an ethylene/Ar/N_2/CO mixture. In the CO-containing feed mixture, the N_2 flow was modified in order to maintain the ethylene pressure at the same value as in the CO-free experiments (P_{ethyl} = 20.5 bar). Figure 11.14 shows the effect of CO on the ethylene conversion over the Ni–SiO_2–Al_2O_3 catalyst at different reaction temperatures, that is, 50°C, 120°C, and 200°C, and CO concentrations, that is, 0.038, 0.8, and 4.0 vol%. As seen in the figure, the presence of CO in the ethylene feed produced a drastic decline of the ethylene conversion from 70% to 95%, depending on reaction temperature, to below 15%. Increasing the CO concentration from 380 ppm to 4.0 vol% had minor effect on the deactivation rate of the catalyst, though the deactivation started the earliest at the highest CO concentration of 4.0 vol%. Such CO-induced catalyst deactivation was seen to be nearly reversible. Thus, reactivation of the poisoned catalyst in flowing dry N_2 at 300°C followed by reaction with a

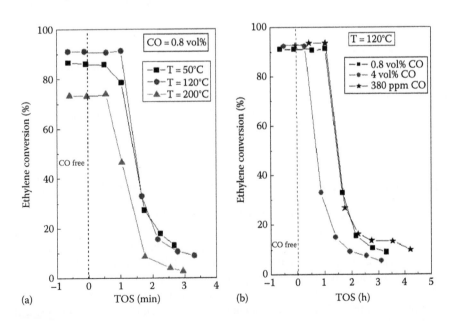

FIGURE 11.14 Ethylene conversion as a function of TOS obtained for a Ni–SiO_2–Al_2O_3 catalyst in the absence (TOS < 0) and presence (TOS > 0) of CO (a) at different temperatures and constant CO concentration of 0.8 vol% and (b) at 120°C and different CO concentration levels. Reaction conditions: P_{total} = 35 bar, P_{ethyl} = 20.5 bar, and WHSV = 2.2 h^{-1}.

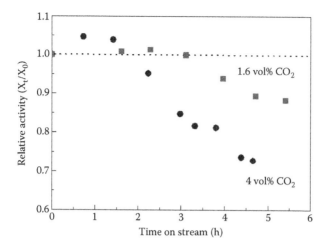

FIGURE 11.15 Influence of CO_2 on the activity of a $Ni-SiO_2-Al_2O_3$ catalyst for the oligomerization of ethylene at 120°C, 35 bar total pressure, 15 bar ethylene partial pressure, and 2.1 h^{-1} WHSV.

CO-free ethylene feed restored about 85%–90% of the initial catalyst activity, suggesting that the loss of activity was likely due to a strong adsorption of CO on the nickel sites.

Besides CO, the presence of CO_2 in the ethylene feed stream was also seen to affect the stability of the catalyst with TOS. In this case, the investigation was performed at T = 120°C, P_{tot} = 35 bar, P_{ethyl} = 15 bar, WHSV = 2.1 h^{-1}, and CO_2 concentrations of 1.6 and 4.0 vol%. The evolution with time of the relative catalyst activity X_t/X_0, where X_t is the ethylene conversion at a reaction time t in the presence of CO_2 and X_0 is the conversion in the absence of CO_2, is shown in Figure 11.15. A gradual loss of the activity with time was evidenced in the presence of CO_2. The poisoning effect of CO_2 was more drastic with increasing its concentration in the feed from 1.6 to 4 vol%. Nonetheless, the detrimental effect of CO_2 on catalyst stability was, at equivalent concentration and reaction conditions, less pronounced than that observed for CO.

The aforementioned results clearly indicated that both CO and CO_2 dramatically reduced the stability with time of the Ni-based catalysts and should, thus, be removed from the ethylene stream before entering the oligomerization reactor.

11.5 CONCLUSIONS

The three nickel-based catalysts developed in the project, namely, Ni-beta, Ni–SiO_2–Al_2O_3, and Ni–Al-MCM-41, displayed high activity and stability during the oligomerization of ethylene under suitable reaction conditions. The stability of these catalysts can be mostly related to the presence of mesopores favoring the mass transport of liquid, and branched, oligomers. The nanocrystalline character of the beta zeolite, with an average crystallite size of 25 nm, and the mesoporosity associated to the interparticle space are the key for the improved stability of the developed Ni-beta catalysts with respect to purely microporous Ni-zeolites reported in previous studies. At a given reaction condition, the activity of Ni-beta and Ni-MCM-41 catalysts was largely determined by the nickel loading. For both catalysts, optimum catalytic performance was obtained at a Ni loading of ca. 5 wt% introduced by incipient wetness impregnation with nickel nitrate solutions. Higher Ni loadings did not result in an improved catalytic activity either due to the formation of inactive bulk-like NiO particles located on the external surface of the beta crystallites in Ni-beta or to partial blockage of pores by very small NiO nanoparticles with a size of ca. 3 nm that are confined within the mesopores in Ni–Al-MCM-41. For silica–alumina-based catalysts, the content and chemical nature of the alumina phase, that is, boehmite, Al_2O_3, and the textural properties were more influential factors than Ni loading (in the 2–6 wt% range).

The activity of Ni-beta correlated well with the concentration of *isolated* Ni^{2+} cations in exchange positions, suggesting that these are the active nickel species for activation and *true oligomerization* of ethylene. An exhaustive in situ FTIR study of CO adsorption and CO-ethylene co-adsorption performed in the context of OCMOL provided further support to this assignment and discarded low-valent Ni^+ ions as the active sites, in contrast to the most generalized opinion. In the case of Ni–Al-MCM-41 catalysts, however, the fact that the maximum activity was attained at Ni loadings (5 wt%) well above the maximum exchange capacity of the Al-MCM-41 support suggested that ion-exchanged Ni cations are not the only active species in these catalysts. This result, which contrasts the general belief, calls for additional studies aiming at elucidating the nature of Ni sites in Ni–Al-MCM-41 systems.

Under conditions of high ethylene conversion (>95%), high selectivity to liquid (C_{5+}) oligomers was achieved over the three optimized Ni-based catalysts. Gasoline-range oligomers (C_5–C_{12}) were the predominant liquid products. For instance, this fraction represented about 80% of the liquids in Ni-beta (5 wt% Ni) at 120°C. Oligomers with more than 6 carbon atoms contained a high proportion of branched products (typically >85%) as a result of acid-catalyzed co-oligomerization and isomerization reactions occurring on the acidic supports. Thus, the C_8–C_{12} fraction obtained on these catalysts could be suitably used, after a simple hydrogenation step, as a high-octane synthetic gasoline component.

Besides the chemical composition of the catalysts, the reaction conditions and, more specifically, the temperature and space velocity had a major influence on the activity and productivity to liquid oligomers. At optimum reaction conditions and catalyst composition, a higher productivity to liquid oligomers was obtained for Ni–Al-MCM-41 as compared to Ni–SiO_2–Al_2O_3 and Ni-beta, the less active. In fact, a productivity to liquids above 6 mmol/(kg$_{cat}$·s) could be sustained in an extended run lasting 60 h over the most active Ni–Al-MCM-41 sample (5 wt% Ni, impregnation) at 120°C, total pressure of 35 bar, ethylene partial pressure of 26 bar, and WHSV of 4 h^{-1}. Therefore, the optimized Ni–Al-MCM-41 catalyst accomplished the productivity and lifetime targets initially established in the OCMOL proposal for the oligomerization subproject.

Finally, tolerance studies performed with a reference Ni–SiO_2–Al_2O_3 catalyst revealed that the catalyst stability with time becomes severely affected by the presence of CO or CO_2 in the ethylene feed. These compounds, and particularly CO, acted as strong poisons for the Ni-based catalyst drastically reducing its lifetime. It is, thus, concluded that both impurities should be removed from the ethylene-rich stream in previous separation steps before entering the oligomerization reactor.

ACKNOWLEDGMENT

This work was undertaken in the context of the project OCMOL (*Oxidative Coupling of Methane followed by Oligomerization to Liquids*). OCMOL is a large-scale collaborative project supported by the European Commission in the Seventh Framework Programme (GA no. 228953). For further information about OCMOL, see http://www.ocmol.eu or http://www.ocmol.com.

REFERENCES

1. T. Mokrani and M. Scurrell, *Catal. Rev. Sci. Eng.* 51 (2009) 1–145.
2. BP Statistical Review of World Energy, June 2013, www.bp.com/statisticalreview, (last accessed May 6, 2014).
3. EU Project OCMOL (Oxidative Coupling of Methane followed by Oligomerization to Liquids), 7th Framework Programme (GA no. 228953), see http://www.ocmol.eu or http://www.ocmol.com, (last accessed May 15, 2014).
4. D.S. McGuiness, *Chem. Rev.* 111 (2011) 2321–2341.
5. V. Hulea and F. Fajula, *J. Catal.* 225 (2004) 213–222.
6. J. Heveling, C.P. Nicolaides, and M.S. Scurrell, *J. Chem. Soc. Chem. Commun.* 173 (1991) 126–127.
7. J. Heveling, C.P. Nicolaides, and M.S. Scurrell, *Appl. Catal. A* 173 (1998) 1–9.
8. J. Heveling, C.P. Nicolaides, and M.S. Scurrell, *Appl. Catal. A* 248 (2003) 239–248.
9. M. Lallemand, O.A. Rusu, E. Dumitriu, A. Finiels, F. Fajula, and V. Hulea, *Appl. Catal. A* 338 (2008) 37–43.
10. M. Lallemand, A. Finiels, F. Fajula, and V. Hulea, *Appl. Catal. A* 301 (2006) 196–201.
11. A.M. Al-Jarallah, J.A. Anabtawi, A.A.B. Siddiqui, A.M. Aitani, and A.W. Alsadoun, *Catal. Today* 14 (1992) 1–121.
12. J.R. Sohn and A. Ozaki, *J. Catal.* 59 (1979) 303–310.
13. J.R. Sohn and A. Ozaki, *J. Catal.* 61 (1980) 29–38.
14. J. Heveling, A. Van Der Beek, and M. De Pender, *Appl. Catal.* 42 (1988) 325–336.
15. F.T.T. Ng and D.C. Creaser, *Appl. Catal. A* 119 (1994) 327–339.
16. M. Tanaka, A. Itadani, Y. Kuroda, and M. Iwamoto, *J. Phys. Chem. C* 116 (2012) 5664–5672.
17. J.R. Sohn, W.C. Park, and S.-E. Park, *Catal. Lett.* 81 (2002) 259–264.
18. M.D. Heydenrych, C.P. Nicolaides, and M.S. Scurrell, *J. Catal.* 197 (2001) 49–57.
19. M. Lallemand, A. Finiels, F. Fajula, and V. Hulea, *Chem. Eng. J.* 172 (2011) 1078–1082.
20. A. Lacarriere, J. Robin, D. Świerczyński, A. Finiels, F. Fajula, F. Luck, and V. Hulea, *Chem. Sus. Chem.* 5 (2012) 1787–1792.
21. L. Bonneviot, D. Olivier, and M. Che, *J. Mol. Catal.* 21 (1983) 415–430.
22. M. Lallemand, A. Finiels, F. Fajula, and V. Hulea, *J. Phys. Chem. C* 113 (2009) 20360–20364.
23. A. Corma, *J. Catal.* 216 (2003) 298–312.
24. J.R. Sohn and J.H. Park, *Appl. Catal. A* 218 (2001) 229–234.
25. R.L. Espinoza, C.J. Korf, C.P. Nicolaides, and R. Snel, *Appl. Catal.* 29 (1987) 175–184.
26. R.L. Espinoza, R. Snel, C.J. Korf, and C.P. Nicolaides, *Appl. Catal.* 29 (1987) 295–303.

27. R.L. Espinoza, C.P. Nicolaides, C.J. Korf, and R. Snel, *Appl. Catal.* 31 (1987) 259–266.
28. J. Heveling, C.P. Nicolaides, and M.S. Scurrell, *Catal. Lett.* 95 (2004) 87–91.
29. J. Heveling and C.P. Nicolaides, *Catal. Lett.* 107 (2006) 117–121.
30. M. Hartmann, A. Pöppl, and L. Kevan, *J. Phys. Chem.* 100 (1996) 9906–9910.
31. T. Cai, D. Cao, Z. Song, and L. Li, *Appl. Catal. A* 95 (1993) L1–L7.
32. T. Cai, *Catal. Today* 51 (1999) 153–160.
33. I.V. Elev, B.N. Shelimov, and V.B. Kazansky, *J. Catal.* 89 (1984) 470–477.
34. A. Martínez, M.A. Arribas, P. Concepción, and S. Moussa, *Appl. Catal. A* 467 (2013) 509–518.

12 Understanding the Mechanisms of Gas to Liquid Processes with the Aid of Quantum Chemistry Tools
Case Studies of the Shilov Reaction and Natural Gas Aromatization

Dorota Rutkowska-Zbik, Renata Tokarz-Sobieraj, and Małgorzata Witko

CONTENTS

12.1 THEORETICAL APPROACH TO MOLECULAR CATALYSIS

Recent advances in computational methods as well as quickly growing performance of computers resulted in the burst of application of computational chemistry tools to study catalysts and catalytic reactions. Most of the examples of application of the modern quantum chemistry methods are based on density functional theory (DFT). The details of the mathematical formulation of the DFT an interested reader may find in a specialized literature. The selection of this formalism is usually due to the fact that it combines high accuracy with relatively low cost of calculations, which is of utmost importance taking into account the usual complexity of the studied catalytic systems. DFT is based on the theorem formulated by Hohenberg and Kohn in 1964, which states that all ground state properties of a system are uniquely determined by its electron density ρ. This approach allows for the simplification of the calculations in comparison to the classical quantum mechanics, which operates in the virtual space of 4N dimensional (where N denotes the number of electrons); DFT deals with an observable ρ, which can be accessed by experiment (x-ray diffraction measurements), and is a function in the real 3D space. Accordingly, DFT allows for the study of the larger systems than traditional quantum mechanical methods. Another important factor explaining the increasing popularity of the DFT method is the inclusion of the electron correlation, rendering it particularly suited for the description of systems containing transition metals, which often play the role in the catalysts.

The employment of the theoretical chemistry methods may help to solve particular problems arising while studying a catalytic system. For understanding the catalysis at the molecular level, one often asks detailed questions about the nature of the reacting species, both the catalyst and the substrates, as well as the more general ones, regarding the relation between the observed properties and the performance of the system as a whole. In such a sense, the theoretical methods may contribute to the rational catalyst design.

It should be borne in mind that with the aid of theoretical modeling, one can approach the following problems and search for replies to the following questions asked:

1. Structure of a catalyst and identification of active sites:
   a. What is the geometry of the catalyst?
   b. What is the electronic state of the catalyst?
   c. How does the active center look like?
   d. Are there any other possible active centers?
   e. How does the interaction between the active phase and its support look like?
2. Elucidation of a reaction mechanism:
   a. What are the key steps of the catalytic reaction?
   b. What is the nature of the transient species?
   c. Are there any alternative routes possible?
   d. What is the rate-determining step?

3. Structure–activity relationship:
   a. What are the factors promoting/tampering the reactivity?
   b. How can the catalyst be tuned to obtain better performance?
   c. How can one modify the catalyst to suppress unwanted reactivity?

In the following, the application of theoretical methods to elucidate the details of the selected examples of methane to liquid processes will be presented. The first one is a low-temperature methane oxidation to methanol on homogeneous Pt complex, known as the Shilov process. The second one is a nonoxidative conversion of methane to aromatics, catalyzed by 3%–5% MoO_3 in ZSM-5-type zeolite.[1–8]

12.2 APPLICATION OF QUANTUM CHEMICAL METHODS TO STUDY THE REACTIVITY OF METHANE

12.2.1 METHANE OXIDATION TO METHANOL IN THE SHILOV PROCESS

In 1972, Shilov reported that the addition of Pt(IV) to the aqueous complex of the $PtCl_4^{2-}$ formula with methane led to the production of the selectively oxidized species, that is, methanol and methyl chloride (Equation 12.1):

$$CH_4 + PtCl_6^{2-} + H_2O(Cl^-) \xrightarrow{PtCl_4^{2-},H_2O} CH_3OH(CH_3Cl) + PtCl_4^{2-} + 2HCl \qquad (12.1)$$

Consequently, a significant fraction of research in the area of the *Shilov chemistry* was concentrated on determining the mechanism of the Shilov reaction. A reasonable mechanistic scheme for the catalytic cycle is present in Figure 12.1.

According to the scheme, the Pt(II) complex reacts with methane to form a Pt(II)–methyl complex. The product is then oxidized to a methyl platinum (IV) species in the second step. Either reductive elimination involving the Pt(IV) methyl group and coordinated water or chloride or, alternatively, nucleophilic attack at the carbon by an external nucleophile (H_2O or Cl^-) was proposed to generate the functionalized product and reduce the Pt center back to Pt(II) in the final step. However, the cycle as pictured earlier is quite general and the intimate mechanism of the individual steps may differ for each particular system.

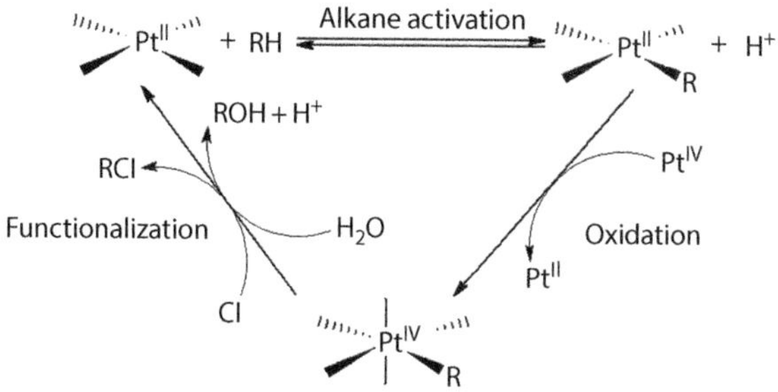

FIGURE 12.1 Proposed reaction mechanism of Shilov process for oxidation of methane.

The most significant advancement for the Shilov system appeared in 1998. Researchers at Catalytica discovered that methane could be selectively converted to methyl bisulfate by a Pt(II) bipyrimidine (bpym) catalyst in concentrated sulfuric acid (Equation 12.2). An impressive 72% yield (89% conversion of methane, 81% selectivity) was observed in this reaction, which formally used SO_3 as the oxidant:

$$CH_4 + 2H_2SO_4 \xrightarrow{\text{(bpym)PtCl}_2} CH_3OSO_3H + 2H_2O + SO_2 \qquad (12.2)$$

The mechanism of the reaction went under investigation, and a general scheme similar to that shown in Figure 12.1 for the Shilov reaction was proposed.

Several articles reported the mechanism of methane oxidation using Pt(II) complex as a catalyst. Methane to methanol transformation employing *C–H activation* chemistry inspired a number of theoretical investigations using DFT. Most of the studies concentrated, however, on various fragments of the mechanism, in full depicted in Figure 12.2.

The most comprehensive approach to the mechanism and energetics of methane to methanol conversion is present in the work done by Paul and Musgrave.[9] This theoretical study established full potential energy profiles for the catalytic cycles of methane to methanol conversion by Catalytica and cisplatin and attributed higher rates for cisplatin compared to Catalytica, based on computed energy barriers for Pt(II) to Pt(IV) oxidation. The calculations focused on four major stages of the catalytic cycle, namely, formation of σ-complex, C–H activation to form the methylated Pt(II) complex, oxidation of the methylated Pt(II) complex to form a Pt(IV) complex, and functionalization of the methyl group to form CH_3OSO_3H through S_N2 nucleophilic attack on the methyl by a bisulfate anion. All quantum chemical calculations were carried out using B3LYP functional, with inclusion of the solvent effects approached for by the CPCM method for H_2SO_4 with a dielectric constant of 98. Detailed theoretical studies were performed for selected Catalytica: (**1**) [Pt(bpym)(Cl)$_2$], (**2**) [Pt(bpym)(Cl)(OSO$_3$H)], (**3**) [Pt(bpymH)(Cl)$_2$]$^+$, (**4**) [Pt(bpymH)(Cl)(OSO$_3$H)]$^+$, and two cisplatin (**5**) [Pt(NH$_3$)$_2$(Cl)$_2$] and (**6**) [Pt(NH$_3$)$_2$(Cl)(OSO$_3$H)] complexes—see Figure 12.3.

Table 12.1 summarizes the relative energies for all major reaction steps. The following discussion of the particularities of the reaction mechanism is divided into four parts, each comprising of one of the major steps of the Shilov reaction, as defined by Musgrave.

12.2.1.1 CH$_4$ Uptake by Pt(II) Complex

C–H activation follows σ-complex formation. Two modes through which σ-complex formation can occur are plausible: (1) dissociation (D) of the chloride ligand and subsequent formation of the σ-complex by association with methane and (2) direct nucleophilic displacement (ND). The calculations indicated that mode (1) is favored over mode (2) for all studied systems. In addition, as seen in Table 12.1, displacing Cl$^-$ in cisplatin ((**2**) and (**3**)) is significantly easier than in Catalytica systems (**1**). This was expected, because the sp^3-N center of NH$_3$ is a better σ-donor than the more electronegative sp^2-N center of bpym and hence significantly weakened the Pt–Cl bond in *trans* position. Also the barrier of HSO$_4$ displacement from (**6**) [Pt(NH$_3$)$_2$(Cl)(OSO$_3$H)] by methane is 26.2 kcal/mol, 4.2 kcal/mol lower than

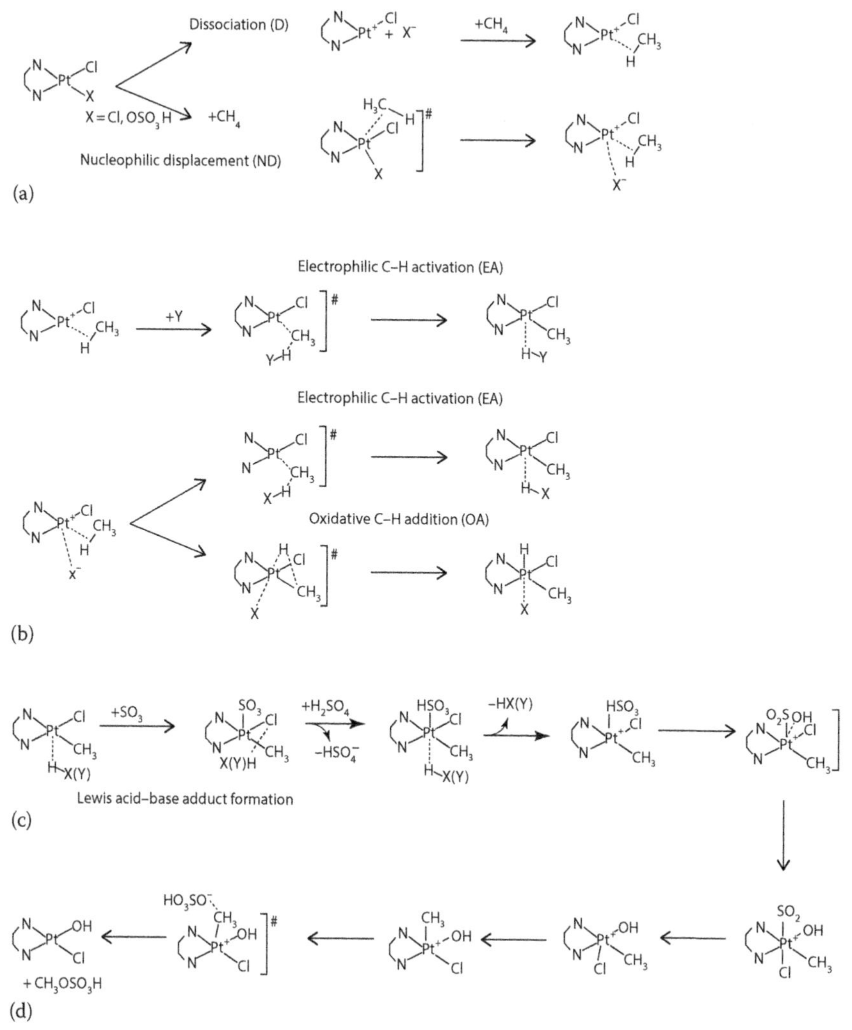

FIGURE 12.2 Schematic representation of the catalytic route for methane to methanol conversion by the Pt(II) catalysts: (a) σ-complex formation, (b) C–H activation, (c) oxidation, and (d) functionalization.

found for (**2**) [Pt(bpym)(Cl)(OSO$_3$H)]. As shown, the role and the position of the ligands were identified as important in the dissociative mechanism. They were also of relevance for assessing the stability of different systems.

Hush et al.[10] demonstrated that ammonia is more tightly bound than water; therefore, the activation energies for the methane–ammonia exchange are expected to be higher than for the analogous methane–water exchange—see Table 12.2. Additionally, by studying the formation of Pt(IV) complexes formed on the oxidative addition pathway, they found that strong *trans*-directing nature of methyl and

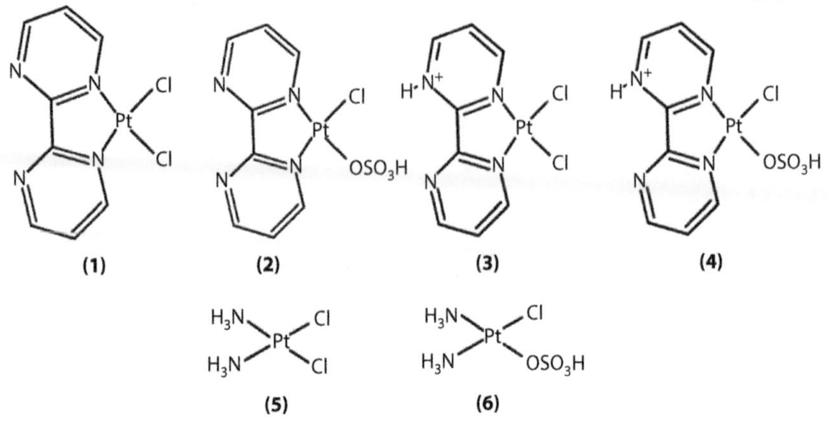

FIGURE 12.3 Structures of the proposed catalysts.

TABLE 12.1
Relative Energies of Different Reaction Steps of Methane Oxidation in the Shilov Reaction (in kcal/mol)

Reaction Step		1	2	3	4	5	6
σ-Complex formation	D	30.1	19.8	29.2	15.2	12.2	2.3
	ND	37.8	30.4	40.7	—	32.1	26.2
C–H activation	EA	40.4	33.1	40.0	30.0	36.4	29.7
	OA	42.5	35.3	42.6	31.6	35.6	27.9
Oxidation of Pt^{II} to Pt^{IV}		44.7	37.5	46.8	35.9	34.9	28.2
Isomerization		31.3	24.0	30.3	19.4	26.0	19.3
S_N2 attack		14.0	7.0	16.3	5.7	14.2	8.0

Note: D, dissociation of chloride; ND, direct nucleophilic displacement; EA, electrophilic C–H activation; OA, oxidative C–H addition.

TABLE 12.2
Gas Phase Binding Energy of Possible Pt(II) Ligands Computed at the B3LYP/ecp-DZ Level (in kcal/mol)

	NH_3	Cl^-	H_2O	CH_4
cis-$PtCl_2(NH_3)_2$	51.5	151	—	7.3
trans-$PtCl_2(NH_3)_2$	64.4	154.4	—	14.5
cis-$PtCl_2(H_2O)_2$	—	166.6	32.1	8.6
trans-$PtCl_2(H_2O)_2$	—	158.6	46.4	21.6

hydride ligands resulted in the formation of the *trans*-Pt(IV) complexes, which were expected to be less stable than *cis*-isomers.

Zhu and Ziegler[11] used DFT theory to investigate complexes formed as a result of $[Pt(Cl)_4]^{2-}$ dissolution in water, followed by their test in dissociative and associative processes of methane uptake. Concentration ratios of the different aqua platinum species depended on the concentration of chloride ions (Cl^-) added to the solution. They found that mainly three complexes $[Pt(Cl)_4]^{2-}$, $[Pt(Cl)_3(H_2O)]^-$, and $[Pt(Cl)_2(H_2O)_2]$ were formed by Cl^-/H_2O ligand substitution. The calculation results slightly favored the dissociative pathway. It was concluded that uptake rather than C–H activation was the rate-determining step under the Shilov reaction conditions. For the dissociative mechanism, $[Pt(Cl)_4]^{2-}$ contributed the most to the overall rate. For the associative mechanism, $[Pt(Cl)_3(H_2O)]^-$ and $[Pt(Cl)_2(H_2O)_2]$ had similar energy barriers, but higher concentration of $[Pt(Cl)_3(H_2O)]^-$ indicated it as the most active species.

12.2.1.2 C–H Activation

12.2.1.2.1 Electrophilic Activation versus Oxidative Addition

It was suggested that two significant routes for C–H activation exist: (1) electrophilic C–H activation (EA) and (2) oxidative C–H addition (OA). EA required that the leaving or dissociated ligand, or an external base (anion), captured the acidic hydrogen from the activated C–H bond, leading to the formation of the Pt(II)–CH_3-methylated complex, whereas oxidative addition led to the formation of a Pt(IV) hydride methyl complex. The calculations showed that energy barrier to oxidative addition is higher than the barrier to electrophilic addition for complexes (**1**), (**2**), (**3**), and (**4**), in agreement with the results of Goddard et al. for complex (**1**).[12] In the case of cisplatin, the barriers associated with C–H activation are notably different. For cisplatin (complexes (**5**) and (**6**)), oxidative addition is favored over EA by only 2.2 kcal/mol, in contrary to the calculations reported previously.

Goddard et al.[12] compared the mechanisms and energetics of methane to methanol reactions taking $[Pt(NH_3)_2(Cl)_2]$ and $[Pt(bpym)(Cl)_2]$ complexes as examples in concentrated sulfuric acid, based on B3LYP functional results. The authors found that reaction profiles were different for ammonia catalyst (Scheme 12.1) compared to the bipyrimidine catalyst (Scheme 12.2):

$$\left[(NH_3)_2\,Pt(Cl)_2\right] \rightarrow \begin{cases} E_A = 33.1 \\ \Delta E = 23.6 \end{cases} \left[(NH_3)_2\,Pt(Cl)_2(CH_4)\right] \rightarrow \begin{cases} E_A = 18.5 \\ \Delta E = 2.6 \end{cases} \left[(NH_3)_2\,Pt(Cl)(CH_3)\right]\ldots HCl$$

(a)

$$\left[(NH_3)_2\,Pt(Cl)_2\right] \rightarrow \begin{cases} E_A = 33.1 \\ \Delta E = 23.6 \end{cases} \left[(NH_3)_2\,Pt(Cl)_2(CH_4)\right] \rightarrow \begin{cases} E_A = 8.5 \\ \Delta E = -2.6 \end{cases} \left[(NH_3)_2\,Pt(H)(Cl)_2(CH_3)\right]$$

(b)

SCHEME 12.1 Reaction between methane and $[Pt(CH_3)_2(Cl)_2]$ (a) following the electrophilic activation pathway and (b) following the oxidative addition pathway. The energies, calculated in sulfuric acid as a solvent, are given in kcal/mol.

$$[(bpym)Pt(Cl)_2] \rightarrow \begin{cases} E_A = 40.7 \\ \Delta E = 30.8 \end{cases} [(bpym)Pt(Cl)_2(CH_4)] \rightarrow \begin{cases} E_A = 4.4 \\ \Delta E = 11.2 \end{cases} [(bpym)Pt(Cl)(CH_3)]...HCl$$

(a)

$$[(bpym)Pt(Cl)_2] \rightarrow \begin{cases} E_A = 40.7 \\ \Delta E = 30.8 \end{cases} [(bpym)Pt(Cl)_2(CH_4)] \rightarrow \begin{cases} E_A = 13.5 \\ \Delta E = -5.1 \end{cases} [(bpym)Pt(Cl)_2(CH_3)]$$

(b)

SCHEME 12.2 Reaction between methane and [Pt(bpym)(Cl)₂] (a) following the electrophilic activation pathway and (b) following the oxidative addition pathway. The energies, calculated in sulfuric acid as a solvent, are given in kcal/mol.

In the case of the [Pt(NH)₃(Cl)₂] complex, energy barriers were equal to 8.5 and 18.5 kcal/mol favoring oxidative addition over electrophilic substitution. The differences between these two mechanisms were larger than those calculated by Paul and Musgrave. In contrast to the amine catalyst, in the case of [Pt(bpym)(Cl)₂], the calculated energy barriers (4.4 and 13.5 kcal/mol) suggested that direct C–H activation was σ-bond metathesis (σBM) rather than oxidative addition. These results led to the general conclusion that C–H breaking mechanism strongly depended on the character of the ligands.

Siegbahn and Crabtree[13] examined the Shilov reaction using *trans*- and *cis*-form of the [M(H₂O)₂Cl₂] species, where M = Pt, Pd, with B3LYP functional.

It was demonstrated (compare Scheme 12.3) that the reaction starts from binding of methane molecule to the metal center and a subsequent methane–water exchange. In the second step, the C–H bond was broken. C–H activation may proceed via H dissociation to a chloride ligand and lead to the formation of a four-center transition state resembling that of σBM. The calculated energy at the transition state was equal to 20.5 kcal/mol (reduced to 16.5 kcal/mol by ZPE and temperature effects) for *trans*-Pt complex, differed only by 0.1 kcal/mol for Pd *trans*-dichlorides. The activation energy was higher by 6.7 kcal/mol for *cis*-Pd than for the *trans*-Pd complex. By passing the low-energy barrier (0.6 kcal/mol), associated with the σBM transition state, the structure was reorganized to contain an (H₅O₂)⁺ ligand with two hydrogen bonds to the chlorides. Additionally, for *cis*-dichloride Pd and Pt complexes, an alternative

$$[(OH_2)_2Pt(Cl)_2](CH_4) \rightarrow \{\Delta E = -3.8 [(OH_2)Pt(Cl)_2(CH_4)](H_2O) \rightarrow$$

$$\begin{cases} E_A = 16.5 \\ \Delta E = 6.7 \end{cases} [(OH_2)Pt(Cl)(ClH)(CH_3)](H_2O) \rightarrow \begin{cases} E_A = 0.6 \\ \Delta E = 12.6 \end{cases} [Pt(Cl)_2(CH_3)]^- (H_5O_2)^+$$

(a)

$$[(OH_2)Pt(Cl)_2(CH_4)](H_2O) \rightarrow \begin{cases} E_A = 23.8 \\ \Delta E = 4.6 \end{cases} [(OH_2)Pt(H)(Cl)_2(CH_3)]$$

(b)

SCHEME 12.3 Reaction between methane and [Pt(Cl)₂(H₂O)₂] (a) following the electrophilic activation pathway and (b) following the oxidative addition pathway. The energies, calculated in sulfuric acid as a solvent, are given in kcal/mol.

reaction pathway, in which the C–H bond dissociated toward the water ligand, was studied. This step required 12.8 and 12.1 kcal/mol (without zero-point and temperature effects) for Pt and Pd, respectively. From this point, the C–H bond could dissociate toward the OH ligand to form water. The obtained high-energy barriers, as compared with C–H activation on Cl ligand, ruled out this reaction pathway. Theoretical calculations showed that the formation of $[Pt^{IV}(H)(OH_2)(Cl)_2(CH_3)]$ complex, in oxidative addition pathway, was only 4.6 kcal/mol above the outer-sphere methane complex. The calculated energy of the transition state, equal to 23.8 kcal/mol (without entropy effects), was comparable to the one calculated for the σBM transition state; therefore, both pathways were difficult to distinguish. However, for the Pd system, this transition state laid 36.0 kcal/mol above the outer-sphere methane complex, suggesting that the oxidative addition pathway was implausible for the studied Pd complex.

Summarizing, the results obtained by Siegbahn and Crabtree suggested that the breaking of the C–H bond in methane occurred in two steps. First, methane is coordinated to the metal; next, the C–H bond is broken via transfer of the hydrogen atom from methane σ-complex, first to the neighboring Cl^- ligand and then to the solvent (water). The total activation energy was computed as 27 kcal/mol and was in very good agreement with the experimental estimate of 28 kcal/mol. An alternative oxidative addition/reductive elimination sequence was also investigated and was found to be competitive with the σBM pathway for Pt(II), but not tenable for Pd(II) complex. The C–H activation was defined as the rate-determining step in this reaction.

Hush et al.[10] showed further differences between *cis*- and *trans*-$[Pt(NH_3)_2(Cl)_2]$ complexes. In the reported calculations carried out using B3LYP functional, a substitution of the NH_3 ligand by CH_4 was considered as the initial step.

Their results, presented in Schemes 12.4 and 12.5, suggested that the highest-energy barrier, computed as 34.2 and 43.9 kcal/mol for cisplatin and transplatin, respectively,

$$\left[Pt(NH_3)_2(Cl)_2\right](CH_4) \rightarrow \begin{cases} G_A = 34.2 \\ \Delta G = 30.5 \end{cases} \left[Pt(Cl)_2(NH_3)(CH_4)\right](NH_3) \rightarrow$$

$$\begin{cases} G_A = -2.1 \\ \Delta G = -8.9 \end{cases} \left[Pt(Cl)_2(NH_3)(CH_4)\right]...(NH_3) \rightarrow \begin{cases} G_A = 4.2 \\ \Delta G = 3.2 \end{cases} \left[Pt(Cl)_2(NH_3)(CH_3)(H)\right]...(NH_3) \rightarrow$$

$$\begin{cases} G_A = 5.3 \\ \Delta G = -6.3 \end{cases} \left[Pt(Cl)_2(NH_3)(CH_3)(H)(NH_3)\right] \rightarrow \begin{cases} G_A = 15.6 \\ \Delta G = -25.9 \end{cases} \left[Pt(Cl)_2(NH)_3(CH_3)\right]^-(NH_4)^+$$

(a)

$$\left[Pt(Cl)_2(NH_3)_2\right](CH_4) \rightarrow \begin{cases} G_A = 34.2 \\ \Delta G = 30.5 \end{cases} \left[Pt(Cl)_2(NH_3)(CH_4)\right](NH_3) \rightarrow$$

$$\begin{cases} G_A = -2.1 \\ \Delta G = -8.9 \end{cases} \left[Pt(Cl)_2(NH_3)(CH_4)\right]...(NH_3) \rightarrow \begin{cases} G_A = 11.0 \\ \Delta G = 8.6 \end{cases} \left[Pt(Cl...H...Cl)(NH_3)(CH_3)\right]...(NH_3)$$

(b)

SCHEME 12.4 Reaction between methane and cisplatin $[Pt(Cl)_2(NH_3)_2]$ (a) following the electrophilic activation pathway and (b) following the oxidative addition pathway. The energies, calculated in water as a solvent, are given in kcal/mol.

$$[Pt(Cl)_2(NH_3)_2](CH_4) \rightarrow \begin{cases} G_A = 43.9 \\ \Delta G = 38.7 \end{cases} [Pt(Cl)_2(NH_3)(CH_4)](NH_3) \rightarrow$$

$$\begin{cases} G_A = 0.3 \\ \Delta G = -8.9 \end{cases} [Pt(Cl)_2(NH_3)(CH_4)]...(NH_3) \rightarrow \begin{cases} G_A = 30.8 \\ \Delta G = 20.9 \end{cases} [Pt(Cl)(Cl...H)(NH_3)(CH_3)]...(NH_3) \rightarrow$$

$$\begin{cases} G_A = 0 \\ \Delta G = -18.0 \end{cases} [Pt(Cl)_2(NH_3)(CH_3)(H)]...(NH_3)$$

$$[Pt(Cl)_2(NH_3)(CH_3)(H)]...(NH_3) \rightarrow \begin{cases} G_A = 17.7 \\ \Delta G = -4.0 \end{cases} [Pt(Cl)_2(CH_3)][(NH_3)...(H)...(NH_3)]^+$$

$$[Pt(Cl)_2(NH_3)(CH_3)(H)]...(NH_3) \rightarrow \begin{cases} G_A = 6.2 \\ \Delta G = -8.8 \end{cases} [Pt(Cl)_2(NH_3)(CH_3)](NH_4)^+$$

(a)

$$[Pt(Cl)_2(NH_3)_2](CH_4) \rightarrow \begin{cases} G_A = 43.9 \\ \Delta G = 38.7 \end{cases} [Pt(Cl)_2(NH_3)(CH_4)](NH_3) \rightarrow$$

$$\begin{cases} G_A = 0.3 \\ \Delta G = -8.9 \end{cases} [Pt(Cl)_2(NH_3)(CH_4)]...(NH_3) \rightarrow \begin{cases} G_A = 10.3 \\ \Delta G = 11.1 \end{cases} [Pt(Cl)_2(NH_3)(CH_3)(H)]...(NH_3) \rightarrow$$

$$\begin{cases} G_A = 1.8 \\ \Delta G = -14.3 \end{cases} [Pt(Cl)_2(NH_3)_2(H)(CH_3)]$$

(b)

SCHEME 12.5 Reaction between methane and transplatin $[Pt(Cl)_2(NH_3)_2]$ (a) following the electrophilic activation pathway and (b) following the oxidative addition pathway. The energies, calculated in water as a solvent, are given in kcal/mol.

corresponded to the transition state for the initial methane–ammonia exchange. In the case of cisplatin, the subsequent C–H activation and proton elimination steps were comparable for the σBM and oxidative addition, suggesting that reaction may occur by both mechanisms. In transplatin, however, the σBM pathway required additional energy of 30.8 kcal/mol needed for H migration to chloride ligand. In contrast to these results, the analogous transition state and product for hydrogen migration in the *trans*-$[PtCl_2(H_2O)_2]$ system, in the study of Siegbahn and Crabtree, were found to be just 16.5 and 6.5 kcal/mol, respectively. Additionally, in contrast to the previous study on exploring oxidative addition mechanism, the authors found a low-energy barrier directly leading to the formation of the $[Pt(NH_3)(Cl)_2(H)(CH_3)]\cdots(NH_3)$ adduct. These results predicted that in the transplatin system, the C–H activation process took over the oxidative addition as the electrophilic activation.

Analogous results, favoring oxidative addition, were obtained by Swang et al.,[14] who studied C–H activation process on $[Pt(tmeda)(CH_3)]^+$ and $[Pt(eda)(CH_3)]^+$ cationic complexes, in which platinum had a reduced coordination number of 3. They found that the formation of σBM required 9.4 kcal/mol, whereas the oxidative addition pathway products were reached by energy barrier of 1.6 kcal/mol. Additionally, by including solvation effect (explicit NF_2H molecule was added to the model), they found that an electron pair donor stabilizes the Pt(IV) complex much more than Pt(II) species. The energy of transition states changed to 13.1 and 14.8 kcal/mol for σBM and OA, respectively.

$$\left[(\text{bpym})\text{Pt}(\text{Cl})\right]^{+} \rightarrow \left\{\Delta E = -9.9\left[(\text{bpym})\text{Pt}(\text{Cl})(\text{CH}_4)\right]^{+} \rightarrow \begin{cases} E_A = 17.3 \\ \Delta E = 10.6 \end{cases}\left[(\text{bpym})\text{Pt}(\text{ClH})(\text{CH}_3)\right]^{+} \rightarrow$$

$$\left\{\left\{\Delta E = 27.8\left[(\text{bpym})\text{Pt}(\text{CH}_3)\right]^{+} + \text{HCl}\right.\right.$$

(a)

$$\left[(\text{bpym})\text{Pt}(\text{Cl})\right]^{+} \rightarrow \left\{\Delta E = -9.9\left[(\text{bpym})\text{Pt}(\text{Cl})(\text{CH}_4)\right]^{+} \rightarrow \begin{cases} E_A = 17.3 \\ \Delta E = 10.6 \end{cases}\left[(\text{bpym})\text{Pt}(\text{ClH})(\text{CH}_3)\right]^{+}\right.$$

(b)

SCHEME 12.6 Reaction between methane and the [Pt(bpym)Cl]⁺ complex (a) following the electrophilic activation pathway and (b) following the oxidative addition pathway. The energies, calculated in water as a solvent, are given in kcal/mol.

The search for the differences between these two suggested reaction pathways was continued in the work done by Ziegler et al.[15,16] The activation of the methane C–H bond was examined on the tricoordinated substituted platinum [Pt(bpym)(X)]⁺ catalysts (X = Cl⁻ or HSO₄⁻) and their diprotonated species. The solvation calculations were performed for sulfuric acid.

The applied computational model predicted the preference for oxidative addition over metathesis (Scheme 12.6), when the catalyst reacted with CH₄. In the first stage, methane coordinated to the metal through a C–H σ-bond, lowering the energy of the systems by −14.5 kcal/mol in the oxidative addition process and −9.9 kcal/mol in the metathesis process. In both cases, the C–H bond then stretched substantially until the transition state was attained, with the energy barriers of 10.0 and 17.3 kcal/mol, respectively. Thus, the oxidative addition product was favored both kinetically and thermodynamically over the metathesis product. Overall, the key C–H activation steps and the formation of the final products were both endothermic and equaled to 5.0 kcal/mol for the oxidative addition process and 10.6 kcal/mol for the metathesis process.

12.2.1.2.2 Influence of Different Ligands on the C–H Activation Process

The fact that the reaction occurs in concentrated sulfuric acid opened up the possibility of formation of new form of active complexes.

Hush et al.[16] studied the thermodynamics of the activation and functionalization steps of the related cisplatin-catalyzed process in sulfuric acid. Their findings demonstrated that the initial reaction of cisplatin with sulfuric acids gave rise to a range of bisulfate complexes, which were next considered as the catalysts in the subsequent reaction with methane. The studied reaction steps, namely, the electrophilic substitution, C–H activation by oxidative addition, and C–H activation by Pt(IV) catalysts, showed that cisplatin [Pt(NH₃)₂(Cl)₂] formed various complexes with sulfuric acid. Among the thirteen investigated systems, the one of [Pt(NH₃)₂(OSO₃H)(H₂SO₄)]⁺ formula was chosen as a good candidate for the active catalyst from a thermodynamic point of view. In the case of the oxidative addition scheme, the calculations were considerably less unequivocal; however, the mechanism was effectively ruled out on the basis that the C–H activation was far too endergonic. The alternative scheme, where the Pt(IV) complex $[Pt(NH_3)_2(Cl)_2(OSO_3H)_2]^+$ was assumed to be the active catalyst, also appeared to be feasible on the basis of thermodynamics.

Gilbert and Ziegler[15,17] found that protonation and double protonation processes of bipyrimidine complex were endothermic with reaction energies of 6.9 and 17.5 kcal/mol:

$$[Pt(bpym)Cl]^+ + H_2SO_4 \rightarrow [Pt(bpymH)Cl]^{2+} + HSO_4^-$$

$$[Pt(bpymH)Cl]^{2+} + H_2SO_4 \rightarrow [Pt(bpymH_2)Cl]^{3+} + 2HSO_4^-$$

These estimates suggested that either the peripheral bipyrimidine nitrogen atoms were not protonated in the Catalytica catalyst or, if they were, the counter ions were probably coordinated to the metal center as well. The larger positive charge increased the Lewis acidity of the platinum center, resulting in a moderately larger exothermicity for the formation of the σ-bond complexes. The oxidative addition product (−12.5 kcal/mol) calculated for $[Pt(bpymH_2)Cl]^+$ complex was indicated as more stable than the metathesis product (−10.9 kcal/mol). Further, the metathesis process was exothermic by 0.3 kcal/mol, while the oxidative addition process was endothermic by 4.1 kcal/mol. As expected, protonating the bipyrimidine ligand enhanced the metathesis pathway compared to the oxidative addition pathway. The values showed a general trend of stabilization for all species when solvation was not included in the modeling.

Another possible consequence of performing methane activation in the fuming sulfuric acid would be the replacement of chloride ligands with bisulfate ligands:

$$[Pt(bpym)Cl_2] + H_2SO_4 \rightarrow [Pt(bpymH)(Cl)(OSO_3H)] + HCl$$

The heat of the reaction was equal to 15.4 kcal/mol. Additionally, it was found that the bisulfate ligand formed more stable bidentate complexes than the [Pt(bpym)(OSO_3H)]^+ complex by 31.4 kcal/mol, in agreement with the results reported by Hush and coworkers. These relatively high energies required to dissociate one oxygen atom providing an explanation for which one must conduct the reaction at 180°C–200°C. The reaction pathways recalculated for bidentate [Pt(bpym)(OSO_3H)]^+ system showed that in oxidative addition, the process was more endothermic (by 9.7 kcal/mol) than the analogous reaction for $[Pt(bpym)(Cl)]^+$. The difference was probably due to the fact that the metal center had less electrons while with OSO_3H^+. Changing the Cl^- ligand to OSO_3H^+ decreased the endothermicity of reaction to 3.3 kcal/mol, making it energetically more favorable than the corresponding oxidative addition process (which has an endothermicity of 14.7 kcal/mol).

Goddard et al.[12] found that ammonia ligands in $[Pt(NH_3)_2(Cl)_2^-]$ were favorably displaced by bisulfate ligands:

$$[(NH_3)_2Pt(Cl_2)] + 2H_2SO_4 \rightarrow [(OSO_3H)_2Pt(Cl)_2] + 2NH_4^+$$

The bidentate form of bisulfate, $[Pt(\eta^2\text{-}OSO_3H)(Cl)_2]^-$ was thermodynamically favored species, whereas the protonated form of bisulfate ligands were highly unfavorable.

In the case of bipyrimidine ligand, the calculations suggested that the singly protonated form $[Pt(bpymH)(Cl)_2]^+$ would be dominant in the real catalytic environment. However, the differently protonated forms were accessible to allow proton shuttling to facilitate the activation and oxidation steps.

These results provided rationale, which should help the screening of possible ligands for the design of new catalysts. The calculations indicated that the influence of a *trans* ligand Y on the Pt–X bond follows in the order $Cl^- > NH_3 > $ (bpym) $>$ $OSO_3H^- > $ (empty site); therefore, it should be easier to dissociate the Pt–Cl bond in $[PtCl_2(NH_3)_2]$ (23.8 kcal/mol) than to dissociate the same bond in $[PtCl_2(bpym)]$ (36.0 kcal/mol). Replacement of the amino ligands with bisulfate ligands was thermodynamically favorable, while replacement of chloride ligands or bpym with bisulfate ligands was not. The Pt–Cl bond was stronger than the Pt–OSO_3H bond, making easier to dissociate the latter.

The other important issue considered in the theoretical calculation is the stability of the various form of Catalytica Pt(II) catalysts.[18] The comparison of NH_3 and bpym systems revealed that $[PtCl_2(NH_3)_2]$ would lose its ammine ligands in hot, concentrated sulfuric acid and the $PtCl_2$ moiety will dimerize and trimerize, leading to the $(PtCl_2)_n$ precipitation. In contrast, $[PtCl_2(bpym)]$ was found resistant to solvent attack, favorably retaining the bpym ligand in hot, concentrated sulfuric acid, in agreement with the experimental findings.

As bpym was protonated, the Pt–N bond strength decreased in the following order: $[PtCl_2(bpym)]$ (87.0 kcal/mol) $> [PtCl_2(bpymH)]^+$ (58.9 kcal/mol) $>$ $[PtCl_2(bpymH_2)]^{2+}$ (26.6 kcal/mol). This explained its role as a σ-donor instead of being a π-acceptor, as it is generally accepted. Both bisulfate forms, $[PtCl_2(OSO_3H)_2]^{2-}$ and $[PtCl_2(\eta^2\text{-}OSO_3H)]^-$, favorably formed dimers and trimers. Protonated bpym catalysts, $[PtCl_2(bpymH_n)]^{n+}$ ($n = 0, 1, 2$), did not favor dimerization and trimerization. Although each Pt(II) kept a local square planar geometry, Pt(II) dimers and trimers adopted bowl structures. The calculations showed that the affinity of bpym for Pt(II) was sufficiently high so that a treatment of bpym in hot, concentrated sulfuric acid would lead to the dissolution of insoluble $(PtCl_2)_n$ and the oxidative dissolution of Pt metal to make a homogeneous solution of the bpym catalyst. No such dissolution occurred in the absence of the bpym ligand, in agreement with the experimental findings.

On the basis of the previous calculations and the experimental data by Zhu and Ziegler,[19] a further study on the mixture of $[Pt(Cl)_2(H_2O)_2]$ and X^- where $X = F^-, Cl^-,$ Br^-, I^-, NO_2^-, CN^- was undertaken. The results of the calculations clearly showed that, within the halogen series, methane uptake was still rate determining and its energy barrier decreased in the order $F^- > Cl^- > Br^- > I^-$, which was in agreement with the increasing *trans*-directing ability throughout this series. As a result, the reactivity of the catalyst increased from $F^- < Cl^- < Br^- < I^-$. For NO_2^- and CN^-, due to their strong *trans*-directing ability, there was a drastic decrease in the energy barrier of methane uptake and a considerable jump in the energy barrier of C–H activation.

The authors noticed also an unusual increase in the energy barrier of methane uptake for CN⁻. In both cases, C–H activation overtook the methane uptake as the rate-determining step. From the data presented in Table 12.1, it was clear that the activity of the catalyst was reduced for X = NO_2^- and CN⁻ due to the high C–H activation energy barrier.

In their next paper,[20] the authors studied the influence of different X ligands on Pt–CH_4 bond and C–H activation barriers in trans-$[Pt(Cl)_2(X)(CH_4)]^-$ and trans-$[Pt(X)_2(Cl)(CH_4)]^-$, where X was, respectively, trans and cis ligands to CH_4. When X was situated in the cis position, the stability of trans-$[Pt(X)_2(Cl)(CH_4)]^-$, in both the ground and transition states, followed the order F⁻ < Cl⁻ < Br⁻ < NO_2^- < I⁻ < CN⁻. For X in trans position to CH_4, the Pt–CH_4 bond energy decreased as F⁻ > Cl⁻ > Br⁻ > I⁻ > NO_2^- > CN⁻, in agreement with the experimentally established trans influence of these ligands. The barrier of C–H activation with CH_4 trans to X follows the order F⁻ < Cl⁻ < Br⁻ < I⁻ < NO_2^- < CN⁻.

The calculations indicated that the trans-directing ligand in the trans-$[Pt(Cl)_2(X)_2]$ complex played a crucial role for the ability of the complex to activate alkanes. Especially, the trans-directing ligand affected the activation energy barriers of methane uptake and C–H bond activation in opposite directions. That is, a strong trans-directing ligand would make the C–H activation rate determining, while a weak trans ligand would make methane uptake rate determining. Either of the two limiting cases would make the overall energy barrier of the reaction too high, resulting in a catalyst with low activity. Therefore, the optimal catalyst should have a moderate trans-directing ligand so that the energy barrier for methane uptake is acceptable but not too weak and the energy barrier of C–H activation is not too high.

In order to gauge the potential of different trans ligands, the authors explored the ability of square planar Pt(II) complexes with different trans (T) and leaving (L) ligands to activate methane. The results of the calculations showed that phosphine as a trans ligand was unlikely to produce an efficient alkane activation catalyst due to high-energy barriers of methane uptake and C–H activation. Replacing the less electronegative P with the more electronegative N should lead to less trans-directing ligands increasing the methane uptake barrier while reducing the C–H activation energy to the point where methane uptake rather than C–H activation became the rate-determining step. Further increase of the electronegativity of ligand T, achieved by replacing N with O, would still lead to a lowering of the C–H activation barriers and an additional increase in the methane uptake activation energy. For all oxygen-based trans ligands, methane uptake was rate determining. Moreover, the energy barrier of C–H cleavage was so small that the location of the corresponding transition states was not possible.

Hush et al.[10] attempted to quantify the importance of such trans influence in the context of methane activation. A detailed comparison of the relative stabilities of such cis and trans isomers was carried out for various systems, where CH_3 and H were in cis or trans positions with respect to each other. The energies of the four isomers with CH_3 and H in cis geometry were within ~7 kcal/mol, and they were by ~32–40 kcal/mol lower than the energy of the fifth isomer with trans CH_3 and H, as expected on the basis of trans influence.

12.2.1.3 Oxidation of Pt(II) to Pt(IV)

Periana and coworkers suggested that the oxidation proceeds through the reaction of the Pt(II)-methylated intermediate with SO_3, the acid anhydride of sulfuric acid. The catalytic cycle is sustained by passing SO_3, which oxidizes Pt(II) to Pt(IV) and is itself reduced to SO_2, through the reaction media.

Hristov and Ziegler[17] proposed that the oxidation by SO_3 proceeds by the following steps: (1) association of SO_3 to the Pt(II) complex; (2) protonation of one of the oxygens on SO_3; (3) stretching of the S–OH bond mediated by Pt, to form a new OH ligand; (4) dissociation of SO_2; and (5) abstraction of OH^- and addition of HSO_4^-, based on the model complexes of the $[Pt(bpym)(Cl)_2]$ and $[Pt(bpym)(CH_3)(Cl)]$ formulae. The energy profiles for the oxidation of the two complexes exhibited a number of differences. First, the SO_3 adduct for the methyl complex was more stable by 6.91 kcal/mol. The protonation at the SO_3 oxygen led to a destabilization of the dichloro complex by 9.2 kcal/mol compared to only 0.4 kcal/mol for the methyl one. In the course of the complete oxidation reaction, SO_3H^+ had to accept two electrons from Pt to reduce the sulfur from oxidation state six to oxidation state four. The calculated internal barrier was basically the same, around 34.6 kcal/mol for both systems. The dissociation of SO_2 stabilized the dichloro complex by 13.6 kcal/mol, while the same step led to the destabilization of only 0.7 kcal/mol for the methyl complex. The differences in the energies for the methyl and dichloro species along the oxidation path could be attributed to the greater electronegativity of Cl^-, which withdrew electrons from Pt to a larger extent than methyl did and thus made the electron transfer to SO_3H^+ more difficult. As shown by the calculations, the oxidation of $[Pt(bpym)(Cl)(CH_3)]$ was exothermic by 22.2 kcal/mol, while the oxidation of the catalyst itself was endothermic by 8.3 kcal/mol. The overall barrier for the oxidation process was 35.1 kcal/mol for the dichloro complex and 15.6 kcal/mol for the methyl complex, which ruled out a two-step oxidation mechanism involving a Pt(IV) dichloro complex and confirmed that the catalyst was stable toward SO_3 even under the harsh conditions applied in the Catalytica process.

Goddard et al. have suggested that Pt(II) to Pt(IV) oxidation proceeds via a proton shuttling mechanism involving an H_2SO_4 molecule interacting with bpym. Extensive thermodynamic studies for various steps of CH_4 to CH_3OSO_3H conversion, C–H activation, and oxidation as well as studies of stability of different Pt complexes (various configurations of Pt with six ligands, namely, Cl^-, NH_3, OSO_3H^-, $\eta^2\text{-}OSO_3H^-$, bpym, bpymH$^+$, and bpymH$_2^{2+}$) led to the following general conclusion: to have a ligand that in its protonated state (at low pH) still can bind strongly to the Pt center was critical to the stability of the catalytic complex in concentrated sulfuric acid. The bipyrimidine ligand (doubly protonated in solution) still had two N centers to bind to the complex. Simple ammines would not live long in acidic media, since it was quite favorable for them to form free RNH_2^+ in solution, and the loss of the ammine ligands would eventually lead to the $PtCl_2$ precipitate and the catalyst death. The favorable ligands should have at least three N π-acid sites with two nitrogens in the right positions to act as bidentate ligands to the Pt complex plus at least one additional N to be protonated in sulfuric acid. Oxidation step was most favorable for the ammine ligand. Since the oxidation step was rate determining, this suggests that the ammine form of the catalyst should be responsible for the short-lived higher catalytic activity before precipitation occurs.

The calculations reported by Paul and Musgrave[9] exhibited that HOMO orbital of methylated Pt(II) complex is characterized by $5dz^2$ configuration. The expected oxidation may occur through removal of these electrons via interaction with SO_3, which was already indicated by Ziegler et al. Methylated Pt(II) complex formed Lewis acid–base adduct with SO_3, in which the electrons from $5dz^2$ orbitals were donated toward the 3p orbitals of S atom in SO_3. Formation of the Lewis acid–base adduct was followed by protonation by H_2SO_4 producing HSO_4^- as a by-product. After protonation, the reaction proceeded through hydroxylation of the Pt center, leading to the formation of Pt(IV). This step involved a high barrier for all examined systems. The more electron-rich Pt(II) center in cisplatin underwent more facile oxidation than the Pt(II) center in Catalytica. The energy of the oxidation transition state was significantly affected by the nature of the ligands. N-sp^3 donors made Pt(II) to Pt(IV) oxidation facile compared to bpym, which is a σ-donor π-acid ligand.

Periana suggested that the Catalytica catalyst got poisoned by the water produced during the catalytic cycle, as catalysis stopped when the concentration of sulfuric acid felt below 90%. This could be explained if the oxidation step was inhibited by water in the reaction medium. In H_2SO_4, H_2O rapidly combined with SO_3, the acid anhydride, by a highly exothermic reaction, and H_2SO_4 could not itself oxidize Pt(II) to Pt(IV).

12.2.1.4 Functionalization to Form CH_3OSO_3H Comprising Isomerization and S_N2 Attack

Hydroxylation was followed by the loss of SO_2. The five-coordinate Pt(IV) center could bind to HSO_4^- from the reaction medium or undergo isomerization to species that had the methyl group in the axial position. The calculations showed that the complexes with methyl group in axial position were more stable than those where axial position was occupied by Cl^-. Isomerization was followed by S_N2 nucleophilic attack by HSO_4^- to form CH_3OSO_3H. In the functionalization phase, a protonation of the OH group can occur, leading to the formation of a bound water molecule. This may further reduce the barrier for the S_N2 attack, as Pt would be more electron deficient by the bound water compared to the hydroxide, and hence the Pt center would serve as a better leaving group. If ligand exchange occurred between the bound water molecules and HSO_4^-, the S_N2 barrier was expected to be lower compared to the hydroxyl bound case, as HSO_4^- was more electron withdrawing than OH.

On the other hand, Goddard et al. presented that the first turnover led to the irreversible loss of chloride ion in the form of HCl. After the functionalization step, $[PtL_2(Cl)(OSO_3H)]$ was regenerated. This became the starting point for the subsequent catalytic cycles, which was found to have a lower barrier for the CH activation step. Weaker binding L ligands enhance C–H activation by stabilizing the formation of a stronger Pt–C bond in the Pt(II)–CH_3 intermediate. Methane ion-pair intermediate complex led to the calculated relative activation barriers for conversion to CH_3 versus loss of CH_4, which were in good agreement with H/D exchange experiments (for bpym). Performed computations suggested also that the reaction pathway involving the close association of X (where X = Cl⁻, OSO_3H^-) kept the intermediate methane complex reactive and able to form the Pt(II)–CH_3 species. Decreasing protonation of bpym favored a more rapid oxidation step. This was reasonable, since the less positive ligand would stabilize the Pt(IV) complex, but too many protonation sites on L may destabilize the oxidation complex. In addition, as the solvent was made less acidic, the oxidation step should be increasingly favorable. Although water was involved in the hydrolysis to convert methyl bisulfate to methanol, its presence in the catalytic cycle had deteriorating effect on the catalyst activity. When more water was generated from the decomposition of H_2SO_3 to SO_2 and H_2O (as the by-product of the oxidation step), the equilibrium between H_2SO_4 and SO_3 (direct oxidant) shifted to reduce the concentration of SO_3 and, hence, inhibited the oxidation step.

12.2.1.5 Alternative Reactivity of the Homogeneous Pt Complex: Methane Oxidation to Acetic Acid

The reactivity of the Pt/Pd metal complexes toward methane in homogeneous systems is not limited to the aforementioned Shilov chemistry. One should have in mind that it covers also alternative reactions as well as alternative metal complexes. One of such processes was reported by Periana, who demonstrated that Pd^{2+} dissolved in concentrated H_2SO_4 catalyzed the direct oxidation of methane to acetic acid in the presence of O_2 and CO at 453 K. The mechanism of this reaction was computationally studied by Bell.[21] Interestingly for the Shilov process, the employed model of the active catalyst was built assuming that Pd(II) is totally solvated by H_2SO_4, so all the ligands in the first coordination sphere are either OSO_3H^- or OSO_3H_2. The calculations indicated that the dominant species in the Pd(II)/H_2SO_4 mixture is [Pd(η^2-OSO_3H)(OSO_3H)(OSO_3H_2)]. In this square planar Pd(II) complex, one of the hydrogen sulfate ligands, η^2-OSO_3H^-, is bound to Pd through two of its oxygen atoms, whereas the other ligands form only one bond between oxygen and the central metal. This species performed a concerted metalation deprotonation (CMD) of methane, in which the formation of the Pd–CH_3 bond formation was concomitant with a proton transfer to the η^2-OSO_3H^- ligand. This step had the highest activation barrier ($\Delta G = 41.5$ kcal/mol) and led to the formation of the [Pd(CH_3)(OSO_3H)(OSO_3H_2)$_2$] system. In the following step, CO was added and inserted into the Pd–CH_3 bond, involving a much lower activation barrier of 14.7 kcal/mol. The final reductive elimination of CH_3COOSO_3H, yielding the final acetic acid product by hydrolysis, required the previous oxidation of Pd(II) to Pd(IV) by H_2SO_4 and O_2. The oxidation and reductive elimination steps were exothermic and involved low-energy barriers. These reported results proved that Pd(II) is capable of promoting methane oxidation in the absence of chloride, amine, and aqua ligands. The calculations also indicated that Pd(II) is reduced to Pd(0) by CO, which causes the death of the catalyst due the formation of a Pd black precipitate.

Finally, also new computational approaches to the discovery of new catalysts were tested. In relation to the Shilov chemistry, one should mention quantum mechanical rapid prototyping (QM-RP) method, which was developed by Goddard. The QM-RP approach profits from the excellent accuracy/cost relationship offered by modern QM calculations (here DFT) and follows a five-step catalyst redesign process:

- Complete and detailed description of the reaction mechanism for a well-known experimental system, in which all intermediates and transition states involved in the mechanism are computed
- Characterization of the catalytic bottlenecks, in which the critical steps involving the largest activation or reaction energies are identified
- Discovery of new catalysts by screening, in which a variety of modifications in the nature of metal, ligand, cocatalyst, or solvents are applied in order to test how they affect the catalytic bottlenecks and other critical factors such as catalyst stability and poisoning
- Catalyst refinement, in which the leads found in the previous step are further improved by fine-tuning of their electronic and steric properties
- Experimental tests, in which the best performing species are finally synthesized and tested in the laboratory

The QM-RP approach was applied to the oxidation of methane by the Catalytica system in the quest for new and more efficient catalysts. One of the new catalysts found in this study was an Ir(III) complex with a pincer NNC ligand, which was predicted to promote C–H activation through energy barriers lower than 30 kcal/mol, without being poisoned by water.

12.2.2   METHANE CONVERSION TO AROMATICS UNDER NONOXIDATIVE CONDITIONS ON MoO₃/ZSM-5

Direct aromatization of methane

$$6CH_4 \rightarrow C_6H_6 + 9H_2$$

was proved to occur on the bifunctional catalyst consisting of ZSM-5 zeolite impregnated by MoO_3. It was observed that during the catalytic process, methane reduces the initial molybdenum(VI) oxide to yield MoO_{3-x}, further molybdenum (oxo)carbide, and molybdenum carbide.[2,22] It seems that molybdenum (oxo)carbide is the active phase responsible for methane dehydrogenation and conversion to ethylene, which, in turn, is aromatized inside the channels of the zeolite.[5,7,23] Despite intensified experimental studies, there were still a lot of unanswered questions regarding the nature and location of the active phase in course of the catalytic reaction. Similarly, the details of the reaction mechanism were not fully understood. In the past 15 years or so, there have been about a number of theoretical studies done mostly with application of density functional methods, which shed some light on the possible structure of the molybdenum species and their location in the zeolitic framework, as well as gave some hints regarding the first stage of methane aromatization, which is its conversion to ethylene.

12.2.2.1   Theoretical Characterization of the Catalyst

The existing theoretical investigations on the structure of the catalyst based on Mo/zeolite refer mainly to pure molybdenum structures: either oxides or carbides.[24–28]

The results of the experimental investigations suggesting that molybdenum phase was highly dispersed[2,3,6,23,29] were the starting point for theoretical consideration as to the composition and geometry structure of the active molybdenum phase. The other hints were geometry indications provided by EXAFS studies. The latter showed that the Mo–Mo distance varied from 3.09 Å (in fresh, calcined catalyst) to 4.10 Å (used catalyst) and Mo–X (X = C or O) distance was between 1.63–1.91 Å (fresh) and 2.00–2.82 Å (used).[3] Further, the existence of carbene-like species (CH_2=Mo) was postulated.[5,29]

12.2.2.1.1   Identification of Mobile Molybdenum Species Entering ZSM-5 Pores

The experimental evidence by MAS NMR, EPR, and TPD proved that molybdenum species were capable of migration into the zeolite pores during catalyst preparation, that is, impregnation and calcination.[30] Theoretical calculations helped to character-ize the structure of the volatile species, which were identified as either monomeric $MnO_2(OH)_2^{31}$ or dimeric $Mo_2O_5^{2+}$ complexes. Both were able to enter into the pore structure, as proved by experimental techniques.[32]

12.2.2.1.2   Modeling of Different Compositions of the Active Phase

The first attempts to model the active phase of the catalyst were made assuming that they were derived from the volatile molybdenum species formed in synthesis conditions, that is, those that were composed of molybdenum and oxygen. It was proposed that they consisted of monomers of tetrahedral complexes of molyb-denum with oxo ligands: MoO_2, MoO_2OH.[33] The comparison of the two models revealed that the first one, where molybdenum atom was able to form two bonds to the zeolite framework, better agreed with experimental findings. In the following, clusters with higher number of molybdenum atoms were proposed: these were a dimer $Mo_2O_5^{2+}$[34] or a cyclic trimer Mo_3O_9.[35] These structures exhibited all relevant surface oxygen species, that is, doubly coordinated Mo=O groups as well as bridg-ing oxygen atoms.

Alternatively, structures arising from the pure molybdenum trioxide, in which Mo exhibits distorted octahedral geometry, were proposed.[36] For such a model, transformations leading to the full carburization of the molybdenum oxide species were proposed. The consecutive replacement of the oxygen atoms by carbon atoms in the bulk structure of MoO_3 was studied. The formation of the C–C as well as CO species was identified, in agreement with the observation that the reduction of the active phase of the catalyst is accompanied by the release of CO and ethylene species. Similar studies were performed for the cluster model in which molybde-num phase was represented by a supermolecule consisting of two octahedra of molybdenum oxide in a geometry cut from the MoO_3 lattice. The oxygen atoms were replaced either by carbon atoms or, in case of the surface oxo species, by the methylidene (CH_2) groups. In the course of the studied process, a barrier-less formation of ethylene and carbon monoxide was observed. Thermodynamic considerations based on the performed calculations revealed that the reduction of molybdenum trioxide clusters by methane is endoenergetic and there was no synergy between the amount of carbon already introduced to the system and the energetic cost of replacing the next oxygen atom by the CH_2 group.

Finally, fully carburized models of the active phase were considered, that is, clusters of molybdenum carbide of Mo_2C_x and Mo_4C_y (where $x = 1$–4, 6; $y = 2$, 4, 6, 8) composition. As shown by the investigation by means of DFT within RPBE/DNP,[37] in all stable clusters, the existence of the Mo–Mo bonds was detected, whose distances varied between 2.10 and 2.90 Å, depending on the Mo–C ratio. When the content of carbon was increased, C_2 and C_4 chains were formed. Their bond distances indicated that they were the precursors of ethylene, in yet another way accounting for the experimental observations.

Knowing that the by-product of methane aromatization is hydrogen, the hydrogenation of $Mo_2C_4^{2+}$ and $Mo_2C_5^{2+}$ was considered.[38] The reaction proved to be exothermic, and the exploration of different possible structures revealed that the models containing CH_2 species would reasonably model the working catalyst. As a result, three models were proposed: monomeric $Mo(CH_2)_2(CH_3)^+$ and dimeric $Mo_2(CH_2)_5^{2+}$ and $Mo_2(CH_2)_4^{2+}$. It should be borne in mind, however, that the presence and location of the hydrogen atoms in the working catalyst was difficult to prove experimentally and should not disturb the conclusions drawn from theoretical results.

12.2.2.1.3   Location of Molybdenum Species on Zeolitic Support

It is postulated that molybdenum species are present on both outer surface and inside the pores of the zeolitic support.[39] Most of the researchers, however, considered models in which molybdenum species were bound at the vicinity of the acidic Brönsted acid sites, in the porous framework of ZSM-5. This assumption was caused by the observation that the introduction of MoO_3 phase into ZSM-5 resulted in the partial loss of zeolite acidity. Consequently, the geometry model of the zeolitic framework was being enlarged starting from the simplest $AlO_2(OH)_2$ cluster, through $Si(OH)_3$–O–$Al(OH)_3$ and $(OH)_3Si$–O–AlO_2–O–$Si(OH)_3$, up to the more elaborate systems in which more silicon and/or aluminum sites were added.

The studies of the MoO_2 and MoO_2OH linked to the single acid site represented by $(OH)_3Si$–O–AlO_2–O–$Si(OH)_3$[35] revealed that molybdenum atoms formed two bonds with zeolite framework preferentially. In the same way, monomeric carbon-containing $Mo(CH_2)_2(CH_3)^+$ cluster also was bound to the support in a bidentate manner.[38] Although the calculated distance between molybdenum atoms and the lattice alumina atom was close enough to induce overlap of their atomic orbitals, Mulliken population analysis showed only a weak coupling between the cluster of the active phase and the support.

Following the indication that the active phase was rather located in the vicinity of the alumina pairs, other models considered the location spot near two Brönsted acid sites. The models were built knowing that 3–6 out of 12 unique tetrahedral sites in ZSM-5 zeolite unit cell were occupied by aluminum in the typical zeolite sample of Si/Al ratio of 15–30—see Figure 12.4. Different possible Al pairs were considered[34] as the binding sites for $Mo_2O_5^{2+}$ clusters. It was concluded that molybdenum phase was bound preferentially at the intersections of ZSM-5 channel systems, at the sites denoted as T6T6 or T6T9. The geometry considerations proved that the active phase cluster may be composed of both monomer and dimer molybdenum species.

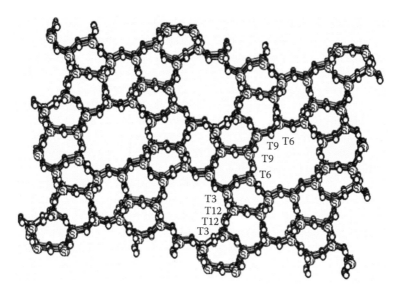

FIGURE 12.4 Location of the T-sites in ZSM-5 zeolite.

The modeling of the carbide-like active phase with the zeolite support was done by Zhou et al.[40] As most of the researchers, they assumed that the active molybdenum species replaced acidic proton and thus were directly bonded to the oxygen atom bridging tetrahedral Si and Al atoms. Their model of the molybdenum phase consisted of a single Mo center in form of $CH_2 = Mo(CH_3)_3$ bound to $Si(OH)_3–O–Al(OH)_3$ fragment, representing the zeolitic framework.

All possible binding locations, that is, surface hydroxyl groups as well as double and single aluminum sites, were considered for binding of the molybdenum carbide nanoparticles.[37] The comparison of the carbide binding energies to different locations in ZSM-5 revealed that the locations close to the aluminum pairs were preferred for clusters where C–Mo ratio was lower than 1.5. Increasing the carbon content made binding to the silicon hydroxyl groups energetically more privileged. Further, binding mode of molybdenum phase to different binding sites was compared. It was shown that molybdenum anchors bidentally when bound to external surface and to the double Al sites. In the case of the single Al sites, the denticity depends on the composition of the molybdenum carbide, but in general decreased with higher carbon content.

Despite different models of the support and use of various compositions of the active phase, one may observe some generalities. The active phase of the fresh catalyst is located at the intersections of the ZSM-5 pores, in the vicinity of the pair of Brönsted acid sites. Upon reduction, when the carbide phase replaces the oxide one, a migration of the molybdenum species toward outer surface is observed. The mobility of the active phase is promoted by increasing C–Mo ratio, which is reflected in the lowering of the number of chemical bonds between the two components of the catalyst (the active phase and the support). This picture adequately explains observed changes in catalyst behavior as seen by experimental techniques.

12.2.2.2   Determination of the Reaction Mechanism

There is a general belief that the MoO_3/ZSM-5 system acts as a bifunctional catalyst. The first stage of the reaction, methane conversion to ethylene, is catalyzed by the molybdenum phase, while the second one, ethylene condensation to benzene, occurs in the zeolite pores and requires the presence of the acidic sites and space constraints imposed by the zeolite framework. Up to date, most, if not all, theoretical efforts were made on the explanation of the first part of the process. The proposed mechanisms were varied, depending of the model of the active phase considered. While a number of studies considered methane activation—see Table 12.3—only a few tried to model the complete reaction pathway.

In the simplest model consisting of MoO_3H^+ species, the first step consisted of methane adsorption on the bare molybdenum atom, followed by the H–H bond formation.[40] The transition step was described as resulting from the electron transfer of methane to the vacant d metal orbitals with the formation of two-electron three-center bond. As a result, methylene group was left on the Mo site. Based on the considerations of the similar mechanisms for more elaborate models, taking into account Brönsted acidity of the support, it was found that strong acidity of the catalyst would favor the adsorption of (numerous) CH_x species and promote formation of the long-chain hydrocarbons, decreasing the selectivity to aromatics.

Alternative route of methane activation on the Mo=O system was identified as the C–H bond addition to the Mo=O group. Such a step resulted in the OH group formation and binding of the methyl moiety to the molybdenum atom. Our results were also in favor of this hypothesis.[41]

A similar conclusion as to the character of the first step of methane activation was drawn by Fu and coworkers,[35] who considered different modes of methane activation on different active sites of the cyclic Mo_3O_9 cluster. The one in which CH_4 is activated through hydrogen abstraction by the oxo group was found to proceed through the lowest-energy barrier.

Zhou et al.[38] considered analogous activation event, catalyzed by the $Mo=CH_2$ group, on their monomeric and dimeric models of carburized molybdenum species. $Mo=CH_2$ polarized the C–H bond in methane, which was dissociated with the consecutive formation of two CH_3 groups bound to molybdenum atom. The second step comprised the H_2 elimination for which hydrogen atoms came from both CH_3 groups. The process led to the C–C coupling and formation of ethylene bound on the catalyst surface. The activation of the second molecule of methane, leading to the dissociation of its C–H bond, was proposed as the third step. Ethylene was desorbed in the final step, followed by the release of molecular hydrogen. This last stage of the reaction was proposed as the rate-determining one.

Our unpublished results indicate that the desorption of ethylene would be a barrierless process, when C_2H_4 was formed as a result of CH_2 coupling from two adjacent molybdenum sites.[41]

Although the complete picture of the mechanism of methane dehydroaromatization on MoO_3/ZSM-5 phase is not fully understood and a number of different active phase models as well as reaction mechanisms were proposed, one may draw some conclusions aiding in the design of the catalyst. It was shown that the efficient

TABLE 12.3

Summary of Theoretical Studies on Methane Activation over Mo/ZSM-5 Catalyst

Model of Active Phase	Model of Support	TS Description	E_{act} (kcal/mol)	Method	References
$Mo(CH_3)_3$	$(OH)_3Si–O–Al(OH)_3$	H–H bond formation in CH_4 bound to the active phase	$\Delta E = 57.3$ $\Delta H = 53.1$	BH and HLYP/Lanl2DZ	[40]
MoO_2	H	H–H bond formation in CH_4 bound to the active phase	$\Delta E = 46.0$ $\Delta H = 42.3$	B3LYP/Lanl2DZ	[40]
$Mo_2O_5^{2+}$	—	C–H addition to Mo=O	$\Delta E = 63.5$	PWC/DNP	[34]
Mo_3O_9	—	C–H addition to Mo=O	$\Delta H = 50.1$ $\Delta G = 80.7$	B3LYP/6-31G**//B3LYP/6-31G	[35]
Mo_3O_9	—	C–H addition to two Mo=O groups (3 + 2 addition)	$\Delta H = 68.6$ $\Delta G = 97.1$	B3LYP/6-31G**//B3LYP/6-31G	[35]
Mo_3O_9	—	C–H addition to two Mo=O groups from adjacent Mo centers (5 + 2 addition)	$\Delta H = 63.2$ $\Delta G = 97.1$	B3LYP/6-31G**//B3LYP/6-31G	[35]
Mo_3O_9	—	Oxenoid insertion to one oxo group	$\Delta H = 69.4$ $\Delta G = 93.4$	B3LYP/6-31G**//B3LYP/6-31G	[35]
Mo_3O_9	—	H abstraction by oxo group	$\Delta H = 45.0$ $\Delta G = 63.4$	B3LYP/6-31G**//B3LYP/6-31G	[35]
Mo_3O_9	—	H abstraction by μ-oxo group	$\Delta H = 63.6$ $\Delta G = 77.7$	B3LYP/6-31G**//B3LYP/6-31G	[35]
$Mo_2(CH_2)_5$	$(OH)_3Al–O–Si(OH)_2–O–Si(OH)_2–O–Al(OH)_3$	C–H addition to Mo=CH_2 group	$\Delta E = 24.2$	B3LYP/6-31G(d,p)	[38]
$Mo_2(CH_2)_4$	$(OH)_3Al–O–Si(OH)_2–O–Si(OH)_2–O–Al(OH)_3$	C–H addition to Mo=CH_2 group	$\Delta E = 34.0$	B3LYP/6-31G(d,p)	[38]
$Mo(CH_2)_3$	$(OH)_3Al–O–Si(OH)_2–O–Si(OH)_2–O–Al(OH)_3$	C–H addition to Mo=CH_2 group	$\Delta E = 27.1$	B3LYP/6-31G(d,p)	[38]

methane activation requires a reduced molybdenum center, a requirement, which is realized during the prereduction period of the process. The catalytic selectivity to obtain aromatics is promoted by binding of the active phase next to the acid sites of the moderate strength.

12.3 CONCLUSIONS

In this chapter, we wanted to show that nowadays modern quantum chemistry tools play a complementary role in catalyst characterization and elucidation of the possible reaction mechanisms. Usually, they are not used as stand-alone methods, but their results are equally treated as those obtained by traditional, and thus already well established, experimental techniques. In the case of the methane conversion to valuable chemicals, this was shown on two different exemplary processes: homogeneous oxidation to methanol and heterogeneous aromatization to benzene. In the Shilov process, theoretical methods actively participated in elucidation of the mechanisms of possible reactions occurring on the active catalyst complex. The influence of the catalyst composition (type of central metal ligands) as well as reaction conditions (type of solvent) was studied. On the one hand, these data helped to understand factors determining observed activity of the system, while on the other hand, they contributed to the better catalyst design and optimization of the process.

In the case of methane aromatization, theoretical methods enabled characterization of the volatile molybdenum species entering zeolitic framework in the course of the catalyst preparation, and their plausible location on the support was proposed. Further, possible transformations of the catalyst active phase during catalytic reaction were described. Finally, the first stage of the catalytic process, which is methane to ethylene transformation, was modeled. The course of the reaction, and in particular its first step, depended on the model of the active phase taken to the calculation.

It always consisted of the C–H bond activation, which in most cases occurred either on Mo=O or Mo=CH$_2$ group. The computed energy barriers varied from 27.1 to 69.4 kcal/mol (taking barrier heights expressed as energies or enthalpies). The next step consisted of the coupling of two methylene groups, to form ethylene precursor, bound to the catalyst. The energetic barrier of this reaction was lower than the energy needed to split C–H bond in methane, irrespectively, of the model used. If the desorption of ethylene was not coupled to the activation of another methane molecule, it proceeded without additional energy input. In the case of such a coupling, the energy cost was considerably higher, and this step became rate determining.

One should have in mind, however, that despite a great effort from theoretical and experimental methods, there still exist a lot of open questions regarding methane reactivity in the processes described above. Scientific community is looking forward to answering these in the nearest future. Our belief is that this will happen with the increasing contribution from the theoretical methods.

REFERENCES

1. Xu, Y., Bao, X., Lin, L. *Journal of Catalysis* 2003, *216*, 386.
2. Ding, W., Li, S., Meitzner, D. G., Iglesia, E. *The Journal of Physical Chemistry B* 2000, *105*, 506.
3. Zhang, J.-Z., Long, M. A., Howe, R. F. *Catalysis Today* 1998, *44*, 293.
4. Wang, L., Tao, L., Xie, M., Xu, G., Huang, J., Xu, Y. *Catalysis Letters* 1993, *21*, 35.
5. Wong, S.-T., Xu, Y., Wang, L., Liu, S., Li, G., Xie, M., Guo, X. *Catalysis Letters* 1996, *38*, 39.
6. Xu, Y., Liu, W., Wong, S.-T., Wang, L., Guo, X. *Catalysis Letters* 1996, *40*, 207.
7. Shu, Y., Ohnishi, R., Ichikawa, M. *Journal of Catalysis* 2002, *206*, 134.
8. Solymosi, F., Szöke, A., Cserényi, J. *Catalysis Letters* 1996, *39*, 157.
9. Paul, A., Musgrave, C. B. *Organometallics* 2007, *26*, 793.
10. Mylvaganam, K., Bacskay, G. B., Hush, N. S. *Journal of the American Chemical Society* 2000, *122*, 2041.
11. Zhu, H., Ziegler, T. *Journal of Organometallic Chemistry* 2006, *691*, 4486.
12. Kua, J., Xu, X., Periana, R. A., Goddard, W. A. 3rd. *Organometallics* 2002, *21*, 511.
13. Siegbahn, P. E., Crabtree, R. H. *Journal of the American Chemical Society* 1996, *118*, 4442.
14. Heiberg, H., Swang, O., Ryan, O. B., Gropen, O. *Journal of Physical Chemistry A* 1999, *103*, 10004.
15. Gilbert, T. M., Hristov, I., Ziegler, T. *Organometallics* 2001, *20*, 1183.
16. Mylvaganam, K., Bacskay, G. B., Hush, N. S. *Journal of the American Chemical Society* 1999, *121*, 4633.
17. Hristov, I., Ziegler, T. *Organometallics* 2003, *22*, 1668.
18. Xu, X., Kua, J., Periana, R. A., Goddard, W. A. 3rd. *Organometallics* 2003, *22*, 2057.
19. Zhu, H., Ziegler, T. *Organometallics* 2007, *26*, 2277.
20. Zhu, H., Ziegler, T. *Organometallics* 2008, *27*, 1743.
21. Chempath, S., Bell, A. T. *Journal of the American Chemical Society* 2006, *128*, 4650.
22. Solymosi, F., Szoeke, A. *Applied Catalysis A: General* 1998, *166*, 225.
23. Jiang, H., Wang, L., Cui, W., Xu, Y. *Catalysis Letters* 1999, *57*, 95.
24. Tokarz-Sobieraj, R., Grybos, R., Witko, M., Hermann, K. *Collection of Czechoslovak Chemical Communications* 2004, *69*, 121.
25. Shi, X.-R., Wang, S.-G., Wang, H., Deng, C.-M., Qin, Z., Wang, J. *Surface Science* 2009, *603*, 851.
26. Shi, X.-R., Wang, J., Hermann, K. *The Journal of Physical Chemistry C* 2010, *114*, 13630.
27. Tokarz-Sobieraj, R., Hermann, K., Witko, M., Blume, A., Mestl, G., Schlogl, R. *Surface Science* 2001, *489*, 107.
28. Witko, M., Tokarz-Sobieraj, R. *Adsorption Science & Technology* 2007, *25*, 583.
29. Xu, Y., Liu, S., Guo, X., Wang, L., Xie, M. *Catalysis Letters* 1995, *30*, 135.
30. Zhang, W., Xu, S., Han, X., Bao, X. *Chemical Society Reviews* 2012, *41*, 192.
31. Zhou, D., Ma, D., Liu, X., Bao, X. *Journal of Molecular Catalysis A: Chemical* 2001, *168*, 225.
32. Borry, R. W., Kim, Y. H., Huffsmith, A., Reimer, J. A., Iglesia, E. *The Journal of Physical Chemistry B* 1999, *103*, 5787.
33. Zhou, D., Ma, D., Liu, X., Bao, X. *The Journal of Chemical Physics* 2001, *114*, 9125.
34. Zhou, D., Zhang, Y., Zhu, H., Ma, D., Bao, X. *The Journal of Physical Chemistry C* 2007, *111*, 2081.
35. Fu, G., Xu, X., Lu, X., Wan, H. *Journal of the American Chemical Society* 2005, *127*, 3989.

36. Rutkowska-Zbik, D., Grybos, R., Tokarz-Sobieraj, R. *Structural Chemistry* 2012, *23*,1417.
37. Gao, J., Zheng, Y., Fitzgerald, G. B., de Joannis, J., Tang, Y., Wachs, I. E., Podkolzin, S. G. *The Journal of Physical Chemistry C* 2014, *118*, 4670.
38. Zhou, D., Zuo, S., Xing, S. *The Journal of Physical Chemistry C* 2012, *116*, 4060.
39. Zheng, H., Ma, D., Bao, X., Hu, J. Z., Kwak, J. H., Wang, Y., Peden, C. H. F. *Journal of the American Chemical Society* 2008, *130*, 3722.
40. Zhou, T., Liu, A., Mo, Y., Zhang, H. *The Journal of Physical Chemistry A* 2000, *104*, 4505.
41. Rutkowska-Zbik, D., Tokarz-Sobieraj, R., Witko, M. Unpublished results.

Section III

Innovative Oxidative and
Nonoxidative GTL Processes

13 Production of High-Octane Fuel Components by Dehydroalkylation of Benzene with Mixtures of Ethane and Propane

Dennis Wan Hussin and Yvonne Traa

CONTENTS

13.1 INTRODUCTION

The main hydrocarbon constituents of natural gas are methane, ethane, and propane. The molar fractions of methane, ethane, and propane typically are in the range of 0.75–0.99, 0.01–0.15, and 0.01–0.10, respectively. In so-called wet gases, larger quantities of higher hydrocarbons are possible (Hammer et al. 2012). Therefore, if remote gas fields are exploited by dehydroaromatization of methane to aromatics (cf. Section 13.3), possible uses for ethane and propane should also be discussed, because flaring ethane and propane causes unnecessary greenhouse gas emissions as well, albeit in lower amounts. This chapter discusses the possibility of producing high-octane fuel components by alkylation of the aromatics produced by dehydroaromatization (cf. Section 13.3) with mixtures of ethane and propane as typically occurring in natural gas. This ensures the valorization of all components of natural gas and allows for the production of liquid fuel, which can be used locally. This is even more attractive, since the concentration of benzene in gasoline is regulated to very low amounts, while alkylaromatics can be blended to gasoline in much larger quantities.

13.2   ALKYLAROMATICS AS FUEL COMPONENTS

One of the most important quality criteria for gasoline is its resistance against autoignition (engine knock), that is, octane quality (Dabelstein et al. 2012). The octane quality of gasoline is described by the octane number. The octane number of gasoline is determined by comparative measurements of its octane quality and that of binary mixtures having variable concentrations of n-heptane (low octane quality) and 2,2,4-trimethylpentane (isooctane, high octane quality). By definition, the octane number of n-heptane is 0 and that of isooctane is 100. The octane numbers of mixtures are given by their percentage by volume of isooctane. Octane numbers > 100 can be determined with lead-containing isooctane or toluene-containing mixtures (Dabelstein et al. 2012). Octane number determination is carried out in single-cylinder, four-stroke test engines specially developed for the purpose, which are used under two different running conditions: The research octane number (RON) describes the knocking performance at low and medium engine speeds, whereas the motor octane number (MON) defines the knock behavior under high speed and load (Dabelstein et al. 2012). As determining octane numbers in test engines was not feasible in our laboratory, we calculated them. Since true octane numbers do not blend linearly, it is necessary to use blending octane numbers in making calculations. Blending octane numbers are based upon experience and are those numbers that, when added on a volumetric average basis, will give the true octane number of the blend (Gary et al. 2007). Because our product compositions are determined in moles, we tested the influence of calculating octane numbers in molar fractions or on a volumetric basis and found that the difference was only in decimal places. Thus, we calculated the RON of our liquid product using Equation 13.1:

$$RON = \sum_{i=1}^{n} x_i \times (\text{blending RON})_i \qquad (13.1)$$

where
   i is the number of liquid components from 1 to n
   x_i is the molar fraction of component i

In Table 13.1, the main constituents of the products from the reaction of aromatics with ethane and propane are listed with their boiling points, their true or actual octane numbers, and their blending octane numbers. It is obvious that it is difficult to determine the true octane numbers of aromatics. When regarding the blending RON numbers, it is noticeable that benzene has a blending RON of 99, which is far lower than the values of *all* alkylaromatics. Alkylaromatics have blending RON values of 120–146 (Owen 1984; Satterfield 1980). Thus, the alkylation of benzene with mixtures of ethane and propane produces high-octane fuel components and at the same

TABLE 13.1

Boiling Points, Actual and Blending Octane Numbers of Hydrocarbons

| Hydrocarbon | Boiling Point (°C) | Octane Number | | | |
| | | Actual | | Blending | |
		RON	MON	RON	MON
n-Butane	0.5	93	—	113	114
2-Methylpropane	−11.7	—	—	122	121
n-Pentane	36	62	62	62	67
2-Methylbutane	27.9	92	90	99	104
Benzene	80.1	>100	>100	99	91
Toluene	110.6	120	>100	124	112
o-Xylene	144.5	>100	>100	120	103
m-Xylene	139.2	117	>100	145	124
p-Xylene	138.5	>100	>100	146	127
Ethylbenzene	136.5	>100	98	124	107
n-Propylbenzene	159.2	>100	98	127	129
Cumene	152.3	>100	99	132	124

Sources: After Owen, K., in: *Modern Petroleum Technology*, 5th edn., Part II, G.D. Hobson (ed.), Wiley, Chichester, U.K., 1984, p. 786; Satterfield, C.N., *Heterogeneous Catalysis in Practice*, McGraw-Hill, 1980, p. 241.

RON, research octane number; MON, motor octane number.

time reduces the toxicity of benzene. The next sections will go into more detail and show the different parameters that influence the octane number of the liquid mixture.

13.3 INFLUENCE OF THE REACTANTS

For preliminary experiments, only two reactants were used (Table 13.2). The alkylation of benzene with ethane is very selective: mainly the primary alkylation product ethylbenzene (see Figure 13.1) is formed with small amounts of the secondary product toluene, which probably originates from the hydrogenolysis of ethylbenzene (Rezai et al. 2009; Vazhnova et al. 2013). Thus, the RON of the liquid product phase without benzene is with a value of 124 rather high. However, since the benzene conversion is low, the RON of the total liquid with unconverted benzene (blending RON, 99; Table 13.1) is only 102. By contrast, the alkylation of benzene with propane is not very selective: propylbenzenes, the primary alkylation products, are only formed in small amounts. Secondary products such as xylenes, ethylbenzene, and toluene (see Figure 13.1) are formed in larger amounts. The largest product fraction consists of butanes besides ethane, pentanes and some methane, which is due to disproportionation/cracking of propane (Ivanova et al. 1999; Wang et al. 2004). This reaction is thermodynamically favored (Traa 2008), which becomes also evident, if ethane and propane are converted without benzene (Table 13.2). In this case, also small amounts of ethylbenzene and toluene are produced by ethane/propane aromatization (Ono 1992). Thus, the higher reactivity of propane as compared to ethane leads to a higher

TABLE 13.2

Influence of the Reactants during the Conversion of Benzene (Bz.), Ethane (Et.), and Propane (Pr.) on the Molar Product Distribution without Hydrogen and on the Calculated RON of the Liquid Phase

Reactant	Et. + Bz.	Bz. + Et. + Pr.	Bz. + Pr.	Et. + Pr.
WHSV (h^{-1})	3.8	3.2	3.5	4.3
$\dot{n}_{Et.}/\dot{n}_{Pr.}/\dot{n}_{Bz.}$	9.4/0/1	8/4.5/1	0/15.1/1	1.77/1/0
Benzene conversion (%)	14	19	18	—
RON without Bz.	124	121	110	99
RON with Bz.	102	104	102	99
Product Distribution				
Methane (%)	0.00	16.29	1.35	17.25
Ethane (%)	—	—	11.68	—
i-Butane (%)	0.00	13.09	18.11	21.91
n-Butane (%)	0.00	34.50	48.43	47.79
i-Pentanes (%)	0.00	2.10	3.41	3.85
n-Pentane (%)	0.00	2.36	3.91	3.45
Benzene (%)	—	—	—	0.86
Toluene (%)	2.00	14.73	7.92	2.71
Ethylbenzene (%)	98.00	8.48	0.85	2.18
Xylenes (%)	0.00	2.07	1.20	0.00
Cumene (%)	0.00	2.01	1.07	0.00
n-Propylbenzene (%)	0.00	4.37	2.08	0.00

Source: After Wan Hussin, D. and Traa, Y., *Energy Fuels*, 28, 3352, 2014, doi:10.1021/ef500333b.
Fixed-bed reactor at 350°C and 6 bar on 1.0Pd-H-ZSM-5 ($n_{Si}/n_{Al} = 35$), TOS = 105 min.

benzene conversion but also to a larger extent of side reactions. The low blending RON of n-pentane (62, Table 13.1) causes the lower RON of the liquid phase. When using a mixture of ethane and propane for the alkylation of benzene, a combination of both product distributions is obtained with the highest RON of 104, while the contribution of the more reactive propane is dominant. Actually, n-butane disposes of a blending RON of 113 (Table 13.1), and limited amounts are added to gasoline. Because of its high vapor pressure, n-butane is good for engine cold start. On the other hand, too high vapor pressures can cause hot-fuel handling problems, so that the addition of n-butane is limited by the vapor pressure specifications (Dabelstein et al. 2012). Therefore, calculating only the RON of the liquid phase is, in fact, a pessimistic approach.

13.4 INFLUENCE OF THE CATALYST

In further experiments, we optimized the catalyst for obtaining a large RON. First of all, the n_{Si}/n_{Al} ratio of the Pd-H-ZSM-5 catalyst was varied. As Figure 13.2 shows, the n_{Si}/n_{Al} ratio of the zeolite has a big influence on the results. The smaller the n_{Si}/n_{Al} ratio or the larger the aluminum content, the larger the ion exchange capacity

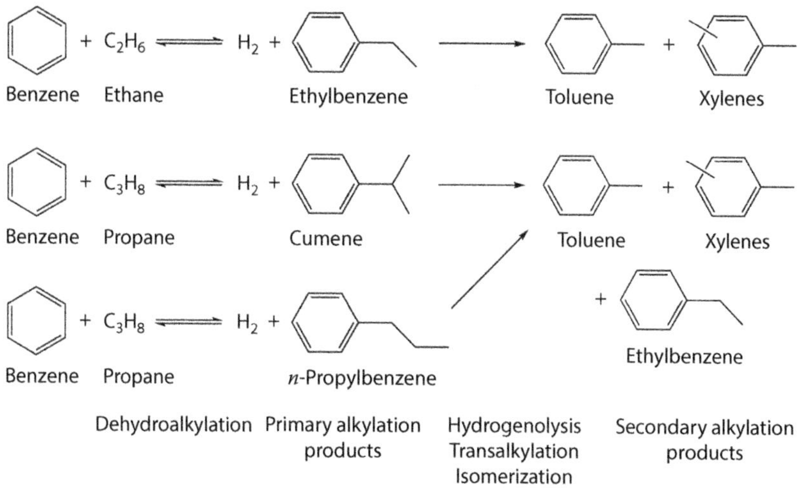

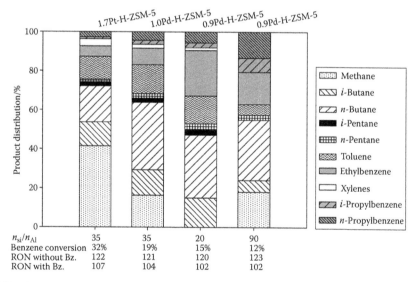

FIGURE 13.1 Reaction scheme showing the dehydroalkylation reaction of benzene with ethane or propane to the primary alkylation products ethylbenzene, cumene and n-propylbenzene as well as secondary reactions such as hydrogenolysis, transalkylation or isomerization leading to the secondary alkylation products toluene, xylenes and ethylbenzene.

FIGURE 13.2 Molar product distribution without hydrogen during the conversion of ethane, propane, and benzene on different catalysts (fixed-bed reactor at 350°C and ≈6 bar, WHSV 3.2–4 h⁻¹, TOS = 105 min, $\dot{n}_{Et.}/\dot{n}_{Pr.}/\dot{n}_{Bz.} = 7.3$–$8.7/4.1$–$5.1/1$).

of the zeolite, and in the case of Pd-H-ZSM-5 catalysts, the larger the concentration of Brønsted acid sites. Since it is presumed that during the alkylation of aromatics with alkanes the alkane is first dehydrogenated to the alkene, which afterward alkylates the aromatic reactant on the acid sites (Caeiro et al. 2006; Traa 2008), the higher concentration of Brønsted acid sites favors not only the alkylation reaction but also other acid-catalyzed side reactions. Therefore, the benzene conversion is highest at low n_{Si}/n_{Al} ratios of 20 and 35. At the high n_{Si}/n_{Al} ratio of 90, the benzene conversion is only 12%, but the RON of the liquid product phase without benzene reaches the highest value of 123. In addition, no secondary xylene products are formed. However, if the benzene conversion is taken into account and the RON of the liquid phase with benzene is calculated, the highest value of 104 is achieved on 1.0Pd-H-ZSM-5 (n_{Si}/n_{Al} = 35), and these maximum yields at intermediate n_{Si}/n_{Al} ratios have been observed before (Bressel et al. 2008).

In the next step, the influence of the noble metal was tested for the zeolite with the intermediate n_{Si}/n_{Al} ratio of 35. The values before the noble metal give the weight percentage of metal as referred to the dry catalyst. The metal contents were chosen so that the molar metal content of the palladium- and the platinum-containing catalyst is the same. As Figure 13.2 shows, the benzene conversion and also the RON of the liquid phase with benzene are with 32% and 107 much higher on the platinum-containing catalyst. Therefore, the following experiments were carried out with 1.7Pt-H-ZSM-5 (n_{Si}/n_{Al} = 35).

13.5 INFLUENCE OF THE CATALYST PRETREATMENT AND THE REACTOR

The dehydroalkylation of aromatics with alkanes is thermodynamically strongly limited. Equilibrium conversions of benzene dehydroalkylation with ethane to ethylbenzene or with propane to cumene/n-propylbenzene amount only to 3%–4% at 600 K (Traa 2008). To overcome these limitations, attempts have been made to shift the equilibrium toward alkylaromatics by selectively removing hydrogen, and thus enhancing the dehydrogenation of the alkane. One approach to this problem is to remove hydrogen from the reaction zone via a hydrogen scavenger, which is mixed with a Pt-H-ZSM-5 catalyst (Smirnov et al. 2000). Another method is to employ a membrane reactor with a hydrogen-selective membrane. We employed a previously described (Rezai and Traa 2008a) tubular packed-bed membrane reactor purchased from REB Research & Consulting with a hydrogen-selective membrane to keep the hydrogen concentration in the reaction zone low. The membrane is 100% selective to hydrogen and presumably consists of a palladium silver alloy.

In order to prevent damage to the membrane from the acidic surface of the catalyst and by coking, the catalyst 1.7Pt-H-ZSM-5 (n_{Si}/n_{Al} = 35) was first precoked according to the method described by Bauer et al. (2001). Methanol was adsorbed on the catalyst at an ambient temperature with a very slow nitrogen flow through a saturator for 24 h. Then under static conditions in nitrogen atmosphere, the sample was slowly heated to 500°C and kept at that temperature for 4 h. On the outer surface, mainly hard coke should form, while inside the pore system, only soft coke can form. Subsequently, the catalyst was treated at 500°C with hydrogen for 24 h. This should remove the soft coke inside the pores while not removing the hard coke on the surface (Bauer et al. 2001).

The precoked catalyst should have been deactivated on the outer surface of the crystals, while maintaining the activity of the active sites inside the pore system. Since the deactivated sites on the surface are no longer available to reactants, shape selectivity should increase because reactants must enter the pore system to reach an active site.

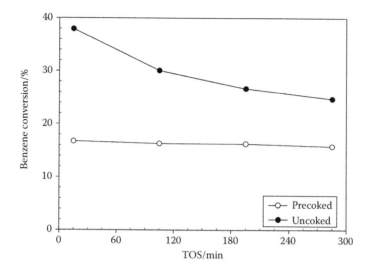

FIGURE 13.3   Benzene conversion as a function of TOS on freshly activated, uncoked and precoked 1.7Pt-H-ZSM-5 (n_{Si}/n_{Al} = 35) catalyst (membrane reactor at 350°C and 6.5 bar, sweep gas: nitrogen [100 cm³ min⁻¹ for the precoked catalyst, 250 cm³ min⁻¹ for the uncoked catalyst], WHSV 3.3–4.0 h⁻¹, $\dot{n}_{Et.}/\dot{n}_{Pr.}/\dot{n}_{Bz.}$ = 9.1–9.2/4.1–5.2/1).

It can be seen that the precoking severely hampers the alkylation performance of the catalyst. Figure 13.3 shows that the catalyst loses almost half of its initial activity by precoking. However, the stability of the catalyst over time on stream (TOS) is greatly improved. (The different sweep gas flow rates used do not have a big influence: see Rezai and Traa (2008), and Section 13.6.) The selectivity to alkylation products does not improve significantly, as seen in Figure 13.4, but the percentage of the primary alkylation products n-propylbenzene, cumene, and ethylbenzene is slightly higher, whereas the percentage of the secondary products toluene and xylenes is slightly lower. Due to the higher benzene conversion of 30% after 105 min TOS, the RON of the liquid phase with benzene is with 107 much higher on the freshly activated, uncoked catalyst than on the precoked catalyst, where the benzene conversion is only 16% and the RON of the liquid phase with benzene is 103. However, the values obtained on the precoked catalyst agree well with the values obtained on an equilibrated catalyst after several days on stream (cf. Section 13.6). Therefore, precoking is an efficient method to quickly and reproducibly make an equilibrated catalyst with a stable TOS behavior. Since so far no harm to the membrane in the reactor could be observed with using uncoked catalyst, further experiments were carried out with fresh catalysts.

In Figure 13.5, the product composition without hydrogen is compared for the fixed-bed reactor and the membrane reactor with freshly activated, uncoked catalysts after 105 min TOS. The reactor type seems to have little influence on the selectivity of the reaction. The combined amounts of primary alkylation products, that is, n-propylbenzene, cumene, and ethylbenzene, correspond to 12% in the fixed-bed reactor and 16% in the membrane reactor. Thus, in the membrane reactor the formation of these primary

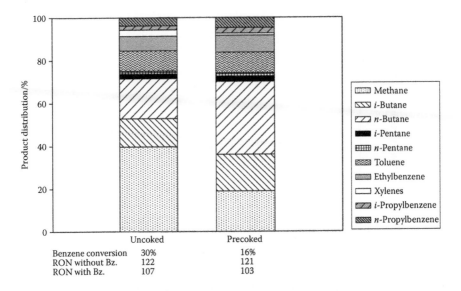

	Uncoked	Precoked
Benzene conversion	30%	16%
RON without Bz.	122	121
RON with Bz.	107	103

FIGURE 13.4 Molar product distribution without hydrogen on freshly activated, uncoked and precoked 1.7Pt-H-ZSM-5 (n_{Si}/n_{Al} = 35) catalyst (membrane reactor at 350°C and 6.5 bar, sweep gas: nitrogen [100 cm^3 min^{-1} for the precoked catalyst, 250 cm^3 min^{-1} for the uncoked catalyst], WHSV 3.3–4.0 h^{-1}, TOS = 105 min, $\dot{n}_{Et.}/\dot{n}_{Pr.}/\dot{n}_{Bz.}$ = 9.1–9.2/4.1–5.2/1).

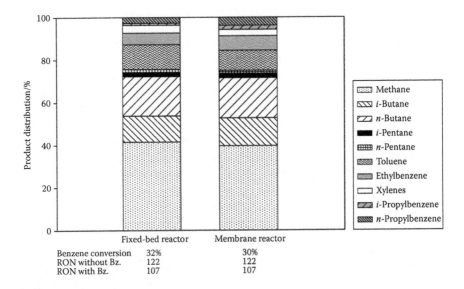

	Fixed-bed reactor	Membrane reactor
Benzene conversion	32%	30%
RON without Bz.	122	122
RON with Bz.	107	107

FIGURE 13.5 Molar product distribution without hydrogen in the fixed-bed reactor and in the membrane reactor on 1.7Pt-H-ZSM-5 (n_{Si}/n_{Al} = 35) catalyst (350°C and 6.7 bar, sweep gas in membrane reactor: nitrogen [250 cm^3 min^{-1}], WHSV = 4 h^{-1}, TOS = 105 min, $\dot{n}_{Et.}/\dot{n}_{Pr.}/\dot{n}_{Bz.}$ = 8.7–9.1/4.1–5.1/1).

alkylation products is favored. The conversions and the RON values are almost the same in both reactors. Since the aim of this study was to maximize the RON, the experimental parameters were not optimized with regard to the membrane effect. A smaller total flow rate would give the produced hydrogen more time to permeate through the membrane. Yet the space time yield would be lowered, which is highly undesired, economically. A calculation of the maximum possible hydrogen flux through the membrane shows that, in principle, all hydrogen produced can permeate through the membrane. In addition, a thermodynamic analysis of the dehydroalkylation reaction has shown that the $\dot{n}_{alkane}/\dot{n}_{aromatic}$ ratio seems to have a big effect as well (Rezai and Traa 2008b). The biggest influence of the membrane was observed at $\dot{n}_{alkane}/\dot{n}_{aromatic}$ ratios around 1. The value of 7 used in an earlier experimental study (Rezai and Traa 2008a) is still in a range with considerable influence (Rezai and Traa 2008b). With our setup for using mixtures of ethane and propane, the lowest $\dot{n}_{ethane+propane}/\dot{n}_{aromatic}$ ratio possible was around 12.5, which is in a range where the membrane reactor effect is only small but still observable, as will be demonstrated in Section 13.6.

13.6 INFLUENCE OF THE TIME ON STREAM AND THE REACTION TEMPERATURE

In order to assess the stability of our 1.9Pt-H-ZSM-5 ($n_{Si}/n_{Al} = 35$) catalyst, experiments over long TOSs and three reaction cycles were performed. As Figure 13.6 shows, the initial conversion significantly decreases from cycle to cycle, while the

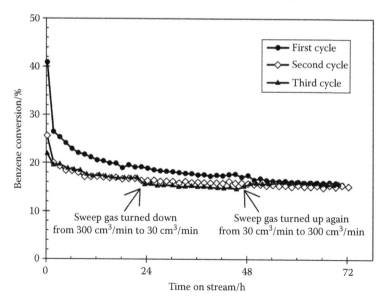

FIGURE 13.6 TOS behavior for benzene conversion on 1.9Pt-H-ZSM-5 ($n_{Si}/n_{Al} = 35$) during three reaction cycles (membrane reactor at 350°C and 6.4 bar, sweep gas: nitrogen [300 cm^3 min^{-1}], WHSV 4.0 h^{-1}, $\dot{n}_{Et.}/\dot{n}_{Pr.}/\dot{n}_{Bz.} = 9.5/5/1$). (From Wan Hussin, D. and Traa, Y., *Energ. Fuels*, 28, 3352, 2014.)

final conversion after 3 days is almost identical for all three reaction cycles (Wan Hussin and Traa 2014). The three reaction cycles were conducted with the same catalyst sample being regenerated with a hydrogen treatment of 24 h between the cycles at 350°C between the first and second cycle, and 400°C between the second and third cycle. The conversion decrease with TOS is explained by the fact that the platinum sites active for hydrogenolysis side reactions deactivate quickly, and then the product distribution equilibrates with primary alkylation products being the main constituents of the liquid phase (Wan Hussin and Traa 2014).

During the third cycle, the sweep gas flow rate was in between reduced by a factor of 10 for 1 day to verify the beneficial effect of hydrogen removal in the membrane reactor. By reducing the sweep gas flow rate, the permeated hydrogen is swept away more slowly, thus reducing the driving force of permeation and the permeation rate. However, the effect of the membrane is rather small here, since the experimental conditions were adapted to maximize the RON and not the membrane effect (Wan Hussin and Traa 2014).

At a lower reaction temperature of 300°C, the initial deactivation of the catalyst is not observed (Figure 13.7). The RON is slightly lower than at 350°C (Table 13.3), due to the lower benzene conversion, while no deactivation and no significant change in product distribution with increasing TOS is detected, as with higher reaction temperatures. The small amount of secondary products at 300°C (not shown) proves that preferably the reaction to primary products occurs at lower temperatures, since the hydrogenolysis activity of the catalyst is low.

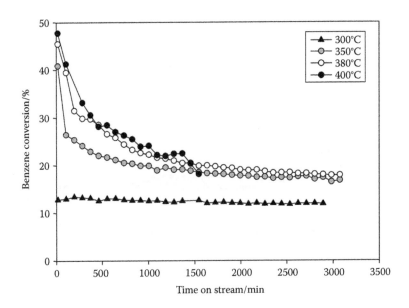

FIGURE 13.7 Temperature dependence of TOS behavior for benzene conversion on 1.9Pt-H-ZSM-5 ($n_{Si}/n_{Al} = 35$) (membrane reactor at 350°C and 6.4 bar, sweep gas: nitrogen [300 cm³ min⁻¹], WHSV 4.0 h⁻¹, $\dot{n}_{Et.}/\dot{n}_{Pr.}/\dot{n}_{Bz.} = 9.5/5/1$).

TABLE 13.3

Influence of the Cycle Number and the Reaction Temperature on the Benzene Conversion, the Yield Y of Alkylaromatics and the Calculated RON of the Liquid Phase during the Conversion of Benzene (Bz.), Ethane, and Propane in the Membrane Reactor on 1.9Pt-H-ZSM-5 (n_{Si}/n_{Al} = 35) after 1 Day on Stream at p = 6.4 bar

Temperature (°C)	300	350 Cycle 1	Cycle 2	Cycle 3	380	400
Benzene conversion (%)	12.7	19.1	16.8	16.9	20.5	20.3
$Y_{alkylaromatics}$ (%)	12.0	19.7	17.2	17.9	21.5	21.6
RON with Bz.	102.6	103.6	103.1	103.1	103.9	104.0

Source: After Wan Hussin, D. and Traa, Y., *Energ. Fuels*, 28, 3352, 2014.

The deactivation observed at higher temperatures between 350°C and 400°C results from the catalysts' loss in hydrogenolysis activity, because especially the amount of typical hydrogenolysis products is strongly reduced after 1 day on stream (Wan Hussin and Traa 2014). Therefore, the reaction temperature has actually no large effect, provided that the equilibrated catalyst after about 1 day on stream is used. Table 13.3 shows that the RON varies only between 102.6 at 300°C and 104.0 at 400°C. In addition, Table 13.3 shows the yield Y of alkylaromatics as the desired products calculated using Equation 13.2:

$$Y_{alkylaromatics} = \frac{\sum n_{alkylaromatics, out}}{n_{benzene, in}} \qquad (13.2)$$

Due to the increasing benzene conversions at increasing reaction temperatures, the yield of alkylaromatics is also increasing from 12.0% at 300°C to 21.6% at 400°C. The aromatization of ethane/propane (Table 13.2) is probably responsible for the fact that the yield of alkylaromatics can be slightly higher than the benzene conversion. Thus, from an economic point of view, operation at a reaction temperature of 400°C is most favorable. Larger amounts of methane formed at 400°C could be recycled into the methane aromatization unit (Wan Hussin and Traa 2014).

13.7 CONCLUSIONS

A viable route to convert ethane and propane from natural gas into valuable liquid, easy-to-transport fuel components was presented. The dehydroalkylation of benzene with mixtures of ethane and propane is an attractive alternative to flaring natural gas in remote locations where the infrastructure for gas transport is not available. Reasonably high conversions to liquid, alkylaromatic products as superior fuel components with high RON values can be achieved.

ACKNOWLEDGMENTS

The research leading to these results received funding from the EU's FP7 through the collaborative project NEXT-GTL under agreement no. 229183. Daniel Geiss is acknowledged for his preliminary work during the NEXT-GTL project. We thank Dr. Ines Kley for helping with the precoking of the catalyst.

REFERENCES

Bauer, F., W. Chen, H. Ernst, S. Huang, A. Freyer, S. Liu, *Micropor. Mesopor. Mater.*, 2001, 47, 67.

Bressel, A., T. Donauer, S. Sealy, Y. Traa, *Micropor. Mesopor. Mater.*, 2008, 109, 278.

Caeiro, C., R.H. Carvalho, X. Wang, M.A.N.D.A. Lemos, F. Lemos, M. Guisnet, F. Ramôa Ribeiro, *J. Mol. Catal. A*, 2006, 255, 131.

Dabelstein, W., A. Reglitzky, A. Schütze, K. Reders, Automotive fuels, in: *Ullmann's Encyclopedia of Industrial Chemistry*, vol. 4, Wiley-VCH, Weinheim, Germany, 2012, p. 425.

Gary, J.H., G.E. Handwerk, M.J. Kaiser, *Petroleum Refining Technology and Economics*, 5th edn., Taylor & Francis Group, Boca Raton, FL, 2007, p. 261.

Hammer, G., T. Lübcke, R. Kettner, M.R. Pillarella, H. Recknagel, A. Commichau, H.-J. Neumann, B. Paczynska-Lahme, Natural gas, in: *Ullmann's Encyclopedia of Industrial Chemistry*, vol. 23, Wiley-VCH, Weinheim, Germany, 2012, p. 739.

Ivanova, I.I., A.I. Rebrov, E.B. Pomakhina, E.G. Derouane, *J. Mol. Catal. A*, 1999, 141, 107.

Ono, Y., *Catal. Rev. Sci. Eng.*, 1992, 34, 179.

Owen, K., in: *Modern Petroleum Technology*, 5th edn., Part II, G.D. Hobson (ed.), Wiley, Chichester, U.K., 1984, p. 786.

Rezai, S.A.S., F. Bauer, U. Decker, Y. Traa, *J. Mol. Catal. A*, 2009, 314, 95.

Rezai, S.A.S., Y. Traa, *Chem. Commun.*, 2008a, 2382.

Rezai, S.A.S., Y. Traa, *J. Membr. Sci.*, 2008b, 319, 279.

Satterfield, C.N., *Heterogeneous Catalysis in Practice*, McGraw-Hill, 1980, p. 241.

Smirnov, A.V., E.V. Mazin, V.V. Yuschenko, E.E. Knyazeva, S.N. Nesterenko, I.I. Ivanova, L. Galperin, R. Jensen, S. Bradley, *J. Catal.*, 2000, 194, 266.

Traa, Y., Non-oxidative activation of alkanes, in: *Handbook of Heterogeneous Catalysis*, 2nd edn., G. Ertl, H. Knözinger, F. Schüth, J. Weitkamp (eds.), Wiley-VCH, Weinheim, Germany, 2008, p. 3194.

Vazhnova, T., S.P. Rigby, D.B. Lukyanov, *J. Catal.*, 2013, 301, 125.

Wang, X., H. Carabineiro, F. Lemos, M.A.N.D.A. Lemos, F. Ramôa Ribeiro, *J. Mol. Catal. A*, 2004, 216, 131.

Wan Hussin, D., Y. Traa, *Energ. Fuels*, 2014, 28, 3352. doi:10.1021/ef500333b.

14 Syngas to Liquids via Oxygenates

Marius Westgård Erichsen,
Juan Salvador Martinez-Espin, Finn Joensen,
Shewangizaw Teketel, Pablo del Campo Huertas,
Karl Petter Lillerud, Stian Svelle,
Pablo Beato, and Unni Olsbye

CONTENTS

14.1 INTRODUCTION

The aim of the OCMOL project is to develop an integrated process for production of liquid fuels from relatively small natural gas wells in a sustainable way, thereby avoiding the current practice of flaring due to unfavorable economics. The oxidative coupling of methane followed by oligomerization to liquids (OCMOL) process, simplified in Figure 14.1, is a fully integrated process characterized by extensive recycling of by-products, including CO_2. The focus of this chapter will be on Step 3, where synthesis gas (syngas) from the methane reforming unit will be converted into a clean-burning gasoline with a low content of aromatics. This target is achieved by combining the processes of synthesis gas conversion to oxygenates (STO) and

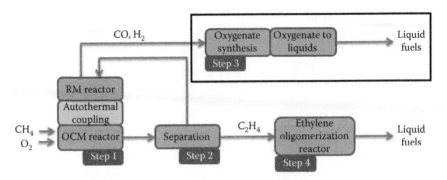

FIGURE 14.1 Block diagram of the OCMOL process, with step 3 (the subject of this chapter) highlighted. OCM, oxidative coupling of methane; RM, reforming of methane.

oxygenates conversion to liquids (OTL). Sections 14.2 and 14.3 will describe each of these two processes, while Section 14.4 provides details of the combined STO–OTL unit and the process engineering behind it.

Commercial routes already exist for the conversion of synthesis gas to gasoline via methanol. However, the OCMOL project introduced two additional challenges to the process:

First, the STO–OTL unit would need to be integrated in the OCMOL process, which implies that the feed for the STO reactor would consist of a blend of unconverted methane, various products, and inert gas and that the pressure of the STO reactor was mainly defined by the preceding steps. Second, the target for the OTL reactor was set to 75% yield of nonaromatic C_{5+} hydrocarbons, a hydrocarbon blend that had not yet been achieved in an OTL process.

14.2 SYNGAS TO OXYGENATES

Methanol has emerged as an important intermediate in the production of petrochemicals and synthetic fuels from syngas. Section 14.3 will describe the production of gasoline from oxygenates such as methanol and dimethyl ether (DME), but such processes are only viable if the production costs of oxygenates are relatively low. A methanol plant with natural gas as feedstock is typically divided into three sections: synthesis gas production, methanol synthesis, and purification. The capital cost of a large-scale methanol plant is significant, and around 60% of the investment is associated with the syngas production. Today, the syngas production is often carried out by steam reforming in tubular reformers. However, autothermal reforming technology (ATR) is becoming preferred since it maximizes the single line capacity and minimizes the investment. ATR combines substoichiometric combustion and catalytic steam reforming to convert natural gas and oxygen into a reactive syngas with a low H_2/CO ratio. Subsequently, the synthesis of methanol is typically carried out on $Cu/ZnO_2/Al_2O_3$ catalysts by means of boiling water reactors (BWRs) [1,2].

The dehydration of methanol yields DME, a compound with properties similar to liquefied petroleum gas (LPG) that can be used to substitute propane as fuel in households and industry. Like LPG, DME is gaseous at normal temperature and pressure, but becomes a liquid when subjected to modest pressure or cooling. With a heat value of 28.4 MJ kg^{-1} and cetane numbers of 55–60, DME can actually be used directly in a diesel engine as a substitute for crude oil-derived diesel fuel [3]. Compared to conventional diesel, DME burns much cleaner in a diesel engine, without significant soot formation. DME is currently being developed as a second-generation biofuel (BioDME), and a first pilot plant based on black liquor gasification is currently in operation in Piteå, Sweden [4].

Another approach to produce oxygenates from synthesis gas is the combination of methanol and DME synthesis, which has been followed by Haldor Topsøe A/S, for their topsoe integrated gasoline synthesis (TIGAS) process [5]. This combined methanol/DME synthesis approach has also been chosen for the OCMOL process, and in the following sections, we will describe some of its characteristics.

14.2.1 THERMODYNAMICS

One of the targets of OCMOL was to design an STO process utilizing the syngas produced in the preceding reforming step (see Figure 14.1), which has a H_2/CO ratio close to 1, with a conversion efficiency above 95%. Thermodynamics play a crucial role to meet this goal. The combined usage of methanol and DME as feed to the gasoline synthesis reactor, like in the TIGAS processes, can be advantageous. This layout exploits the favorable thermodynamics of syngas conversion into oxygenates as shown in Figure 14.2 [5]. The combination of methanol and DME synthesis provides an important enhancement to the conversion of syngas, enabling a remarkable efficiency at pressures significantly below those required for methanol synthesis. However, the equilibrium conversion is still limited and recycle is required in order to reach the targeted 95% syngas conversion levels.

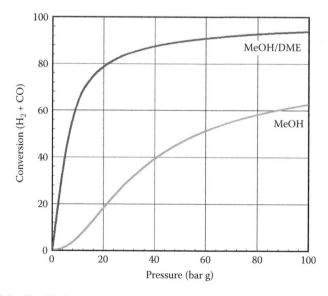

FIGURE 14.2 Equilibrium conversion of syngas versus pressure of synthesis gas to methanol/CO_2 and to methanol/DME/CO_2, respectively. Synthesis gas composition (mol%) H_2/CO/CO_2 = 51/48/1. Temperature = 250°C.

The conversion of syngas to methanol and DME proceeds via three main reactions:

$$CO + 2H_2 \leftrightarrow CH_3OH \quad \Delta H = -90.7 \text{ kJ mol}^{-1} \tag{14.1}$$

$$CO + H_2O \leftrightarrow CO_2 + H_2 \quad \Delta H = -41.1 \text{ kJ mol}^{-1} \tag{14.2}$$

$$2CH_3OH \leftrightarrow CH_3OCH_3 + H_2O \quad \Delta H = -23.6 \text{ kJ mol}^{-1} \tag{14.3}$$

$$\text{Net reaction} \quad 3H_2 + 3CO \rightarrow CH_3OCH_3 + CO_2 \tag{14.4}$$

Combining the synthesis of methanol and DME not only improves the process efficiency, but also induces flexibility in terms of synthesis gas composition. Figure 14.3 illustrates the effect of syngas composition on the product distribution at equilibrium. At CO-rich conditions, the water gas shift reaction induces a strong enhancement of conversion because water, formed in the dehydration step (Equation 14.3), is almost completely shifted by reaction with CO, forming H_2 and CO_2 (Equation 14.2). The net reaction becomes essentially that hydrogen and carbon monoxide form DME and carbon dioxide (Equation 14.4).

The combined methanol/DME yield is particularly high at H_2/CO ratios close to one, which is similar to that obtained in the OCMOL process reforming step. At 35–50 atm pressure and H_2/CO = 1, thermodynamics permit close to 35 mol% oxygenate production, most of which is DME.

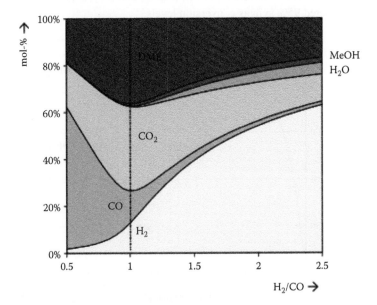

FIGURE 14.3 Effect of synthesis gas H_2/CO ratio on the thermodynamic equilibrium distribution during combined methanol/DME synthesis. Temperature = 240°C, pressure = 35 atm.

14.2.2 CATALYSTS

Traditional catalysts for the conversion of synthesis gas into MeOH and DME are based on Cu/ZnO in combination with an acid component such as γ-Al$_2$O$_3$, silica–alumina, or zeolites [6–10]. Technology for methanol synthesis on the active copper sites (Equation 14.1) is well known, and recently, researchers have focused on the dehydration of methanol to DME over the acid sites of the catalyst (Equation 14.3). For this purpose, zeolites are typically preferred due to their Brønsted acidity. On the other hand, the strongly acidic sites of the zeolites might also provoke the formation of higher hydrocarbons (see Section 14.3) and lead to coking of the catalyst. For this reason, a careful adjustment of acid strength is required. Numerous studies have investigated how the type and strength of the acid sites influence the reaction in terms of CO conversion, selectivity to DME, and catalyst lifetime. For instance, García-Trenco and Martínez [7,11] observed that methanol dehydration was mostly driven by the number of strong Brønsted acid sites, while the presence of strong Lewis sites was not crucial. The preparation methodology is also a controversial point that has been extensively studied. The two catalytic functions can be prepared together as a unique material (composite catalyst), or they can be prepared separately and mechanically mixed afterward (hybrid catalyst).

In this work, Topsøe's portfolio of proprietary DCK-10 series of composite catalysts was applied using a molar feed composition close to the near-ideal blend coming from the reformer unit (Figure 14.1) at 50 bar. Figure 14.4a shows the effluent composition of the optimized catalyst in a bench scale unit at 50 bar when using a molar feed composition H$_2$/CO/CO$_2$/Ar = 48/46/3/3. The once-through syngas conversion was 83%, with DME concentrations between 25 and 30 mol% in the product stream. With the addition of a modest recycle stream, the targeted 95% conversion should be possible to obtain.

The selectivity toward oxygenates over the optimized catalyst, compared to other hydrocarbons, was also studied in detail, as shown in Figure 14.4b. Combined

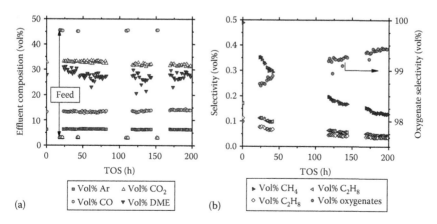

FIGURE 14.4 (a) Effluent composition versus TOS for validation test at 50 atm. Conditions: T_{inlet} = 250°C (H$_2$/CO/CO$_2$/Ar = 48:46:3:3), NHSV = 2000. (b) Hydrocarbon by-product formation and overall product selectivity (excluding CO$_2$); 50 bar, 250°C, and NHSV = 2000 h^{-1}.

methanol/DME selectivities higher than 99% were observed after the first 100 h of testing. The main by-products were methane, ethane, and propane, but selectivity to each of these was always below 0.5%.

14.2.3 Process Optimization

As shown in Section 14.2.1, the conversion of syngas into oxygenates is exothermic, with an enthalpy of −246 kJ per mol DME (Equation 14.4 in Section 14.2.1). Thus, providing efficient heat removal is mandatory in order to achieve high conversions. The use of BWRs is preferential, but slurry bed processes have also been proposed. A BWR consists of multiple tubular reactors surrounded by boiling water on the shell side, ensuring that the heat released during the methanol/DME synthesis is efficiently transformed into medium-pressure steam and keeping the catalyst temperature at a suitable level. Thereby, this system provides a high catalyst utilization, low recirculation, and low by-product formation.

Emphasis was placed on optimizing the synthesis of oxygenates via recycling unconverted reactants. Process studies varying the ratio of recycle to makeup gas (R/M) for different reaction pressures and product separation temperatures were essential to determine the most viable conditions for synthesis of oxygenates. Figure 14.5 illustrates that an R/M ratio close to 1 is sufficient to reach high conversion and that further increase in the R/M ratio does not lead to significant gains. As expected from thermodynamic and kinetic analysis, higher pressures result in higher conversion. A steady-state conversion value close to the targeted 95% is reached when an R/M ratio of 1 is used at 50 bar.

The separation temperature of the products formed in the STO unit was also studied at different R/M ratios (Figure 14.6). The syngas conversion increases by decreasing the separation temperature, as products are more efficiently separated. Below 5°C and with an R/M ratio of 1 or higher, 95% syngas conversion can be obtained.

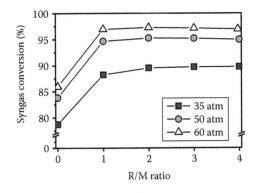

FIGURE 14.5 Syngas conversion versus R/M ratio, effect of pressure. T = 240°C, approach to equilibrium MeOH/WGS/DME = 20°C/20°C/200°C. Product separation performed at 5°C.

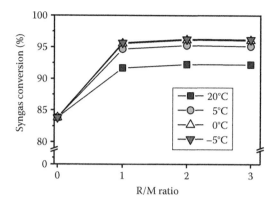

FIGURE 14.6 Syngas conversion versus R/M ratio: effect of separation temperature at 50 atm pressure. T = 240°C, approach to equilibrium, MeOH/WGS/DME = 20°C/20°C/200°C.

14.2.4 CONCLUSION/SUMMARY OF STO STUDY

This section has described the optimization of the catalytic conversion of syngas into oxygenates. The received feed gas composition, characterized by low H_2/CO ratios and a moderate synthesis pressure (50 atm), provides an efficient layout for the synthesis of oxygenates. Syngas conversion levels above 80% are achieved per pass and with a modest recycle to makeup ratio of 1, the conversion levels are above the targeted 95%. The product separation is carried out efficiently below 5°C, resulting in a product liquid phase composed of DME and CO_2. Due to the high solubility of CO_2 in DME, the separation step provides a suitable means of rejecting CO_2 from the synthesis loop. Thus, the gaseous phase is recycled directly back to the oxygenate synthesis reactor without an additional CO_2 removal unit, significantly reducing capital costs.

14.3 OXYGENATES TO LIQUIDS

In the 1970s, researchers at Mobil discovered that methanol could be converted to hydrocarbons over zeolite catalysts [12,13]. After attempting to react isobutane with methanol over H-ZSM-5, they found that even though isobutane was not consumed, a mixture of alkanes and aromatics similar to high octane gasoline was produced. Due to the Arab oil embargo and subsequent oil crisis, alternative energy sources were already being sought [14,15]. As methanol may be produced with high energy efficiency via any carbon-based feedstock, this discovery sparked an extensive research effort and considerable commercial interest in the methanol to hydrocarbon (MTH) reaction.

The stoichiometry of the acid-catalyzed reaction from MTHs can be represented by the general reaction equation:

$$CH_3OH \xrightarrow{\ H^+\ } CH_2 + H_2O$$

SCHEME 14.1   General scheme of the MTH reaction. (From Chang, C.D., *Catal. Rev.*, 25, 1, 1983.)

CH_2 in this equation represents a range of both aliphatic and aromatic hydrocarbons. The exact product distribution can be varied through changes in process conditions and by the use of shape-selective zeolite catalysts. The general reaction path consists of an initial equilibration of methanol to DME and water, before this mixture reacts further to form alkenes. These alkenes then react further to form alkanes, aromatics, and larger alkenes, as shown in Scheme 14.1.

The reaction is strongly exothermic, and this makes control and removal of reaction heat a major factor in process design. The amount of heat released during the reaction depends on the exact product distribution [16], and the dehydration of methanol to DME accounts for a significant fraction of the total reaction heat. For this reason, some processes use a mixture of methanol and DME as feed for the MTH reactor.

Several processes based on the reaction have been developed, and the first commercial natural gas to gasoline plant utilizing Mobil's methanol-to-gasoline (MTG) process was brought on-stream in New Zealand in 1986 [17]. Haldor Topsøe also developed an alternative gasoline technology, TIGAS [5]. This process combined methanol, DME, and gasoline synthesis from syngas in a single loop, featuring a high conversion of the syngas feed (see Section 14.2.1). Schematic representations of these two processes are displayed in Figure 14.7. In addition to this, Mobil developed

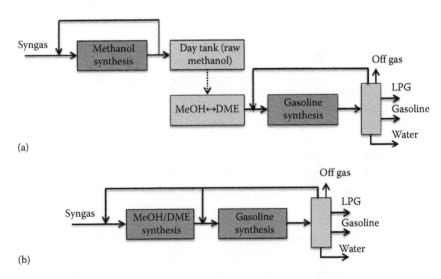

FIGURE 14.7   (a) Mobil's MTG process and (b) Topsøe's TIGAS process.

a fluid-bed process producing either gasoline or light alkenes over H-ZSM-5 based on process conditions [18]. The alkenes produced from this process could also react in a second step to produce gasoline and distillate fuels in the so-called mobil olefins to gasoline and distillates (MOGD) process [19].

Shortly after the construction of the MTG plant in New Zealand, oil prices plummeted and the gasoline plant shut down in the mid-1990s. Nevertheless, methanol conversion to hydrocarbons has remained an important research topic both in academia and industry, and now that oil prices have risen, the process is again being commercialized. Also, commercialization is taking place on a large scale in China for production of light olefins from coal. So far, plants based on three different technologies have been constructed: the Lurgi methanol to propylene (MTP) process [20], the UOP/Norsk Hydro (now UOP/INEOS) methanol to olefin (MTO) process [21], and the Dalian methanol to olefin (DMTO) process [22]. Of these, Lurgi's process utilizes H-ZSM-5 in a parallel fixed-bed setup with feed injection between beds and product recycle to maximize propene yields, while the latter two processes utilize the narrow-pore zeotype catalyst H-SAPO-34 in fluid-bed operation to produce a mixture of ethene and propene.

The processes described earlier show the flexibility inherent in the MTH chemistry. For example, H-ZSM-5 is a catalyst both for production of gasoline (MTG, TIGAS) at high pressure and moderate temperature and propene (with gasoline as by product) at higher temperatures and lower pressures (MTP). On the other hand, catalysts with different pore structures may also be used to obtain different product selectivity. An example of this is the use of H-SAPO-34 in the MTO processes, which produces ethene and propene very selectively.

Figure 14.8 shows the pore structures of the two commercial MTH catalysts, H-ZSM-5 and H-SAPO-34. Where the former catalyst has a 3D channel system of medium-sized intersecting pores, H-SAPO-34 comprises large voids connected by apertures too small to allow diffusion of branched hydrocarbons. These two

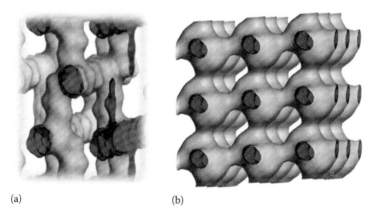

(a)   (b)

FIGURE 14.8 Pore structures of the commercial zeotype catalysts (a) H-ZSM-5 and (b) H-SAPO-34. The crystal framework has been removed for clarity, and the inner channel surface is drawn in blue.

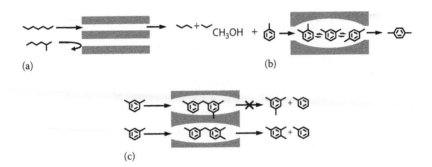

FIGURE 14.9 The three classes of shape selectivity: (a) reactant selectivity, (b) product selectivity, and (c) transition state selectivity. (Adapted from Csicsery, S.M., *Pure Appl. Chem.*, 58, 841, 1986.)

structures nicely illustrate the concept of shape selectivity, which results from the uniform pores present in the zeolite frameworks. As these pores are of molecular dimensions, the size of the pores determines which chemical species fit inside and what reactions can occur inside the materials. Shape selectivity is usually divided into three differing types [23–26]: reactant, product, and (restricted) transition state shape selectivities. These are schematically illustrated in Figure 14.9.

In reactant selectivity (Figure 14.9a), bulky reactants are prevented from diffusing into the pore system, while smaller molecules diffuse more easily into the pores. The larger molecules will thus be unable to reach an active site and will not react. Product shape selectivity (Figure 14.9b) is observed when some of the product molecules are too large to diffuse out of the pore system. This is often observed in zeolites featuring large internal cavities connected by narrow apertures such as in H-SAPO-34. Large molecules can be formed in the cavities but must react further to smaller species before they can diffuse out of the structure. In (restricted) transition state shape selectivity (Figure 14.9c), neither the product nor the reactant is hindered from diffusing into or out of the zeolite, but the available space cannot accommodate certain transition states. It is worth noting that when the dimensions of the pores and reacting molecules are similar, even subtle changes in pore structure or reactant molecules can cause large differences in diffusivities and reactivity [24].

Typical product compositions such as the MTG, MTP, and MTO processes are displayed in Figure 14.10. However, the relative amounts of products can be altered even further than what is shown in Figure 14.10 by including a second reaction step. For instance, the MTO process can incorporate a cracking step, such as the UOP/total petrochemical olefin cracking process, to convert by-product C_{4+} alkenes into ethene and propene [27], or propene can be oligomerized in a second step to form gasoline and distillate as in the MOGD process [19].

The gasoline produced in the MTG process over H-ZSM-5 typically contains 25–35 vol% aromatics [28], while the target of the OCMOL OTL process was an aromatic-free gasoline. It is possible to produce gasoline with a lower content of aromatics (2% [29]) in the MOGD process, but there is currently no available process to make gasoline range alkenes in a single step. To accomplish this, new catalysts and

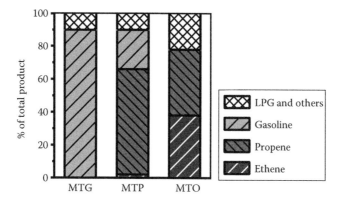

FIGURE 14.10 A comparison of the product fractions produced by the MTG, MTP, and MTO processes.

a high degree of insight into the reaction mechanisms of zeolite-catalyzed methanol conversion were required. Sections 14.3.1 and 14.3.2 are dedicated to a description of MTH research up to and including the current state of the art in reaction mechanisms and how this insight led to the discovery of a catalyst selective toward C_{5+} alkenes.

14.3.1 EVOLUTION TOWARD THE DUAL CYCLE MECHANISM

Ever since the discovery of the MTH reaction, the reaction mechanisms have been studied and debated [15,16,30–34]. In 1979, Chen and Reagan proposed that the formation of higher olefins in MTH reaction was autocatalytic [35], and several works published in the early 1980s proposed indirect rather than direct coupling of C_1 species. For instance, Dessau and Lapierre [36,37] proposed that the reaction over H-ZSM-5 was driven by a continuous cycle of alkene methylation and cracking, as illustrated in Scheme 14.2.

Both Langner [38] and Mole et al. [39,40] also proposed indirect mechanisms based on cycles of methylation and dealkylation, but involving cyclic or aromatic intermediates, around the same time. At present, there is a strong consensus that

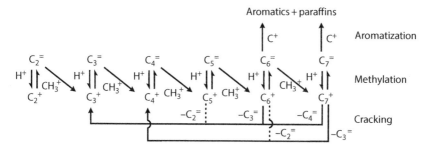

SCHEME 14.2 The methylation/cracking mechanism proposed by Dessau. (Adapted from Dessau, R.M., *J. Catal.*, 99, 111, 1986.)

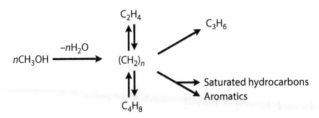

SCHEME 14.3   The hydrocarbon pool mechanism proposed by Dahl and Kolboe. (Adapted from Dahl, I.M. and Kolboe, S., *J. Catal.*, 149, 458, 1994.)

direct C_1–C_1 coupling reactions are insignificant compared to the rate at which trace impurities of C_{2+} compounds initiate the reaction [41–43].

In the mid-1990s, Dahl and Kolboe carried out isotopic labeling experiments by cofeeding alkene precursors (ethanol, propanol) and ^{13}C-methanol over an H-SAPO-34 (CHA) catalyst. Analysis of the effluent showed that most of the products were formed exclusively from methanol under the applied reaction conditions [44–46]. Hence, the concept of a *hydrocarbon pool* was proposed. While their proposal shared many similarities with previous works, this schematic concept had a greater immediate influence than the works of the previous decade [30]. The original hydrocarbon pool model, as shown in Scheme 14.3, assumed that methanol was continuously added to a pool of adsorbed hydrocarbons, which successively eliminated light alkenes.

The initial hydrocarbon pool was given an overall stoichiometry $(CH_2)_n$, and the chemical structure was not specified [44–46]. Thus, the concept of the hydrocarbon pool could cover all intermediates in the proposed indirect mechanisms from the previous decade. The groups of Haw and Kolboe simultaneously concluded that polymethylbenzenes were the main hydrocarbon pool species in H-SAPO-34 (CHA) [47–49]. Additional evidence for the hydrocarbon pool mechanism in H-ZSM-5 (MFI), H-SAPO-34 (CHA), and H-SAPO-18 (AEI) was also provided by Hunger et al. [50–52]. Later studies of the MTH reaction in zeolite H-beta (*BEA) cemented the importance of polymethylbenzene intermediates also in this catalyst [53,54].

After a long period focusing on aromatic intermediates in the MTH reaction, steady-state isotope transient studies of the H-ZSM-5 (MFI) catalyst revealed that aromatics did not act as intermediates for all alkenes formed [55,56]. This finding gave rise to the dual cycle concept, which states that the hydrocarbon pool proceeds through two partly separated cyclic reaction mechanisms, as shown in Scheme 14.4. One of these cycles (the alkene cycle) involves methylation and cracking of alkenes in a similar manner to what was previously proposed by Dessau [36,37]. A main difference from the proposal by Dessau is that ethene formation from the alkene cycle is assumed to be negligible. The arene cycle involves continuous methylation of aromatic molecules and their subsequent dealkylation. The mechanism for dealkylation has not yet been fully elucidated, but isotopic labeling results at moderate temperature suggest that a systematic ring expansion and/or contraction is involved [53,57,58]. A possible reaction pathway is the paring reaction proposed by Sullivan et al. [59].

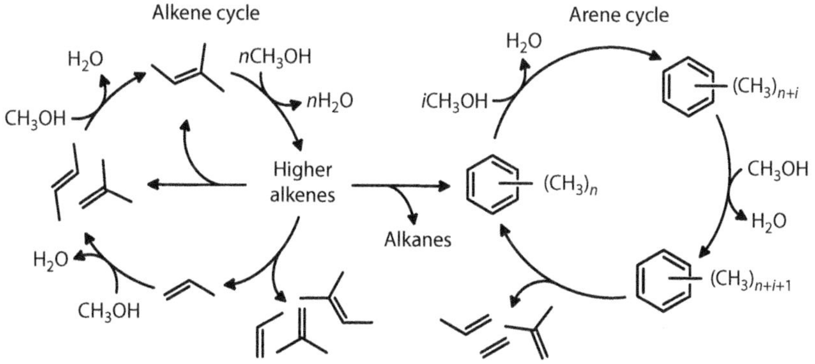

SCHEME 14.4 Suggested dual cycle concept for methanol conversion over zeotype catalysts. The relative importance of each cycle, as well as the exact structure of intermediates, depends on the catalyst employed and the process conditions. Thus, not all products shown here are observed in all systems. (Reproduced from Westgård Erichsen, M. et al., *Catal. Today*, 215, 216, 2013.)

The dual cycle proposal initiated a series of similar studies over different catalysts with the aim of relating catalyst structure to product selectivity. Several studies have shown that pore size is an important parameter determining which of the two cycles is favored. In general, it has been found that the arene cycle is more favored in large-pore than in medium-pore catalysts [60–64]. However, it has also been suggested that large-pore zeolites favor the alkene cycle at high pressure and low temperatures [65,66] or if the catalyst has a low acid strength [57,58].

14.3.2 Selectivity Control through Fundamental Insight

The introduction of the dual cycle concept raised an interesting fundamental question of whether it was possible to run one cycle independently of the other [55,56]. As both cycles are active in the large-pore zeolite H-beta (although the arene cycle is favored) [61], attention turned to whether a catalyst with smaller pores than H-ZSM-5 would suppress formation of aromatics and force the alkene cycle to operate on its own. In a rare example of rational catalyst design, this question was answered by studying the unidimensional narrow 10-ring zeolite H-ZSM-22 (TON). This catalyst was indeed found to strongly favor the alkene cycle while suppressing the formation of aromatic products [67,68]. The resultant product spectrum was rich in C_{3+} alkenes, with a high fraction of branched and disbranched C_{5+} products and a very low amount of aromatics. A GC–MS chromatogram of a typical product distribution from H-ZSM-22 is shown in Figure 14.11. Similar GC–FID chromatograms are published elsewhere [67].

Even though the product distribution from H-ZSM-22 is highly interesting, the catalyst displays a low activity for methanol conversion compared to other zeolites such as H-ZSM-5. In fact, H-ZSM-22 was previously tested for methanol conversion by Cui et al. [69] and reported to be inactive for conversion of methanol alone, although the same authors later reported activity for alkene methylation reactions [70]. Teketel et al. [68], and later Li et al. [71,72], found that a low weight hourly

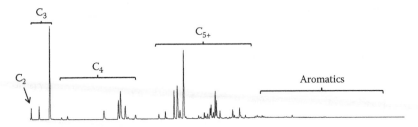

FIGURE 14.11 A representative GC–MS chromatogram of H-ZSM-22 effluent. Note the near empty aromatic region.

space velocity (WHSV) was a prerequisite for appreciable conversion of methanol in H-ZSM-22. This low activity of H-ZSM-22 was also further explored in a recent work by Janssens et al. [73], where it was found that the critical contact time (i.e., the contact time needed in order for the autocatalytic reaction to dominate the reaction rate) is much higher than in H-ZSM-5. The reason behind the low activity of H-ZSM-22 is unknown, but it is in line with the lower rate constants reported by Hill et al. [74] for propene methylation over H-FER (a zeolite with nearly as small pores as H-ZSM-22) compared to H-ZSM-5.

Another issue with H-ZSM-22 is its fairly rapid deactivation, which is illustrated in Figure 14.12 by plots of methanol conversion versus time on stream (TOS) at different temperatures and the corresponding conversion capacities before deactivation. While the total methanol conversion capacity per gram catalyst in the material tested by Teketel et al. [68] is lower than for H-SAPO-34 [75], which is used commercially, the lower amount of acid sites in typical H-ZSM-22 samples means that the conversion capacity per acid site in the catalyst is higher. A comparison with the very coke-resistant catalyst H-ZSM-5 containing the same amount of acid sites

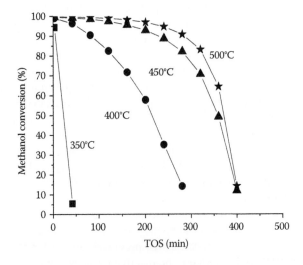

FIGURE 14.12 Methanol conversion over H-ZSM-22 versus TOS at different temperatures at WHSV = 2 gg_{cat}^{-1} h^{-1}. (Reproduced from Teketel, S. et al., *ChemCatChem*, 1, 78, 2009.)

performed by Janssens et al. [73] estimated the methanol conversion capacity of H-ZSM-22 to be roughly 1/20 of H-ZSM-5.

14.3.3 Catalyst Comparison

The OCMOL target for OTL was to identify a catalyst that would give 65% selectivity to gasoline range (C_{5+}) aliphatic hydrocarbons. A further target was to obtain at least 75% selectivity to C_{5+} aliphatic hydrocarbons by process optimization. Although the first target was already demonstrated for H-ZSM-22 [68], further catalyst optimization was attempted by synthesizing and screening a series of 1D 10-ring catalysts with similar channel dimensions. The results of this study have been reported previously [76], and only the main findings will be presented here. Figure 14.13 illustrates the channel systems of the investigated materials. The main channels of the four 1D 10-ring zeolites are of comparable size, but the shape of the pore differs, and some of the catalysts possess side pockets protruding from the main channel. H-ZSM-22 has slightly zigzagging channels measuring 4.6 Å × 5.7 Å at the apertures. H-ZSM-23 has slightly smaller channels than H-ZSM-22, which are teardrop shaped with dimensions 4.5 Å × 5.2 Å, but the shape of the pores creates small side pockets. H-ZSM-48 has nearly symmetrical straight channels with dimensions 5.6 Å × 5.3 Å [77], which are the largest channels of any of the four catalysts described here.

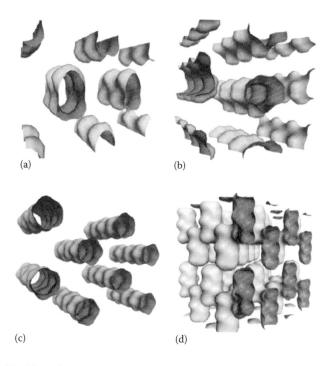

(a)   (b)

(c)   (d)

FIGURE 14.13 Illustrations of channel systems of (a) H-ZSM-22, (b) H-ZSM-23, (c) H-ZSM-48 and (d) H-EU-1 catalysts. (Adapted from Teketel, S. et al., *ACS Catal.*, 2, 26, 2012.)

TABLE 14.1
List, Descriptions, and Characterization Data for the 1D 10-Ring Catalysts Employed

Topology (Material)	10-Ring Channel System			Crystal Size (μm)	Si/Al	BET Area (m² g⁻¹)
	Main Channel Size	Shape	Side Pocket			
TON (H-ZSM-22(C))	4.6 Å × 5.7 Å	Elliptical	None	1–2	48	196
TON (H-ZSM-22(H))	4.6 Å × 5.7 Å	Elliptical	None	2–3	30[b]	207
MTT (H-ZSM-23)	4.5 Å × 5.2 Å	Teardrop	Very small	<1	33	115
MRE (H-ZSM-48)[a]	5.3 Å × 5.6 Å	Cylindrical	None	1–2	52	275
EUO (H-EU-1)	4.1 Å × 5.4 Å	Zigzag	6.8 × 5.8 × 8.1 Å (12 rings)	<1	30	420

Source: Adapted from Teketel, S. et al., *ACS Catal.*, 2, 26, 2012.

For the H-ZSM-22 samples, "C" denotes the commercial and "H" the homemade sample.

[a] Disordered structure.

[b] Si/Al ratios have been determined by NH_3 TPD for all samples except H-ZSM-22(H), for which it was determined by ICP-AES.

H-EU-1 has channels with dimensions 4.1 Å × 5.4 Å and large 12-ring side pockets. The 12-ring side pockets are 6.8 Å × 5.8 Å wide and 8.1 Å deep.

Characterization data for the tested materials are shown in Table 14.1. All materials were single phase and crystalline, exhibited fairly high BET surface areas, and have particle sizes ranging from <1 to 2–3 μm. The Si/Al ratios of the catalysts, as determined from NH_3 TPD, were within a limited range of 30–52 (see Table 14.1 for details). On the basis of these similarities, it appears reasonable to assign major selectivity differences to differences in topologies. Two different samples of H-ZSM-22 were employed in the study, denoted (C) for commercial (supplied by Zeolyst International) and (H) for homemade, respectively.

Product yield versus conversion plots obtained during gradual deactivation for the five catalysts at 400°C are presented in Figure 14.14. Previous investigations of H-ZSM-5 have concluded that deactivation does not influence the selectivity of the catalysts for the MTH reaction [78,79], and this was confirmed to be valid for H-ZSM-22, H-ZSM-23, and H-EU-1 as well (H-ZSM-48 was tested at one WHSV only, so the possible deactivation influence on this structure is inconclusive) [76]. The change in selectivity with reaction time is therefore regarded as a change with contact time.

Concentrating first on H-ZSM-22 and H-ZSM-23, very similar yield versus conversion data were observed for the two structures. C_{5+} hydrocarbon yields up to 55%, corresponding to selectivities between 40 and 70 C%, were observed depending on the conversion. Moreover, the selectivity toward aromatic hydrocarbons was inferior to 2% at all measured conversions, and the main by-products over H-ZSM-22 and H-ZSM-23 were C_2–C_4 alkenes. The product yield versus conversion curves for all products were rather linear in the 0%–80% conversion range and then changed slope in the upward or downward direction. In general, an upward slope

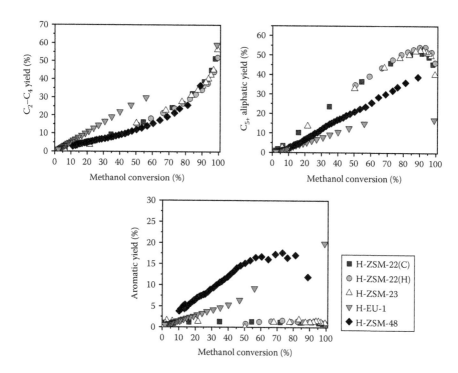

FIGURE 14.14 Yields of various hydrocarbon products versus methanol conversion at 400°C and WHSV = 2 gg_{cat}^{-1} h^{-1} for the catalysts investigated. For the H-ZSM-22 samples, "C" denotes the commercial and "H" the homemade sample.

change at increasing conversion suggests that the product is formed by secondary reactions, while a downward slope change with increasing conversion suggests that the product is further converted into other products. As such, the slope changes observed suggest that C_{5+} aliphatic products are further converted to lighter alkenes over H-ZSM-22 and H-ZSM-23 at the high conversion. It is also worth noting that an optimal C_{5+} aliphatic yield of 53%–55% was observed for both topologies (and both batches of H-ZSM-22 catalyst) at 90%–95% methanol conversion. A change in product distribution at high methanol conversions has been observed also for other topologies, and it was suggested that the alkene interconversion reactions are suppressed in the presence of methanol, possibly due to competitive adsorption at the Brønsted acid sites [80].

The total selectivity toward gasoline range (C_{5+}) hydrocarbons is slightly higher (between 55 and 75 C%) over H-ZSM-48 than over H-ZSM-22 and H-ZSM-23 but includes 20–40 C% selectivity for aromatic products. Formation of considerable amounts of aromatics over H-ZSM-48 makes this material poorly suited for production of aromatic-free gasoline. The most probable reason for the difference is the slightly larger channel diameter of this topology, strongly suggesting that even subtle changes in channel dimensions may have an important impact on product selectivity.

H-EU-1 displayed a significantly lower selectivity for C_{5+} hydrocarbons than the other catalysts, ranging between 35 and 45 C%, including ~15 C% selectivity for aromatics. This decreased selectivity to gasoline range hydrocarbons was accompanied by a higher selectivity toward C_3 and C_4 hydrocarbons. A plausible explanation for the unexpected product distribution over H-EU-1 zeolite, which has the narrowest 10-ring channel, might be the involvement of the 12-ring side pockets on the outer surface of the crystal. Large aromatics should not be able to diffuse through the narrow 10-ring channels, but if they are formed in side pockets open to the exterior of the crystals, the 10-ring channel would not be involved. A similar phenomenon has previously been suggested for MCM-22 zeolite [64,81–84]. Since the H-EU-1 crystals employed here were small (see Table 14.1 or Ref. [76] for details), reactions occurring on the external surface area may be more pronounced than in the other samples.

Figure 14.15 displays the product yields versus conversion for four of the five catalysts at 450°C. H-ZSM-22(H) is not included, but as selectivity was nearly identical to H-ZSM-22(C) at 400°C, it is assumed to behave similarly also at 450°C. Focusing on the initial yields at high methanol conversion for H-ZSM-22 and H-ZSM-23, an increase in reaction temperature from 400°C to 450°C led to a slight decrease in C_{5+} yield, accompanied by an increase in the selectivity for lighter fractions. This may be ascribed to both an increased tendency toward cracking of heavier hydrocarbons to lighter fragments at high temperatures and thermodynamic effects. At lower methanol conversion (<95%), product yields over H-ZSM-22 and H-ZSM-23 were similar to those obtained at 400°C. For H-ZSM-48 and H-EU-1, an increase in the yield of

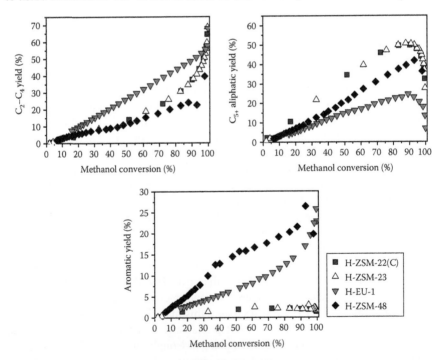

FIGURE 14.15 Yields of various hydrocarbon products versus methanol conversion at 450°C and WHSV = 2 gg_{cat}^{-1} h^{-1} for the catalysts investigated.

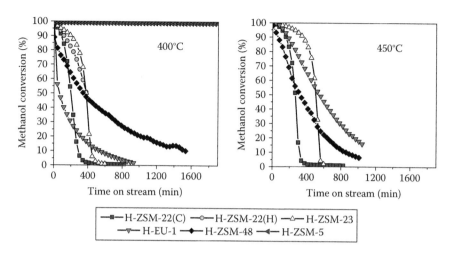

FIGURE 14.16 Methanol conversion as a function of TOS at 400°C and 450°C; reaction carried out at WHSV = 2 gg_{cat}^{-1} h^{-1}.

aromatic products was observed at high methanol conversion, but otherwise, the yields were similar to those obtained at 400°C.

The second issue to be addressed in the comparison of 1D 10-ring structures was catalyst activity. Screening tests at WHSV of 2 gg_{cat}^{-1} h^{-1} or larger showed that a WHSV of 4 gg_{cat}^{-1} h^{-1} or less was required to achieve an initial conversion near 100% for all tested catalysts. Figure 14.16 displays methanol conversion versus TOS plots for H-ZSM-22 (homemade and commercial), H-ZSM-23, H-ZSM-48, and H-EU-1 at 400°C and 450°C at WHSV = 2 gg_{cat}^{-1} h^{-1} and compares it to a typical H-ZSM-5 sample tested at the same WHSV and temperatures. It was observed that all samples deactivated significantly in less than 400 min on stream, which is very rapid when compared to H-ZSM-5. An increase in space velocity was found to lead to even more rapid deactivation, with WHSV above 4 gg_{cat}^{-1} h^{-1} leading to near instant deactivation.

The deactivation curves observed for H-ZSM-22, H-ZSM-23, and H-ZSM-48 displayed a shape typical for the MTH reaction, with a period of full conversion followed by methanol breakthrough and rapid deactivation. This curve shape is typical of catalytic systems where the reactant (in this case methanol) is strongly involved also in reactions leading to deactivation (coke formation) and where the active segment of the catalyst bed progresses downward through the reactor until it reaches the end of the catalyst bed and deactivation becomes apparent [73,78]. H-EU-1, on the other hand, displayed a different conversion versus TOS behavior, which might suggest a different deactivation mechanism, possibly involving coke-forming reactions from the products alone.

H-ZSM-22 and H-ZSM-23 were tested at various WHSV at 400°C, and both the deactivation rate of the active catalyst segment and the total methanol conversion capacity were determined from t_{50} (the time required before deactivation from 100% to 50% conversion) versus contact time plots, according to the method employed by Janssens et al. [73] (methanol conversion capacity is defined as the total gram

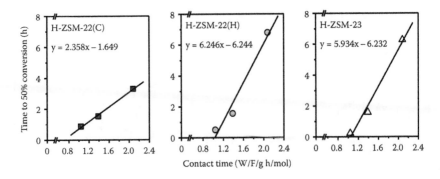

FIGURE 14.17 Catalytic lifetime to 50% conversion as a function of contact time with a methanol concentration of 13% in He at 400°C.

TABLE 14.2

Total Methanol Conversion, Defined as the Gram Amount of Methanol Converted to Products per Gram Amount of Catalyst over H-ZSM-22 and H-ZSM-23

	Total Conversion Capacity at Different Temperatures			
Material	350°C	400°C	450°C	500°C
H-ZSM-22(C)	1.1	6.5 (9.8)	8.5	11.0
H-ZSM-22(H)	0.7	12 (26.0)	10.5	11.2
H-ZSM-23	0.3	11.6 (24.7)	16.0	9.3

Values in parenthesis are calculated according to the method described in Ref. [73].

---

amount of methanol converted to hydrocarbons per gram amount of catalyst before complete deactivation [85,86].) The plots are shown in Figure 14.17, and the resulting conversion capacities are reported in Table 14.2, together with methanol conversion data determined by integration of single methanol conversion versus TOS plots at 350°C–500°C (as in Figure 14.16).

The data in Table 14.2 show that the methanol conversion capacity of the home-made and commercial sample of H-ZSM-22 differed by a factor of 2 at 400°C. The reason for this difference is unknown, but implies that synthesis method and the exact physical characteristics of the catalyst are important for the lifetime of the catalyst, even if the product selectivities do not change. H-ZSM-23 displayed a conversion capacity in the upper range of the H-ZSM-22 batches, suggesting that even the selectivity toward deactivating species is similar for these two catalyst structures.

Furthermore, the data in Table 14.2 and the high critical contact time (Figure 14.17) strongly suggest that the conversion capacity of H-ZSM-22 and H-ZSM-23 can be increased if the autocatalytic reaction is initiated at the start of the catalyst bed. This possibility was further explored by the addition of alkenes to the reactant feed, and this is described in Section 14.3.4.2.

The identity and distribution of the compounds retained within the catalyst pores during the MTH reaction were analyzed using GC–MS by dissolving the catalyst

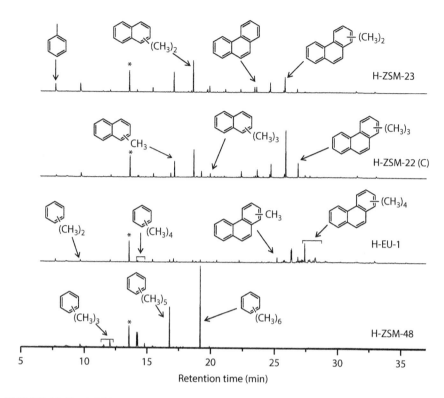

FIGURE 14.18 MTH reaction over H-ZSM-22 (C), H-ZSM-23, H-EU-1, and H-ZSM-48; GC/MS total ion chromatogram of the hydrocarbon extracts of the deactivated catalysts. Reaction carried out at 400°C and WHSV = 2 gg_{cat}^{-1} h^{-1}. All the peaks are normalized relative to the standard C_2Cl_6 peak, indicated by the asterisk (*) in the chromatogram. (From Teketel, S. et al., *ACS Catal.*, 2, 26, 2012.)

with 15% HF and extracting the organic phase with CH_2Cl_2. The results are shown in Figure 14.18. Previous GC–MS analysis of retained coke species [67], and computational modeling [71] of them, in H-ZSM-22 suggested that species such as polymethylated benzenes and naphthalenes were responsible for the deactivation by blocking the zeolite channel openings, rather than acting as active centers for alkene production, and this explanation is plausible also for the very similar H-ZSM-23 catalyst.

For H-ZSM-22 and H-ZSM-23 catalysts, very similar hydrocarbons were detected as retained species. However, the concentration of relatively heavy species was slightly higher over H-ZSM-22 (C) than H-ZSM-23. Even heavier species were trapped in the H-EU-1 pore structure compared with H-ZSM-22 and H-ZSM-23, and this is probably related to the large side pockets in EU-1 (Table 14.1): a strong correlation between cavity size and formation of polyaromatic species has previously been reported for 3D 10-ring topologies [87]. Moreover, the distribution of retained species is substantially shifted toward heavier compounds, strongly suggesting the involvement of the more spacious 12-ring side pockets during the reaction. A very different chromatogram was obtained for H-ZSM-48, where polymethylbenzenes were the dominant species detected.

14.3.4 OTL YIELD OPTIMIZATION

The data presented in Section 14.3.3 show that among the tested structures, only H-ZSM-22 and H-ZSM-23 meet the OCMOL targets for product selectivity (low-aromatic gasoline). Moreover, it was found that H-ZSM-23 and H-ZSM-22 possess similar activity and methanol conversion capacity. Further process optimization with the aim of achieving the ultimate OCMOL target of 75% net selectivity toward C_{5+} alkenes was therefore carried out using H-ZSM-22 (C).

14.3.4.1 High-Pressure Operation

It has previously been shown that a higher reactant pressure leads to higher selectivity toward C_{5+} hydrocarbons over H-ZSM-5 at 350°C (see Table 14.3) and over H-ITQ-13 [88–90].

An attempt was therefore made to perform the MTG reaction at elevated pressure also over H-ZSM-22 (C). The experiment was performed at 450°C, with total pressure either 1 or 30 bar, respectively. The results are shown in Figure 14.19. Contrary

---

TABLE 14.3

Product Selectivities Obtained at 100% Methanol Conversion during the MTH Reaction over H-ZSM-5 (PZ-2/100H from Zeolyst International) at 350°C at Ambient and Elevated Pressure

P(MeOH) (atm)	P(Tot) (atm)	WHSV (gg_{cat}^{-1} h^{-1})	C_1	C_2	C_3	C_4	C_{5+}	Refs.
					Selectivity (C%)			
0.04	1	1.1	0.1	4.8	31	24	40	[89]
4.2	15	0.75	<1	<1	9	17	72	[90]

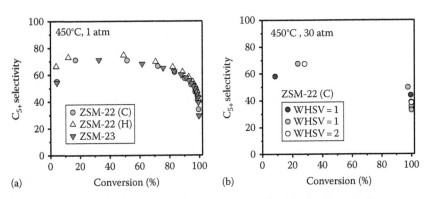

(a) (b)

FIGURE 14.19 C_{5+} selectivity as a function of methanol conversion during reaction at 450°C at atmospheric pressure (a, WHSV = 2 gg_{cat}^{-1} h^{-1}) and at 30 atm (b, WHSV = 1 and 2 gg_{cat}^{-1} h^{-1}) over H-ZSM-22 (C). The mole fraction of methanol in the 1 and 30 atm experiments was 13% and 5.6%, respectively. The data at high pressure were collected from three experiments: one at WHSV = 2 gg_{cat}^{-1} h^{-1}, with the first analysis after 10 min on stream, and two experiments at WHSV = 1 gg_{cat}^{-1} h^{-1}, where the first analysis was performed after either 10 or 50 min on stream.

to what was expected, the pressure increase did not lead to significant changes in product selectivities for the C_{5+} products. Furthermore, a more rapid deactivation was observed at 30 atm compared to 1 atm pressure. The total conversion capacity at 30 atm was determined to be approximately 2.2 g/g_{cat} at WHSV = 2 $gg_{cat}^{-1} h^{-1}$ and 5.2 or 6.3 gg_{cat}^{-1} (based on two different experiments) at WHSV = 1 $gg_{cat}^{-1} h^{-1}$. A decrease in conversion capacity at high pressure was also found previously for H-ITQ-13 and may be caused by an increased selectivity to aromatic products that remain trapped inside the catalyst.

14.3.4.2 Product Recycle

C_2–C_4 alkenes constitute the main by-products in the MTG process over H-ZSM-22 and H-ZSM-23 (see Figures 14.14 and 14.15). Recycling of those alkenes could potentially lead to enhanced net C_{5+} alkene yields. Therefore, co-conversion tests of methanol with C_2–C_4 alkenes (fed as the corresponding alcohols, alone or in mixtures) were performed over H-ZSM-22 as a model study of product recycle. The full study will be published elsewhere [91] and only the main findings will be reported here.

Results obtained when cofeeding methanol with isopropanol (47:1 partial pressure ratio) over H-ZSM-22 at 400°C and a methanol WHSV of 2 $gg_{cat}^{-1} h^{-1}$ are shown in Figure 14.20. Separate experiments where isopropanol was fed alone under otherwise similar conditions showed that the alcohol was rapidly converted to its alkene analogue. The curves obtained for only methanol feed or cofeeding with isopropanol were similar at all conversion levels (Figure 14.20). Furthermore, it was found that the amount of propene detected in the reactor effluent was much lower than the amount of propanol in the feed, indicating a net conversion of the added propene to hydrocarbons. When the methanol to isopropanol ratio in the feed was decreased to 10.4 (not shown), similar product selectivities for the methanol only and coconversion cases were observed only at close to full methanol conversion, while enhanced C_{5+} yields were observed for the coconversion case compared to the

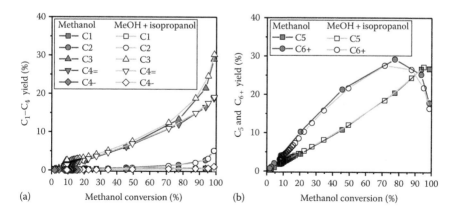

FIGURE 14.20 (a) Product yields of C_1–C_4 hydrocarbons and (b) C5+ hydrocarbons during conversion of methanol (closed symbols) and a 102.8:2.2 mixture of methanol and isopropanol (open symbols) over an H-ZSM-22 catalyst at 400°C. (Adapted from Teketel, S. et al., manuscript in preparation.)

methanol only case at lower conversions (obtained during deactivation). At lower than 85% methanol conversion, the propene concentration in the effluent was lower than the isopropanol content in the feed, demonstrating net conversion of isopropanol to hydrocarbons in the reactor [91]. Similar results were obtained for ethanol and isobutanol cofeeds, although ethanol showed lower conversion to hydrocarbons (except its corresponding alkene, ethene) than the higher alcohols. Isotopic labeling experiments, where ^{13}C-methanol was cofed with each of the unlabeled alcohols at a methanol to alcohol pressure ratio of 10.4, showed two regimes of product formation: at high conversion (fresh catalyst), random labeling of the products was observed, suggesting that product formation was dominated by the hydrocarbon pool. However, at low conversion (after significant deactivation), methylation of the cofed alkene was seen to dominate product formation, leading to predominantly mono- and dilabeled alkene products [91]. Together, the cofeed experiments suggest that C_2–C_4 product recycle may increase the net C_{5+} alkene yield in the MTG process over H-ZSM-22 but that an enrichment of ethene in the product stream may be foreseen, due to its lower reactivity compared to the C_3–C_4 alkenes.

In addition to their advantageous effect on C_{5+} product yields, a C_2–C_4 alcohol cofeed was observed to enhance the lifetime and methanol conversion capacity of the H-ZSM-22 catalyst [91]. An example is shown in Figure 14.21, where methanol was coreacted with individual C_2–C_4 alcohols. The reason for the enhanced lifetime observed when adding C_2–C_4 alcohol to the methanol feed is not straightforward, but two possible explanations may be suggested: First, the critical contact time of the catalyst may be reduced by the presence of alcohols/alkenes in the feed, leading to a larger active fraction of the catalyst bed (see Section 14.3.3) [91]. Zeolite-based

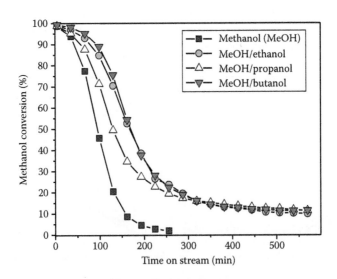

FIGURE 14.21 Conversion versus TOS for methanol alone and for mixtures of methanol and either ethanol, isopropanol, or isobutanol at 400°C and WHSV = 2 gg_{cat}^{-1} h^{-1} with respect to methanol. In the coconversion studies, about 6% of the carbon atoms in the reactant stream originate from higher alcohols.

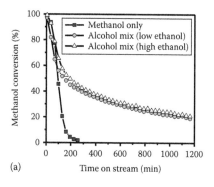
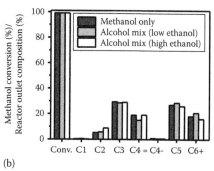

(a)   Time on stream (min)   (b)

FIGURE 14.22 Methanol conversion versus (a) TOS and (b) product selectivity at 99% conversion and WHSV (methanol) = 2 $gg_{cat}^{-1} h^{-1}$ for pure CH_3OH feed and for $CH_3OH/C_2H_5OH/i\text{-}C_3H_7OH/t\text{-}C_4H_9OH$ = 310:6:16:9 or 350:15:13:7 mixtures over an H-ZSM-22 catalyst at 400°C.

methanol conversion using a fixed-bed reactor was reported to have zone ageing along the catalyst bed, and in some cases, the very first layer of the catalyst bed is almost inert toward methanol conversion [78,92,93]. Such an effect is suggested from the graphs in Figure 14.21, where the slope of the conversion versus TOS curves is similar, but the methanol breakthrough point is displaced toward longer times on stream when C_2–C_4 alcohols are cofed. Second, it may be speculated that the presence of higher alcohols in the feed may alter the coking rate of the catalyst, although no evidence for this has been found.

Figure 14.22 shows the conversion versus TOS and the product selectivities at 99% conversion for coreaction of methanol with a mixture of C_2–C_4 alcohols. As ethanol is less reactive than the higher alkenes, mixtures with both low and high ethanol content were investigated. The joint cofeed of the C_2–C_4 alcohols barely affected the product selectivity at very high methanol conversion, thus confirming that recycling of C_2–C_4 alkenes in the reactor effluent may lead to an overall increase in gasoline production, with close to 100% C_{5+} alkene selectivity in the MTH process over H-ZSM-22 at 400°C. Overall, the coconversion studies suggested that recycling of light hydrocarbons during the MTH process over 1D 10-ring catalysts will increase the lifetime of the catalyst before deactivation and thus improve the total methanol conversion capacity.

14.3.5 Conclusion and Outlook of OTL Study

As an overall conclusion to the OTL study, H-ZSM-22 and H-ZSM-23 seem equally attractive as catalysts for the conversion of methanol and DME to aromatic-free gasoline. Both catalysts give an optimum of 53%–55% per pass yield of C_{5+} alkenes at 400°C–450°C, which could be increased substantially by cofeeding C_2–C_4 alcohols (simulating a by-product recycle), and the production of aromatic products was very low (<2% yield). However, both catalysts pose the same challenges concerning low catalyst activity and rapid deactivation under OTL conditions. Therefore, further optimization work is required in order for the OTL process to compete with a conventional H-ZSM-5-based MTG processes.

On the materials side, the activity and methanol conversion capacity of H-ZSM-22 and H-ZSM-23 need to be further optimized. Recent studies have pointed to catalyst morphology as a key factor for activity and conversion capacity optimization of zeolite catalysts. In particular, an increase in the accessible fraction of the zeolite crystal by nanosized crystals or hierarchical pore systems has been highlighted [94–97]. For H-ZSM-22 and H-ZSM-23 that usually crystallize as needlelike structures, a change in catalyst morphology toward shorter channels would be particularly advantageous. Finally, studies of other zeolites have pointed to acid site density as an important parameter for catalyst deactivation [98,99]. Thus, increasing the Si/Al ratio by steaming and/or acid treatment might also lead to enhanced methanol conversion capacity for the selected catalyst structures. Exchanging the H-ZSM-22 and H-ZSM-23 zeolites with zeotypes that offer lower acid strength and possibly sites that prevent coke formation could lead to further OTL process improvements on a longer timescale.

14.4   STO–OTL PROCESS OUTLINE AND CONCLUSIONS

The synthesis of gasoline from syngas as part of the OCMOL process is feasible by integrating STO with OTL, as shown schematically in Figure 14.23. The products from the oxygenate reactor are directed to a separator where mainly DME and CO_2 are condensed and subsequently sent to the OTL reactor. A recycle stream of CO and hydrogen is also implemented in the layout with the aim of increasing the conversion of syngas. To further reduce the concentration of reactants, the oxygenate-rich stream is mixed with a recycle stream from the post-OTL process separation step. This recycle stream contains a high fraction of light alkenes, which may react to form higher hydrocarbons and improve the gasoline yield. As reported in Section 14.3.4.2 and in Ref. [91], the addition of light alkenes improves both the gasoline yield and prolongs the lifetime of the catalyst.

Since the conversion of oxygenates into hydrocarbons is a strongly exothermic reaction, and less than 100% conversion gives the highest gasoline yields over H-ZSM-22 or H-ZSM-23 catalysts, the use of a fluidized bed reactor would be beneficial. Fluidization of the catalyst provides an excellent means for controlling the reaction temperature, ensuring homogeneous reaction conditions along the catalytic bed, and minimizing the risk for a runaway or hot spots, in addition to allowing stable operation at less than 100% conversion. Further, the application of a fluidized

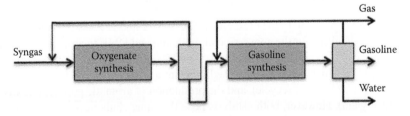

FIGURE 14.23   Simplified process layout involving liquid product intermediate. Oxygenate synthesis purge stream adds to the main (liquid) feed for OTL synthesis.

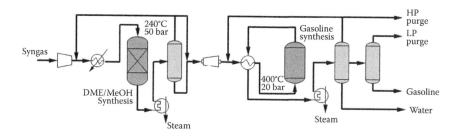

FIGURE 14.24 Process flow scheme of the integrated synthesis of gasoline from syngas through oxygenates.

bed process enables continuous regeneration of the catalyst particles in a separate regenerator, which is particularly relevant for catalysts suffering from rapid deactivation such as H-ZSM-22 and H-ZSM-23.

It should be noted that carbon dioxide is an inert diluent in the OTL step, reducing the concentration of oxygenates and acting as a heat sink during the exothermic gasoline synthesis reaction. The oxygenate-rich stream is further diluted by mixing with a recycle stream from the subsequent separation process as illustrated in Figure 14.23 to ensure that the temperature rise in the adiabatic reactor does not exceed 60°C–70°C thereby limiting secondary cracking reactions. Typical inlet and outlet temperatures in the reactor (in the case of H-ZSM-23 as the OTL catalyst) are in the range 390°C–410°C and 450°C–470°C, respectively. To increase the energy efficiency of the process and limit the reaction temperature, steam can be generated by exchange with the heat released during the reaction.

A process layout is shown in Figure 14.24, including the integration of oxygenate synthesis from synthesis gas. The syngas is fed at 240°C and 50 atm in order to achieve a syngas conversion around 95% at an R/M ratio of about 1. The exit gas is cooled to 5°C enabling a sufficient product separation of a recycle stream rich in unreacted syngas and a product stream containing DME, methanol, CO_2, and minor amounts of CO, CH_4, and H_2.

It has been experimentally demonstrated that in the case of H-ZSM-22, the reaction pressure does not lead to significant changes in the product selectivity, but it does negatively affect the lifetime of the catalyst before deactivation. When considering equipment size, a higher pressure is usually preferred. However, high pressure during the gasoline HP separation step also induces the dissolution of light olefins into the liquid product, and these will be lost after depressurization. Even though the use of low pressures results in larger recycling flows, it was found that a separation pressure of 20 atm represented a reasonable compromise between equipment size and gasoline yield. Figure 14.25 shows how the pressure in the HP separator affects the gasoline yield in terms of wt% of C_{4+} and the content of light olefins in the product stream coming out from this separator. Furthermore, the high-pressure purge contains a significant amount of methane, which can be diverted to methane reforming unit, intensifying the concept of recycling by-products that characterizes the OCMOL process.

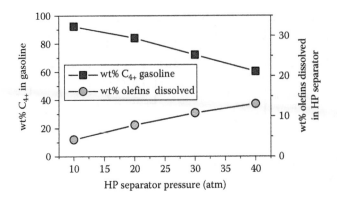

FIGURE 14.25   wt% C_{4+} in gasoline final product (dark squares) and wt% light alkenes dissolved in the liquid product in an HP separator (light gray circles) versus pressure. Effect of separation pressure at $T = 5°C$.

A key point in order to make the process viable is the stability toward regeneration of the H-ZSM-22/23 catalysts. Since a high pressure was found to increase the deactivation rate, the catalyst will need to be recycled very frequently.

In summary, the full conversion of syngas via DME/MeOH with light olefin recycled into higher hydrocarbons can give a gasoline product with C_{5+} and C_{4+} yields superior to 70 and 80 wt%, respectively, of the total hydrocarbon product. In addition, the final product stands out by its low content in aromatics. The topology of H-ZSM-22 and H-ZSM-23 inhibits the formation of aromatic compounds, which makes this process very attractive for producing environmentally friendly gasoline [67].

REFERENCES

1. L.J.C. Jens Rostrup-Nielsen, *Concepts in Syngas Manufacture*, Imperial College Press, London, U.K., 2011.
2. K. Aasberg-Petersen, C. Stub Nielsen, I. Dybkjr, Very large scale synthesis gas production and conversion to methanol or multiple products, *Stud. Surf. Sci. Catal.*, 167 (2007), 243–248.
3. Y. Ohno, M. Omiya, Coal conversion into dimethyl ether as an innovative clean fuel, in: *12th ICCS Coal Conversion into DME*, 2003.
4. F. Granberg, H. Nelveg, I. Landälv, *Proceedings of the International Chemical Recovery Conference*, Williamsburg, VA, Vol. 2, 2010, pp. 179–186.
5. J. Topp-Jørgensen, Topsøe integrated gasoline synthesis—The tigas process, *Stud. Surf. Sci. Catal.* 36 (1988), 293–305.
6. D. Mao, W. Yang, J. Xia, B. Zhang, Q. Song, Q. Chen, Highly effective hybrid catalyst for the direct synthesis of dimethyl ether from syngas with magnesium oxide-modified HZSM-5 as a dehydration component, *J. Catal.*, 230 (2005), 140–149.
7. A. García-Trenco, A. Martínez, Direct synthesis of DME from syngas on hybrid CuZnAl/ZSM-5 catalysts: New insights into the role of zeolite acidity, *Appl. Catal. A Gen.*, 411–412 (2012), 170–179.
8. A. García-Trenco, A. Vidal-Moya, A. Martínez, Study of the interaction between components in hybrid CuZnAl/HZSM-5 catalysts and its impact in the syngas-to-DME reaction, *Catal. Today*, 179 (2012), 43–51.

9. J. Ereña, R. Garoña, J.M. Arandes, A.T. Aguayo, J. Bilbao, Effect of operating conditions on the synthesis of dimethyl ether over a CuO-ZnO-Al2O3/NaHZSM-5 bifunctional catalyst, *Catal. Today*, 107–108 (2005), 467–473.

10. T. Takeguchi, K.I. Yanagisawa, T. Inui, M. Inoue, Effect of the property of solid acid upon syngas-to-dimethyl ether conversion on the hybrid catalysts composed of Cu-Zn-Ga and solid acids, *Appl. Catal. A Gen.*, 192 (2000), 201–209.

11. A. García-Trenco, A. Martínez, The influence of zeolite surface-aluminum species on the deactivation of CuZnAl/zeolite hybrid catalysts for the direct DME synthesis, *Catal. Today*, 227 (2014), 144–153.

12. C.D. Chang, A.J. Silvestri, Conversion of methanol and other O-compounds to hydrocarbons over zeolite catalysts, *J. Catal.*, 47 (1977), 249–259.

13. S.L. Meisel, J.P. Mccullough, C.H. Lechthaler, P.B. Weisz, Gasoline from methanol in one-step, *Chemtech*, 6 (1976), 86–89.

14. F.J. Keil, Methanol-to-hydrocarbons: Process technology, *Micropor. Mesopor. Mater.*, 29 (1999), 49–66.

15. S. Kvisle, T. Fuglerud, S. Kolboe, U. Olsbye, K.P. Lillerud, B.V. Vora, Methanol-to-hydrocarbons, H. Ertl, F. Knözinger, F. Schüth and J. Weitkamp (eds), in: *Handbook of Heterogeneous Catalysis*, Wiley-VCH Verlag GmbH & Co. KGaA, (2008), 2950–2965.

16. C.D. Chang, Hydrocarbons from methanol, *Catal. Rev.*, 25 (1983), 1–118.

17. C.D. Chang, The New Zealand gas-to-gasoline plant: An engineering tour de force, *Catal. Today*, 13 (1992), 103–111.

18. H.R. Grimmer, N. Thiagarajan, E. Nitschke, Conversion of methanol to liquid fuels by the fluid bed Mobil process (a commercial concept), *Stud. Surf. Sci. Catal.*, 36 (1988), 273–291.

19. A.A. Avidan, Gasoline and distillate fuels from methanol, *Stud. Surf. Sci. Catal.*, 36 (1988), 307–323.

20. H. Koempel, W. Liebner, Lurgi's methanol to propylene (MTP®) report on a successful commercialisation, *Stud. Surf. Sci. Catal.*, 167 (2007), 261–267.

21. B.V. Vora, T.L. Marker, P.T. Barger, H.R. Nilsen, S. Kvisle, T. Fuglerud, Economic route for natural gas conversion to ethylene and propylene, *Stud. Surf. Sci. Catal.*, 107 (1997), 87–98.

22. J. Liang, H. Li, S. Zhao, W. Guo, R. Wang, M. Ying, Characteristics and performance of SAPO-34 catalyst for methanol-to-olefin conversion, *Appl. Catal.*, 64 (1990), 31–40.

23. P.B. Weisz, Molecular shape selective catalysis, *Pure Appl. Chem.*, 52 (1980), 2091–2103.

24. S.M. Csicsery, Catalysis by shape selective zeolites—Science and technology, *Pure Appl. Chem.*, 58 (1986), 841–856.

25. B. Smit, T.L.M. Maesen, Towards a molecular understanding of shape selectivity, *Nature*, 451 (2008), 671–678.

26. T.F. Degnan, The implications of the fundamentals of shape selectivity for the development of catalysts for the petroleum and petrochemical industries, *J. Catal.*, 216 (2003), 32–46.

27. J.Q. Chen, A. Bozzano, B. Glover, T. Fuglerud, S. Kvisle, Recent advancements in ethylene and propylene production using the UOP/Hydro MTO process, *Catal. Today*, 106 (2005), 103–107.

28. S. Yurchak, Development of Mobil's fixed-bed methanol-to-gasoline (MTG) process, *Stud. Surf. Sci. Catal.*, 36 (1988), 251–272.

29. S.A. Tabak, S. Yurchak, Conversion of methanol over ZSM-5 to fuels and chemicals, *Catal. Today*, 6 (1990), 307–327.

30. J.F. Haw, W.G. Song, D.M. Marcus, J.B. Nicholas, The mechanism of methanol to hydrocarbon catalysis, *Acc. Chem. Res.*, 36 (2003), 317–326.

31. M. Stocker, Methanol-to-hydrocarbons: Catalytic materials and their behavior, *Micropor. Mesopor. Mater.*, 29 (1999), 3–48.

32. U. Olsbye, S. Svelle, M. Bjørgen, P. Beato, T.V.W. Janssens, F. Joensen, S. Bordiga, K.P. Lillerud, Conversion of methanol to hydrocarbons: How zeolite cavity and pore size controls product selectivity, *Angew. Chem., Int. Ed.*, 51 (2012), 5810–5831.

33. K. Hemelsoet, J. Van der Mynsbrugge, K. De Wispelaere, M. Waroquier, V. Van Speybroeck, Unraveling the reaction mechanisms governing methanol-to-olefins catalysis by theory and experiment, *ChemPhysChem*, 14 (2013), 1526–1545.

34. S. Ilias, A. Bhan, Mechanism of the catalytic conversion of methanol to hydrocarbons, *ACS Catal.*, 3 (2013), 18–31.

35. N.Y. Chen, W.J. Reagan, Evidence of auto-catalysis in methanol to hydrocarbon reactions over zeolite catalysts, *J. Catal.*, 59 (1979), 123–129.

36. R.M. Dessau, On the H-Zsm-5 catalyzed formation of ethylene from methanol or higher olefins, *J. Catal.*, 99 (1986), 111–116.

37. R.M. Dessau, R.B. Lapierre, On the mechanism of methanol conversion to hydrocarbons over Hzsm-5, *J. Catal.*, 78 (1982), 136–141.

38. B.E. Langner, Reactions of methanol on zeolites with different pore structures, *Appl. Catal.*, 2 (1982), 289–302.

39. T. Mole, G. Bett, D. Seddon, Conversion of methanol to hydrocarbons over Zsm-5 zeolite—An examination of the role of aromatic-hydrocarbons using carbon-13-labeled and deuterium-labeled feeds, *J. Catal.*, 84 (1983), 435–445.

40. T. Mole, J.A. Whiteside, D. Seddon, Aromatic co-catalysis of methanol conversion over zeolite catalysts, *J. Catal.*, 82 (1983), 261–266.

41. D. Lesthaeghe, V. Van Speybroeck, G.B. Marin, M. Waroquier, Understanding the failure of direct C–C coupling in the zeolite-catalyzed methanol-to-olefin process, *Angew. Chem., Int. Ed.*, 45 (2006), 1714–1719.

42. D.M. Marcus, K.A. McLachlan, M.A. Wildman, J.O. Ehresmann, P.W. Kletnieks, J.F. Haw, Experimental evidence from H/D exchange studies for the failure of direct C–C coupling mechanisms in the methanol-to-olefin process catalyzed by HSAPO-34, *Angew. Chem. Int. Ed.*, 45 (2006), 3133–3136.

43. W.G. Song, D.M. Marcus, H. Fu, J.O. Ehresmann, J.F. Haw, An oft-studied reaction that may never have been: Direct catalytic conversion of methanol or dimethyl ether to hydrocarbons on the solid acids HZSM-5 or HSAPO-34, *J. Am. Chem. Soc.*, 124 (2002), 3844–3845.

44. I.M. Dahl, S. Kolboe, On the reaction mechanism for hydrocarbon formation from methanol over SAPO-34.2. Isotopic labeling studies of the co-reaction of propene and methanol, *J. Catal.*, 161 (1996), 304–309.

45. I.M. Dahl, S. Kolboe, On the reaction-mechanism for propene formation in the MTO reaction over SAPO-34, *Catal. Lett.*, 20 (1993), 329–336.

46. I.M. Dahl, S. Kolboe, On the reaction-mechanism for hydrocarbon formation from methanol over Sapo-34.1. Isotopic labeling studies of the co-reaction of ethene and methanol, *J. Catal.*, 149 (1994), 458–464.

47. B. Arstad, S. Kolboe, The reactivity of molecules trapped within the SAPO-34 cavities in the methanol-to-hydrocarbons reaction, *J. Am. Chem. Soc.*, 123 (2001), 8137–8138.

48. B. Arstad, S. Kolboe, Methanol-to-hydrocarbons reaction over SAPO-34. Molecules confined in the catalyst cavities at short time on stream, *Catal. Lett.*, 71 (2001), 209–212.

49. W.G. Song, J.F. Haw, J.B. Nicholas, C.S. Heneghan, Methylbenzenes are the organic reaction centers for methanol-to-olefin catalysis on HSAPO-34, *J. Am. Chem. Soc.*, 122 (2000), 10726–10727.

50. M. Hunger, M. Seiler, A. Buchholz, In situ MAS NMR spectroscopic investigation of the conversion of methanol to olefins on silicoaluminophosphates SAPO-34 and SAPO-18 under continuous flow conditions, *Catal. Lett.*, 74 (2001), 61–68.

51. M. Seiler, U. Schenk, M. Hunger, Conversion of methanol to hydrocarbons on zeolite HZSM-5 investigated by in situ MAS NMR spectroscopy under flow conditions and on-line gas chromatography, *Catal. Lett.*, 62 (1999), 139–145.

52. M. Seiler, W. Wang, A. Buchholz, M. Hunger, Direct evidence for a catalytically active role of the hydrocarbon pool formed on zeolite H-ZSM-5 during the methanol-to-olefin conversion, *Catal. Lett.*, 88 (2003), 187–191.

53. M. Bjørgen, U. Olsbye, D. Petersen, S. Kolboe, The methanol-to-hydrocarbons reaction: insight into the reaction mechanism from [C-12]benzene and [C-13]methanol coreactions over zeolite H-beta, *J. Catal.*, 221 (2004), 1–10.

54. A. Sassi, M.A. Wildman, H.J. Ahn, P. Prasad, J.B. Nicholas, J.F. Haw, Methylbenzene chemistry on zeolite HBeta: Multiple insights into methanol-to-olefin catalysis, *J. Phys. Chem. B*, 106 (2002), 2294–2303.

55. M. Bjørgen, S. Svelle, F. Joensen, J. Nerlov, S. Kolboe, F. Bonino, L. Palumbo, S. Bordiga, U. Olsbye, Conversion of methanol to hydrocarbons over zeolite H-ZSM-5: On the origin of the olefinic species, *J. Catal.*, 249 (2007), 195–207.

56. S. Svelle, F. Joensen, J. Nerlov, U. Olsbye, K.-P. Lillerud, S. Kolboe, M. Bjørgen, Conversion of methanol into hydrocarbons over Zeolite H-ZSM-5: Ethene formation is mechanistically separated from the formation of higher alkenes, *J. Am. Chem. Soc.*, 128 (2006), 14770–14771.

57. M. Westgård Erichsen, S. Svelle, U. Olsbye, H-SAPO-5 as methanol-to-olefins (MTO) model catalyst: Towards elucidating the effects of acid strength, *J. Catal.*, 298 (2013), 94–101.

58. M. Westgård Erichsen, S. Svelle, U. Olsbye, The influence of catalyst acid strength on the methanol to hydrocarbons (MTH) reaction, *Catal. Today*, 215 (2013), 216–223.

59. R.F. Sullivan, C.J. Egan, G.E. Langlois, R.P. Sieg, A new reaction that occurs in the hydrocracking of certain aromatic hydrocarbons, *J. Am. Chem. Soc.*, 83 (1961), 1156–1160.

60. M. Bjørgen, F. Joensen, K.-P. Lillerud, U. Olsbye, S. Svelle, The mechanisms of ethene and propene formation from methanol over high silica H-ZSM-5 and H-beta, *Catal. Today*, 142 (2009), 90–97.

61. S. Svelle, U. Olsbye, F. Joensen, M. Bjørgen, Conversion of methanol to alkenes over medium- and large-pore acidic zeolites: Steric manipulation of the reaction intermediates governs the ethene/propene product selectivity, *J. Phys. Chem. C*, 111 (2007), 17981–17984.

62. W. Song, H. Fu, J.F. Haw, Supramolecular origins of product selectivity for methanol-to-olefin catalysis on HSAPO-34, *J. Am. Chem. Soc.*, 123 (2001), 4749–4754.

63. B.P.C. Hereijgers, F. Bleken, M.H. Nilsen, S. Svelle, K.P. Lillerud, M. Bjorgen, B.M. Weckhuysen, U. Olsbye, Product shape selectivity dominates the methanol-to-olefins (MTO) reaction over H-SAPO-34 catalysts, *J. Catal.*, 264 (2009), 77–87.

64. M. Bjørgen, S. Akyalcin, U. Olsbye, S. Benard, S. Kolboe, S. Svelle, Methanol to hydrocarbons over large cavity zeolites: Toward a unified description of catalyst deactivation and the reaction mechanism, *J. Catal.*, 275 (2010), 170–180.

65. J.H. Ahn, B. Temel, E. Iglesia, Selective homologation routes to 2,2,3-Trimethylbutane on solid acids, *Angew. Chem., Int. Ed.*, 48 (2009), 3814–3816.

66. D.A. Simonetti, J.H. Ahn, E. Iglesia, Mechanistic details of acid-catalyzed reactions and their role in the selective synthesis of triptane and isobutane from dimethyl ether, *J. Catal.*, 277 (2011), 173–195.

67. S. Teketel, U. Olsbye, K.P. Lillerud, P. Beato, S. Svelle, Selectivity control through fundamental mechanistic insight in the conversion of methanol to hydrocarbons over zeolites, *Micropor. Mesopor. Mater.*, 136 (2010), 33–41.

68. S. Teketel, S. Svelle, K.P. Lillerud, U. Olsbye, Shape-selective conversion of methanol to hydrocarbons over 10-ring unidirectional-channel acidic H-ZSM-22, *ChemCatChem*, 1 (2009), 78–81.

69. Z.-M. Cui, Q. Liu, W.-G. Song, L.-J. Wan, Insights into the mechanism of methanol-to-olefin conversion at zeolites with systematically selected framework structures, *Angew. Chem. Int. Ed.*, 45 (2006), 6512–6515.

70. Z.-M. Cui, Q. Liu, Z. Ma, S.-W. Bian, W.-G. Song, Direct observation of olefin homologations on zeolite ZSM-22 and its implications to methanol to olefin conversion, *J. Catal.*, 258 (2008), 83–86.

71. J. Li, Y. Wei, Y. Qi, P. Tian, B. Li, Y. He, F. Chang, X. Sun, Z. Liu, Conversion of methanol over H-ZSM-22: The reaction mechanism and deactivation, *Catal. Today*, 164 (2011), 288–292.

72. J. Li, Y. Wei, G. Liu, Y. Qi, P. Tian, B. Li, Y. He, Z. Liu, Comparative study of MTO conversion over SAPO-34, H-ZSM-5 and H-ZSM-22: Correlating catalytic performance and reaction mechanism to zeolite topology, *Catal. Today*, 171 (2011), 221–228.

73. T.V.W. Janssens, S. Svelle, U. Olsbye, Kinetic modeling of deactivation profiles in the methanol-to-hydrocarbons (MTH) reaction: A combined autocatalytic–hydrocarbon pool approach, *J. Catal.*, 308 (2013), 122–130.

74. I.M. Hill, S.A. Hashimi, A. Bhan, Corrigendum to "kinetics and mechanism of olefin methylation reactions over zeolites", *J. Catal.*, 291 (2012), 155–157.

75. F. Bleken, M. Bjorgen, L. Palumbo, S. Bordiga, S. Svelle, K.P. Lillerud, U. Olsbye, The effect of acid strength on the conversion of methanol to olefins over acidic microporous catalysts with the CHA topology, *Top. Catal.*, 52 (2009), 218–228.

76. S. Teketel, W. Skistad, S. Benard, U. Olsbye, K.P. Lillerud, P. Beato, S. Svelle, Shape selectivity in the conversion of methanol to hydrocarbons: The catalytic performance of one-dimensional 10-ring zeolites: ZSM-22, ZSM-23, ZSM-48, and EU-1, *ACS Catal.*, 2 (2012), 26–37.

77. J.L. Schlenker, W.J. Rohrbaugh, P. Chu, E.W. Valyocsik, G.T. Kokotailo, The framework topology of Zsm-48—A high silica zeolite, zeolites, 5 (1985), 355–358.

78. F.L. Bleken, T.V.W. Janssens, S. Svelle, U. Olsbye, Product yield in methanol conversion over ZSM-5 is predominantly independent of coke content, *Micropor. Mesopor. Mater.*, 164 (2012), 190–198.

79. T.V.W. Janssens, A new approach to the modeling of deactivation in the conversion of methanol on zeolite catalysts, *J. Catal.*, 264 (2009), 130–137.

80. S. Teketel, M. Westgård Erichsen, F. Bleken, S. Svelle, K.P. Lillerud, U. Olsbye, Shape selectivity in zeolite catalysis. The methanol to hydrocarbons (MTH) reaction, *Catalysis*, 26 (2014), 173–211.

81. S. Inagaki, K. Kamino, M. Hoshino, E. Kikuchi, M. Matsukata, Textural and catalytic properties of MCM-22 zeolite crystallized by the vapor-phase transport method, *Bull. Chem. Soc. Jpn.*, 77 (2004), 1249–1254.

82. S.L. Lawton, M.E. Leonowicz, R.D. Partridge, P. Chu, M.K. Rubin, Twelve-ring pockets on the external surface of MCM-22 crystals, *Micropor. Mesopor. Mater.*, 23 (1998), 109–117.

83. R. Ravishankar, D. Bhattacharya, N.E. Jacob, S. Sivasanker, Characterization and catalytic properties of zeolite MCM-22, *Microporous Mater.*, 4 (1995), 83–93.

84. W. Souverijns, W. Verrelst, G. Vanbutsele, J.A. Martens, P.A. Jacobs, Micropore structure of zeolite MCM-22 as determined by the decane catalytic test reaction, *J. Chem. Soc. Chem. Commun.*, 14 (1994), 1671–1672.

85. M. Bjørgen, S. Kolboe, The conversion of methanol to hydrocarbons over dealuminated zeolite H-beta, *Appl. Catal. A Gen.*, 225 (2002), 285–290.

86. O. Mikkelsen, S. Kolboe, The conversion of methanol to hydrocarbons over zeolite H-beta, *Micropor. Mesopor. Mater.*, 29 (1999), 173–184.

87. F. Bleken, W. Skistad, K. Barbera, M. Kustova, S. Bordiga, P. Beato, K.P. Lillerud, S. Svelle, U. Olsbye, Conversion of methanol over 10-ring zeolites with differing volumes at channel intersections: Comparison of TNU-9, IM-5, ZSM-11 and ZSM-5, *Phys. Chem. Chem. Phys.*, 13 (2011), 2539–2549.

88. W. Skistad, S. Teketel, F. Bleken, P. Beato, S. Bordiga, M. Nilsen, U. Olsbye, S. Svelle, K. Lillerud, Methanol conversion to hydrocarbons (MTH) over H-ITQ-13 (ITH) zeolite, *Top. Catal.*, 57 (2014), 143–158.

89. B.-T.L. Bleken, D. Wragg, B. Arstad, A. Gunnæs, J. Mouzon, S. Helveg, L. Lundegaard et al., Unit cell thick nanosheets of zeolite H-ZSM-5: Structure and activity, *Top. Catal.*, 56 (2013), 558–566.

90. F.L. Bleken, K. Barbera, F. Bonino, U. Olsbye, K.P. Lillerud, S. Bordiga, P. Beato, T.V.W. Janssens, S. Svelle, Catalyst deactivation by coke formation in microporous and desilicated zeolite H-ZSM-5 during the conversion of methanol to hydrocarbons, *J. Catal.*, 307 (2013), 62–73.

91. S. Teketel, U. Olsbye, P. Beato, K.P. Lillerud, S. Svelle, Manuscript in preparation.

92. M. Kaarsholm, F. Joensen, J. Nerlov, R. Cenni, J. Chaouki, G.S. Patience, Phosphorous modified ZSM-5: Deactivation and product distribution for MTO, *Chem. Eng. Sci.*, 62 (2007), 5527–5532.

93. H. Schulz, "Coking" of zeolites during methanol conversion: Basic reactions of the MTO-, MTP- and MTG processes, *Catal. Today*, 154 (2010), 183–194.

94. M. Choi, K. Na, J. Kim, Y. Sakamoto, O. Terasaki, R. Ryoo, Stable single-unit-cell nanosheets of zeolite MFI as active and long-lived catalysts, *Nature*, 461 (2009), 246–249.

95. M.S. Holm, E. Taarning, K. Egeblad, C.H. Christensen, Catalysis with hierarchical zeolites, *Catal. Today*, 168 (2011), 3–16.

96. D. Verboekend, J. Perez-Ramirez, Design of hierarchical zeolite catalysts by desilication, *Catal. Sci. Technol.*, 1 (2011), 879–890.

97. J. Perez-Ramirez, C.H. Christensen, K. Egeblad, C.H. Christensen, J.C. Groen, Hierarchical zeolites: Enhanced utilisation of microporous crystals in catalysis by advances in materials design, *Chem. Soc. Rev.*, 37 (2008), 2530–2542.

98. J.W. Park, S.J. Kim, M. Seo, S.Y. Kim, Y. Sugi, G. Seo, Product selectivity and catalytic deactivation of MOR zeolites with different acid site densities in methanol-to-olefin (MTO) reactions, *Appl. Catal. A Gen.*, 349 (2008), 76–85.

99. J. Liu, C.X. Zhang, Z.H. Shen, W. Hua, Y. Tang, W. Shen, Y.H. Yue, H.L. Xu, Methanol to propylene: Effect of phosphorus on a high silica HZSM-5 catalyst, *Catal. Commun.* 10 (2009), 1506–1509.

15 Solar-Aided Syngas Production via Two-Step, Redox-Pair-Based Thermochemical Cycles

Christos Agrafiotis, Martin Roeb, and Christian Sattler

CONTENTS

15.1 INTRODUCTION

Conventional, fossil, liquid hydrocarbon fuels will continue to cover a major portion of the ever-increasing world energy requirements for the foreseeable future, basically due to their demand in transportation and the existing relevant extensive infrastructure. In this perspective, the direct use of nonfossil, synthetic liquid fuels (SLFs) attracts a lot of interest. Currently, the term *synthetic fuel (synfuel)* refers to a liquid fuel produced at commercial scale from low-energy-content carbonaceous sources, such as coal, natural gas, oil shale, tar sand, and other biomass, that are upgraded at the expense of additional energy, also obtained from the combustion of fossil fuels. The terms gas to liquid (GTL) and coal to liquid (CTL) refer to processes to convert natural gas (or other gaseous hydrocarbons) and coal, respectively, into longer-chain liquid hydrocarbons such as gasoline or diesel fuel. Syngas is a gas mixture that contains varying amounts of CO and H_2 whose exothermic conversion to fuel and other products has been commercially practiced since a long time ago, for example, via the Fischer–Tropsch (FT) technology, and which can be also used as a source of pure hydrogen and carbon monoxide [1,2]. Thus, in fact, hydrogen and syngas are the basic raw materials to produce SLFs and chemicals via industrially available processes. These procedures can be rendered more attractive and environmentally friendlier when combined with a renewable energy source, such as solar energy. Indeed, it is generally accepted that only solar-driven technologies offer a permanent solution to both oil independence and climate change due to the unmatched magnitude and availability of solar resource [3]. When solar energy is employed for the production of the raw materials for the synthesis of such synthetic fuels, the latter are characterized with the term *solar fuels*. In the broad sense, this term can contain in addition to *solar hydrogen* synthetic liquid hydrocarbons and alcohols produced from reactions between H_2 and CO that have originated from solar-aided dissociation processes as well as metal powders obtained by solar thermal reduction (TR) of metal oxides [4].

Figure 15.1 depicts a partial listing of the various feedstocks and solar energy variances that can be employed to produce such fuels. In this perspective, there are basically three pathways for producing syngas with the aid of solar energy: photochemical/photobiological, thermochemical, and electrochemical [6–8]. The first route makes direct use of solar photon energy for photochemical and photobiological processes. The thermochemical route uses solar heat at high temperatures supplied by concentrated solar power (CSP) systems—that is, special mirror assemblies that track the sun. These systems provide concentrated solar radiation as the energy source for performing high-temperature reactions that produce syngas from

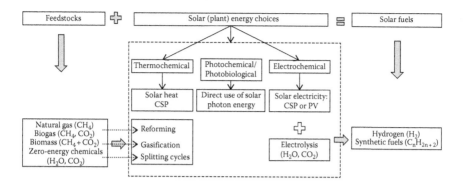

FIGURE 15.1 Partial listing of the various feedstocks and solar energy variances that can be employed to produce solar liquid hydrocarbon fuels. (Adapted from Trainham, J.A. et al., *Curr. Opin. Chem. Eng.*, 1, 204, 2012. With permission; Wegeng, R.S. and Mankins, J.C., *Acta Astron.*, 65, 1261, 2009. With permission.)

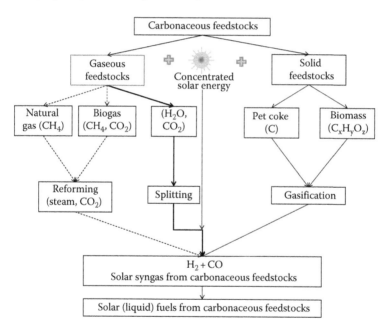

FIGURE 15.2 Transformation routes of various fossil and nonfossil fuels to *solar syngas* via high-temperature reactions and concentrated solar energy. (From Agrafiotis, C. et al., *Renew. Sustain. Energ. Rev.*, 29, 656, 2014. With permission.)

transformation of various fossil and nonfossil fuels via different routes (Figure 15.2) such as natural gas steam reforming, gasification of solid carbonaceous materials like coal or biomass, or water splitting (WS) to hydrogen and oxygen and reaction of H_2 with CO_2 via the reverse water–gas shift (RWGS) reaction to produce syngas [9,10]. In addition, CSP systems can be employed alternatively to photovoltaics for the renewable WS indirectly, for example, by supplying the (solar thermal) electricity for high-temperature electrolysis of steam or of steam/CO_2 mixtures [11,12] according to the third electrochemical route in Figure 15.1. In the next step, the most promising liquid fuels to be generated from solar syngas are methanol, dimethyl ether (DME), and FT diesel [13–15].

Among the thermochemical routes to solar syngas of Figure 15.2, solar reforming has the advantage of lower temperatures required compared to water/carbon dioxide thermochemical splitting processes. However, the latter route employs CO_2 as a reactant and, in this perspective, lies within the broad, increasing R&D effort on developing effective solutions for reusing and *valorizing* atmospheric CO_2 as a carbon-containing raw material for the production of fuels, chemicals, and materials rather than currently treating it as a waste [12]. A review on solar-aided reforming (dashed line) has been recently published by the present authors [16], whereas coal/biomass steam gasification (dotted line) has been covered excessively in a series of publications [17,18]. The present work reviews the activities on solar-aided syngas production via water/carbon dioxide thermochemical splitting processes (solid line).

15.2 TWO-STEP THERMOCHEMICAL CYCLES FOR SYNGAS SYNTHESIS: REDOX CHEMISTRY

The so-called WS thermochemical cycles were proposed initially for the production of hydrogen from the splitting of water. The single-step thermal dissociation of water (thermolysis)—the conceptually simplest reaction to split water at a first glance—is impractical due to the very high temperatures needed (above 2500 K) as well as the problem of effectively separating H_2 and O_2 to avoid explosive mixtures. Thermochemical cycles are a series of consecutive chemical reactions (≥ 2), their *net* sum being the splitting of H_2O to H_2 and O_2, where the maximum-temperature (endothermic) step takes place at a temperature lower than that of the single water decomposition chemical reaction, bypassing in addition the H_2/O_2 separation problem. Nevertheless, since they involve a highly endothermic step, they need the input of external energy that can be provided by a source of high-temperature process heat. A revival in the research and development of thermochemical cycles has taken place in the past few years with the driving force being the production of hydrogen as a greenhouse-gas-free energy carrier to fulfill the requirements of the Kyoto Protocol. In parallel, in contrast to the work conducted 30–40 years ago, concentrated solar radiation has now become more important as a heat source than nuclear heat.

Among the various thermochemical cycles, of particular interest are two-step thermochemical cycles based on oxide redox-pair systems. Such cycles operate on the principle of transition between the oxidized (higher-valence, MeO_{ox}) and reduced (lower-valence, MeO_{red}) form of an oxide of a metal exhibiting multiple oxidation states [19]. In this concept, during the first, higher-temperature, endothermic step, the higher-valence oxide state of a metal exhibiting multiple oxidation states is subjected to TR (Reaction 15.1), that is, under the supply of external heat, it releases a quantity of oxygen and transforms to its lower-valence state. In a subsequent, lower-temperature, exothermic, WS (Reaction 15.2) step, the reduced (lower-valence) oxide is oxidized to the higher-valence state by taking oxygen from water and producing hydrogen, establishing hence a cyclic operation, according to the generalized reaction scheme as follows:

$$MeO_{oxidized} + (\Delta H)_1 \rightarrow MeO_{reduced} + \frac{1}{2}O_2(g) \tag{15.1}$$

$$MeO_{reduced} + H_2O\ (g) \rightarrow MeO_{oxidized} + H_2(g) + (\Delta H)_2 \tag{15.2}$$

An advantage of the oxide redox-pair-based, two-step thermochemical cycles versus other cycles employed like the sulfur–iodine one [19] is that the same thermochemical principle can be exploited for carbon dioxide (CO_2) splitting (CDS) as well, for the production of CO, according to the following reaction scheme [20]:

$$MeO_{oxidized} + (\Delta H)_1 \rightarrow MeO_{reduced} + \frac{1}{2}O_2(g) \tag{15.3}$$

$$MeO_{reduced} + CO_2(g) \rightarrow MeO_{oxidized} + CO(g) + (\Delta H)_3 \tag{15.4}$$

The acknowledgment of this flexibility and of the fact that the H_2 produced via Reaction 15.2 and the CO produced via Reaction 15.4 can then be combined leading to syngas that can be further used via the established FT technology to create a number of different product streams, including diesel-like fuel, alcohols, and other chemicals, has lead various researchers to propose the previously described *solar syngas* production scheme in solar reactors [20]. The approach is conceptually simple since the TR step is common for both WS and CDS (Reactions 15.1 and 15.3, respectively). Therefore, a particular redox material and a respective thermochemical reactor can be used for both WS and CDS either separately from each other, that is, to produce a stream of H_2 and a separate stream of CO, or even simultaneously—when steam and CO_2 are co-fed above the redox material—to produce syngas in one step. This generic scheme is depicted in Figure 15.3.

Obviously, the oxidized form of the redox material with the metal cation at its highest valence can dissociate either to a lower-metal-valence oxide or all the way to the respective metal (of zero valence). It is of no surprise therefore that both such kinds of systems have been considered for hydrogen/syngas generation via the cycles mentioned earlier: either metal oxide/metal systems (e.g., ZnO/Zn) or metal oxide/metal oxide pairs with typical examples being oxides of a single multivalent metal such as Fe_3O_4/FeO, Mn_3O_4/MnO, and the more recently tested CeO_2/Ce_2O_3. The first such systems proposed to be run with the aid of solar energy were those of iron oxides Fe_3O_4/FeO—magnetite/wüstite [22]. The first solar furnace experiments for the TR of magnetite were performed in the early 1980s at the solar facilities of Laboratoire PROcédés, Matériaux et Energie Solaire (PROMES) in Odeillo, France [23–25]. However, the temperature needed for TR was higher than the melting points of Fe_3O_4 (1808 K) and FeO (1643 K), resulting in liquid Fe_3O_4 and FeO phases and causing a rapid decrease of the iron oxide surface area and thus a deactivation of the material. Therefore, research efforts shifted toward alleviating these issues. Indeed, subsequent studies have shown that partial substitution of iron cations in the magnetite's lattice by a transition (e.g., Mn, Co, Ni, or Zn),

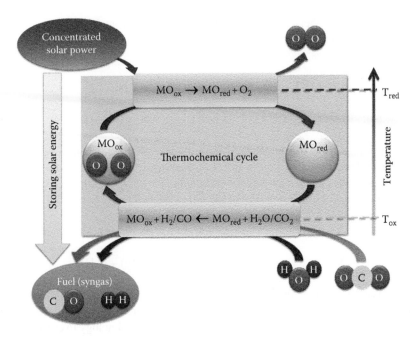

FIGURE 15.3 Schematic of the solar-aided, two-step, redox-oxide-pair-based, WS/CDS thermochemical cycle for syngas production. (Adapted from Roeb, M. et al., *Materials*, 5, 2015, 2012.)

or an alkaline earth (e.g., Mg) metal, forms mixed metal iron oxides of the form $(Fe_{1-x}M_x)_3O_4$—ferrites—that can be reduced at lower temperatures than the pure magnetite while their reduced mixed wüstite phase $(Fe_{1-x}M_x)O$ is still capable of splitting water [26]. Ferrites belong to the family of spinel materials that together with the perovskites are two of the most known families of oxides containing multiple metal cations that can demonstrate various valence states. Spinels and perovskites are of the general formulae $A^{+2}B^{+3}_2O_4$ and $A^{+2}B^{+4}O_3$, respectively; however, more than one cations of the same valence can occupy the A or B sites, for instance, $(A_x,B_{1-x})^{+2}(C_y,D_{2-y})_2^{+3}O_4$ and $(A_x,B_{1-x})^{+2}(C_y,D_{2-y})^{+4}O_3$ are respective typical formulae with two metal cations at each site. In addition, both single- and multimetal, multivalent oxides such as Fe_3O_4, CeO_2, spinels, and perovskites form a wide range of stable, nonstoichiometric compounds, denoted as $Fe_3O_{4-\delta}$, $CeO_{2-\delta}$, $AB_2O_{4-\delta}$, and $ABO_{3-\delta}$ respectively. All such nonstoichiometric materials and families can operate in a reaction scheme similar to (15.1) and (15.2) or (15.3) and (15.4). Therefore, the reactions can represent much more complex situations like nonstoichiometric/solution phases or multicomponent oxide materials [27] like schemes (15.5) and (15.6) and (15.7) and (15.8) for the exemplary cases of a mixed ferrite and ceria, respectively:

$$(A_xB_{1-x})Fe_2O_4 \rightarrow (A_xB_{1-x})Fe_2O_{4-\delta} + (\delta/2)O_2 \tag{15.5}$$

$$(A_xB_{1-x})Fe_2O_{4-\delta} + \delta H_2O(g) \rightarrow (A_xB_{1-x})Fe_2O_4 + \delta H_2 \tag{15.6}$$

$$CeO_2 \rightarrow CeO_{2-\delta} + (\delta/2)O_2 \qquad (15.7)$$

$$CeO_{2-\delta} + \delta H_2O(g) \rightarrow CeO_2 + \delta H_2 \qquad (15.8)$$

The high TR temperatures may cause melting and/or evaporation of volatile compounds, reactant loss or composition changes, activity reduction, or side reactions with the reactor materials. Therefore, when lower TR temperatures are selected to avoid such phenomena, the reduction reactions proceed rather like schemes (15.5) through (15.8). In other words, the reduction product does not contain all the reducible metal cations at their lower valence state but only a smaller percentage of them since δ is usually much smaller than 1. This has as a direct result the evolution of much less oxygen during the TR step and the consequent generation of less H_2/CO at the subsequent WS/CDS step, thus an overall lower cycle efficiency. An advantage of this on the other hand is that one can make use of the percentage of the bivalent metal(s) present in the single/mixed oxide phase and *oscillate* during the redox cycle between the reduced and the oxidized structure without undergoing phase transformations (such as to the lower-valence metal oxide or to the respective metal) that can cause structure disruption.

Thermodynamic calculations have been performed to determine potentially attractive redox systems and the respective temperature ranges required, initially for WS [28–31] and subsequently for CDS [27] on such single-metal oxides that can be reduced either to their lower-valence oxide or to the respective metal, such as Fe_3O_4, Mn_3O_4, Co_3O_4, Nb_2O_5, WO_3, MoO_2, SiO_2, SnO_2, In_2O_3, CdO, ZnO, and CeO_2. These studies have identified a limited number among these oxides being of practical interest. In addition, such calculations provide the upper limits of expected efficiencies since, in most cases, consider full reduction-oxidation of all the metal cations to the lower or higher valence respectively – which is not the case as already pointed out. This issue has been addressed by the research group of SANDIA National Labs (SNL) Albuquerque, NM, U.S.A. for the FeO/Fe_3O_4 redox pair [32]. The TR temperatures of the oxides exhibiting high hydrogen yields during hydrolysis can be lowered by reducing the oxygen's partial pressure. Thermodynamic calculations have identified the magnitude of oxygen partial pressure effects on the equilibrium of the TR step and have shown the prohibitive thermodynamics for performing this step under air atmosphere. In practice, most experimental setups and reactors use an inert gas to sweep the oxygen out of the reaction chamber and keep the atmosphere under a low oxygen partial pressure, but recycling inert gas imposes additional energy penalties to the overall process. An alternative option is to reduce the metal oxide under vacuum, eliminating the need for purge gas [33].

Materials identified as promising are synthesized and tested in lab scale usually in the form of powders, under nonsolar-provided heat. Typical temperatures required for the high-temperature step of these cycles range from 1100 to 2300 K. Laboratory tests involve WS, CDS, or co-splitting to experimentally determine the respective conditions, product yields, material performance, and stability. Such tests are performed either in thermogravimetric analyzers (TGA) or in typical electrically or IR-heated high-temperature furnace test rigs, coupled to mass spectrometer (MS) or suitable gas analysis (e.g., CO, CO_2, oxygen) equipment, where the gaseous products can be monitored as a function of cyclic operating temperature and surrounding atmosphere (in addition to the oxide's weight change in the case of TGA).

15.3 REDOX-PAIR-BASED, WS/CDS THERMOCHEMICAL CYCLES

All the various redox-pair categories have been initially explored extensively for WS/TR thermochemical cyclic operation, and the relevant issues can be found in several review articles [6,19,34,35]. Investigations among the most promising of them extended recently to include CDS; these are described in more detail in the following.

15.3.1 Volatile Cycles

The temperatures required for the TR reactions are high, often exceeding the boiling temperatures of the reduced species; thus *volatile* redox pairs employed in two-step splitting cycles commonly exhibit a solid-to-gas phase transition of the other-than-oxygen product in the reduction step. This phase transition is thermodynamically beneficial for the process because a high entropy gain is obtained. On the other hand, significant challenges occur due to the recombination of the decomposition reaction product with oxygen back to the initial reactant in the product gas stream. All such cycles thus share common characteristics with the main problem being the need of fast quenching of the decomposition products to avoid recombination of them back to the oxidized form.

15.3.1.1 ZnO/Zn Cycle

Among the many metal oxide/metal systems having been considered for the two-step WS process [36], the ZnO/Zn system is characterized by a combination of favorable properties: Zn is nonprecious and has low atomic weight, thus comparatively high energy content per mass. However, the decomposition temperature of ZnO is approximately 2300 K, whereas Zn melts at 692 K and has a boiling point of 1180 K. This system has been studied for quite some time, in particular from researchers from the group at Swiss Federal Institute of Technology/Paul Scherrer Institute (ETH/PSI), Switzerland [37]. Their work originally focused mostly on the application of this cycle for WS and hydrogen generation [38] but, more recently, has been extended to include investigations on CDS as well. The group considered thermodynamically the ZnO/Zn and the Fe_3O_4/FeO thermochemical cycles using concentrated solar energy and determined maximum theoretical solar-to-chemical energy conversion efficiencies of 39% and 29% respectively [63]. The reaction rate law and Arrhenius parameters for the reduction of ZnO into Zn and $\frac{1}{2}O_2$ were derived for the case of directly solar irradiated ZnO pellets [40]. The product mixture needs to be quenched to avoid recombination. As a result, the product leaving the solar thermolysis reactor/quencher, that is, the feed for the oxidation step, generally contains a substantial amount of ZnO observed to be as low as 6 mol% and as high as 85 mol% depending on reaction conditions and inert gas/Zn(g) dilution ratio.

The second step of these cycles involves the oxidation of the elementary metal in H_2O/CO_2 where H_2/CO is generated and the metal oxide is recovered and recycled.

This step need not be solar-aided, and since the two steps are decoupled, the production of H_2 and/or CO can be carried out on demand and round the clock at convenient sites and independent of solar energy availability [41]. Nonsolar exothermic oxidation of Zn by H_2O and/or CO_2 to form fuel (H_2 and/or CO) has been investigated [42], showing that the presence of inert ZnO affects the oxidation kinetics and the final conversion of Zn.

15.3.1.2 SnO_2/SnO Cycle

The group of PROMES has proposed and systematically investigated the SnO_2/SnO cycle. In this cycle, the solar-aided TR step consists of the reduction at approximately 1873 K of SnO_2 into gaseous SnO under atmospheric pressure—since its boiling temperature is 1800 K—and O_2 [43], followed by a nonsolar exothermic hydrolysis of SnO(s) to form H_2 and SnO_2(s) at about 873 K. Therefore, the solar step faces the same recombination problems, with the product consisting of a mixture of SnO and SnO_2 [44]. Quenching devices as well as reduced partial pressure of oxygen were employed to suppress this recombination. On the other hand, the dissociation rate of SnO_2 is high, and it is less dependent on the quenching rate than the dissociation rate of ZnO, whereas SnO reactivity with O_2 was shown to be lower than Zn reactivity in the high-temperature dissociation zone [45]. Thus, the solar step encompasses the formation of SnO-rich nanopowders that can be hydrolyzed efficiently in the temperature range of 800–900 K with a H_2 yield over 90%, higher than that of respective Zn nanoparticles ($\approx 55\%$), which, however, exhibited a faster hydrolysis rate [46].

This cycle has been studied for CDS as well. Solar-produced SnO-rich nanopowders were reoxidized separately by water and CO_2 in a TGA apparatus. SnO conversion approaching 90% was reported, while CO_2 reduction required significantly higher temperatures than H_2O reduction for reaching the same conversion. Parametric studies have shown as a global trend that both the conversion and the reaction rate improved with increasing temperature and the reacting gas mole fraction. However, the SnO disproportionation reaction had an influence on the H_2O and CO_2 reduction rates at temperatures above 773 K [47]. The reactivity of tin-based species was studied via TGA in order to elucidate the phenomena occurring during heating and subsequent reoxidation in a H_2O or CO_2 atmosphere to produce H_2 or CO. The activity of SnO and Sn/SnO_2 with H_2O was found much higher than that with CO_2 at a given temperature in the range of 823–923 K. The simultaneous splitting of H_2O and CO_2 was not favorable, given the higher reactivity of tin species with H_2O than with CO_2; the splitting of CO_2 required significantly higher temperatures (1073 K) to reach complete particle conversion [48].

15.3.2 Nonvolatile Cycles

Such cycles employ redox-pair oxides that remain condensed during the whole process, bypass the recombination issue of volatile cycles, and include today a wide variety of single- as well as multimetal, multivalent, metal oxide families. The material aspects have been comparatively presented in recent publications by researchers of the German Aerospace Center (DLR), Cologne [21], and of the SNL group [49].

15.3.2.1 Ferrite Cycle

The concept of using oxide pairs of multivalent metals for thermochemical production of hydrogen via WS first appeared in the literature in 1997 where cycling between Fe_3O_4 and FeO under the pair of Reactions 15.1 and 15.2 earlier was proposed [22], introducing the notion of metal oxides as thermochemical reaction media and a new strategy for thermochemical fuel production [50]. The first such systems studied for solar-aided WS were these of oxide pairs of multivalent metals such as iron oxides Fe_3O_4/FeO [33,51] or Mn_3O_4/MnO [25] with which the *proof of concept* of the particular process was demonstrated. However, even though such reaction schemes lead to high theoretical efficiencies, practical problems occur during their actual implementation. For instance, the thermochemistry of Fe_3O_4/FeO pair of reactions is such that (under standard pressures) the first is favored at temperatures above 3100 K and the latter at temperatures around 1000–1200 K. The extremely high temperatures required for the reduction step, exceeding the melting temperatures of both FeO and Fe_3O_4, render this specific cycle impracticable.

To lower the temperature of the reduction step, substitution of Fe by more easily reduced metals has been pursued. It has been reported that partial substitution of Fe in the magnetite phase by Mn, Ni, or Zn forms mixed metal oxides of the type MFe_2O_4 (mixed ferrites) that are more reducible and require moderate, upper (TR) operating temperatures. A whole series of such ferrites has been tested experimentally as well as studied thermodynamically initially for WS, including $MnFe_2O_4$ [52], $ZnFe_2O_4$ [53,54], $NiFe_2O_4$ [55–58], $CoFe_2O_4$ [59], $Ni_{0.5}Mn_{0.5}Fe_2O_4$ [26,60], $Mn_{0.5}Zn_{0.5}Fe_2O_4$ [61], and other cation stoichiometries [32,62,63]. Comparative testing of ferrite powders has identified the temperature ranges recommended for cyclic operation as well as the effects of process parameters on hydrogen/oxygen yields [55,58,64]. Their TR temperatures are still high ($\approx$1600–1700 K)—an important drawback since it can cause significant sintering of the metal oxide. Attempts to tackle this problem have been realized by supporting the redox reagent on high-temperature and stable ZrO_2 fine particles or supports [65–67].

Such testing of iron oxides and ferrites has been recently extended to include CDS as well, either separately [68,69] or simultaneously with WS. The SNL group investigated experimentally the capability of the iron oxide/yttria-stabilized zirconia (Fe_2O_3/YSZ) materials with respect to WS and CDS [70–72] employing temperature-programmed reduction and oxidation in a TGA between 1673 and 1373 K, respectively, and in situ, high-temperature x-ray diffraction (XRD). Both CDS and WS were demonstrated over multiple cycles. It has also been reported that Fe ions dissolved within the YSZ lattice are more *redox active* than nondissolved ones leading to higher oxygen yields. In addition to thermodynamic comparisons between the Zn/ZnO and FeO/Fe_3O_4 systems mentioned earlier, the ETH/PSI group also performed comparative TGA studies on simultaneous H_2O and CO_2 reduction over the same materials [73]. They reported that during the interface-controlled regime for both Zn and FeO, H_2O exhibited higher reaction rates with the solids compared to CO_2 and a strong dependency between the H_2O/CO_2 molar ratio of the input gases and the H_2/CO molar ratio of the product gases over the temperature ranges investigated. The research group of PROMES studied co-splitting of water and carbon dioxide over FeO (produced via solar TR of Fe_3O_4) in a TGA configuration and also reported that H_2 production was favored over CO production with the WS reaction being responsible for over 80% of the global FeO conversion [74]. In a recent work by the group of Aerosol and Particle Technology Laboratory (APTL), of CERTH, Thessaloniki, Greece, Ni-ferrite was shown to successfully split simultaneously CO_2 and H_2O under a co-feeding mode of operation to CO and H_2 with a H_2/CO ratio close to 1.40 with the two splitting reactions proceeding with nearly the same rate indicating a similar mechanism at the particular experimental conditions tested [75].

With respect to material composition, the current consensus seems to be that among the many ferrite materials tested for the targeted application, Zn-containing ones exhibit Zn-volatilization problems and Mn-containing ones face stability problems under air atmosphere at high temperatures [54,58,76–78]. These facts practically leave only $NiFe_2O_4$ and $CoFe_2O_4$ (and their combined stoichiometries) as the most *robust* among the ferrites, capable to operate reliably at the real conditions of a solar-aided process.

15.3.2.2 Ceria Cycle

The CeO_2/Ce_2O_3 pair, extensively used in automotive emission exhaust after-treatment as an oxygen storage system [79], has emerged as a possible redox pair in 2006, when the group of PROMES has investigated it for both WS and CDS [80]. Dissociation of CeO_2 to Ce_2O_3 was achieved via a solar reactor at pressures of 100–200 mbar and at temperatures over 2220 K where CeO_2 was already in the molten state.

Based on the nonstoichiometry of such compounds described previously, a research group at California Institute of Technology (Caltech), United States, proposed a similar ceria-based cycle based on the nonstoichiometric reduction of ceria ($CeO_2 \rightarrow CeO_{2-\delta}$) not involving melting and reported a higher activation energy for CO_2 dissociation (to CO and O_2) than for H_2O dissociation [10]. The research group of ETH/PSI modeled oxygen nonstoichiometry of several doped cerium oxide–based materials as a function of temperature and oxygen partial pressure [81]. Above 1200 K, ceria was found to react more efficiently with H_2O and CO_2 as the dopant concentration is increased.

Studies of SNL addressing comparatively the thermodynamics of the WS and CDS processes and considering CeO_2 as a redox material [9] indicated that at any temperature below about 1800 K, reduction of CO_2 to CO by Ce_2O_3 is thermodynamically favored and furthermore, at temperatures greater than 1100 K, CO_2 reduction is more favored than H_2O reduction. Similar efforts were extended to ferrites [82]. Research with ceria has thus shifted toward this direction of performing the reduction at temperatures below its melting point. The same group of PROMES studied the TR of CeO_2 by TGA in N_2 stream at 1773 K, whereas WS was performed in a separate packed bed reactor between 973 and 1318 K [83]. A plethora of dopants, basically lanthanides and alkaline earths, has been tested to improve the redox/thermal stability characteristics [81,84–86]. A consensus on the beneficial effect of—anyway well-known from automotive emission control—blending of CeO_2 with ZrO_2 with additions of La_2O_3 to induce thermal stability seems to have been established [87,88], but rather modest results have been achieved in terms of further doping with other cations [49].

Ceria offers several benefits for thermochemical cycling over alternative metal oxide systems: oxygen-deficient ceria shows very good reactivity with water and satisfactory H_2 production yield. However, is not without its own set of challenges, the most significant being the high TR temperatures required—higher than those of ferrites—to achieve significant reduction efficiency. At such temperatures—higher than 1800 K—partial sublimation of ceria can occur that gradually decreases the reduction yield [49,83].

15.3.2.3 Hercynite Cycle

A mixed cobalt ferrite–hercynite system ($CoFe_2O_4/FeAl_2O_4$) has been proposed by researchers from the University of Colorado, United States, as an effective WS-TR system [89]; the hercynite occurred due to the reaction between $CoFe_2O_4$ and Al_2O_3 (the ferrite was deposited on an Al_2O_3 support) at the elevated temperatures employed during testing of the ferrite in cyclic experiments. In this case, the redox-pair reactions are as follows:

$$CoFe_2O_4 + 3Al_2O_3 \rightarrow CoAl_2O_4 + 2FeAl_2O_4 + \tfrac{1}{2}O_2 \qquad (15.9)$$

$$CoAl_2O_4 + 2FeAl_2O_4 + H_2O \rightarrow CoFe_2O_4 + 3Al_2O_3 + H_2 \qquad (15.10)$$

Very low TR temperatures (1213 K, ~200 K lower than that of $CoFe_2O_4$) have been reported attributed to a reaction between the ferrite and Al_2O_3 resulting in the formation of the stable aluminates $FeAl_2O_4$ and $CoAl_2O_4$ that is thermodynamically more favorable than the formation of solid solutions or vacancies. Additionally, $CoFe_2O_4/Al_2O_3$ was capable of being cycled between 1473 (TR) and 1273 K (WS) producing significant amounts of H_2 with no obvious changes in H_2 conversion. The studies were extended to CDS where under either 75 or 600 Torr total pressure, a $CoFe_2O_4$-coated Al_2O_3 material was capable of producing appreciable amounts of CO after TR at a temperature as low as 1633 K, approximately 100°–150° lower than values reported for ferrites or CeO_2 with consistent oxidation behavior up to 23 TRs [90].

15.3.2.4 Perovskite Cycle

Despite the fact that perovskites are also nonstoichiometric compounds well known for their reversibility in delivering and picking up oxygen at high temperatures from their applications in fuel cells, they did not until recently attract significant attention in relation to a redox-pair reaction scheme similar to those discussed earlier for WS/CDS. Thermal dissociation studies of perovskites appeared in the literature only in 2013 when the ETH/PSI group investigated thermodynamically and experimentally lanthanum strontium manganite (LSM)/$La_{1-x}Sr_xMnO_{3-\delta}$ for both WS and CDS [91]. TR results at 1273 K have corroborated the higher O_2 yield of perovskites compared to that of ceria; however, on the other hand, perovskites were characterized by incomplete reoxidation from CO_2 at 1073, 1173, and 1273 K. A subsequent study from SNL [92] on lanthanum strontium aluminates ($La_{1-x}Sr_xMn_yAl_{1-y}O_{3-\delta}$) on WS and CDS reported also much higher reduction extent than ceria, achieved in addition at about 300 K lower temperature as well as multicyclic capability between 1623 (TR) and 1273 K (WS or CDS).

15.4 COUPLING REDOX CHEMISTRY TO SOLAR ENERGY: HEAT TRANSFER ISSUES

15.4.1 SOLAR CONCENTRATION SYSTEMS

Once a redox composition system has been selected, the next step is its coupling to solar energy. The first step of a solar-powered process requires the reflection and concentration of direct insolation using collectors/heliostats. Large-scale concentration of solar energy is accomplished at pilot and commercial solar thermal power plants (STPPs) aimed at the production of electricity from the sun's rays with four kinds of optical configuration systems using movable reflectors (mirrors) that track the sun, namely [93], parabolic trough (PT) collectors; linear Fresnel (LF) reflector systems; power towers, also known as central receiver (CR) systems; and dish–engine (DE) systems that provide in the listed order increasingly higher solar concentrations and increasingly higher process temperatures. The solar energy is concentrated on a focal point/area by the mirrors providing thus medium-to-high temperature heat. A heat exchanger (receiver) is located in the concentration field of the radiation, its task being to *trap* (absorb) the concentrated solar radiation and transfer it to a heat transfer fluid (HTF) (air, water, or molten salt) at the highest possible temperatures. In solar thermochemical applications instead of a *plain* receiver, the concentrated solar radiation is focused on a receiver–reactor where chemical reactions are performed. At this stage, the problem is shifted from chemistry issues to heat transfer ones, and the solar reactor type becomes important. In the particular case, since process temperatures of several hundred up to 2300 K are required to drive oxide-based thermochemical cycles, only solar towers and dishes (Figure 15.4) remain as the technology of choice. Furthermore, dishes are considered only in the developmental stage, since, considering that hydrogen/syngas production plants require a certain size, most research and development work in this field was concentrated on implementing thermochemical cycles with solar towers. Predevelopment tests under *solar* heat are performed progressively in scale, starting by solar simulator or solar furnace facilities.

15.4.2 SOLAR RECEIVER–REACTOR CONCEPTS

15.4.2.1 Directly and Indirectly Irradiated Receivers

One generic categorization of solar receivers is according to the mechanism of transferring the solar heat to the heat transfer fluid: directly and indirectly heated ones. The characteristics of both these receiver kinds have been described in detail in a recent review on solar natural gas reforming by the present authors [16]. Indirectly irradiated receivers (IIRs) consist of absorbing surfaces exposed to the concentrated solar radiation, with heat conducted across their walls to the thermal fluid. The simplest examples are conventional tubular receivers in the interior of which the heat transfer fluid is moving in a direction vertical to that of the incident solar radiation. Alternatively, directly irradiated receivers (DIRs) make use of fluid streams or solid particles directly exposed to the concentrated solar radiation; they are also called *volumetric* receivers since they enable the concentrated solar radiation to penetrate and be absorbed within the entire volume of the absorber. In different designs, the absorber is either a stationary matrix

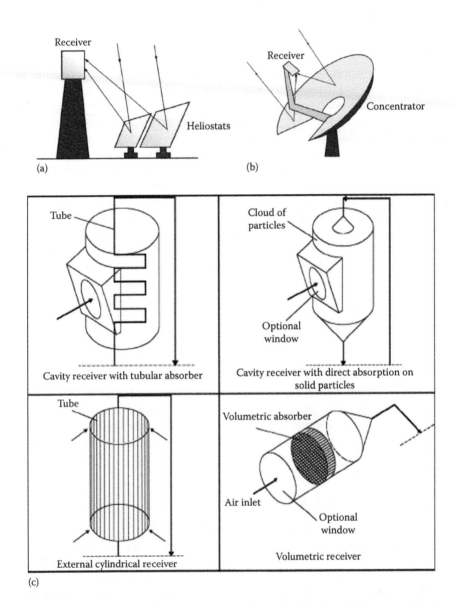

FIGURE 15.4 Schematic of operation of two typical point-focusing CSP systems: (a) solar tower; (b) solar dish (From Roeb, M. et al., *Materials*, 5, 2015, 2012); and (c) examples of solar receiver configurations—cavity receiver with tubular panel, cavity receiver with direct absorption on particle-laden flow, external receiver with tubular panels, and volumetric receiver with porous absorber. (From Romero, M. and Steinfeld, A., *Energ. Environ. Sci.*, 5, 9234, 2012. With permission.)

(grid, wire mesh, foam, honeycomb, etc.) or moving (usually solid) particles. In all cases, the solar receiver is designed so as to approach a blackbody in its capability to trap incident solar radiation by making use of cavities, black-painted tube panels, or volumetric porous absorbers [93]. There are basically two design options: external and cavity-type receivers. In a cavity receiver, the incident concentrated solar radiation enters through a small aperture into a well-insulated enclosure containing the absorber. Because of multiple reflections among the cavity's inner walls, the cavity approaches a blackbody absorber. Cavities are constrained angularly in contrast to external receivers that can be cylindrically shaped. Schematic examples of such configurations of cavity, external, and volumetric receivers are shown in Figure 15.4c [93].

15.4.2.2 *Structured* versus *Nonstructured* Reactors

For the efficient design and operation of solar receiver–reactors, concepts from *traditional* chemical reactor engineering should be combined with ways to achieve efficient heating of the reactor via concentrated solar irradiation. The situation in the case of redox pairs resembles that of catalytic reactions between gases where reactant gas species react on the surface of a solid catalyst to be converted to useful gaseous products. In the *traditional* nonsolar chemical engineering, such catalytic reactor types can be distinguished in two broad categories depending on whether the catalyst particles are distributed randomly or are *arranged* in space at the reactor level: the first category includes packed and fluidized catalytic beds; the second includes the so-called *structured* catalytic systems like honeycomb, foam, and membrane catalytic reactors, all three of them being free of randomness at the reactor's level [94–96].

In a direct analogy to *conventional* catalytic applications, both these configurations—redox oxide particles as well as porous structures—can be contained either within solar-heated tubular receivers (IIRs) or being directly exposed to solar irradiation (DIRs). Therefore, both receiver kinds can be *transformed* and adapted to operate as solar chemical receiver–reactors (IIRRs or DIRRs, where the last "R" stands for reactors) where chemical reactions can take place in an efficient and elegant manner. There are, however, fundamental differences between *traditional* and *solar* chemical engineering, the most important being that the solid reactant (metal oxide) is not a *catalyst* present in much smaller quantities than those of the gaseous reactants, but a reactant itself, with nonnegligible mass, that gets progressively depleted during the course of the reaction having to be replenished. Therefore, it has to be fed constantly into the reactor if the reactions are to be performed in a continuous mode, or, alternatively, practical ways have to be *invented* for its in situ regeneration. In addition, the reactor, being either a packed/fluidized bed or a honeycomb/foam, has to incorporate as much of the redox oxide solid reactant as possible on one hand to maximize volumetric product yield and on the other hand to avoid the *waste* of external energy in heating chemically inert materials. The higher-temperature TR step that needs to be solar aided requires temperature levels (1600–1900 K, depending on the redox material used) imposing rather challenging reactor operation conditions. Requirements in relation to reactor construction, in particular considering the harsh thermal conditions and chemical atmospheres having to be faced, are superimposed to those for the redox material together with issues of solar absorbance and resistance against thermal shock and fatigue.

15.4.3 TEMPERATURE-SWING OPERATION AND HEAT RECUPERATION ISSUES

Another issue is that the TR and the (H_2O/CO_2) splitting reactions are favored by different conditions [97]. TR is thermodynamically favored by high temperatures and low oxygen partial pressures, whereas splitting by low temperatures and high partial pressures of H_2O/CO_2. However, the splitting reaction kinetics needs to be considered; to ensure satisfactory reaction rates requires sufficiently high temperatures— but not so high to induce the TR simultaneously to splitting. Therefore, the splitting reactions should, in principle, take place at lower temperature levels (1000–1300 K) than the TR ones. In other words, the complete cyclic operation has to be carried out under a *temperature-swing* mode of about 400 K, whereas in parallel to this, the gaseous feed to the metal oxide has to swing between H_2O/CO_2 and inert purge gas. This is a common problem for all single- and mixed-oxide redox systems mentioned earlier causing complications with respect to both reactor design and oxide material handling between the two stages.

This temperature swing induces additional issues relevant to efficient heat utilization between the two reaction steps. The sensible heat available after the higher-temperature TR step has somehow to be recovered and reused effectively; heat rejection at the TR temperature levels of 1600–1900 K needs to be avoided as far as possible because the associated heat losses can render the efficiency of the system detrimentally low for the economics of the whole process. The importance of such heat recuperation has been extensively stressed [98,99]. Whereas the *material-related* problem in this respect is the low extent of TR, in solar-aided operation, the heat recuperation, which is a property of the reactor design, comes also into play. In this respect, the various reactor designs proposed and implemented have stemmed from the quest for effective heat recuperation concepts.

Reactor designs have implemented different technical solutions to meet the aforementioned technical requirements of the thermochemical cycle made up of two process steps performed at different temperature levels with different heat demands. All such concepts will be presented in the following. Actually, due to the fact that when using IIRRs, problems relevant to the resistance to heat transfer and the tube materials temperature limit the actual heat flux to the reaction site (inner region of the tube), there are only few examples of such reactors proposed and employed for such systems. The majority of reactors are DIRRs that use solid particles or structures directly exposed to the concentrated solar radiation. In these cases, since the working fluid at the thermal decomposition stage is not air, the receiver–reactors must be equipped with a transparent window, which allows concentrated light to enter the receiver while isolating the working gas from ambient air and providing for reactor operation under nonatmospheric pressures if needed [100].

15.5 SOLAR REACTORS EMPLOYED FOR BOTH STEPS OF THERMOCHEMICAL WS/CDS CYCLES

Based on the aforementioned discussion, we can initially distinguish between two kinds of solar reactors according to the distinction of redox cycles to volatile and nonvolatile. The common characteristic of all volatile cycles proposed is that during the TR phase, a gaseous mixture of the reduced phase (this being either an oxide or a metal) and oxygen occurs, which needs special treatment to avoid its recombination back to the original oxidized form—a fact that precludes the direct combination with the other cycle step, that is, the *oxidation* (via steam or CO_2) reactor. All such systems therefore consist of two separate reactors, one that performs the TR step from which a condensed reduced phase is obtained together with gaseous oxygen and a second one where this condensed phase is oxidized (not necessarily solar aided) and performs the required WS/CDS step. In this respect, the cyclic operation can be decoupled to two stages: one diurnal for TR and one nocturnal when syngas is produced. Thus, solar reactors for such systems have been designed to perform only the higher-temperature TR step. These reactors will not be presented here for reasons of brevity.

Cycles employing nonvolatile redox pairs that remain condensed during the whole process bypass the recombination issue encountered in volatile cycles. Furthermore, these materials offer more possibilities concerning the reactor concept and process design for the reduction step due to the possible use of either particle receiver–reactors or monolithic structures, such as honeycombs, foams, or fins. Such reactors are designed to perform both cycle steps, but different concepts are implemented for this goal; one fundamental distinction has to do with whether such reactors employ moving or only stationary parts. The main approaches introduced are following.

15.5.1 Nonstructured Directly Irradiated Receiver–Reactors

15.5.1.1 Packed Bed Reactors

A first such packed bed reactor consisting of a small quartz tube (2 cm diameter) was tested at the solar furnace of PSI in 1995 [26], placed in the focus of the solar furnace with a secondary concentrator behind it to provide a uniform irradiation of the tube. $Ni_{0.5}Mn_{0.5}Fe_2O_4$ powder mixed with Al_2O_3 grains was used as reactive particle bed. During TR, Ar was passed through the packed bed, and afterward, a mixture of Ar and steam was introduced. H_2 and O_2 evolution was observed. Even though this is the first reported solar-aided production of hydrogen from ferrites with TR temperatures below their melting point, nevertheless, this concept was not followed up.

15.5.1.2 Spouted Bed Reactors

The research group of Niigata University in Japan has set forth the concept of an internally circulating fluidized bed reactor (or, in a more precise reactor terminology, a *spouted* bed reactor) to be combined with a beam-down solar concentrating system, in a series of publications [56,57,101,102]. A schematic of the operating concept and a photograph of such a reactor operating while irradiated by a solar simulator are shown in Figure 15.5a. The redox oxide particles are circulated through an internal annulus from the bottom to the top of the reactor and are exposed to concentrated solar radiation entering from a window at the top, at the uppermost points of their trajectories. The idea is that internal circulation will, among others, inhibit sintering and agglomeration of the redox oxide particles avoiding hence a pulverization process after the TR step. On the other hand, sequential performance of the two steps of the cycle would be possible in a single reactor by switching the feed gas between

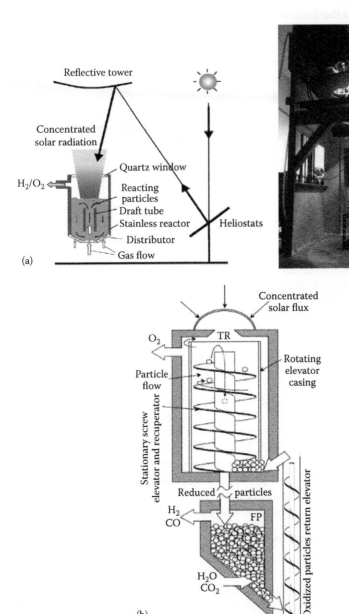

FIGURE 15.5 Solar reactors employing moving powder beds: (a) the spouting bed reactor of Niigata University, Japan, schematic of the operation concept and actual photograph of such a reactor operating while irradiated by a solar simulator. (From Gokon, N. et al., *Int. J. Hydr. Energ.*, 36, 4757, 2011. With permission.); (b) operating principle schematic of the moving packed bed redox solar reactor proposed by Sandia Labs, Bucknell University, and Arizona State University. (From Ermanoski, I. et al., *J. Sol. Energ. Eng.*, 135, 031002, 2013.)

an inert gas (N_2) for the TR step to steam/CO_2 for the WS/CDS step. Such chemical reactors were tested for TR of unsupported $NiFe_2O_4$ as well as supported $NiFe_2O_4$/ ZrO_2 particles on a laboratory scale using solar-simulating Xe-beam direct irradiation. First tests involved only the TR reaction, removing the reduced powder and performing in another nonsolar rig the WS reaction, whereas subsequently both steps were tested in the same spouted bed reactor [57]. During these tests, the temperature at the surface of the fluidized particle bed during TR step reached 1773–1873 K in the draft tube region and 1373–1523 K in the annulus region. Approximately 35% of the supported $NiFe_2O_4$ was reported to have been converted to the reduced phase and subsequently completely reoxidized with steam at 1373 K to generate hydrogen, remaining in powder form without sintering and agglomerating during Xe-beam irradiation over 30 min.

15.5.1.3 Moving Packed Bed Reactors

Recently, another reactor concept, based on a continuously moving packed bed of redox oxide particles, was proposed by a consortium of U.S. researchers from SNL, Bucknell University, and Arizona State University [103]. A schematic of its operating principle is shown in Figure 15.5b. The oxide particles in their oxidized state are transported by a vertical screw elevator/feeder with a rotating casing to the top of a solar tower, where concentrated solar radiation enters through a window-covered aperture, directly heating and thermally reducing them. This step produces gas phase oxygen, which is pumped away from the chamber. The central idea is that of effective heat recuperation since the packed bed of the reduced particles produced is then supposed to move downward via a connecting tube in a counterflow arrangement with respect to the oxidized particles moving upward, essentially preheating them. Heat transfer via conduction, from the hot (reduced) particles to the colder (oxidized) particles, is augmented by the extended surface area of the conveyor auger. In this respect, the whole reactor consists of three sections: a TR chamber at the top, a recuperator (solid–solid heat exchanger) in the middle, and a fuel production chamber (at the bottom). In the fuel production chamber, the particles are exposed to reactant gases (H_2O or CO_2), reducing them to fuel products (H_2 or CO). The mix of reactants and products (H_2O/H_2 or CO_2/CO) is removed from the chamber, and the reoxidized particles empty into a return elevator that brings them to the inlet of the recuperator/ elevator to continue the cyclic process. Advantages claimed are spatial separation of pressures, temperature, and reaction products in the reactor; solid–solid sensible heat recovery between reaction steps; continuous on-sun operation; and direct solar illumination of the working material. However, this concept has not been experimentally implemented and tested as yet.

15.5.2 STRUCTURED DIRECTLY IRRADIATED RECEIVER–REACTORS WITH NONMOVING PARTS

Such reactors stem from the traditional gas–solid catalytic chemical reactors, where ceramic porous supports, chemically inert with respect to the targeted reactions, are coated with a catalyst material capable to catalyze them. The evolution of such solar receiver–reactors proceeded along this developmental path for some time, that is, by depositing layers of redox oxide materials upon ceramic supports capable of absorbing concentrated solar irradiation and developing the temperatures required for the TR step. As experience with such systems was accumulated, it was at some point realized that, on one hand, possible side reactions of the redox coating with the support could have an adverse effect on the desired reactions and that, on the other hand, the much larger thermal mass of the redox-inert support material had a detrimental effect on the reactor's thermal efficiency. In this perspective, a self-supported, shaped, porous structure—honeycomb, foam, or fin—incorporating the maximum possible amount of a redox oxide in its structure instead of a thin coating layer (and even better, entirely made out of it) would introduce minimal thermal mass not directly involved with the reaction process into the reactor and consequently higher volumetric product yields and enhanced efficiency. Indeed, recent efforts have been shifted in this direction; the relevant developments are presented in the following.

15.5.2.1   Reactors Based on Ceramic Honeycombs

The first examples of honeycomb reactors for solar-aided chemistry applications can be traced back to 1989 when researchers at Weizmann Institute of Science (WIS), Israel, have deposited Rh on alumina and cordierite honeycombs and irradiated it in their solar furnace to catalyze the reaction of CO_2 methane reforming [104,105]. However, such honeycomb reactors have not been scaled up until 2006 when the HYDROSOL research group (including among others the research group of APTL, Greece, and the present authors) introduced the concept of monolithic, honeycomb solar reactors for performing redox-pair cycles for the production of hydrogen from the splitting of steam [106]. The reactor has no moving parts and is based on the incorporation of active redox-pair powders as coatings on multichanneled monolithic honeycomb structures of SiC capable of achieving and sustaining high temperatures when irradiated with concentrated solar irradiation. The operating concept is shown in Figure 15.6a: when steam passes through the solar reactor, the coating material splits water vapor by *trapping* its oxygen and leaving in the effluent gas stream pure hydrogen. In a subsequent step, the oxygen-*trapping* coating is thermally reduced by increasing the amount of solar heat absorbed by the reactor. Such redox-material-coated honeycombs (Figure 15.6b) have achieved continuous solar operation WS–TR cycles. The issue of continuous production of solar hydrogen in a single solar receiver–reactor has been resolved with a modular dual-chamber fixed honeycomb absorber design and implementation shown in Figure 15.6c [107]. One part of modules splits water while the other is being thermally reduced; after completion of the reactions, the thermally reduced modules are switched to the splitting process and vice versa by switching the feed gas [108]. Due to its modularity and the lack of movable parts, this design is amenable to straightforward scale-up and can be effectively coupled with a solar tower platform facility for continuous mass production of hydrogen. Indeed, such a modular, dual-chamber, ferrite-coated-honeycomb HYDROSOL reactor has been scaled up to the 100 kW level, coupled on the solar tower facility of Plataforma Solar de Almeria, Spain, and achieved continuous solar-operated WS—regeneration cycles demonstrating the *proof of concept* of the design (Figure 15.6d and e) [109].

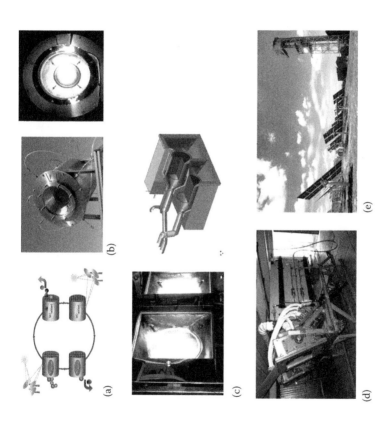

FIGURE 15.6 Solar reactors employing honeycombs. The HYDROSOL solar reactor technology evolution: (a) operating concept; (b) first single-chamber reactor assembled and in operation. (From Agrafiotis, C. et al., *Sol. Energy*, 79, 409, 2005. With permission.); (c) the continuous solar hydrogen production operation concept and the first dual-chamber reactor in operation. (From Roeb, M. et al., *Int. J. Hydrogen Energy*, 34, 4537, 2009. With permission.); (d) the 100 kW$_{th}$ scale dual-chamber reactor on the top of the PSA solar tower facility. (From Roeb, M. et al., *Sol. Energ.*, 85, 634, 2011. With permission.) coupled with (e) the solar field producing two focal points. (Courtesy of DLR) *(Continued)*

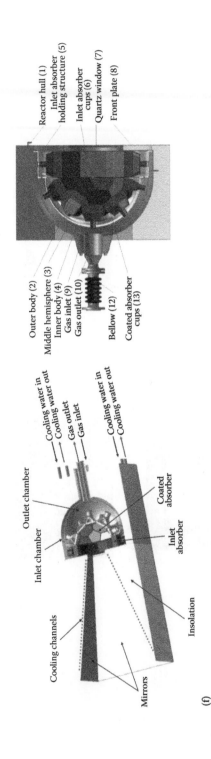

FIGURE 15.6 (CONTINUED) Solar reactors employing honeycombs. The **HYDROSOL** solar reactor technology evolution: (f) schematic of the 100 MW reactor module with secondary concentrator and of the absorber/reactor module designs. (Left: Courtesy of DLR; right: From Houaijia, A. et al., *Sol. Energ.*, 97, 26, 2013. With permission.)

In such a facility, the different heat demands for the two process stages were realized not by moving the reactors but by adjusting the flux density on each module when the status of the cycle is switched from regeneration to splitting and vice versa, via partitioning the heliostat field and providing two focal spots with independent power modulation (Figure 15.5e) [110].

An optimization of the reactor shape has been carried out to reduce the quite high reradiation losses due to the high temperatures and the large exposed absorber surface area revealed by experiments and simulations [111]. A new reactor design has been proposed [112] where the overall shape of the absorber is close to a hemisphere and a suitable secondary reflector is included as well (Figure 15.6f). The reactor–receiver consists of two parts: a *receiver flat* part made of nonredox, square-shaped honeycombs at the front plate of the reactor (just behind the quartz window) and a *domed* part at the rear, which is the reactor part consisting of the redox-coated modules. The introduction of a spherical shape of the absorber and a suitable secondary reflector ensures a more homogeneously distributed solar flux and therefore a more homogeneous temperature distribution than that of the previous *flat design* version. Furthermore, the whole reactor setup and all components were designed in a way allowing easy maintenance and replacement of parts, in particular of the individual absorber monoliths.

The approach of directly fabricating the redox powders into monolithic structures was adopted from the SNL group even though designed to be integrated into another reactor design concept (to be discussed later). They employed robocasting—an SNL-developed technique for free-form processing of ceramics to manufacture monolithic structures [59]. Physical mixtures of ferrite powder with various supports were fabricated directly into small-scale 3D lattice structured monoliths with dense and porous rods after firing at 1700 K with design objectives of high geometric surface area, thermal shock resistance, and allowance for light penetration. Photographs of such items are shown in Figure 15.7a [113]. Such cast pellets made of 1:3 $Co_{0.67}Fe_{2.33}O_4/$ YSZ, $Co_{0.67}Fe_{2.33}O_4/Al_2O_3$, and $Co_{0.67}Fe_{2.33}O_4/TiO_2$ were tested for cyclic TR/WS in a typical, laboratory-scale, nonsolar reheated flow reactor between approximately 1673 K (TR) and 1273 K (WS). Structural integrity could be maintained over successive cycles; however, only the first composition was proved capable of cyclic WS/TR operation, probably due to the reaction of the ferrite with the support to iron spinels (e.g., $FeAl_2O_4$ and Fe_2TiO_4) in the other two cases. The work was continued along the same approach with $(CeO_2)_{0.25}(ZrO_2)_{0.75}$ monoliths cast in a similar fashion, which were shown capable of WS as well as CDS and in fact exhibiting greater yield of CO relative to H_2.

Exploring the same issue of maximum incorporation of redox oxide in the receiver–reactor's structure, a U.S. research consortium from SNL, University of Arizona, and Missouri University of Science and Technology proceeded in the coextrusion of zirconia–iron oxide honeycomb substrates to be tested for solar-based thermochemical CDS [114]. The test material was formed from a 75 vol% YSZ–25 vol% Fe_2O_3 mixture processed into honeycombs of cell sizes 2.4, 4.0, or 6.0 mm in diameter, by the polymer-based coextrusion of ceramics. The extruded specimens were sintered at 1673 K before finally tested in a laboratory-scale test

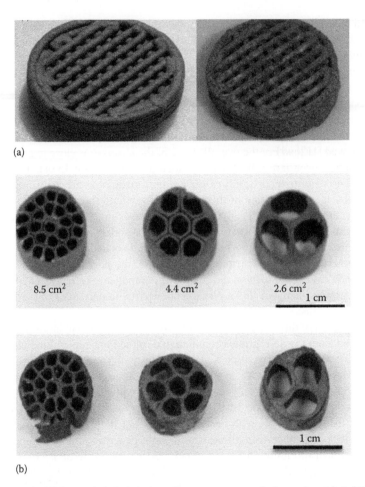

(a)

(b)

FIGURE 15.7 SNL-produced honeycomb structures used for solar-aided WS/CDS: (a) robocast cobalt ferrite/zirconia cylindrical pellets (≈15 mm diameter × 5.4 mm thick) tested for cyclic TR/WS, as-cast and after 31 TR and water oxidation cycles (From Diver, R.B. et al., *J. Sol. Energ. Eng.*, 130, 041001, 2008); (b) coextruded YSZ–Fe₂O₃ honeycombs of SNL's consortium: prior to (top) and after (bottom) thermochemical CDS testing. (From Walker, L.S. et al., *Energ. Fuels*, 26, 712, 2011. With permission.)

furnace for CDS. The TR step was performed at 1743 K under He and the CDS step at temperatures of 1173, 1273, 1373, and 1473 K. It was reported that the melting of FeO at high temperature allows for full reduction of the substrate system in a very short time period (<60 min), making it highly reactive for CO_2 generation cycles. The honeycomb substrates survived 6–10 cycles of melting at high temperature without significant deformation or structural failure (Figure 15.7b).

15.5.2.2 Reactors Based on Ceramic Foams

Such reactors based on ceramic foams were the first structured reactors tested in solar chemistry applications—in fact, for solar-aided syngas production via CO_2 (dry) methane reforming in a 100 kW directly irradiated volumetric receiver–reactor mounted on a solar thermal concentrating dish by SNL and DLR [115] between 1987 and 1990. However, such foam-based reactors were not used until recently for solar-aided *splitting* reactions (either WS or CDS). The first study on ceramic foams coated with redox materials for WS is reported by the group of Niigata University, Japan [116], where a ceramic foam device made of magnesia partially stabilized zirconia (MPSZ) was coated with Fe_3O_4 and c-YSZ particles and examined as a thermochemical WS device. The two cycle steps were performed in independent reactor rigs: WS under irradiation of a Xe lamp at 1073 K and TR in an electrically heated tube furnace at 1373 K with various Fe_3O_4 loading amounts. Throughout 32 cycles of the two-step WS reaction, hydrogen production was successfully continued; however, at that point, the YSZ/MPSZ foam device was cracked and broken (Figure 15.8a). The work was continued with $m-ZrO_2$-suported $NiFe_2O_4$ and Fe_3O_4 powders [117], and eventually the most reactive foam device of $NiFe_2O_4/m-ZrO_2/MPSZ$ was tested in a windowed single reactor using solar-simulated Xe-beam irradiation producing successfully hydrogen for 20 cycles [118]. The group is conducting a solar demonstration project of thermochemical WS at Inha University in Korea using a larger $NiFe_2O_4/m-ZrO_2$-coated MPSZ foam device with a 5 kW_{th} dish concentrator (Figure 15.8b).

The ETH/PSI group in collaboration with the group of Caltech developed a solar reactor based on reticulated porous ceramic foams manufactured entirely from the redox material—cerium oxide (CeO_2) in the particular case—for the splitting of carbon dioxide via thermochemical redox cycles. Actually this has occurred as a culmination of their work employing initially pellets, then porous cylinders, fibers, and eventually foams made entirely of the particular redox material. The evolution of their technology of ceria-based solar cavity reactors for two-step, solar-driven thermochemical production of fuels is depicted schematically in Figure 15.8c through e.

FIGURE 15.8 Solar reactors employing ceramic foams. (a and b) Niigata University: (a) from left to right, photographs of the noncoated MPSZ foam, the Fe_3O_4/YSZ-coated MPSZ foam before testing, after 11 cycles and after 32 cycles when cracked (From Gokon, N. et al., *Energy*, 33, 1407, 2008. With permission.); (b) the 5 kW_{th} Inha University's dish concentrator in Korea and the foam solar reactor installed. (From Gokon, N. et al., *Int. J. Hydrogen Energy*, 36, 2014, 2011. With permission.) *(Continued)*

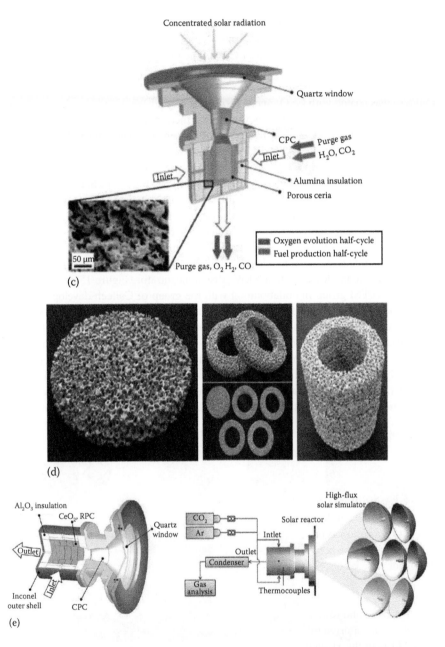

FIGURE 15.8 (CONTINUED) Solar reactors employing ceramic foams. (c–e) Technology evolution of ETH's ceria-based reactors: (c) schematic of operating principle of the first solar reactor containing a porous ceria cylinder. (From Chueh, W.C. et al., *Science*, 330, 1797, 2010.); (d) CeO$_2$ reticulated parts; and (e) schematic of the solar reactor configuration employing ceria foams and experimental test setup at ETH's high-flux solar simulator. (From Furler, P. et al., *Energ. Fuels*, 26, 7051, 2012. With permission.)

In their first such experiments, they employed CeO_2 and $Sm_{0.15}Ce_{0.85}O_{1.925}$ annealed for 3 h at 1773 K to perform separately WS and CDS [50]. In the former case, the materials were shaped in the form of 65% porous pellets loaded into a tubular reactor inside an infrared furnace and operated for 400 cycles. With respect to the TR step, upon heating to 1773 K under an inert atmosphere ($p_{O_2} = 10^{-5}$ atm), release of oxygen was observed, whereas ceria reduction at 1873 K resulted in higher oxygen yield. Hydrogen production was observed by introducing steam at 1073 K with the total amount of hydrogen produced implying complete reoxidation of the reduced ceria. The first solar campaign was reported by the same group [20] employing a cavity receiver containing a porous monolithic ceria cylinder directly exposed to concentrated solar radiation impinging on its inner walls for the (separate) two-step, solar-driven thermochemical WS and CDS. TR of ceria was performed at ≈1873 K and CDS at ≈1173 K. An analogous set of experiments was performed for H_2O dissociation for 500 WS cycles, where the most obvious feature was the much faster rate of fuel production than that of O_2 release. In a subsequent work [119], they employed a solar cavity receiver containing porous ceria felt directly exposed to concentrated thermal radiation provided by a solar simulator. The TR of ceria was performed at 1800 K and reoxidation with a gas mixture of H_2O and CO_2 at 1100 K, producing syngas with H_2/CO molar ratios from 0.25 to 2.34. Even though CeO_2 sublimation was observed, 10 consecutive H_2O/CO_2 gas splitting cycles have been performed. Based on the aforementioned discussion, they have prepared and tested in lab-scale solar–thermal high-temperature chemistry configuration (1873 K) reticulated porous ceramic foams manufactured entirely from cerium oxide (CeO_2—Figure 15.8d) for the splitting of CO_2 via thermochemical redox cycles [120]. Currently, the consortium is working toward the optimization of the solar reactor for the combination of the two thermochemical splitting cycles for solar syngas synthesis targeted to the production of liquid aviation fuels.

15.5.3 STRUCTURED DIRECTLY IRRADIATED RECEIVER–REACTORS WITH MOVING PARTS

15.5.3.1 Rotary-Type Reactors

In 2006, the research group from the University of Tokyo, Japan, introduced the concept of a rotary-type reactor in which a cylindrical rotor coated with redox-pair materials is rotating between two chambers (Figure 15.9a), one where the WS reaction (H_2-generation reaction cell) and one where the TR reaction (O_2-releasing reaction cell) are performed under solar irradiation [121]. Such a cylindrical rotor reactor of 40 mm diameter coated separately with CeO_2 and mixed ferrites ($Ni_{0.5}Mn_{0.5}Fe_2O_4$) was fabricated and tested under infrared lamp irradiation [122]. Successive evolution of H_2 gas was reported under temperatures of 1623 K in the O_2-releasing reaction cell and 1273 K in the H_2-generation reaction cell for CeO_2. Also, repetition of the two-step WS process was achieved using the (Ni, Mn) ferrite with reported optimum reaction temperatures of the O_2-releasing and H_2-generation reactions as 1473 and 1173 K. A scaled-up version was manufactured (Figure 15.9a, right) with $0.8CeO_2$–$0.2ZrO_2$ solid solution as the redox material. Preliminary tests of the reactor were performed using a solar simulator of concentrated Xe lamp beams between 1473 and 1773 K demonstrating cyclic H_2 production without sintering of the redox material [123].

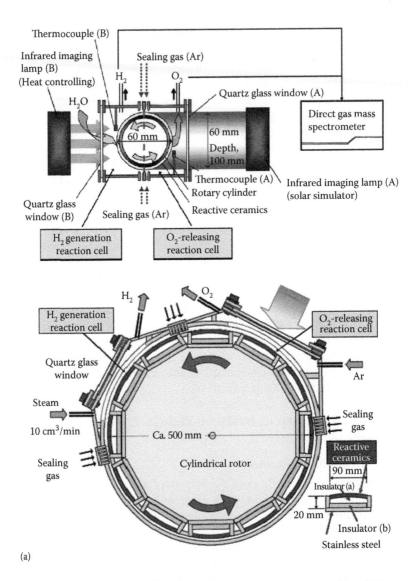

FIGURE 15.9 Solar reactors employing moving parts for the two-step process. (a) The rotary-type solar reactor of the University of Tokyo, Japan: schematic outlines of laboratory-scale (top) and scaled-up, 500 mm cylindrical rotor diameter reactor (bottom). (From Kaneko, H. et al., *Energ. Fuels*, 21, 2287, 2007. With permission.) (b–e) SNL's counter-rotating-ring receiver reactor recuperator (CR5): (b) schematic illustration of operating principles for WS.

(Continued)

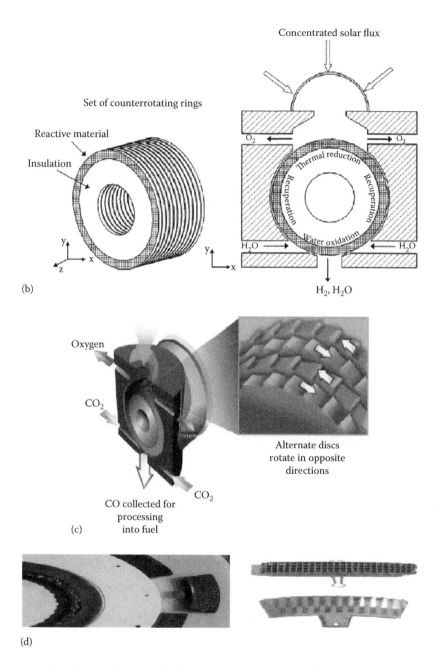

FIGURE 15.9 (CONTINUED) Solar reactors employing moving parts for the two-step process. (b–e) SNL's counter-rotating-ring receiver reactor recuperator (CR5): (c) schematic illustration of operating of operating principles for CDS. (d) Robocast reactant ring assembly and prototype reactant fin segments. (From Diver, R.B. et al., *J. Solar Energy Eng.*, 130, 041001-1, 2008.) (*Continued*)

(e)

FIGURE 15.9 (CONTINUED)   Solar reactors employing moving parts for the two-step process. (e) Photo of the CR5 taken shortly after the conclusion of a successful four-ring test (with ceria)—ring segments recovered after the test showing ceria fins (outer periphery of the disk), zirconia carrier segments (adjacent white ring) and metallic central hub (gray inner ring). (From Miller, J.E. et al., Final Report, Reimagining liquid transportation fuels: Sunshine to petrol, Sandia Report, Sandia National Laboratories, Albuquerque, NM, 2012. With permission.)

This reactor was eventually tested in the solar tower facility of CSIRO, Newcastle, Australia, but results of the campaign have not been published as yet.

15.5.3.2 The CR5 Reactor

At the same year (2006), the research group at SNL proposed their own concept of a rotating reactor, the so-called counter-rotating-ring receiver reactor recuperator (CR5) for producing hydrogen and oxygen from water [62,124]. The key feature of the CR5 is a stack of counter-rotating rings or disks that are outfitted with fins around the circumference that are constructed of a redox metal oxide. The principle of its operation was conceived as follows: as the rings rotate, the reactive material passes from a solar-irradiated high-temperature TR receiver/reactor to a relatively low-temperature hydrolysis reactor where the reactant material performs a WS reaction, and then back again (Figure 15.9b). That is, the fins are cyclically heated and reduced and then cooled and oxidized. Each ring rotates in the opposite direction to its neighbor at a rotational speed on the order of one RPM or less. As the oxidized redox material in the fins leaves the WS reactor and enters the recuperator, it *sees* hotter fins leaving the TR reactor on both sides. In the recuperator, it heats up as the neighboring fins moving in the opposite direction cool.

This group proposed combining the two splitting cycles (of water and carbon dioxide) with the CR5 reactor (Figure 15.9c) into a project named *Sunshine to Petrol* (S2P) targeted to splitting carbon dioxide into oxygen and carbon monoxide with the latter to be used afterward for the production of hydrogen or of liquid synthetic fuels (e.g., methanol, diesel) [125]. The ultimate goal of the S2P project was that the CR5 reactor would become a building block for the synthesis of liquid combustible fuels. From an engineering design standpoint, the working redox oxide must be amenable to fabrication into physical forms that can be integrated into the engine design concept. This led the group initially to the approach of directly fabricating the reactant fin sections for the CR5 prototype by the same robocasting technique mentioned already earlier [59]. Photographs of such items are shown in Figure 15.9d [113]. However, the bulk of the on-sun testing effort with the CR5 reactor was performed with thin ceria fins fabricated by a casting/lamination/laser cutting methodology, captured within a zirconia-based carrier ring that is in turn captured by a metal hub (Figure 15.9e). During operation, argon is injected into the reduction side of the CR5, while CO_2 is provided to the oxidation side. In operation with 4 and 8 ceria-finned rotating rings, CDS was demonstrated successfully; in general, the CO yield increased with increasing reduction temperature and increasing CO_2 (and total) flow rate. The efficiency significantly increased when the reduction temperature was increased from 1723 to 1823 K, but the increase was less pronounced when increasing the temperature to 1893 K. However, when the scaled-up version of 22 fins was tested, operation ceased shortly, mainly since several of the rings ceased to rotate due to cracking/separation of zirconia wedge segments at or near the metal/zirconia connection points [126,127].

15.6 CURRENT DEVELOPMENT STATUS AND FUTURE PROSPECTS

The acknowledgment of the fact that the experience and technology accumulated with solar-aided, two-step, redox-pair-based WS thermochemical cycles for hydrogen production could be *transferred* to CDS as well and essentially culminate to the synthesis of liquid hydrocarbons from solar energy, water, and (waste) carbon dioxide has created many aspirations and research efforts worldwide. Taking into account the fact that solar energy is currently unmatched in its magnitude and availability as well as undisputedly scalable to any future energy demand, this is indeed a fascinating perspective worth exploring. However, it should not be underestimated that despite large efforts and significant progress during the last years, hydrogen production from such solar-aided, WS thermochemical cycles, even though having been already successfully demonstrated at bench and pilot scale, has not yet solved crucial technical challenges and is still far from being commercially exploitable. As the present authors have already mentioned, "...The reactions involved are on the edge of being feasible and practicable because high temperatures are needed and the gas-solid reactions cannot be carried out continuously (the reactive solid is consumed as the reaction proceeds)..." [97]. Technical barriers have to do both with the redox material's chemistry and the solar energy exploitation issues that were discussed earlier.

With respect to material issues, a first distinction can be made according to whether they are employed in volatile or nonvolatile cycles. The volatility of Zn/SnO is not only a fundamental thermodynamic advantage but also a major technical drawback: it means that gaseous mixtures of Zn/SnO and O_2 result from ZnO/SnO_2 decomposition. All such products reported so far consist, in a lesser or higher degree, of a mixture of the reduced/oxidized state of the system. Another issue, common for all reactors that perform only the TR step, is that their integration with their *oxidizing* (splitting) counterparts and relevant heat recuperation still needs to be demonstrated.

Therefore, it seems that the choices are limited within performing the cycles below the melting point of the redox pair, keeping it practically in the solid state throughout the process. Even though screening among nonvolatile materials has narrowed the selection among few material families, one such family combining several desirable properties seems not to have been discovered yet. Most of the materials operating in this mode are limited by the small width of the nonstoichiometry swing δ. Between the two most studied systems, ferrites and ceria, the former have demonstrated higher levels of reduction at lower temperatures. The higher TR temperatures of ceria necessitate the use of special reactor materials. On the other hand, ceria is characterized by much faster gas splitting kinetics. The high oxygen yield of perovskites at moderately high temperatures perhaps can provide a future solution.

Additional concerns come into play when redox chemistry has to be coupled with solar energy. With respect to solar receiver–reactors, it is rather reasonable to expect that due to the relatively high temperatures to be encountered, directly irradiated receiver–reactors have an advantage over indirectly irradiated ones. Furthermore, from the various worldwide research approaches involving structured reactors with nonmoving parts, it can be inferred that the majority of them converge in the concept of fabricating, if possible, the entire receiver–redox reactor body—that being either honeycomb or foam—from the redox material itself. There are arguments in favor of each structure, and in this perspective, the kind of the porous structure is not so crucial for solar operation but has to be selected rather with ease-of-manufacture criteria.

The low efficiency of such reactors remains their main problem. The implementation of efficient heat recuperation is still a major issue of concern. Many alternative solar reactor design and relevant simulations have predicted enhanced heat recuperation and reactor efficiency, but when it comes to implementation, technically complicated concepts have not proved yet to have enough potential for eventual scale-up to commercially exploitable levels. For instance, to the best of the authors' knowledge, none of the proposed and tested structured reactors that include moving parts from a *hotter* region to a *colder* one to enhance such recuperation has been successfully scaled up so far; issues on mechanical vibrations of rotating parts and sealing at high temperatures became fatal during prolonged operation and have caused eventual damage of reactor parts in all cases. In a similar argumentation, techniques proposing the circulation and transport of hot solid particles between higher- and lower-temperature reactor regions with the aid of mechanical parts, sometimes in combination with vacuum conditions, have not been tested even at laboratory scale yet, and they are likely to face similar problems when implemented in large scale. The difficulties in scaling up such concepts should not be overlooked. In contrast, cascades employing nonmoving porous structures are in principle scalable to any commercial-relevant level [128] due to their inherent modularity characteristics. Perhaps a compromise between efficiency and simplicity in operation combined with scalability is necessary at this stage.

In conclusion, it can be said that even though thermodynamic and process efficiency studies have shown that such cycles meet the metric of potential viability compared to *benchmark* technologies like solar-aided electrolysis, further research efforts are needed for the achievement of these targets in practice. The two main research tasks are the improvement of solar interfaces and integrated heat recovery schemes on the one hand and solving the main material-related issues and providing the right functional materials at reasonable costs on the other hand. Solutions to those tasks will decisively help solar thermal processes to achieve the role of significantly contributors to carbon-free and sustainable hydrogen and syngas production on large scale, using only solar energy, carbon dioxide, and water as clean and abundant sources for those fuels.

ACKNOWLEDGMENTS

Dr. C. Agrafiotis would like to thank the European Commission for funding his staying at DLR during this work, within the project "FP7-PEOPLE-2011-IEF, 300194" under the Intra-European Fellowship, Marie Curie Actions of the 7th Framework Programme.

REFERENCES

1. Rostrup-Nielsen JR. Syngas in perspective. *Catalysis Today* 2002;71:243–247.
2. Rostrup-Nielsen JR, Sehested J, Nørskov JK. Hydrogen and synthesis gas by steam- and CO_2 reforming. *Advances in Catalysis* 2002;47:65–139.
3. Trainham JA, Newman J, Bonino CA, Hoertz PG, Akunuri N. Whither solar fuels? *Current Opinion in Chemical Engineering* 2012;1:204–210.
4. Steinfeld A, Palumbo R. Solar thermochemical process technology. In: Meyers RA (ed.) *Encyclopedia of Physical Science and Technology*. Academic Press, New York, NY, U.S.A., 2001, pp. 237–256.
5. Wegeng RS, Mankins JC. Space power systems: Producing transportation (and other chemical) fuels as an alternative to electricity generation. *Acta Astronautica* 2009;65:1261–1271.
6. Steinfeld A. Solar thermochemical production of hydrogen—A review. *Solar Energy* 2005;78:603–615.
7. Koumi Ngoh S, Njomo D. An overview of hydrogen gas production from solar energy. *Renewable and Sustainable Energy Reviews* 2012;16:6782–6792.
8. Meier A, Sattler C. Solar Fuels from Concentrated Sunlight. *Solar Paces Report* 2009, pp. 1–37.
9. Miller JE. Initial case for splitting carbon dioxide to carbon monoxide and oxygen. Sandia Report SAND2007-8012, 2007.
10. Chueh WC, Haile SM. Ceria as a thermochemical reaction medium for selectively generating syngas or methane from H_2O and CO_2. *ChemSusChem* 2009;2:735–739.
11. Graves C, Ebbesen SD, Mogensen M, Lackner KS. Sustainable hydrocarbon fuels by recycling CO_2 and H_2O with renewable or nuclear energy. *Renewable and Sustainable Energy Reviews* 2011;15:1–23.
12. Centi G, Perathoner S. Towards solar fuels from water and CO_2. *ChemSusChem* 2010;3:195–208.
13. Dry ME. High quality diesel via the Fischer–Tropsch process—A review. *Journal of Chemical Technology and Biotechnology* 2001;77:43–50.
14. Bartholomew CH, Farrauto RJ. *Fundamentals of Industrial Catalytic Processes*, 2nd edn. Hoboken, NJ: Wiley, 2006.
15. Huber GW, Iborra S, Corma A. Synthesis of transportation fuels from biomass: Chemistry, catalysts, and engineering. *Chemical Reviews* 2006;106:4044–4098.
16. Agrafiotis C, von Storch H, Roeb M, Sattler C. Solar thermal reforming of methane feedstocks for hydrogen and syngas production—A review. *Renewable and Sustainable Energy Reviews* 2014;29:656–682.
17. Trommer D, Noembrini F, Fasciana M, Rodriguez D, Morales A, Romero M et al. Hydrogen production by steam-gasification of petroleum coke using concentrated solar power—I. Thermodynamic and kinetic analyses. *International Journal of Hydrogen Energy* 2005;30:605–618.
18. Piatkowski N, Wieckert C, Weimer AW, Steinfeld A. Solar-driven gasification of carbonaceous feedstock—A review. *Energy and Environmental Science* 2011;4:73–82.
19. Kodama T, Gokon N. Thermochemical cycles for high-temperature solar hydrogen production. *Chemical Reviews* 2007;107:4048–4077.

20. Chueh WC, Falter C, Abbott M, Scipio D, Furler P, Haile SM et al. High-flux solar-driven thermochemical dissociation of CO_2 and H_2O using nonstoichiometric ceria. *Science* 2010;330:1797–1801.

21. Roeb M, Neises M, Monnerie N, Call F, Simon H, Sattler C et al. Materials-related aspects of thermochemical water and carbon dioxide splitting: A Review. *Materials* 2012;5:2015–2054.

22. Nakamura T. Hydrogen production from water utilizing solar heat at high temperatures. *Solar Energy* 1977;19:467–475.

23. Tofighi A. Sibieude F. Note of the condensation of the vapor phase above a melt of iron oxide in a solar parabolic concentrator. *International Journal of Hydrogen Energy* 1980;5:375–381.

24. Tofighi A, Sibieude F. Dissociation of magnetite in a solar furnace for hydrogen production. Tentative production evaluation of a 1000 kW concentrator from small scale (2 kW) experimental results. *International Journal of Hydrogen Energy* 1984;9:293–296.

25. Sibieude F, Tofighi A, Ambriz J. High temperature experiments with a solar furnace: The decomposition of Fe_3O_4, Mn_3O_4, CdO. *International Journal of Hydrogen Energy* 1981;7:79–88.

26. Tamaura Y, Steinfeld A, Kuhn P, Ehrensberger K. Production of solar hydrogen by a novel, 2-step, water-splitting thermochemical cycle. *Energy* 1995;20:325–330.

27. Meredig B, Wolverton C. First-principles thermodynamic framework for the evaluation of thermochemical H_2O- or CO_2-splitting materials. *Physical Review B* 2009;80:245119-1-8.

28. Abanades S, Charvin P, Flamant G, Neveu P. Screening of water-splitting thermochemical cycles potentially attractive for hydrogen production by concentrated solar energy. *Energy* 2006;31:2805–2822.

29. Bilgen E, Ducarroir M, Foex M, Sibieude F. Use of solar energy for direct and two-step water decomposition cycles. *International Journal of Hydrogen Energy* 1977;2:251–257.

30. Bilgen E, Bilgen C. Solar hydrogen production using two-step thermochemical cycles. *International Journal of Hydrogen Energy* 1982;7:637–644.

31. Lundberg M. Model calculations on some feasible two-step water splitting processes. *International Journal of Hydrogen Energy* 1993;18:369–376.

32. Allendorf MD. Two-step water splitting using mixed-metal ferrites: Thermodynamic analysis and characterization of synthesized materials. *Energy and Fuels* 2008;22:4115–4124.

33. Charvin P, Abanades S, Flamant G, Lemort F. Two-step water splitting thermochemical cycle based on iron oxide redox pair for solar hydrogen production. *Energy* 2007;32:1124–1133.

34. Perkins C, Weimer AW. Solar-thermal production of renewable hydrogen. *AIChE Journal* 2009;55:286–293.

35. Xiao L, Wu S-Y, Li Y-R. Advances in solar hydrogen production via two-step water-splitting thermochemical cycles based on metal redox reactions. *Renewable Energy* 2012;41:1–12.

36. Steinfeld A, Kuhn P, Reller A, Palumbo R, Murray J, Tamaura Y. Solar-processed metals as clean energy carriers and water-splitters. *International Journal of Hydrogen Energy* 1998;23:767–774.

37. Palumbo RLJ, Boutin O, Elorza Ricart E, Steinfeld A, Möller S, Weidenkaff A, Fletcher EA, Bielicki J. The production of Zn from ZnO in a high-temperature solar decomposition quench process—I. The scientific framework for the process. *Chemical Engineering Science* 1998;53:2503–2517.

38. Steinfeld A. Solar hydrogen production via a two-step water splitting thermochemical cycle based on Zn/ZnO redox reactions. *International Journal of Hydrogen Energy* 2002;27:611–619.

39. Gálvez ME, Loutzenhiser PG, Hischier I, Steinfeld A. CO_2 Splitting via two-step solar thermochemical cycles with Zn/ZnO and FeO/Fe_3O_4 redox reactions: Thermodynamic analysis. *Energy and Fuels* 2008;22:3544–3550.

40. Möller S, Palumbo R. Solar thermal decomposition kinetics of ZnO in the temperature range 1950–2400 K. *Chemical Engineering Science* 2001;56:4505–4515.

41. Steinfeld A. Thermochemical production of syngas using concentrated solar energy. In: *Annual Review of Heat Transfer*. Chen G, Prasad V, Jaluria Y, Karni J (eds.), Begell House Publishers, Danbury, CT, U.S.A., 2012, pp. 255–275.

42. Stamatiou A, Steinfeld A, Jovanovic ZR. On the effect of the presence of solid diluents during Zn oxidation by CO_2. *Industrial and Engineering Chemistry Research* 2013;52:1859–1869.

43. Charvin P, Abanades S, Lemont F, Flamant G. Experimental study of SnO_2/SnO/Sn thermochemical systems for solar production of hydrogen. *AIChE Journal* 2008;54:2759–2767.

44. Abanades S, Charvin P, Lemort F, Flamant G. Novel two-step SnO_2/SnO water-splitting cycle for solar thermochemical production of hydrogen. *International Journal of Hydrogen Energy* 2008;33:6021–6030.

45. Chambon M, Abanades S, Flamant G. Solar thermal reduction of ZnO and SnO_2 characterization of the recombination reaction with O_2. *Chemical Engineering Science* 2010;65:3671–3680.

46. Chambon M, Abanades S, Flamant G. Kinetic investigation of hydrogen generation from hydrolysis of SnO and Zn solar nanopowders. *International Journal of Hydrogen Energy* 2009;34:5326–5336.

47. Abanades S. CO_2 and H_2O reduction by solar thermochemical looping using SnO_2/SnO redox reactions: Thermogravimetric analysis. *International Journal of Hydrogen Energy* 2012;37:8223–8231.

48. Leveque G, Abanades S, Jumas J-C, Olivier-Fourcade J. Characterization of two-step tin-based redox system for thermochemical fuel production from solar-driven CO_2 and H_2O splitting cycle. *Industrial and Engineering Chemistry Research* 2014;53:5668–5677.

49. Miller JE, McDaniel AH, Allendorf MD. Considerations in the design of materials for solar-driven fuel production using metal-oxide thermochemical cycles. *Advanced Energy Materials* 2014;4:1300469.

50. Chueh WC, Haile SM. A thermochemical study of ceria: Exploiting an old material for new modes of energy conversion and CO_2 mitigation. *Philosophical Transactions of the Royal Society A: Mathematical, Physical and Engineering Sciences* 2010; 368:3269–3294.

51. Steinfeld A, Sanders S, Palumbo R. Design aspects of solar thermochemical engineering—A case study: Two-step water-splitting cycle using the Fe_3O_4/FeO redox system. *Solar Energy* 1999;65:43–53.

52. Tamaura Y, Ueda Y, Matsunami J, Hasegawa N, Nezuka M, Sano T et al. Solar hydrogen production by using ferrites. *Solar Energy* 1999;65:55–57.

53. Aoki H, Kaneko H, Hasegawa N, Ishihara H, Suzuki A, Tamaura Y. The $ZnFe_2O_4$/(ZnO + Fe_3O_4) system for H_2 production using concentrated solar energy. *Solid State Ionics* 2004;172:113–116.

54. Kaneko H, Kodama T, Gokon N, Tamaura Y, Lovegrove K, Luzzi A. Decomposition of Zn-ferrite for O_2 generation by concentrated solar radiation. *Solar Energy* 2004; 76:317–322.

55. Agrafiotis C, Zygogianni A, Pagkoura C, Kostoglou M, Konstandopoulos AG. Hydrogen production via solar-aided water splitting thermochemical cycles with nickel ferrite: Experiments and modeling. *AIChE Journal* 2013;59:1213–1225.

56. Gokon N, Takahashi S, Yamamoto H, Kodama T. Thermochemical two-step water-splitting reactor with internally circulating fluidized bed for thermal reduction of ferrite particles. *International Journal of Hydrogen Energy* 2008;33:2189–2199.

57. Gokon N, Mataga T, Kondo N, Kodama T. Thermochemical two-step water splitting by internally circulating fluidized bed of $NiFe_2O_4$ particles: Successive reaction of thermal-reduction and water-decomposition steps. *International Journal of Hydrogen Energy* 2011;36:4757–4767.

58. Fresno F, Fernandez-Saavedra R, Gomez-Mancebo MB, Vidal A, Sanchez M, Rucandio MI et al. Solar hydrogen production by two-step thermochemical cycles: Evaluation of the activity of commercial ferrites. *International Journal of Hydrogen Energy* 2009;34:2918–2924.

59. Miller JE, Allendorf MD, Diver RB, Evans LR, Siegel NP, Stuecker JN. Metal oxide composites and structures for ultra-high temperature solar thermochemical cycles. *Journal of Materials Science* 2008;43:4714–4728.

60. Tamaura Y, Kojima M, Sano T, Ueda Y, Hasegawa N, Tsuji M. Thermodynamic evaluation of water splitting by a cation-excessive (Ni, Mn) ferrite. *International Journal of Hydrogen Energy* 1998;23:1185–1191.

61. Inoue M, Hasegawa N, Uehara R, Gokon N, Kaneko H, Tamaura, Y. Solar hydrogen generation with $H_2O/ZnO/MnFe_2O_4$ system. *Solar Energy* 2004;76:309–315.

62. Miller JE, Evans LR, Stuecker JN, Allendorf MD, Siegel NP, Diver RB. Materials development for the CR5 solar thermochemical heat engine. *Proceedings of ISEC 2006 ASME International Solar Energy Conference*, Denver, CO, July 8–13, 2006, pp. 311–320.

63. Kojima M, Sano T, Wada Y, Yamamoto T, Tsuji M, Tamaura Y. Thermochemical decomposition of H_2O to H_2 on cation-excess ferrite. *Journal of Physics and Chemistry of Solids* 1996;57:1757–1773.

64. Fresno F, Yoshida T, Gokon N, Fernandez-Saavedra R, Kodama T. Comparative study of the activity of nickel ferrites for solar hydrogen production by two-step thermochemical cycles. *International Journal of Hydrogen Energy* 2010;35:8503–8510.

65. Kodama T, Nakamuro Y, Mizuno T. A two-step thermochemical water splitting by iron-oxide on stabilized zirconia. *Journal of Solar Energy Engineering* 2006;128:3–7.

66. Gokon N, Mizuno T, Nakamuro Y, Kodama T. Iron-containing yttria-stabilized zirconia system for two-step thermochemical water splitting. *Journal of Solar Energy Engineering* 2008;130:11016–11018.

67. Scheffe JR, McDaniel AH, Allendorf MD, Weimer AW. Kinetics and mechanism of solar-thermochemical H_2 production by oxidation of a cobalt ferrite–zirconia composite. *Energy and Environmental Science* 2013;6:963–673.

68. Ma LJ, Chen LS, Chen SY. Study on the cycle decomposition of CO_2 over $NiCr_{0.08}Fe_{1.92}O_4$ and the microstructure of products. *Materials Chemistry and Physics* 2007;105:122–126.

69. Ma LJ, Chen LS, Chen SY. Studies on redox H_2-CO_2 cycle on $CoCr_xFe_{2-x}O_4$. *Solid State Sciences* 2009;11:176–181.

70. Coker EN, Ambrosini A, Rodriguez MA, Miller JE. Ferrite-YSZ composites for solar thermochemical production of synthetic fuels: In operando characterization of CO_2 reduction. *Journal of Materials Chemistry* 2011;21:10767–10776.

71. Coker EN, Ohlhausen JA, Ambrosini A, Miller JE. Oxygen transport and isotopic exchange in iron oxide/YSZ thermochemically-active materials via splitting of $C(^{18}O)_2$ at high temperature studied by thermogravimetric analysis and secondary ion mass spectrometry. *Journal of Materials Chemistry* 2012;22:6726–6732.

72. Coker EN, Rodriguez MA, Ambrosini A, Miller JE. Thermochemical cycle for H_2 and CO production: Some fundamental aspects. *Proceedings of 15th Solar PACES International Symposium*, Berlin, Germany, 2009.

73. Stamatiou A, Loutzenhiser PG, Steinfeld A. Solar syngas production via H_2O/CO_2-splitting thermochemical cycles with Zn/ZnO and FeO/Fe_3O_4 redox reactions. *Chemistry of Materials* 2010;22:851–859.

74. Abanades S, Villafan-Vidales HI. CO_2 and H_2O conversion to solar fuels via two-step solar thermochemical looping using iron oxide redox pair. *Chemical Engineering Journal* 2011;175:368–375.

75. S. Lorentzou, G. Karagiannakis, C. Pagkoura, Zygogianni. A, Konstandopoulos AG. Thermochemical CO_2 and CO_2/H_2O splitting over $NiFe_2O_4$ for solar fuels synthesis. *Proceedings of 19th SolarPACES Conference*, Las Vegas, NV, 2013.

76. Agrafiotis CC, Pagkoura C, Zygogianni A, Karagiannakis G, Kostoglou M, Konstandopoulos AG. Hydrogen production via solar-aided water splitting thermochemical cycles: Combustion synthesis and preliminary evaluation of spinel redox-pair materials. *International Journal of Hydrogen Energy* 2012;37:8964–8980.

77. Tsuji M, Togawa T, Wada Y, Sano T, Tamaura Y. Kinetic study of the formation of cation-excess magnetite. *Journal of Chemical Society Faraday Transactions* 1995;91:1533–1538.

78. Rosmaninho M, Herreras S, Lago R, Araujo M, Navarro R, Fierro J. Effect of the partial substitution of Fe by Ni on the structure and activity of nanocrystalline $Ni_xFe_{3-x}O_4$ ferrites for hydrogen production by two-step water-splitting. *Nanoscience and Nanotechnology Letters* 2011;3:705–716.

79. Trovarelli A, Fornasiero P. *Catalysis by Ceria and Related Materials*. World Scientific, Publishing Co. Pte. Ltd., Singapore, 2013.

80. Abanades S, Flamant G. Thermochemical hydrogen production from a two-step solar-driven water splitting cycle based on cerium oxides. *Solar Energy* 2006;80:1611–1623.

81. Scheffe JR, Steinfeld A. Thermodynamic analysis of cerium-based oxides for solar thermochemical fuel production. *Energy and Fuels* 2012;26:1928–1936.

82. Miller JE, Evans LR, Siegel NP, Diver RB, Gelbard F, Ambrosini A et al. Summary report: Direct approaches for recycling carbon dioxide into synthetic fuel. Technical Report, Sandia National Laboratories, Albuquerque, NM, 2009.

83. Abanades S, Legal A, Cordier A, Peraudeau G, Flamant G, Julbe A. Investigation of reactive cerium-based oxides for H_2 production by thermochemical two-step water-splitting. *Journal of Materials Science* 2010;45:4163–4173.

84. Lee C, Meng Q-L, Kaneko H, Tamaura Y. Solar hydrogen productivity of ceria–scandia solid solution using two-step water-splitting cycle. *Journal of Solar Energy Engineering* 2013;135:011002.

85. Meng Q-L, Lee C, Shigeta S, Kaneko H, Tamaura Y. Solar hydrogen production using $Ce_{1-x}Li_xO_{2-\delta}$ solid solutions via a thermochemical, two-step water-splitting cycle. *Journal of Solid State Chemistry* 2012;194:343–351.

86. Meng Q-L, Lee C, Ishihara T, Kaneko H, Tamaura Y. Reactivity of CeO_2-based ceramics for solar hydrogen production via a two-step water-splitting cycle with concentrated solar energy. *International Journal of Hydrogen Energy* 2011;36:13435–13441.

87. Le Gal A, Abanades S. Catalytic investigation of ceria-zirconia solid solutions for solar hydrogen production. *International Journal of Hydrogen Energy* 2011;36:4739–4748.

88. Le Gal A, Abanades S. Dopant incorporation in ceria for enhanced water-splitting activity during solar thermochemical hydrogen generation. *The Journal of Physical Chemistry C* 2012;116:13516–13523.

89. Scheffe JR, Li J, Weimer AW. A spinel ferrite/hercynite water-splitting redox cycle. *International Journal of Hydrogen Energy* 2010;35:3333–3340.

90. Arifin D, Aston VJ, Liang X, McDaniel AH, Weimer AW. $CoFe_2O_4$ on a porous Al_2O_3 nanostructure for solar thermochemical CO_2 splitting. *Energy and Environmental Science* 2012;5:9438–9443.

91. Scheffe JR, Weibel D, Steinfeld A. Lanthanum–strontium–manganese perovskites as redox materials for solar thermochemical splitting of H_2O and CO_2. *Energy and Fuels* 2013;27:4250–4257.

92. McDaniel AH, Miller EC, Arifin D, Ambrosini A, Coker E, O'Hayre R et al. Sr-and Mn-doped $LaAlO_{3-\delta}$ for solar thermochemical H_2 and CO production. *Energy and Environmental Science* 2013;6:2424–2428.

93. Romero M, Steinfeld A. Concentrating solar thermal power and thermochemical fuels. *Energy and Environmental Science* 2012;5:9234–9245.

94. Heck RM, Farrauto RJ, Gulati ST. *Catalytic Air Pollution Control*. Van Nostrand Rheinhold, New York, NY, U.S.A., 1995.

95. Cybulski A, Moulijn JA. *Structured Catalysts and Reactors*. CRC Press, Boca Raton, FL, U.S.A., 2005.

96. Twigg MV, Richardson JT. Fundamentals and applications of structured ceramic foam catalysts. *Industrial and Engineering Chemistry Research* 2007;46:4166–4177.

97. Roeb M, Sattler C. Isothermal water splitting. *Science* 2013;341:470–471.

98. Lapp J, Davidson J, Lipiński W. Efficiency of two-step solar thermochemical non-stoichiometric redox cycles with heat recovery. *Energy* 2012;37:591–600.

99. Siegel NP, Miller JE, Ermanoski I, Diver RB, Stechel EB. Factors affecting the efficiency of solar driven metal oxide thermochemical cycles. *Industrial and Engineering Chemistry Research* 2013;52:3276–3286.

100. Ávila-Marín AL. Volumetric receivers in solar thermal power plants with central receiver system technology: A review. *Solar Energy* 2011;85:891–910.

101. Kodama T, Enomoto S, Hatamachi T, Gokon N. Application of an internally circulating fluidized bed for windowed solar chemical reactor with direct irradiation of reacting particles. *Journal of Solar Energy Engineering* 2008;130:014504.

102. Gokon N, Takahashi S, Yamamoto H, Kodama T. New solar water-splitting reactor with ferrite particles in an internally circulating fluidized bed. *Journal of Solar Energy Engineering* 2009;131:011007–011009.

103. Ermanoski I, Siegel NP, Stechel EB. A new reactor concept for efficient solar-thermo-chemical fuel production. *Journal of Solar Energy Engineering* 2013;135:031002.

104. Levy M, Rosin H, Levitan R. Chemical reactions in a solar furnace by direct solar irra-diation of the catalyst. *Journal of Solar Energy Engineering* 1989;111:96–97.

105. Levy M, Rubin R, Rosin H, Levitan R. Methane reforming by direct solar irradiation of the catalyst. *Energy* 1992;17:749–756.

106. Agrafiotis C, Roeb M, Konstandopoulos AG, Nalbandian L. Solar water splitting for hydrogen production with monolithic reactors. *Solar Energy* 2005;79:409–421.

107. Roeb M, Monnerie N, Schmitz M, Sattler C, Konstandopoulos A, Agrafiotis C et al. Thermo-chemical production of hydrogen from water by metal oxides fixed on ceramic substrates. *Proceedings of the 16th World Hydrogen Energy Conference*, Lyon, France, 2006.

108. Roeb M, Sattler C, Klüser R, Monnerie N, De Oliveira L, Konstandopoulos A et al. Solar hydrogen production by a two-step cycle based on mixed iron oxides. *Journal of Solar Energy Engineering* 2006;128:125–133.

109. Roeb M, Säck JP, Rietbrock P, Prahl C, Schreiber H, Neises M et al. Test operation of a 100 kW pilot plant for solar hydrogen production from water on a solar tower. *Solar Energy* 2011;85:634–644.

110. Roeb M, Neises M, Säck J-P, Rietbrock P, Monnerie N, Dersch J et al. Operational strat-egy of a two-step thermochemical process for solar hydrogen production. *International Journal of Hydrogen Energy* 2009;34:4537–4545.

111. Neises M, Goehring F, Roeb M, Sattler C, Pitz-Paal R. Simulation of a solar receiver-reactor for hydrogen production. *ASME Conference Proceedings*, San Francisco, CA, July 19–23, 2009, pp. 295–304.

112. Houaijia A, Sattler C, Roeb M, Lange M, Breuer S, Säck JP. Analysis and improvement of a high-efficiency solar cavity reactor design for a two-step thermochemical cycle for solar hydrogen production from water. *Solar Energy* 2013;97:26–38.

113. Diver RB, Miller JE, Allendorf MD, Siegel NP, Hogan RE. Solar thermochemical water-splitting ferrite-cycle heat engines. *Journal of Solar Energy Engineering* 2008;130:041001–041008.

114. Walker LS, Miller JE, Hilmas GE, Evans LR, Corral EL. Coextrusion of zirconia–iron oxide honeycomb substrates for solar-based thermochemical generation of carbon monoxide for renewable fuels. *Energy and Fuels* 2011;26:712–721.

115. Buck R, Muir JF, Hogan RE. Carbon dioxide reforming of methane in a solar volumetric receiver/reactor: The CAESAR project. *Solar Energy Materials* 1991;24:449–463.

116. Gokon N, Hasegawa T, Takahashi S, Kodama T. Thermochemical two-step water-splitting for hydrogen production using Fe-YSZ particles and a ceramic foam device. *Energy* 2008;33:1407–1416.

117. Gokon N, Murayama H, Nagasaki A, Kodama T. Thermochemical two-step water splitting cycles by monoclinic ZrO_2-supported $NiFe_2O_4$ and Fe_3O_4 powders and ceramic foam devices. *Solar Energy* 2009;83:527–537.

118. Gokon N, Kodama T, Imaizumi N, Umeda J, Seo T. Ferrite/zirconia-coated foam device prepared by spin coating for solar demonstration of thermochemical water-splitting. *International Journal of Hydrogen Energy* 2011;36:2014–2028.

119. Furler P, Scheffe JR, Steinfeld A. Syngas production by simultaneous splitting of H_2O and CO_2 via ceria redox reactions in a high-temperature solar reactor. *Energy and Environmental Science* 2012;5:6098–6103.

120. Furler P, Scheffe J, Gorbar M, Moes L, Vogt U, Steinfeld A. Solar thermochemical CO_2 splitting utilizing a reticulated porous ceria redox system. *Energy and Fuels* 2012;26:7051–7059.

121. Kaneko H, Fuse A, Miura T, Ishihara H, Tamaura Y. Two-step water splitting with concentrated solar heat using rotary-type reactor. *Proceedings of 13th Solar PACES International Symposium*, Seville, Spain, 2006.

122. Kaneko H, Miura T, Fuse A, Ishihara H, Taku S, Fukuzumi H et al. Rotary-type solar reactor for solar hydrogen production with two-step water splitting process. *Energy and Fuels* 2007;21:2287–2293.

123. Kaneko H, Lee C-I, Ishihara T, Ishikawa Y, Hosogoe K, Tamaura Y. Solar H_2 production with rotary-type solar reactor in international collaborative development between Tokyo Tech (Japan) and CSIRO (Australia). *Proceedings of 15th Solar PACES International Symposium*, Berlin, Germany, 2009.

124. Diver RB, Miller JE, Allendorf MD, Siegel NP, Hogan RE. Solar thermochemical water-splitting ferrite-cycle heat engines. *Proceedings of ISEC 2006 ASME International Solar Energy Conference*, Denver, CO, 2006.

125. Kim J, Miller JE, Maravelias CT, Stechel EB. Comparative analysis of environmental impact of S2P (Sunshine to Petrol) system for transportation fuel production. *Applied Energy* 2013;111:1089–1098.

126. Miller JE, Allendorf MA, Ambrosini A, Chen K, Coker EN, Dedrick DE et al. Final report, Reimagining liquid transportation fuels: Sunshine to petrol. Sandia report, Sandia National Laboratories, Albuquerque, NM, 2012.

127. Miller JE, Allendorf MA, Ambrosini A, Coker EN, Diver RB, Ermanoski I et al. Development and assessment of solar-thermal activated fuel production: Phase 1 Summary. Sandia Report, Sandia National Laboratories, Albuquerque, NM, 2012.

128. Graf D, Monnerie N, Roeb M, Schmitz M, Satter C. Economic comparison of solar hydrogen generation by means of thermochemical cycles and electrolysis. *International Journal of Hydrogen Energy* 2008;33:4511–4519.

16 Membrane-Assisted CPO
A New Sustainable Syngas Process

G. Iaquaniello, A. Salladini, E. Palo,
Salvatore Abate, and Gabriele Centi

CONTENTS

16.1 INTRODUCTION

Syngas production is one of the main base processes for chemical industry, playing a major role in quite a few different processes ranging from ammonia, methanol, aldehydes, and gas to liquid (GTL) applications [1–3]. All these processes are energy intensive and require different H_2/CO ratios. Decreasing the energy and raw material intensity in these processes is a necessary direction to move to a more sustainable and competitive chemical production, which well integrates also with the general objective of enabling the use for both energy and chemical utilizations of unexploited resources, particularly remote strained natural gas or shale gas that cannot be connected to a pipeline network.

The European project "Innovative Catalytic Technologies & Materials for Next Gas to Liquid Processes (NEXT-GTL)" aimed to explore different options and alternative routes for the utilization of these remote strained natural gas or shale gas [3]. One of these lines regarded the development of novel sustainable routes for the production of syngas to be then converted by the GTL processes. The novel process scheme investigated regarded the possibility to improve the performances by integrating the use of membranes into a process scheme based on the use of catalytic partial oxidation (CPO) scheme.

As commented later, CPO was mainly developed for H_2 production. The use of CPO, rather than conventional scheme, shows various advantages (as commented in the following), particularly for small-/medium-scale productions, a specific issue for using remote strained natural gas or shale gas [3].

We were interested to investigate the use of this CPO process for syngas rather than for H_2 production and especially analyze the possibility of further improvement, especially in terms of reduction of energy consumption, by introducing the use of membranes in the process scheme. The use of membranes for H_2, CO_2, and O_2 separation was evaluated, although the use of the first type of membranes was proven to be more successful and will be especially discussed here.

This chapter reports some aspects of the key activity performed by KT in the frame of the cited NEXT-GTL project on the development and testing of a semi-industrial scale unit for CPO reaction assisted by H_2-separation membrane modules and discusses the state of the art on membrane-assisted CPO process.

16.1.1   CATALYTIC PARTIAL OXIDATION AT SHORT CONTACT TIME

The CPO of methane is an alternative technology (to steam [SR] and autothermal reforming [ATR]) that has gained interest in recent years [4], particularly the short contact time catalytic partial oxidation (SCT-CPO). CPO was indicated as a promising new solution compared with the most consolidated technologies (SR and ATR) due to the fact that the process is carried out in very small reactors having a very high flexibility toward reactant flow and feedstock composition variations, by performing conversion and selectivity values higher than those predicted by the thermodynamic equilibrium at the reactor exit temperature [5].

The SCT catalytic reactions are realized by flowing a premixed reactant stream inside a small reactor containing an incandescent catalytic bed. The chemical processes occur in a few milliseconds inside a high temperature and thin (<1 μm) solid–gas interphase zone surrounding the catalyst particles and do not propagate into the gas phase. These conditions favor the formation of primary reaction products and allow the utilization of several HC feedstock (also containing S and/or aromatic compounds). These conditions also allow a wide flexibility with respect to variations of compositions and flows of reactant mixtures. SCT-CPO is particularly suited for coupling to GTL process.

Eni has been long active in the development of SCT-CPO technologies for producing hydrogen/synthesis gas [5–8]. Three types of technologies were developed: (1) air-blown SCT-CPO, (2) enriched air-/oxygen-blown SCT-CPO of gaseous hydrocarbons and/or light compounds with biological origin, and (3) enriched air-/oxygen-blown SCT-CPO of liquid hydrocarbons. While the air-blown partial oxidation of gaseous hydrocarbons provided the feedstock for the ammonia synthesis, the O_2-blown partial oxidation of hydrocarbons provided H_2 or synthesis gas.

In 2001, Snamprogetti (the engineering company of the Eni group) and Haldor Topsoe A/S successfully operated the first pilot plant in Houston, TX, and in 2005, Eni Tecnologie realized and operated a second multipurpose plant in Milazzo, Sicily. Recently, the license rights on a nonexclusive basis for the commercialization of SCT-CPO-based processes for H_2/synthesis gas production from light hydrocarbons with production capacity lower than 5000 N-m³/h of H_2 or 7500 N-m³/h of syngas were assigned to two external companies. These activities evaluate the utilization of SCT-CPO for matching the variable hydrogen demand in several contexts of oil refining operation.

The SCT-CPO process is commercialized (actually by Rosetti under Eni license) and shows advantages, particularly for (1) debottlenecking of existing steam methane reformers in hydrogen-consuming industries, (2) modularized and skid-mounted plants for small and medium industrial hydrogen consumers, (3) hydrogen and syngas from fossil feedstocks for clean energy applications with CO_2 recovery, and (4) *green hydrogen* from renewable feedstocks (CO_2 neutral). There are thus many attractive features, particularly in light to moving to a more sustainable chemical production. Figure 16.1 shows

(a)

(b)

(c)

FIGURE 16.1   (a) SCT-CPO process module (street transportable). (b) Skid-mounted SCT-CPO unit (for medium–small hydrogen consumer). (c) SCT-CPO pilot plant for industrial hydrogen production (99.999%). (From Rosetti company website, www.rosetti.it/fileadmin/docs/Tecnoloy/ROSETTI_hydrogen-A4.pdf.)

some picture of a module (street transportable) and SCT-CPO pilot plant for industrial hydrogen production (99.999%).

Competitive advantages of SCT-CPO technology include

- Small-size reaction systems (two orders of magnitude compared to traditional reforming)
- Simplified plant complexity
- Feedstock flexibility: from natural gas, light hydrocarbons with the possibility of treating bio-derived organic fluids (without major equipment modifications)
- Oxidant flexibility: possibility of using air, enriched air, and oxygen (to be evaluated case by case)
- Operating flexibility: simple and easy to operate, short startup/shutdown
- Design flexibility: enhanced energy recovery with integrated steam production and quench systems available
- No flue gas emissions: easy CO_2 recovery from compressed process gas
- Modular construction with reduced plant footprint and short installation time

16.1.2 CATALYSTS FOR CPO

Some recent reviews have discussed the performance of different noble and non-noble transition metals for partial oxidation of methane. Enger et al. [9] have discussed in detail the (1) thermodynamics and heat and mass transport aspects, (2) reaction mechanism, and (3) characteristics of CPO catalysts. The comparison of elementary reaction steps on Pt and Rh illustrates that a key factor to produce hydrogen as a primary product is a high activation energy barrier to the formation of OH. Another essential property for the formation of H_2 and CO as primary products is a low surface coverage of intermediates, such that the probability of O–H, OH–H, and CO–O interactions is reduced. York et al. [10] have also reviewed methane partial oxidation for syngas production

Although Co- and Ni-based catalysts have good performances, particularly for SCT-CPO, it is preferable to use noble-metal-based catalysts, particularly Rh in low amounts, but the support has a critical role, because it must be thermally stable at high temperatures (about 1100°C) and especially must avoid the sintering of the noble-metal nanoparticles when subjected to these high temperatures.

Tanaka et al. [11] have recently shown how the addition of an optimum amount of CeO_2 (Ce/Rh = 4) to 1.0% Rh/MgO promoted the CPO of methane to synthesis gas. The addition of CeO_2 also suppressed the temperature increase at the catalyst bed inlet during the CPO of methane. The Rh–CeO_2/MgO (Ce/Rh = 4) has the ability to maintain a more reduced state in the CPO of methane, and this may be related to the high ability to activate methane by the synergy of Rh metal surface and partially covering Ce species. The catalyst bed of the CPO of methane consists of the initial oxidation zone and the subsequent steam reforming zone. The catalyst with higher resistance to the oxidation of Rh gives larger zone for the steam reforming, and it also enables the overlap of the exothermic oxidation zone with the endothermic steam reforming zone.

The same group [12] have also investigated the modifying effects of Fe, Co, or Ni over 1 wt% Rh/MgO on the catalytic performance in the CPO of methane. The optimum amount of Co (Co/Rh = 1) or Ni (Ni/Rh = 1.5) addition to Rh/MgO enhances the CH_4 conversion and selectivities to H_2 and CO, while the addition of Fe to Rh/MgO decreased the properties.

Cimino and Lisi [13] have instead investigated the effect of sulfur poisoning of Rh active sites during the CPO of methane. The effect of the type of alumina support and the partial substitution of Rh with either Pt or Pd in order to enhance the sulfur tolerance of the catalyst were also investigated.

Enger et al. [9] have instead reported the potential of inactive Ni- and Ni–Co aluminate spinels prepared at high temperatures (1393 K) as precursors for the design of catalysts for CPO and steam methane reforming (SMR). By exposing the aluminate spinel to hydrogen at 1073 K for 2 h, the inactive spinel was restructured to an active catalyst with excellent initial stability (20–40 h). The hydrogen treatment enabled the growth of supported nanosized (15–25 nm) metal particles. Methane conversion during CPO over unreduced $NiAl_2O_4$ did not exceed the empty reactor conversion (6% at 1073 K). Thus, the reduction with H_2 was critical for obtaining an active phase capable of activating methane and catalyzing the CPO and SMR reactions. Once activated by high temperature reduction, close to equilibrium yields were obtained over the in situ reduced catalysts.

16.1.3 CPO TECHNOLOGY

Both air and oxygen may be used as oxidant in a CPO reactor. Experiments with CPO and air as oxidant have been conducted at the Topsøe pilot plant in Houston, Texas. Methane conversion levels close to the equilibrium of the methane steam reforming reaction were observed.

Even if some small–medium case commercial applications already exist, CPO technology cannot be still considered a full commercial technology. One of the issues of this route is the highly flammable mixture that may ignite at temperatures above 250°C. Therefore, the reactants may not be preheated at high temperatures, resulting in high natural gas and oxygen consumption since part of the feed has to be burned to generate the heat required to achieve the reaction temperature. This increases both the oxygen consumption and the natural gas consumption as shown in Table 16.1.

A higher oxygen consumption increases the air separation unit investment. The increased oxygen consumption and the potential safety problems related to the premixing of oxygen and hydrocarbon feed are some of the issues that limit the commercialization of CPO for large-scale production of synthesis gas, although there are other advantages in terms of compactness (thus better retrofitting) and possibility of easy recovery of CO_2.

TABLE 16.1

Relative Oxygen and Natural Gas Consumption for CPO- and ATR-Based GTL Front Ends for the Production of Hydrogen and Carbon Monoxide

Reactor Technology	ATR	CPO	ATR	CPO
S/C ratio	0.6	0.6	0.3	0.3
HC feed temp. reactor inlet (°C)	650	200	650	200
O_2 consumption (relative)	100	121	97	114
NG consumption (relative)	100	109	102	109

Source: Adapted from Zeppier, M. et al., *Appl. Catal. A Gen.*, 387, 147, 2010.

Notes: An adiabatic prereformer is located upstream of the ATR. CO_2 is introduced before the partial oxidation reactor at 200°C in an amount to give $H_2/CO = 2.0$. Pressure: 25 bar abs. Oxygen temperature: 200°C. Exit temperature: 1050°C.

16.1.4 H_2 Dense Membranes

The incorporation of H_2 dense membrane, based on a thin Pd-based dense film over a ceramic or metallic support, in H_2 production process (particularly, based on CPO) is economically advantageous, even if it still remains a problem on membrane long-term stability. KT is actively developing this route [15–18]. The plant scheme is shown in Figure 16.2 and consists of two-step reformer and membrane modules (RMM).

An RMM test plant with a hydrogen capacity of 20 N-m³/h has been designed and built to investigate the performance of this innovative architecture at an industrial level [17]. A major benefit of the proposed RMM configuration is the shift in the chemical equilibrium of the steam reforming reactions by removing

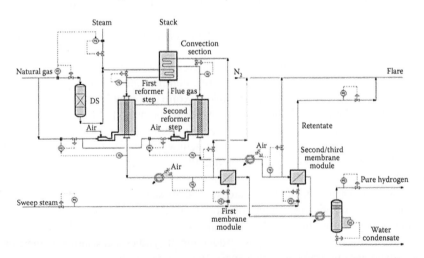

FIGURE 16.2 Process scheme of KT of prototypal plant for steam reforming of natural gas in an RMM plant. (Reproduced from De Falco, M. et al., *J. Membr. Sci.*, 368, 264, 2011. With permission.)

the hydrogen produced at high temperatures, thanks to the integration of highly selective Pd-based membranes. In this way, the process can operate at a lower thermal level (below 650°C in comparison to the 850°C–950°C temperature needed in traditional plants).

Four types of Pd-based membranes, three already installed and one yet to be assembled, with an active area in the range 0.12–0.4 m², were tested in order to compare performance in terms of permeated hydrogen flux. Moreover, a noble-metal catalyst supported on a SiC foam catalyst is placed inside the reactor in order to improve thermal transport inside the reforming tubes. About 1000 operating hours and more than 70 heating and cooling cycles were performed (Figure 16.3). The average H_2 permeability for membranes tested are calculated and compared, and permeability expressions are reported. An overall feed conversion of 57.3% was achieved at 600°C, about 26% higher than what can be achieved in a conventional reformer at the same temperature, thanks to the integration of selective membranes.

The 20 N-m³/h RMM installation makes it possible to completely understand the potential of selective membrane application in industrial high-temperature chemical processes and represents a unique installation worldwide.

The difference between conventional technology production costs and this hybrid technology production costs is shown in Figure 16.4 reporting H_2 production cost vs. yearly depreciation rate [18]. Under the considered conditions and economic scenario, the proposed hybrid scheme is able to reduce production costs by at least 13% compared to conventional steam reforming technology + cogeneration unit and by almost 30% compared to conventional technology. A wider acceptance and use of membranes will increase the already interesting gap and further lower the production costs of hydrogen.

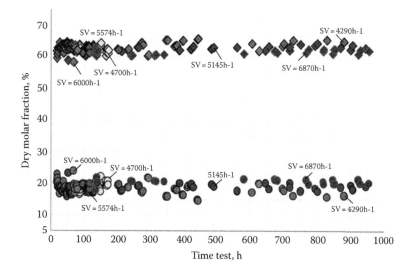

FIGURE 16.3 Methane and hydrogen content in the dry product along 960 h operation in pilot plant for steam reforming of natural gas in an RMM plant. (Adapted from De Falco, M. et al., *J. Membr. Sci.*, 368, 264, 2011.)

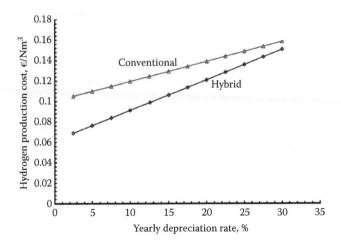

FIGURE 16.4 Production cost vs. yearly depreciation rate for a new membrane reforming MRR concept to convert natural gas to hydrogen and electricity, in comparison with conventional tubular steam reforming technology. (Adapted from Iaquaniello, G. et al., *Int. J. Hydrogen Energ.*, 33, 6595, 2008.)

16.2 CHARACTERISTICS OF THE SEMI-INDUSTRIAL PLANT FOR MEMBRANE-ASSISTED CPO FOR SYNGAS PRODUCTION

The potential economic advantages of CPO against SR in several different scenarios and for different capacity were assessed [19,20]. The possibility, for instance, to exploit the CPO stage in a process scheme routed to bulk chemical production is extensively reported [20]. Of course, the benefits derived by the use of such concept are strongly dependent on the gain for energy efficiency and oxygen consumption, since when CPO reactor is fed with pure oxygen, this cost could be limiting for the overall syngas unit cost production.

To this regard, over the last years, Capoferri et al. [21] proposed a novel scheme with a series of CPO reactors each working at lower temperature (in the 760°C–800°C temperature range), compared to conventional CPO technology, the temperature being a critical parameter with a major impact on the overall energy efficiency of the proposed scheme. Furthermore, each reactor was followed by a membrane for hydrogen separation by syngas mixture. This type of arrangement, enabling for lower reaction temperature, meets the demand for less stressing operating conditions for CPO catalyst, thus increasing its durability on time on stream tests. Such concept was already applied by De Falco et al. in steam reforming reactions [22].

Further improvement of this CPO-based scheme was relevant to the replacement of the first steps of CPO reaction with a preformer step (Figure 16.5), thus enabling for an overall reduction of 10% of the variable operating costs with respect to conventional CPO-technology-based plants [23]. The experimental campaign was carried out at the KT facility previously built [17] and rearranged for this purpose.

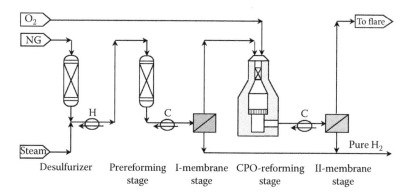

FIGURE 16.5 Simplified membrane-assisted CPO reforming for syngas and H_2 production; C = cooler, H = heating.

The plant has been designed to be operated both in a standalone configuration and in an integrated one. In the former approach, CPO reactor is directly fed by oxygen and a mixture of steam and natural gas under a controlled steam-to-carbon ratio, while in the latter, it is coupled to a low-temperature reformer and fed by retentate coming from membrane separation. The aim of the former approach is to test the potentiality of CPO reactor and its good performance. Designed capacity in terms of pure hydrogen is 20 N-m³/h.

Semi-industrial plant is served by natural gas derived from town grid and available at a pressure of 12 barg. Fraction used as process gas is desulfurized on activated carbon bed prior to being mixed with superheated steam and routed to steam reforming or directly to CPO section. Boiler feed water is produced by a demi-water package based on two stages of reverse osmosis coupled to a finishing step of ionic–anionic exchange bed (final conductivity = 0.2–0.4 μS/cm), while process steam is produced in a dedicated hot oil boiler.

A mixture of steam and natural gas is preheated in the convective section of reformer and fed to reforming reactor working at an exit temperature of 550°C–600°C. Syngas is slightly cooled to approximately 450°C–480°C before entering the first membrane module where a permeate and a retentate stream are produced. The latter, depleted of hydrogen, is routed to the CPO reactor together with a stream of pure oxygen from cylinders. Hot syngas exiting the CPO reactor is cooled down to 450°C–480°C in a gas–gas exchanger and sent to the second membrane separator for a further hydrogen recovery step. The last retentate in the pilot assembly is routed to the flare. The overall pilot plant covers an area of about 1000 m²; a view of constructed pilot plant is shown in Figure 16.6.

Sampling system allows monitoring composition around steam reformer, membrane separation, and CPO stages. The analyzing system is based on non-dispersive infra-red (NDIR) analyzers (Uras 14, ABB) for real-time measurements of CH_4, CO, and CO_2, while the concentration of H_2 is performed with a thermoconductivity analyzer (Caldos 17, ABB). The composition of permeate streams are performed by a gas chromatograph that allows to detect impurities of CH_4, CO, and CO_2 in order of ppm. Residual oxygen at the outlet of CPO reactor is monitored by a field-mounted oxygen transmitter based on a fuel cell sensor (OxyTrans) measuring O_2 concentration in the range 0–10,000 ppm.

FIGURE 16.6   View of the semi-industrial plant for membrane-assisted CPO located in Chieti (Italy) at the facility of KT.

16.2.1   CPO Reactor

Catalytic tests were carried out on a CPO reactor whose main dimensions are reported in Table 16.2, while a schematic configuration together with the on-site mounted view are reported in Figure 16.7.

Pure oxygen is fed from the upper section of reactor at room temperature and mixed with a preheated stream of steam and natural gas. All components in contact with pure oxygen were properly cleaned for oxygen service. Mixing is performed directly inside the reactor through an uncoated SiC foam (20 ppi), wrapped with high-temperature ceramic fiber in order to reduce bypass of reactant gases between monolith and reactor wall. Oxygen distributor is realized by three tubes inserted in the mixing foam in order to assure a premixing within the bulk of foam. Static mixer plays also the role of radiant thermal shield.

As reported by Livio et al. [24], the introduction of a front heat shield in an SCT-CPO reactor may help to reduce radiation from the glowing front face of the catalytic bed thus allowing to moderate temperatures of the hot entry region. The upward heat recirculation due to radiation creates an excess enthalpy environment

TABLE 16.2

CPO Reactor Characteristics

Main CPO Reactor Characteristics	Dimensions
Overall reactor length, L	1.5 m
Reactor diameter, ND	4 in.
Reactor material	INCOLOY 800HT
Catalytic bed diameter, D_c	45 mm
Catalytic bed length, L_c	360–500 mm

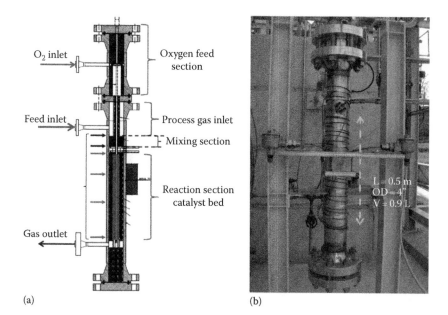

O₂ inlet

Oxygen feed
section

Feed inlet

Process gas inlet

Mixing section

Reaction section
catalyst bed

Gas outlet

L = 0.5 m
OD = 4"
V = 0.9 L

(a)   (b)

FIGURE 16.7   CPO reactor (a) design and (b) construction.

at the catalyst entrance, which increases reactants temperature by this way extending flammability limits of mixture. A proper design of mixing section is thus
required in order to restrict reactions at the catalyst surface and prevent ignition of
the reactant mixtures.

To check where ignition point occurs, the reactor is equipped with an axial mul-
tipoint thermocouple monitoring temperature profile along the static mixer as well
as along the catalyst bed. The latter plays a major role for a stable operation of SCT-
CPO reactor in order to properly manage hot spot temperatures and reduce sintering
phenomena of metal particles in the oxidation zone.

To assure autothermal conditions and prevent reactor wall from overheating, the
catalytic bed is internally insulated with high-density ceramic fiber. Moreover, external metal temperature is monitored by a continuous spiral thermocouple in order to
assure safety operation and prevent hot spot due to leaks in the insulation layer. To
further reduce thermal loss, the reactor is also externally insulated with high-density
ceramic fiber (120 mm). Two couples of electrical cartridge heaters are installed at
the inlet of the catalyst bed and switched on only during heating cycle to adjust the
inlet gas temperature. The reactor was properly designed to accommodate the catalyst both in the form of pellets and monoliths.

A programmable logic controller (PLC) is used to manage safety interlocks
and control system of the overall pilot plant. Regarding CPO section, emergency
shutdown system (ESD) activates safety scenarios in case of too much temperature in the mixing and/or along the catalytic bed, as well as on the reactor wall.
Distributed control system (DCS) allows controlling and registering main process
variables such as temperature, pressure, and flow rates. Figure 16.8 reports DCS
control graphic page dedicated to CPO section.

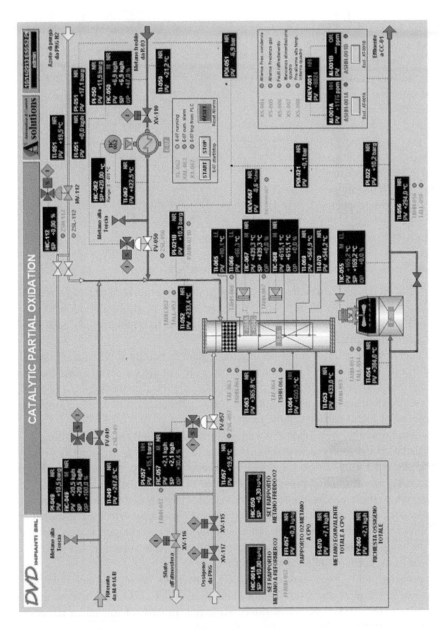

FIGURE 16.8   DCS graphic page for CPO section.

16.3 PERFORMANCES OF THE SEMI-INDUSTRIAL PLANT

We summarize here some of the key performances obtained in the semi-industrial plant for membrane-assisted CPO for syngas production described in the previous section. More extensive results will be reported elsewhere [25]. Separation module installed on the first stage realizes a total area of 0.4 m² through 13 ceramic supported membranes having a palladium thickness of 2.5 μm. Permeance in order of 30 N-m³/hm² bar$^{0.5}$ at 400°C and a permeate purity over 99.95% were obtained.

Figure 16.9 shows the product composition on dry basis at steady-state conditions as measured at the outlet of reformer, membrane separation, and CPO reactor for an exit reforming temperature of 590°C, a hydrogen recovery factor (HRF) of 40%, and an oxygen carbon ratio at the inlet of CPO equal to 0.28. The latter is evaluated referring oxygen to CO and CH_4 in the retentate stream.

An evident reduction in hydrogen content takes place on the retentate stream due to the hydrogen recovery performed by first-stage membrane, while the composition of other components grow up due to the fact that the mixture becomes more concentrated. With respect to the nonintegrated architecture, depending on the reformer exit temperature and HRF performed by membrane module, CPO is fed by a mixture rich in hydrogen and methane together with a consistent amount of CO_2. This kind of feed, in the case of monolith, completely modifies thermal behavior, while for pellets, no substantial change may be observed.

Figure 16.10 shows a comparison in terms of oxygen consumption between the standalone and the integrated configuration. In order to have a direct comparison of oxygen consumption, the latter has been referred to the natural gas at the inlet

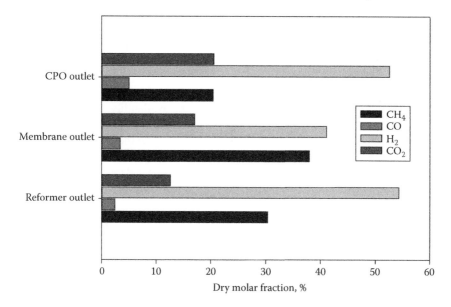

FIGURE 16.9 Composition on dry basis at the outlet of reformer, membrane, and CPO reactor at steady-state conditions. Reformer outlet $T = 600°C$; $P = 10$ barg; S/C = 2.7; HRF = 39%; $O_2/(CH_4 + CO) = 0.28$. (Adapted from Iaquaniello, G. et al., manuscript in preparation.)

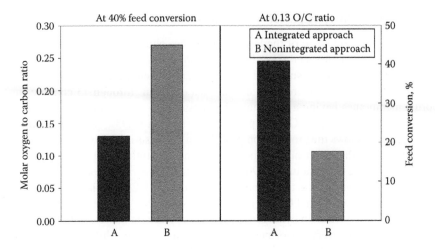

FIGURE 16.10 Comparison between standalone and integrated configuration. *Left*: molar oxygen-to-carbon ratio necessary to have a 40% feed conversion. *Right*: feed conversion obtained for a molar oxygen-to-carbon ratio of 0.13. P = 10 barg; S/C = 2.7; GHSV = 53,000 h^{-1}. (Adapted from Iaquaniello, G. et al., manuscript in preparation.)

of the reaction scheme. For a total feed conversion of 40%, for example, the integrated architecture allows to reduce oxygen consumption over 50%. It means a lower operating temperature for CPO section with a consequent difference in the exit gas composition. The reduction of oxygen consumption plays a major role in the economics of process even if additional capital costs are associated to the introduction of a prereformer stage.

16.4 CONCLUSIONS

The utilization of natural gas, although a grand challenge in catalysis for a long time, has received a large impulse from one side from the availability of unconventional gas resources that have depressed gas prices and altering oil–gas price parity, creating new opportunities to use them as raw material, and from the other side from new scientific discoveries and sometimes breakthrough developments. However, still performances are often far from the possibility of exploitation and thus the old technologies via syngas production will play a relevant role also in the future, particularly when further improvement could be achieved.

We have shown here that there is still space to improve this traditional route, for example, in developing SCT-CPO routes to H$_2$ or syngas, better if coupled to the use of novel H$_2$ separation membranes. The first area considered is for small-scale applications, where the established (conventional) routes are less attractive, but they can be successfully applied in perspective also to large-scale processes. There are, however, still materials issues, in terms of membranes.

We have discussed some aspects of a novel process scheme for syngas production from natural gas, developed and implemented on semi-industrial scale in the frame of EU project NEXT-GTL. The process scheme includes the presence of prereforming stage followed by a membrane for hydrogen separation, a CPO step, and a further step of syngas purification by membrane. The proposed process scheme could enable for a reduction of 10% of the variable operating cost with respect to conventional CPO technology.

The implementation on semi-industrial scale involved the catalytic tests at a capacity of 20 N-m³/h of hydrogen. Different types of catalysts in the form of pellets or honeycomb monolith were tested, as well as the CPO reactor both in a standalone configuration, with a dedicated natural gas stream fed to the reactor, and in an integrated configuration, with the hydrocarbon feedstock represented by the retentate from the membrane separation unit. Further details will be reported elsewhere [25], but this contribution gives a glimpse of the relevant advantages possible by this novel process scheme.

The comparison of the CPO reactor performance in standalone and integrated configurations evidenced how for a total feed conversion of 40% the integrated architecture performs better from an economic point of view, enabling for a reduction in oxygen consumption over 50%. This is a very relevant result, taking into account that oxygen consumption is a key factor for both energy economy and process competitiveness.

ACKNOWLEDGMENT

This work was realized in the frame of the European Project "Innovative Catalytic Technologies & Materials for Next Gas to Liquid Processes, NEXT-GTL."

REFERENCES

1. Liu K, Song C, Subramani V (Eds.). *Hydrogen and Syngas Production and Purification Technologies*, John Wiley & Sons, Hoboken, NJ, 2010.
2. Rostrup-Nielsen J, Christiansen LJ. *Concepts of Syngas Manufacture*, World Scientific, Singapore, Singapore, 2011.
3. Centi G, Perathoner S. Recent developments in gas-to-liquid conversion and opportunities for advanced nanoporous materials, in *Nanoporous Materials: Advanced Techniques for Characterization, Modeling, and Processing*, Kanellopoulos, N. (Ed.), CRC Press, 2011, Chapter 14, pp. 481–511.
4. Specchia S. Hydrocarbons valorisation to cleaner fuels: H_2-rich gas production via fuel processors, *Catal Today* 2011, *176*, 191–196.
5. Basini L, Aasberg-Petersen K, Guarinoni A, Ostberg M. Catalytic partial oxidation of natural gas at elevated pressure and low residence time, *Catal Today* 2001, *64*, 9–20.
6. Basini L. Issues in H_2 and synthesis gas technologies for refinery, GTL and small and distributed industrial needs, *Catal Today* 2005, *106*, 34–40.
7. Basini L. Industrial perspectives in H_2 generation through short contact time—Catalytic partial oxidation technologies, *AIChE Annual Meeting, Conference Proceedings*, Philadelphia, PA, November 16–21, 2008, pp. 841/1–841/8.
8. Guarinoni A, Ponzo R, Basini L. Hydrogen production with short contact time—Catalytic partial oxidation of hydrocarbons and oxygenated compounds: Recent advances in pilot- and bench-scale testing and process design, *DGMK Tagungsbericht* 2010, *3*, 125–132.
9. Enger BC, Lodeng R, Walmsley J, Holmen A. Inactive aluminate spinels as precursors for design of CPO and reforming catalysts, *Appl Catal A Gen* 2010, *383*, 119–127.

10. York APE, Xiao TC, Green MLH, Claridge JB. Methane oxyforming for synthesis gas production, *Catal Rev—Sci Eng* 2007, *49*, 511–560.

11. Tanaka H, Kaino R, Okumura K, Kizuka T, Tomishige K. Catalytic performance and characterization of Rh–CeO$_2$/MgO catalysts for the catalytic partial oxidation of methane at short contact time, *J Catal* 2009, *268*, 1–8.

12. Tanaka H, Kaino R, Nakagawa Y, Tomishige K. Comparative study of Rh/MgO modified with Fe, Co or Ni for the catalytic partial oxidation of methane at short contact time. Part II: Catalytic performance and bed temperature profile, *Appl Catal A Gen* 2010, *378*, 187–194.

13. Cimino S, Lisi L. Impact of sulfur poisoning on the catalytic partial oxidation of methane on rhodium-based catalysts, *Ind Eng Chem Res* 2012, *51*, 7459–7466.

14. Zeppier M, Villa PL, Verdone N, Scarsella M, De Filippis P. Kinetic of methane steam reforming reaction over nickel- and rhodium-based catalysts, *Appl Catal A Gen* 2010, *387*, 147–154.

15. Abate S, Centi G, Perathoner S, Genovese C, Iaquaniello G, Lollobattista E. Performances and stability of Pd-alloy thin film/ceramic membranes for H$_2$ separation in natural gas reforming gases, *Preprints—Am Chem Soc Div Petrol Chem* 2008, *53*(1), 85–87.

16. Abate S, Centi G, Perathoner S, Iaquaniello G. Thin-film membranes downstream to the reactor for novel energy-efficient processes in H$_2$ production, *Preprints Symp—Am Chem Soc Div Fuel Chem* 2010, *55*(2), 222–223.

17. De Falco M, Iaquaniello G, Salladini A. Experimental tests on steam reforming of natural gas in a reformer and membrane modules (RMM) plant, *J Membr Sci* 2011, *368*, 264–274.

18. Iaquaniello G, Giacobbe F, Morico B, Cosenza S, Farace A. Membrane reforming in converting natural gas to hydrogen: Production costs: Part II, *Int J Hydrogen Energ.* 2008, *33*, 6595–6601.

19. Basini L, Iaquaniello G. Process for the production of hydrogen starting from liquid and gaseous hydrocarbons and/or oxygenated compounds also derived from biomasses, Patent WO2011072877 (A1), 2011.

20. Iaquaniello G, Antonetti E, Cucchiella B, Palo E, Salladini A, Guarinoni A, Lainati A, Basini L. Natural gas catalytic partial oxidation: A way to syngas and bulk chemicals production, in *Natural Gas—Extraction to End Use*, Gupta SB (Ed.), InTech Publisher, 2012, Chapter 12, DOI: 10.5772/48708. Available from: http://www.intechopen.com/books/natural-gas-extraction-to-end-use/natural-gas-catalytic-partial-oxidation-a-way-to-syngas-and-bulk-chemicals-production. Accessed July, 2014.

21. Capoferri D, Cucchiella B, Mangiapane A, Abate S, Centi G. Catalytic partial oxidation and membrane separation to optimize the conversion of natural gas to syngas and hydrogen, *ChemSusChem* 2011, *4*, 1787–1795.

22. De Falco M, Iaquaniello G, Salladini A, Reformer and membrane modules for methane conversion experimental assessment and perspectives of an innovative architecture, *ChemSusChem* 2011, *4*, 1157–1165.

23. Salladini A, Palo E, De Falco M, Iaquaniello G. Process intensification in membrane assisted steam reforming at semi industrial scale, in *WHTC2013*, Shanghai, China, September 25–28, 2013.

24. Livio D, Donazzi A, Beretta A, Groppi G, Forzatti P. Optimal design of a CPO-Reformer of light Hydrocarbons with Honeycomb catalyst: effect of frontal heat dispersions on the temperature profiles, *Top Catal* 2011, *54*, 866–872.

25. Iaquaniello G, Salladini A, Palo E, Centi G. Catalytic partial oxidation coupled with membrane as a way to optimize syngas manufacturing, manuscript in preparation.

17 Methane Direct Aromatization Process from an Industrial Perspective

Technical Catalyst and Conceptual Design for Reactor and Process

Jens Aßmann, Rainer Bellinghausen, and Leslaw Mleczko

CONTENTS

17.1 METHANE DIRECT AROMATIZATION: INTRODUCTION

In the next decades, chemical industry is expecting to undergo significant changes. This development will be driven by the changing market demands, increasing pressure on the reduction of the CO_2 footprint, and finally transformation in the feedstock base. Currently, chemical industry is almost solely dependent on crude oil. It suffers from the volatility of the fuel market since many chemicals are side products of the fuel-oriented petrochemical industry. For example, the availability of benzene depends on the development of the gasoline market. Since soot emissions are correlated to the aromatic content in gasoline, refineries strive to reduce production of aromatic hydrocarbons. Finally, the limited availability of crude oil together with the growing wealth and mobility in the new economies like China and India causes chemical industry to look for alternatives [1]. In principle, three options for alternative feedstocks are available: coal, biomass, and natural gas. Unfortunately, when using coal, in general, CO_2 footprint also increases. Therefore, with exception of China, coal is systematically losing its importance as chemical feedstock. Biomass-based economy was the research hype of the last decade. Unfortunately, almost all practiced technologies based on sugar face the moral question *food* or *fuel/chemicals*. On the other hand, lignin-based technologies have not fulfilled the expectations. It is therefore a general agreement that natural gas will be the next replacement for crude. This statement is not new. The growing role of natural gas as chemical feedstock has been postulated since the last three decades. However, in spite of the rising prices for crude oil, its replacement in the chemical industry is marginal. The reasons for this development are multitudinous. First of all, the price for natural gas was correlated with the one for crude. It happened just lately that due to the shell gas market, these prices decoupled. Furthermore, technologies for cheap conversion of natural gas to base chemicals are still not available. Natural gas is the primary feedstock for the synthesis of gas-based processes. For example, in the last few decades, world-scale methanol units have been built. Technologies exist to convert methanol gas to olefins and aromatics. However, this multistep processes are economically viable only in a very large scale. Also the first large-scale gas to liquid (GTL) plants have been commissioned. Again, also the Fischer–Tropsch-based route is commercially viable in a large scale and in the places where natural gas is very cheap.

Against this background, there is large interest to develop technologies for direct conversion of methane to valuable chemicals. Direct conversion means lower investment costs. This would allow utilization of stranded gas. The direct oxidation of methane suffers from very low yields. A quite significant effort has been spent on the development of the oxidative coupling of methane (OCM). However, also this technology faces barriers with respect to the yield. Obviously, the maximum yield in this technology is limited to around 25%. In novel concepts no more yield increase is targeted but carbon efficiency [2]. Furthermore, consecutive conversion steps are necessary to utilize side products. This makes the OCM technology more complex. One of the interesting alternatives for the direct conversion of methane is its dehydroaromatization. This route is given by Equation 17.1. With this, single-stage benzene that is the valuable feedstock for chemical and polymer industry can be made:

$$6CH_4 \leftrightarrow C_6H_6 + 9H_2 \quad \Delta H_R = +523 \text{ kJ/mol} \tag{17.1}$$

This route would be significantly simpler than the syngas route that includes several high-pressure and high-temperature steps (Figure 17.1).

Unfortunately, the thermodynamics is not favorable. The strong increase of gas volume significantly limits equilibrium conversion (see Figure 17.2). In the temperature range between 600°C and 800°C that is technically feasible, conversion is significantly lower than 25%. The reaction is strongly endothermic; therefore, heat supply becomes the real challenge. These advantages are compensated by the high selectivity to aromatics, which is for the technically interesting conditions far above 80%. Beside benzene as the major aromatic product, by-products like naphthalene and toluene are produced. However, the high selectivity to aromatics is limited to the kinetically controlled regime since the thermodynamic equilibrium leads to carbon and hydrogen.

In spite of this unfavorable thermodynamics, the methane direct aromatization (MDA) reaction has been extensively studied since it allows conversion of methane to benzene that is an important feedstock for chemical and polymer industry. For the overview of the earlier studies, we refer to [4–7]. In the meantime, there is a good understanding of the chemistry of this reaction [4]. It has been found that in order to obtain a high productivity, a heterogeneous catalyst is necessary. In all studies, the best performance exhibited zeolite loaded with a metal, like Mo, Rh, and Ga. From the cost and performance perspective, Mo has been the preferred metal. Within an induction period, molybdenum transforms under reaction conditions to carbide and

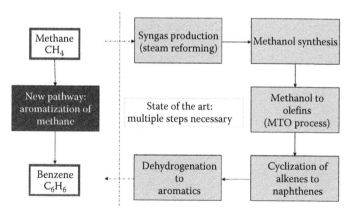

FIGURE 17.1 MDA vs. syngas route via several consecutive processes.

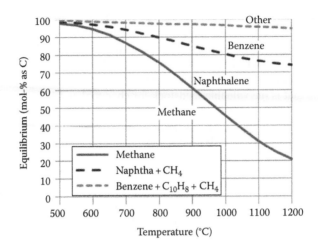

FIGURE 17.2 Equilibrium compositions for MDA reaction at atmospheric pressure and different temperatures acc. to Wang et al. [3].

is responsible for methane activation [8–10]. Protons in zeolites play a crucial role (Si/Al = 10–40) in cyclization. Finally, channels of zeolite should be in similar size relative to benzene molecule. Therefore, ZSM-5 zeolite that has pores of 5.4 Å × 5.6 Å exhibited superior performance compared to other materials. Slightly later, Mo/MCM-22 catalysts were also shown to be competitive to Mo/ZSM-5 catalysts. Especially, the benzene selectivity could be increased above 80% [11]. Acidic and catalytic properties and the structure of the working catalysts were analyzed in detail applying fourier transform infrared (FTIR) spectroscopy [12], in situ Raman spectroscopy, as well as x-ray absorption spectroscopy (XAS) [13,14].

The major challenge for the chemical performance of this catalyst is its deactivation in the timescale of minutes. In order to cope with this deactivation, a fluidized-bed reactor has been proposed. Following the same concept like in the fluid catalytic cracking (FCC) process [15], a two-reactor arrangement has been proposed (Figure 17.3).

In the synthesis reactor, the dehydroaromatization reaction takes place. The catalyst deactivates quite fast, but it is continuously removed from the synthesis reactor to the second one, which is also operated as a fluidized bed. In the second reactor that is called regenerator carbon or precisely saying coke that is deposited on the catalyst is burned off. This process not only regenerates the catalyst but also generates the heat that is transported with the catalyst to the synthesis reactor where it is utilized in the endothermic MDA reaction. It is the matter of the reaction conditions if the synthesis reactor is operated as bubbling bed or as a riser. Higher yields have been reported for the fluidized-bed reactor compared to the performance of the fixed bed [16]. However, problems with the mechanical stability of the catalyst applied in the fluidized bed have been reported. This basic configuration has been the subject of various modifications. For example, a two-stage fluidized-bed has been proposed. In the first stage, oxygen has been co-fed to generate heat [17]. It is also possible to run in the first stage the OCM reaction [18].

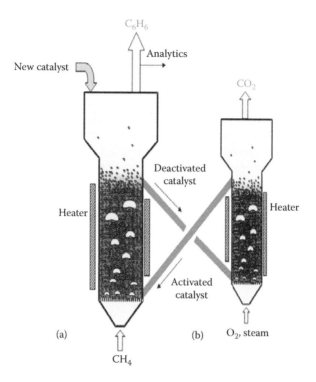

FIGURE 17.3   Two-reactor arrangement: (a) MDA reaction and (b) catalyst regenerator.

This arrangement allows generation of heat and of higher hydrocarbons that are easier to aromatize. Oxygen can be also fed into the fluidized bed by means of a membrane [19]. Or finally, a membrane for the H_2 removal can be inserted in the fluidized bed for shifting the equilibrium [20].

In spite of the large number of scientific publications, scale-up activities have not been reported. It is also not easy to assess the economic potential of the MDA technology. Within this study, major challenges and requirements toward an industrial process have been addressed. The primary goal was to develop a catalyst that will be suitable for application in a fluidized-bed reactor and to test chemical performance of this catalyst. Furthermore, a fluidized-bed reactor that will allow a stable operation with respect to the catalytic stability and heat supply has been conceptually designed. Finally, economic evaluation of various process options has been performed.

17.2   SCALE-UP STUDY OF MDA CATALYST

17.2.1   DEVELOPMENT OF TECHNICAL CATALYST

In various laboratory studies, bifunctional catalysts like Mo/ZSM-5 catalyst were investigated. Different loadings and also promoters were tested to increase yield toward aromatics, mainly benzene. Spivey and Hutchings reviewed recent developments in the field of catalyst composition and transformation of Mo into molybdenum carbide species [5]. However, these studies do not consider the additional needs of a technical catalyst like mechanical stability and manufacturing concepts, which still can provide the catalyst for reasonable low costs. Herein, we describe an approach toward the development of a technical catalyst for its application in a fluidized-bed reactor. Fluidized-bed reactors require mechanically stable particles of defined size, in general between 50 and 300 µm. In order to minimize attrition of the catalyst during operation, typically, binders are added during the synthesis procedure. Generally, alumina is known to deliver mechanical stable zeolite catalysts. However, a strong interaction between binder and molybdenum could result in an unselective catalyst. Thus, the development of a technical catalyst was performed according to the following strategy:

1. Selection of binder material that provides sufficient mechanical stability and that reveals low affinity toward molybdenum precursor salt solutions.
2. Develop a suitable granulation process such as a spray-drying procedure to produce zeolite-binder composite material in kg scale.
3. Impregnation of the corresponding composite material with ammonium molybdate to deliver the *complete* technical catalyst.
4. Optimization of zeolite-to-binder mass ratio to achieve best compromise between mechanical stability and benzene yield.
5. Verification of the catalyst lifetime and of the potential regeneration procedures.

17.2.1.1 Synthesis Recipe

17.2.1.1.1 Binder

There are several binders that are generally used in order to prepare mechanically stable zeolite particles. Commercial suppliers often use alumina or aluminum phosphate. However, it has been found that alumina has high tendency to adsorb Mo from an aqueous ammonium molybdate solution. In order to evaluate this potential drawback, the following experiments were performed. The impregnated ZSM-5 catalyst that was not yet calcined was dissolved in water, and the solid was then filtered off. Knowing the molybdenum concentration in the aqueous phase, different binders were added in separate experimental trials and the suspensions were stirred for constant time period. The binders were filtered off and the Mo concentration in the remaining aqueous phase was again analyzed by inductively coupled plasma optical emission spectrometry (ICP-OES). Table 17.1 shows the amount of Mo that was adsorbed on the binder materials. High percentage means that the preliminary dissolved Mo was recovered onto the binder material, which indicates a strong interaction between Mo in aqueous precursor salt solutions and the corresponding binder. If the affinity to the binder material is high, the impregnation of the zeolite catalyst will be not highly selective. The results reveal that Al_2O_3 cannot be used as binder due to its strong affinity toward Mo species. ZrO_2 and TiO_2 would be a better choice but SiO_2 revealed the best result, that is, the lowest affinity toward molybdenum uptake.

Similar experiments were described by Honda et al. [21] who investigated the adsorption of MoO_3 solution in water at 80°C. They could demonstrate that Al_2O_3

TABLE 17.1

Results: Molybdenum Uptake by Different Binder Types

Binder	Mo Uptake from Mo Salt Solution (%)	Recovery Amount of Predissolved Mo on Binder
Al_2O_3	100	High
$AlPO_4$	12	Average
SiO_2/Al_2O_3	100	High
SiO_2	3	Low
ZrO_2	19	Average
TiO_2	20	Average

reveals a 225-fold higher affinity for MoO_3 uptake than SiO_2. The adsorption of MoO_3 onto the Zeolite ZSM-5 was shown to be slightly higher than on SiO_2. Finally, we could conclude that the application of SiO_2-based binder materials should deliver a technical catalyst that allows a selective Mo impregnation of the zeolite and prevents adsorption onto the binder itself.

17.2.1.1.2   Spray Drying

Spray-drying synthesis technology allows the production of catalyst particles with the required size of >50 and <300 μm for application in fluidized-bed reactors. Alternatively, extrudates could be crushed, but this method yields small particles with significantly lower mechanical stability as shown in stability tests (compare 17.2.2.1). For the spray-drying procedure, SiO_2 was used in the form of a fresh sol (acidic aqueous SiO_2 solution [pH = 2] with 6 wt.% solids). The fresh sol is taken directly from a production plant and must be processed within 24 h to prevent agglomeration of the colloidal solution. The quality of the HZSM-5 catalyst powder was controlled, purified from partially clumped particles if needed, and homogenized before using them. The HZSM-5 is dispersed into the fresh sol and further promoted as a suspension into the spray-drying tower. Binder contents of 10 and 17 wt.% were targeted.

In Table 17.2, synthesis parameters are listed. It points out that not only the targeted binder content differentiates both synthesis procedures but also the solid

TABLE 17.2

Spray-Drying Synthesis: Binder and Zeolite Amounts Used

Binder fraction (%)	17	10
Solid fraction in *fresh sol* (wt.%)	6	6
Target weight of catalyst (kg)	5	5
Mass HZSM-5 (kg)	4.15	4.50
Mass *fresh sol* (kg)	14.17	8.33
Total mass (zeolite + *fresh sol*) (kg)	18.32	12.83
Solid fraction of suspension (wt.%)	27.30	38.96

fraction within the suspension used for the spray-drying step. The latter can directly affect the particle size of the product. The suspensions were fed at a rate of 6–8 kg/h into the spray-drying tower (with ca. 5 m height). They were sprayed with an atomizing pressure of 1 bar (N_2) through a single hole. The drying is carried out in cocurrent with an inlet temperature of 160°C and an outlet temperature of 80°C.

17.2.1.2   Characterization of Technical Catalyst

The catalyst particles (zeolite/binder composite) obtained by the spray-drying synthesis procedure were fractionated using adequate sieve sizes. Figure 17.4 shows the particle-size distribution after sieving catalysts, one with 10 wt.% and another one with 17 wt.% binder content. It can be seen that a significantly larger proportion of particles in the range between 50 and 100 μm were found for the catalyst with 17 wt.% binder content. However, for both catalysts, the particle-size distribution is rather narrow in the range between 100 and 200 μm with minor amounts up to 300 μm size.

Scanning electron microscope (SEM) pictures (Figure 17.5) confirm the particle-size distribution obtained from the sieve analysis and reveal in addition that the zeolite/binder composite materials are mainly characterized by spherical geometry. However, Figure 17.6 illustrates that the proportion of nonuniform particles is significantly higher for the 17 wt.% binder containing sample.

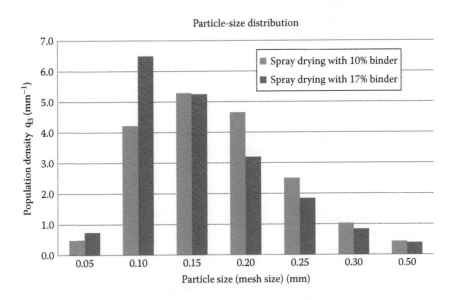

FIGURE 17.4   Comparison of particle-size distribution of spray-dried catalyst with 10 and 17 wt.% binder amount, respectively.

FIGURE 17.5   SEM picture of spray-dried particles with 10 wt.% binder (a) and 17 wt.% binder fraction (b).

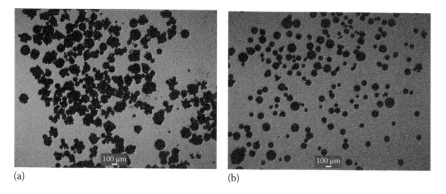

FIGURE 17.6   Optical microscope images of the spray-dried particles with 17 wt.% binder (a) and 10 wt.% binder portion (b).

17.2.2 PERFORMANCE OF TECHNICAL CATALYST

17.2.2.1 Mechanical Stability

To evaluate the mechanical stability of the catalysts containing 10 and 17 wt.% SiO_2 binder, two approaches were applied:

1. Long-term fluidization at room temperature (the so-called cold-flow experiments) and monitoring the weight loss from time to time to determine attrition losses
2. Severe treatment of the particles in an aqueous dispersion within an ultrasonic bath and measuring average diameter of the particles by laser diffraction analysis

In cold-flow experiments, the fluidization velocity was chosen to be a factor of five times higher compared to the minimum fluidization velocity of the studied catalyst.

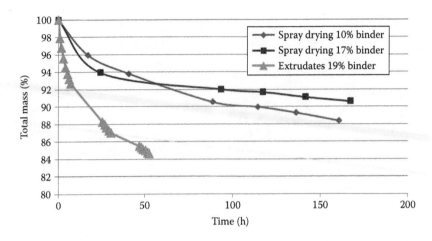

FIGURE 17.7 Stability test in cold-flow experiments: spray-dried catalyst in comparison to crushed catalyst extrudates.

At specific time intervals, the weight loss due to abrasion of the catalyst particles was measured, and thus a mass loss rate was determined. These rates are compared in Figure 17.7 for both spray-dried catalysts. For comparison, commercial zeolite extrudates were crushed down to the similar particle size (100–300 µm) and also operated under cold-flow conditions to monitor the weight loss as a function of time. From the illustration in Figure 17.7, it becomes obvious that both spray-dried catalysts reveal a higher mechanical stability than the commercial zeolite particles derived from the extrudates. The very high weight loss rate in the initial phase that has been observed for all catalysts is caused by the entrainment of fines and reduction of the catalyst roughness. Considering only the time of operation after the induction period (30 min), the following mass loss rates could be calculated: 0.02%/h at 17 wt.% and 0.04%/h at 10 wt.% binder content, respectively. In contrast, the mass loss rate of the crushed extrudate is 0.15%/h.

Using ultrasonic treatment, it was possible to accelerate degradation of the particles and to discriminate also both spray-dried catalysts from each other. The samples were treated differently long with ultrasound before particle-size distribution was measured by laser diffraction (compare Figure 17.8). Plotting the ratio of the corresponding d50 values and the d50 value of the untreated sample as a function of the ultrasonic treatment duration (Figure 17.9) reveals significant differences in mechanical stability for both spray-dried catalysts and benchmark materials.

Through this experiment, the rather poor mechanical stability of the extrudate with 19 wt.% binder content observed in the cold-flow measurements has been confirmed. Another commercial extrudate (H-ZSM 5 plus an unknown type of binder) was also not competitive to the spray-dried particles.

A significant difference can be seen between the two spray-dried particles, too. The average particle diameter of particles with 10 wt.% binder content is shrinking much faster compared with the catalyst containing a higher amount of

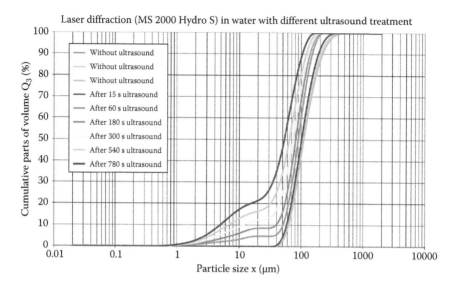

FIGURE 17.8   Particle-size distribution of spray-dried samples with 17 wt.% binder after ultrasonic treatment for different time intervals.

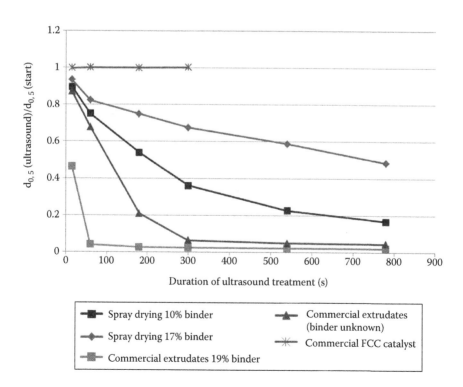

FIGURE 17.9   Comparison of stability of spray-dried catalysts and commercial catalysts.

binder (17 wt.%). Outstanding mechanical stability was observed when testing a standard FCC catalyst applied in advance in a circulating fluidized-bed reactor. For this type of catalyst, however, Al_2O_3 is used as a binder, which leads to a significantly higher mechanical stability compared with SiO_2-based binder. As outlined earlier, alumina cannot be applied as binder for Mo/zeolite catalysts. It can be concluded that the spray-dried catalysts containing SiO_2-based binder material reveal sufficient strength for operation in a fluidized-bed reactor. Higher amounts of binder increase the stability but might not be advantageous with respect to product selectivity as discussed in the upcoming section.

17.2.2.2 Catalytic Performance Test on Bench Scale in Fluidized-Bed Reactor

As mentioned before, binder materials for zeolites behave not completely inert. Considering the benzene formation rates, the difference between a binder-free and a binder-containing catalyst is clearly visible. In Figure 17.10, it can be seen that the binder-free catalyst for the entire duration of the reaction has a higher benzene formation rate and the deactivation is slower than that of the binder-containing catalysts. In comparison, both binder-containing catalysts reveal the same maximum in benzene formation rate. However, the reaction rate decreases faster at higher binder content. Although the applied binder SiO_2 is less catalytically active than Al_2O_3 species, it has still a negative impact. The strong decrease in the benzene formation rate for the catalyst with 17 wt.% binder content is accompanied by a decrease in the benzene selectivity over the complete reaction time (see Figure 17.11). This has a significant difference to binder-free catalysts that exhibited a constant selectivity in fixed-bed experiments. In contrast, the catalyst with 10 wt.% binder component shows only a negligible loss of the benzene selectivity with respect to the binder-free catalyst. In contrast, the coke selectivity for the catalyst with 17 wt.% binder

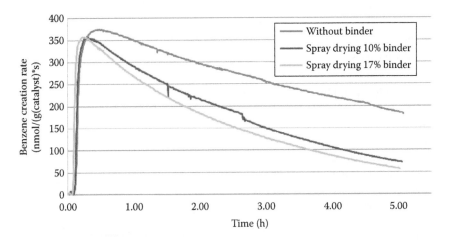

FIGURE 17.10 Comparison of benzene formation rates for catalysts with different binder fractions.

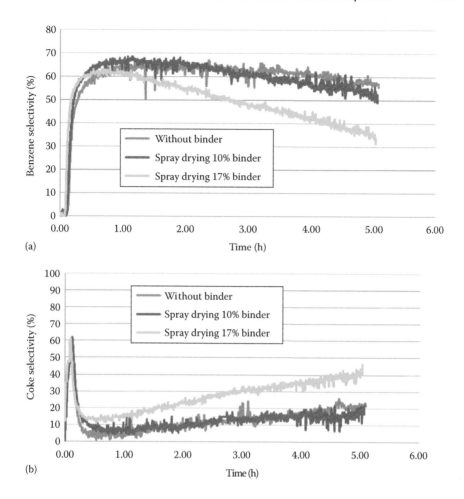

FIGURE 17.11 Comparison of (a) benzene and (b) coke selectivity for catalysts with different binder fractions.

content is higher and increases faster while the coke selectivity for the catalyst with 10 wt.% binder does not differ that much from that of the binder-free catalyst.

It is obvious that although a binder with low affinity to molybdenum was selected, the influence of the binder on catalyst performance is still visible via product selectivity during MDA. To achieve similar performance as shown for powder-based binder-free catalysts, the binder content has to be decreased down to a level of 10 wt.%. Higher binder contents cause a decrease in benzene selectivity and accelerate coking of the catalyst but offer an improved mechanical strength. Finally, the optimized catalyst always represents a trade-off between maximum selectivity in MDA reaction and mechanical stability. Our study indicates that binder contents between 10 and 20 wt.% should be considered for the technical catalyst.

17.2.2.3   Development of the Regeneration Procedure

The formation of carbonaceous deposits is understood as a major reason for catalyst deactivation. Within several studies, the different nature of coke species was investigated applying temperature-programmed oxidation and/or thermal gravimetric analysis with the coked Mo/ZSM-5 or Mo/MCM-22 catalysts, respectively [22–24]. From a theoretical point of view, there are four different ways of removing coke:

1. Burning coke by oxygen: $C + O_2 \rightarrow CO_2$
2. Hydrogenation of coke to volatile compounds (e.g., methane): $C + 2H_2 \Leftrightarrow CH_4$
3. Transforming coke into carbon monoxide: $C + CO_2 \Leftrightarrow 2CO$
4. Coke gasification by steam: $C + H_2O \Leftrightarrow CO + H_2$

Applying CO as co-feed, L. Wang et al. [25] could demonstrate a stabilized conversion of methane and high aromatic selectivity. Others describe the combination of methane CO_2 reforming with the dehydroaromatization reaction [26] or $H_2 + CO_2$ switching experiments for coke removal [27].

Herein, we briefly describe our results obtained when following routes (1) and (2). Hydrogenation of coke to volatile species was carried out at the reaction temperature of 700°C. Catalytic regeneration by pure hydrogen for 3 h could not restore catalytic activity. Though small amounts of methane could be detected by mass spectroscopy, transmission electron microscopy showed that still carbon nanotube–like structures could be found on the catalyst surface. In addition, x-ray photoelectron spectroscopy showed that the amount of coke was reduced by only ca. 25%. Overall, reductive regeneration via hydrogen could not be shown as useful technology for complete catalyst regeneration, although it would have been attractive to use the aromatization *by-product* hydrogen for catalyst regeneration.

Thus, oxidative regeneration was investigated. *Burning coke* is a standard technology in crude oil refinery. In addition, oxygen (air) is rather cheap. The drawback of this technology is the fact that molybdenum oxide has a relatively low sublimation point. Thus, care must be taken in order to prevent temperatures higher than 500°C. Since coke oxidation is an exothermic process, local hotspots need to be avoided. Thus, one of the tasks was to find an appropriate oxygen concentration. As a result of an experimental optimization, an oxygen concentration of ~4 vol.% at 500°C was found to be promising. In addition, sintering of molybdenum oxide due to high temperatures also needs to be avoided. We found out that reactivation of the oxidized catalyst can be successfully carried out by heating the catalyst in methane to the reaction temperature. Heating the catalyst in nitrogen until reaction temperature before introducing methane has a negative influence on benzene selectivity and is thus not an appropriate method to reactivate the spent catalyst. Using the optimized procedure, a catalytic performance can be achieved, which is even better than that of the fresh catalyst as shown in Figure 17.12. As can be seen in Figure 17.12, six reactions (regeneration cycles) could be performed without loss of productivity, that is, the rate of benzene formation.

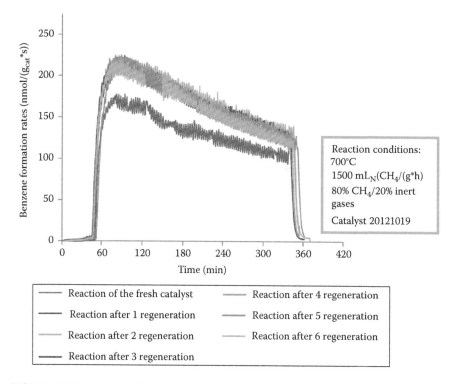

FIGURE 17.12 Benzene formation rate of the fresh Mo/(zeolite + silica) catalyst in comparison to the performances of the spent catalyst regenerated several times in a gas mixture containing 4 vol.% O_2 in N_2.

17.3 DESIGN OF TECHNICAL REACTOR

17.3.1 REACTOR MODEL

In the MDA process, several reactions have to be regarded (see Table 17.3). In addition to the gas-phase reactions, also reactions resulting from the catalyst transformation should be taken into account. They are important to close energy balances, for example, heat generation due to the catalyst regeneration. On the basis of experiments in the laboratory reactor, a global kinetics have been developed.

17.3.2 MODE OF OPERATION AND SIZE

For an initial design and choice of a reactor type, the assumptions listed in Table 17.4 were made according to the first results in lab trials and expectations for further developments. For these conditions, the main design data listed in Table 17.5 can be derived. According to the primary concept, the endothermic MDA reaction should be driven with the heat supplied from the catalyst regeneration. However, the heat demand is so high that catalyst regeneration solely, for example, assuming the maximal temperature of 800°C for the catalyst, less than 1 MW can be transported with that catalyst

TABLE 17.3
Regarded Reaction Enthalpies

Reaction

$$6CH_4 \leftrightarrow C_6H_6 + 9H_2 \qquad \Delta H_R = 530 \text{ kJ/mol}$$
$$10CH_4 \leftrightarrow C_{10}H_8 + 16H_2 \qquad \Delta H_R = 895 \text{ kJ/mol}$$
$$2CH_4 \leftrightarrow 2H_2 + C_2H_4 \qquad \Delta H_R = 201 \text{ kJ/mol}$$
$$CH_4 \leftrightarrow 2H_2 + C_{Graphite} \qquad \Delta H_R = 74 \text{ kJ/mol}$$

Catalyst formation

$$2CH_4 + 2MoO_{3solid} \leftrightarrow CO_2 + 4H_2O + Mo_2C_{solid} \qquad \Delta H_R = 410 \text{ kJ/mol}$$
$$4O_2 + Mo_2C_{solid} \leftrightarrow CO_2 + 2MoO_{3solid} \qquad \Delta H_R = -2016 \text{ kJ/mol}$$

Regeneration

$$C_{Graphite} + O_2 \leftrightarrow CO_2 \qquad \Delta H_R = -393 \text{ kJ/mol}$$
$$2H_2 + C_{Graphite} \leftrightarrow CH_4 \qquad \Delta H_R = -74 \text{ kJ/mol}$$
$$CO_2 + C_{Graphite} \leftrightarrow 2CO \qquad \Delta H_R = 172 \text{ kJ/mol}$$

Heat supply

$$2H_2 + O_2 \leftrightarrow 2H_2O \qquad \Delta H_R = -484 \text{ kJ/mol}$$
$$CH_4 + 2O_2 \leftrightarrow CO_2 + 2H_2O \qquad \Delta H_R = -803 \text{ kJ/mol}$$

TABLE 17.4
Assumptions for MDA Reaction

Benzene capacity	500,000 t/annum
Production time	8,000 h/annum
Methane conversion	11%
Benzene selectivity	70%
Naphthalene selectivity	15%
C_2H_x selectivity	10%
C—selectivity	5%
VHSV	3,000 m³/h/t catalyst
Catalyst lifetime	8 h
Catalyst bulk density	600 kg/m³
Reaction temperature	750°C

circulation but 190 MW is required. Accordingly, additional heat source is necessary. Finally, the catalyst regeneration study has shown that such a high temperature would cause irreversible catalyst deactivation due to the loss of Mo. Another critical issue is the low conversion. This has impact on the reactor size and, in turn, on captial expenditure (CAPEX). Due to the negative effect on equilibrium conversion rates, the pressure was limited to 4 bar, abs. Therefore, in order to achieve the desired capacity and keep the reactor size as small as possible, high gas flow rates are necessary. In this assessment, gas velocity was limited to 1.3 m/s. This velocity is slightly higher compared to the velocities applied in the stationary fluidized beds for catalytic transformation, for example, selective oxidation reactions are operated at gas velocities of about 0.8 m/s [15]. It is in the lower range of gas velocities applied in the circulating fluidized

TABLE 17.5
Main Design Data for MDA Reactors

Methane feed	1000 t/h	
Catalyst holdup	500 t	
Catalyst exchange rate	65 t/h	
Reaction heat	190 MW	
Fluidization conditions	Case A	Case B
Particle size	~1 mm	~50 µm
Min fluidization velocity umf (cm/s)	25	~0.3
Superficial gas velocity u (m/s)	2.5	1.35
Velocity ratio u/umf	10	500
Reactor diameter (m)	10	10
Number of reactors	2	4

beds and is much lower than the one in the risers, that is, much lower than in FCC reactors where gas velocity might be as high as 10 m/s. Unfortunately, gas velocity in fluidized-bed reactors is coupled with bed porosity. With increasing gas velocity, porosity increases from 0.6–0.7 in bubbling fluidized beds to 0.99 in the risers. Therefore, for riser reactors, very active catalysts are necessary, while catalyst deactivation on the timescale of seconds might be tolerated. When comparing catalytic activity expressed by the space-time yield in Table 17.4 and the residence time of a few hours before reactivation, a bubbling bed seems to be the preferred reactor design. In this reactor type, hydrodynamics is controlled by the gas velocity and a particle diameter. In order to cover the whole range of possible solutions, two cases with particle diameters between 50 and 1000 µm were studied, even when both cases would be hard to realize (compare Table 17.5). For the smallest particle, the ratio of superficial gas velocity to minimum fluidization velocity amounts to 500. Thus, the entrainment of particles would be very high and a complex fines separation would be necessary in order to keep the catalyst in the reactor. On the other hand, particles with 1000 µm are very large for catalytic applications in fluidized beds. Smaller particles are usually applied in order to avoid effect of the intraparticle mass transport and to minimize attrition.

Due to the high gas velocity, a reactor with the diameter of 10 m is necessary in order to achieve the nameplate capacity. Of course, also smaller reactors could be used, for example, with a diameter of 5 m, the number of reactors will rise by a factor of 4. The unexpanded bed height of the catalyst amounts to 3–6 m, depending on the number of reactors. Even if assuming a very high expansion of $H/H_0 = 3$, the total reactor height will not exceed 20 m. In case of VHSV of 1500 m^3/h/t catalyst, the bed height will be doubled. In that case, a minor expanded bed can limit the overall reactor height so that the 20 m will not be exceeded.

17.3.3 REACTOR DESIGN: HEAT SUPPLY

As already indicated, heat supply is one of the major challenges for the reactor design. Original concept based on the interconnected fluidized beds as a mean for catalyst regeneration and heat supply faces obstacles due to the insufficient energy

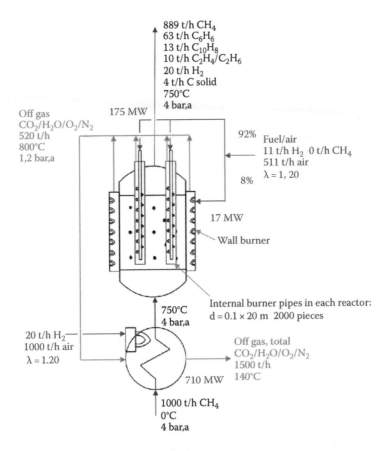

FIGURE 17.13 Fluidized-bed reactor heated by a large number (ca. 2000) of internal burner pipes and additional wall burners. Methane feed is preheated via heat exchanger using hot off-gas coming from internal burner pipes plus additional heat provided by H_2 combustion.

supply and limitations resulting from the catalyst stability. Therefore, several alternative designs have been challenged. One of the options utilizes membrane walls and the immersed heat exchanger that are supplied with heat generated by burners. This concept is illustrated in the flow sheet presented in Figure 17.13. In this design, heat can be supplied homogeneously by using a number of burners that are assigned to the tubes of the heat exchangers. It is a similar design to the one used in a number of highly exothermic reactions like selective oxidations or ammoxidations where a tubular heat exchanger is immersed in a fluidized bed [15]. The major issue in this design is the operation of a large number of small burners in order to avoid local hot spots that might be critical from the mechanical integrity point of view. However, new developments like pore or catalytic burners should allow homogeneous distribution of heat sources and very high heat transfer coefficients on the tube side. In this design, it has been assumed that the off-gas temperature is 800°C, same temperature as the wall temperature of the heating pipes. A heat transfer coefficient of the shell side of 300 W/m² K was assumed. It is a usual value for heat transfer coefficient between a fluidized bed and immersed tubes. The heat transfer on hot gas side was not regarded in detail as it is supported by flame radiation in case of burner pipes or can be accelerated by high enough hot gas temperatures.

As expected, heat transfer over the outer reactor walls is not sufficient. Only 8% of the necessary heat can be transported. Additionally, about 40 km pipes with diameter of 100 mm are necessary to supply the heat. Thus, the number of pipes is not that high: if 20 m length can be dipped into the fluidized bed, only 2000 pipes are necessary in 2–4 reactors. As fuel, the hydrogen created by the MDA reaction can be applied. It turns out that 11 t/h of hydrogen is necessary and 20 t/h is generated. In these calculations, it has been assumed that methane feed is heated up to reaction temperature in an upstream located gas heater. The required energy is about 3.5 times higher than the energy that is necessary for the MDA reaction alone. This energy should be recovered in large extent from the outlet reactor gas.

As an alternative, a more conventional design fluidized bed is heated by an immersed heat exchanger. However, the heat is supplied by hot off-gas from an upstream combustion process. The corresponding process scheme is shown in Figure 17.14. In calculations, it was assumed that the inlet off-gas temperature should not exceed 1000°C and the outlet off-gas temperature is 800°C. Then 4000 t/h off-gas is necessary and about 82 t/h methane as fuel is needed. In order to transport this huge gas flow, a pressure of 10 bar and a gas velocity in pipes of 50 m/s were assumed. In this design, 1000 parallel pipes are necessary. Thus, each pipe needs to be dipped about 4.0 m in the fluidized bed so that each pipe can drive once up and down in the reactor. In this case, the off-gas from the heating system of the reactor has enough enthalpy to heat the inlet methane feed up to 750°C. However, there is still another huge amount of energy in the product gas leaving the reactor. This energy has to be recovered separately.

These evaluations show that alternative concepts for the heat supply for the highly endothermic MDA reaction are technically feasible. Safety issues have not been analyzed so far, for example, what happens if the heating pipes start to leak. In case of the off-gas heating with separate incinerator, the off-gas has some amount of oxygen that could burn together with the methane of the MDA process and heat up the reactor significantly. Therefore, the case with internal burners might be more suitable because here, the overstoichiometric air supply can be minimized.

17.3.4 MEMBRANE REACTOR

The two major challenges of the MDA reaction, that is, low conversion and heat supply, can be addressed by application of a membrane reactor. Theoretical study of fixed-bed membrane reactor [12] has shown that the removal of H_2 across ceramic membranes can lead to almost complete CH_4 conversion at ~730°C at practical reactor residence times of ca. 100 s. This performance requires catalytic reactors that use thin dense ceramic films (10–100 μm) in order to achieve high permeations and practical reactor diameters.

However, this theoretical potential is hard to realize with the existing materials. A dense $SrCe_{0.95}Yb_{0.05}O_{3-\alpha}$ thin-film membrane with the thickness of ca. 2 μm has been tested [28]. At 720°C, about 16% of generated hydrogen has been removed.

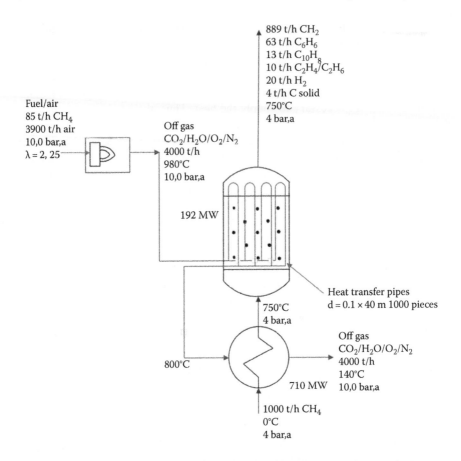

FIGURE 17.14 Fluidized-bed reactor heated by hot off-gas (980°C) from an incinerator. After leaving the heat exchangers within the reactor (ca. 1000 pipes), the partially cooled off-gas (800°C) is used for preheating methane feed in an upstream heat exchanger.

This resulted in mode increase of conversion. H_2 removal also led to slightly higher deactivation rates due to the increased cocking. Furthermore, hydrogen permeability decreases due to the gradual and slight reduction of the stoichiometric perovskite. Alternatively, Pd-coated alloy tubular membranes with high hydrogen selectivity have been tested [29–31]. Also in this case, hydrogen removal caused increase in conversion but accelerated deactivation. However, the major disadvantage of the Pd membrane is its low thermal stability. All experiments were performed at temperature up to 600°C. Therefore, the conversion level obtained with the membrane was much lower than at higher temperatures without membranes. With Pd–Ag composite membranes supported on porous stainless steel, stable operation at temperatures up to 700°C has been obtained [32]. With this membrane, CH_4 conversion up to 16% has been achieved. Like with other membranes, an accelerated deactivation has been observed. However, by means of periodic switching between operation with and without permeation, a metastable operation could be achieved.

Application of a fixed-bed reactor needs a periodic operation with periodic catalyst regeneration. Continuous operation can be achieved by immersing bundle of membranes in a fluidized bed. The membrane fluidized-bed reactor has been already studied for reforming reactions (e.g., [33]). The removal of hydrogen from the MDA reaction is combined with its combustion on the tube side of the membrane tubes. Thereby, the membrane tubes have double function, that is, they drive the MDA reaction above the equilibrium and supply heat to the fluidized bed. However, this multifunctionality implicates some limitations since only a third of the hydrogen created by MDA is necessary to deliver the heat if air supply is stoichiometric and heat loss is neglected. In simulations, a stoichiometric ratio of air to hydrogen of 1.2 and about 10% heat loss were assumed. This limits hydrogen consumption for heat supply. Hydrogen removal can be further driven by applying two sets of tubes: first set for pure removal and the second for removal and combustion. However, complete hydrogen removal has also disadvantages. It reduces partial pressure of H_2, which is the driving force for the permeation rate. In consequence, membranes with large surface areas are necessary. Furthermore, it has been reported that with decreasing partial pressure of hydrogen, deactivation rate increases. In order to get additional flexibility, also unconverted methane can be used to support combustion on the tube side of the membranes (see Figure 17.15). The previously described concept of multifunctional membranes has been roughly assessed. The effect of the hydrogen removal on the methane conversion is presented in Figure 17.16. In order to achieve 50% conversion of

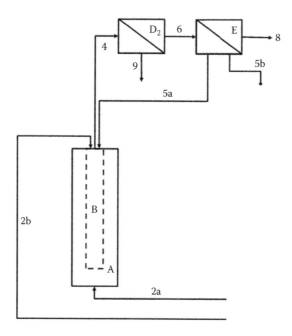

FIGURE 17.15   Reactor with burning zone B separated by membrane for removal of H_2 and CH_4 from MDA reaction zone A. A: MDA reaction zone of reactor; B: combustion zone; D_2: aromatic separation; E: gas separation; 2a: CH_4 feed; 2b: air supply; 4: product gas; 6: off-gas; 5a: CH_4; 5b: H_2 and C_{2+}; 8: H_2O/CO_2; 9: aromatics.

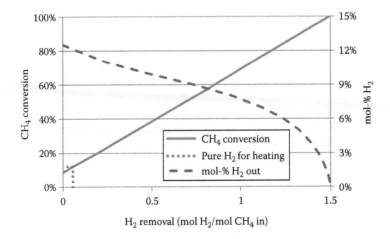

FIGURE 17.16 Equilibrium conversion of methane (straight line) when created hydrogen is removed: complete conversion of CH_4 is reached when all produced H_2 is removed, that is, 1.5 mol H_2/mol CH_4. In this case, the H_2 concentration in outlet (dashed line) is zero. Removal of H_2 that is needed as fuel for heating the reactor by a combustion reaction is marked by dotted line.

methane, about 1 mol of H_2 per converted mole of methane has to be removed. When the reactor is operated at 4 bar and assuming the membrane surface of 13,000 m², that is, the necessary surface for heat transfer, with a permeability of 10^{-7} mol/s/m²/Pa, only 2.5% of the generated hydrogen can be removed by the membrane. In order to remove 33% of hydrogen generated at about 10% conversion, the surface increases to 200,000 m². Even in this case, the increase of methane conversion is small (see dotted mark in Figure 17.16). The main challenge for this concept is still the availability of highly permeable membranes that are stable at high temperatures.

17.4   PROCESS CONCEPTS

17.4.1   PROCESS OVERVIEW

The complete MDA process consists of several functional blocks: the reaction section, the gas–liquid (G/L) separation where the aromatics are removed, a gas separation for hydrogen removal before recycling the unconverted methane and lower hydrocarbons back to the reactor, and a liquid separation for various aromatics (see Figure 17.17).

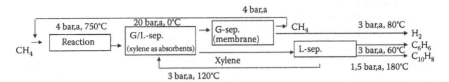

FIGURE 17.17 Block process scheme—including downstream separation processes: reaction, gas–liquid (G/L) separation for removal of aromatics, gas (G) separation for removal of hydrogen, and liquid (L) separation to separate the various aromatics.

For the reaction, the pressure was chosen as a compromise between high gas density and thus high mass flow rate and a tolerable reduction of equilibrium conversion of methane: 4 bar, abs was chosen for the synthesis reactor. However, this pressure has to be increased to 20 bar in gas separation where a membrane process has been chosen (Figure 17.18). As high pressure also enhances the gas–liquid separation, the pressure is raised directly downstream the reaction block. The large differences occurred for the temperatures. The reactor is operated at high temperature that amounts to 750°C, but the low temperatures are required for minimizing the electrical energy for compression and for effective absorption of aromatics. Therefore, heat integration is important for achieving a high energetic efficiency. Methane entering the reactor is preheated by the outlet gas. The product has to be further cooled and compressed up to 20 bar before it enters the absorber column. o-Xylene is proposed as an absorbent for aromatics with the aim to absorb homologues and to avoid freezing of benzene and naphthalene by dissolution. Absorbent and aromatics that leave the column at the bottom are separated by distillation. All light gases, mainly unconverted methane and hydrogen, leave the absorber at the top and are directed toward gas/gas separation unit. Liquid/liquid separation is carried via two distillation columns. In the first column, benzene is separated from o-xylene and naphthalene. In the second column, o-xylene is separated from naphthalene and recycled back to the absorber. The challenge for the gas/gas separation is given by the need to separate small amounts of hydrogen from excess methane. Because of the low conversion rates of methane, a recycle of the unconverted methane and lower hydrocarbons is necessary. Therefore, hydrogen needs to be removed as higher ratio of hydrogen will reduce the conversion of methane in the reaction step. Different options for the gas/gas separation unit like membranes, pressure swing adsorption (PSA), and cryogenic separation have been considered.

In reforming technology, cold boxes are widespread for separation of hydrocarbons. In this case, however, a very small hydrocarbon, methane, needs to be liquefied as the hydrogen is removed as gas. And the amount of methane is fairly high as about 90% of the feed is not converted. A huge cooling energy at very low temperatures (~−90°C) would be necessary in the cold box. Therefore, another separation technology was chosen: the membrane technology. For that technology, high-pressure differences are helpful especially the pressure difference in partial pressure of hydrocarbon. A pressure of 20 bar was chosen for the off-gas feed. This pressure level is also advantageous for the absorption process for removal of the aromatic liquids; the pressure rise is arranged upstream the G/L separation process directly after cooling down the reaction gas as explained earlier.

The complete model of the process is shown in Figure 17.18. An ASPEN simulation software for the process was done in order to determine the necessary feed of auxiliaries, electric power supply, and heating/cooling energy. The results of this calculation were transferred to the economic evaluation of the process. It turned out that the gas separation is pretty expensive as very large membranes are necessary. Therefore, an alternative process was created without gas recycle. As the remaining process steps for this alternative process are very similar to the corresponding process steps in Figures 17.15 and 17.18, no special process sheet for the new process was worked out. Instead, the gas recycle rate was substituted by a feed of fresh methane.

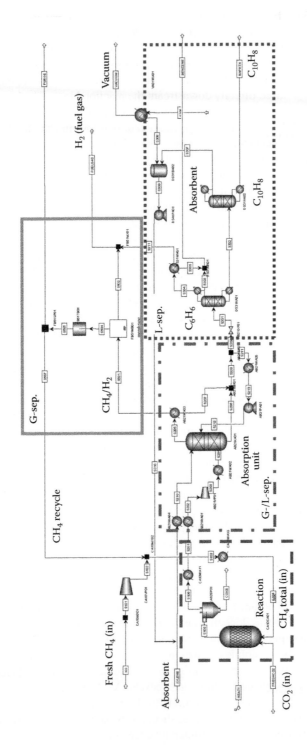

FIGURE 17.18 Process scheme—including downstream separation processes. The process blocks as introduced in Figure 17.17 are marked by the dashed frame (reaction), dot-dashed frame (G/L separation), dotted frame (L separation), and solid frame (G separation).

17.4.2 REACTIVE CH$_4$/H$_2$ SEPARATION

In order to avoid the cost-intensive separation of CH$_4$ and H$_2$, alternative process concepts were developed. The idea behind is to replace hydrogen separation by selective conversion. For this purpose, hydrogenation of carbon dioxide, which is also called the Sabatier reaction, was selected (Equation 17.2). This concept allows utilization of CO$_2$ that is mostly available in the associated gas. Furthermore, it has positive effect on the emissions of greenhouse gases. Finally, a part of the benzene has its roots in CO$_2$ and therefore it can be considered as *green* benzene.

$$CO_2 + 4H_2 = CH_4 + 2H_2O \quad \Delta H_R = -165 \text{ kJ/mol} \quad (17.2)$$

The Sabatier reaction is well known and it is already applied on the industrial scale [34]. Usually, this reaction is performed over Ni-based catalysts at temperatures above 400°C (e.g., [35]). Temperature is important since CO$_2$ and H$_2$ conversion are thermodynamically limited. In order to achieve high conversions, the Sabatier reactor should be operated at low temperature. That means that a very active catalyst is necessary. With noble metal catalysts like Ru or Rh, the temperature can be lowered to the range 200°C–300°C (e.g., [36]). Due to the large temperature difference between the Sabatier and MDA reactors, it is not possible to couple these steps thermally. The integrated process that includes reactive separation and heat integration is presented in Figure 17.19. All external heat was generated by combustion of methane. It turned out that the overall supply of methane in this process was smaller than in the basic configuration. That means that about 20% of carbon molecules from CO$_2$ are converted to benzene.

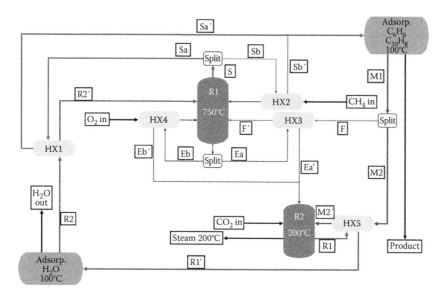

FIGURE 17.19 Process scheme for a coupled MDA/methanation process. R1: MDA process; R2: methanation process; HX1-5: heat exchangers; R1, R1': CH$_4$ + H$_2$O; R2, R2': recycled CH$_4$ for MDA; Sa, Sa', Sb, Sb': MDA product gas; M1: recycle gas CH$_4$ + H$_2$ + CxHy; F, F': fuel for heating MDA process; Ea, Ea', Eb, Eb': combustion gas (H$_2$O + CO$_2$) from MDA heating; M2, M2': H$_2$ feed for methanation.

17.5   PROCESS ECONOMICS

Economic assessment was done on the basis of laboratory performance data obtained under optimized reaction conditions.

CH_4 conversion = 11% best case
Benzene yield = 9.1% best case
Benzene selectivity = 83% best case

Furthermore, a plant size with a capacity of 100,000 t/annum benzene (=12.5 t/h) was chosen. For simplification, products like ethylene and toluene were ignored for the process assessment as well as for the economic estimation.

Case studies were calculated for two different locations: Europe and U.S. Gulf coast. Although all utilities like electricity rate, steam, chilled water, or ammonia (used as cooling medium) differ slightly, the major difference is given by the natural gas price. For Europe, the natural gas prices were considered to be a factor of about four times larger compared to U.S. Gulf coast. Furthermore, it was assumed that by-products such as naphthalene and the steam from cooling down the products can be applied completely and provide credits for the process.

The cost estimate (Figure 17.20) clearly indicates the strong advantage for the U.S. Gulf coast scenario. Whereas for Europe the net benzene manufacturing costs are estimated to be larger than the benzene market price, the cost estimate for the process with internal recycling of methane at the U.S. Gulf coast looks promising. In this case, the depreciation, that is the capital costs, are comparable with the raw material costs.

Figure 17.21 shows the capital cost breakdown. Capital costs for distillation and absorption are rather small. Surprisingly, capital costs for gas/gas separation based

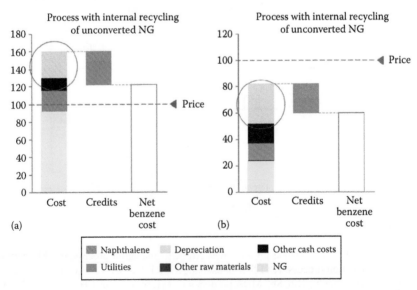

FIGURE 17.20   Cost estimate for MDA reaction for (a) European sites and (b) U.S. Gulf coast.

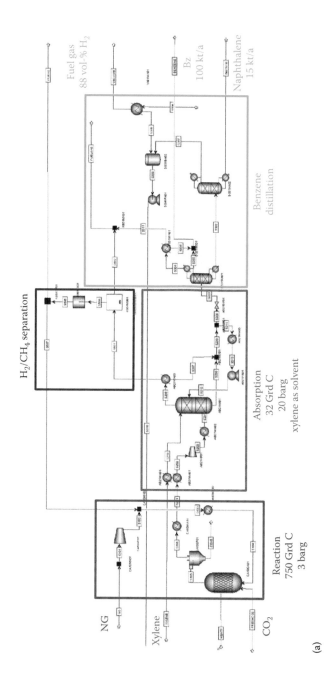

FIGURE 17.21   (a) Process flow sheet highlighting four sections: reaction (red), absorption (blue), distillation (green) and gas separation (yellow). *(Continued)*

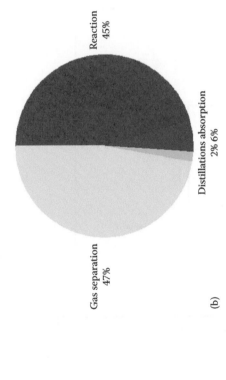

(b)

FIGURE 17.21 (CONTINUED) (b) Capital cost breakdown (percentage) for marked process sections.

on the membrane modules were estimated to be on the same level as capital costs for the reactor/regeneration unit with approximately 47% of total capital costs.

It is obvious that the MDA technology is an interesting option for the utilization of stranded gas that has a very low price. For these locations, where no other options for gas utilization are available, MDA should be already now an economically viable option. Economic analysis shows also further development directions. The high costs for separation is partly a *chicken-egg* situation. Polymer membranes are expensive because there are not many applications for them. When generating a market, their price can be potentially reduced. This would have a significant impact on capital costs for the separation section. The largest potential for further improvement is in the MDA reactor. Definitely, any measure that increase conversion per pass will significantly reduce the costs for the reactor and, in consequence, also for the gas separation.

17.6 CONCLUSIONS

In spite of many years of MDA research, majority of the research studies has been focused on catalysis and reaction engineering. Scale up, process development, and economics still have not been discussed. In this study, major challenges toward commercialization have been addressed. The primary challenge is given by the development of the mechanically stable catalyst for application in a fluidized bed that does not undergo fast cocking. It has been found that usually used Al-binders are not suitable for the MDA reaction. A recipe for the technical catalyst based on the SiO_2 binder has been developed. This catalyst was extensively tested in a bench-scale fluidized bed. It has shown sufficient mechanical stability. However, it exhibits slightly lower catalytic performance, that is, lower yields and faster deactivation have been observed. The study confirmed that the primary challenge for the MDA process is regeneration of the cocked catalyst. The widely proposed concept of two interconnected fluidized beds as reaction engineering solution for the catalyst regeneration and heat supply can't be directly applied for the MDA reaction. This solution is not suitable for catalyst regeneration since in the oxidative atmosphere and temperatures above 500°C, the catalyst cannot be reactivated due to the volatile MoO_3 specie nor energy balance can be closed when regenerating the catalyst significantly below the reaction temperature. Alternative designs with the external heat supply have been conceptually studied. A fluidized-bed reactor with immersed heat exchanger with internal methane combustion in the tubes of this heat exchanger seems to be technically feasible. Obviously, a membrane fluidized-bed reactor with the multifunctional membranes that have the function to remove and to combust hydrogen has the highest potential. However, no suitable membranes are still available. Nevertheless, economic evaluation indicates the MDA technology might economically be viable for location with a low price for natural gas; certainly, it should be an interesting option for conversion of stranded gas. This technology might be especially interesting for conversion of natural gas containing carbon dioxide in high concentration. By combining MDA with the Sabatier reaction, CO_2 can be converted to *green* benzene.

REFERENCES

1. W. Keim, M. Röper, Change in the raw material base, 2010. Available: http://www. dechema.de. Accessed March 1, 2014.
2. J.W. Thybaut, G.B. Marin, C. Mirodatos, Y. Schuurman, A.C. van Veen, V. Sadykov, H. Pennemann, R. Bellinghausen, L. Mleczko, *Chem. Ing. Tech.*, 2014, xx, 8.
3. D. Wang, J.H. Lunsford, M.P. Rosynek, *J. Catal.*, 1997, 169, 341–358.
4. Z. Tian, Y. Xu, L. Lin, *Chem. Eng. Sci.*, 2004, 59, 1745–1753.
5. J.J. Spivey, G. Hutchings, *Chem. Soc. Rev.*, 2014, 43, 792.
6. Z.R. Ismagilov, E.V. Matus, L.T. Tsikoza, *Energy Environ. Sci.*, 2008, 1, 526–541.
7. Y. Xu, X. Bao, L. Lin, *J. Catal.*, 2003, 216, 386–395.
8. C. Bouchy, I. Schmidt, J.R. Anderson, C.J.H. Jacobsen, E.G. Derouane, S.B. Derouane-Abd Hamid, *J. Mol. Catal. A Chem.*, 2000, 163, 283–296.
9. S.B. Derouane-Abd Hamid, J.R. Anderson, I. Schmidt, C. Bouchy, C.J.H. Jacobsen, E.G. Derouane, *Catal. Today*, 2000, 63, 461–469.
10. D. Ma, Y. Shu, M. Cheng, Y. Xu, X. Bao, *J. Catal.*, 2000, 194, 105–114.
11. Y. Shu, D. Ma, X. Bao, Y. Xu, *Catal. Lett.*, 2000, 70, 67.
12. Z. Sobalik, Z. Tvaruzkova, B. Wichterlova, V. Fila, S. Spatenka, *Appl. Catal. A Gen.*, 2003, 253, 271–282.
13. W. Li, G.D. Meitzner, R.W. Borry, E. Iglesia, *J. Catal.*, 2000, 191, 373–383.
14. W. Ding, S. Li, G.D. Meitzner, E. Iglesia, *J. Phys. Chem. B*, 2001, 105, 506–513.
15. D. Kunii, O. Levenspiel, *Fluidization Engineering*, Butterworth-Heinemann, Oxford, U.K., 1991.
16. B. Cook, D. Mousko, W. Hoelderich, R. Zennaro, *Appl. Catal. A Gen.*, 2009, 365, 34–41.
17. M.P. Gimeno, J. Soler, J. Herguido, M. Menendez, *Ind. Eng. Chem. Res.*, 2010, 49, 996–1000.
18. K. Skutil, A. Taniewski, *Fuel. Process. Technol.*, 2006, 87, 511–521.
19. Z. Cao, H. Jiang, H. Luo, S. Baumann, W.A. Meulenberg, J. Aßmann, L. Mleczko, Y. Liu, J. Caro, *Angew. Chem. Int. Ed.*, 2013, 52, 13794–13797.
20. L. Li, R.W. Borry, E. Iglesia, *Chem. Eng. Sci.*, 2002, 57, 4595–4604.
21. K. Honda, X. Chen, Z.-G. Zhang, *Appl. Catal. A Gen.*, 2008, 351, 122–130.
22. H. Liu, L. Su, H. Wang, W. Shen, X. Bao, Y. Xu, *Appl. Catal. A Gen.*, 2002, 236, 263–280.
23. H. Liu, T. Li, B. Tian, Y. Xu, *Appl. Catal. A Gen.*, 2001, 213, 103–112.
24. D. Ma, D. Wang, L. Su, Y. Shu, Y. Xu, X. Bao, *J. Catal.*, 2002, 208, 260–269.
25. L. Wang, R. Ohnishi, M. Ichikawa, *J. Catal.*, 2000, 190, 276–283.
26. S. Yao, L. Gu, C. Sun, J. Li, W. Shen, *Ind. Eng. Chem. Res.*, 2009, 48, 713–718.
27. Y. Shu, H. Ma, R. Ohnishi, M. Ichikawa, *Chem. Comm.*, 2003, 86–87.
28. Z. Liu, L. Li, E. Inglesia, *Catal. Lett.*, 2002, 82, 175–180.
29. O. Rival, B.P.A. Grandjean, C. Guy, A. Sayari, F. Larachi, *Ind. Eng. Chem. Res.*, 2001, 40, 2212–2219.
30. M.C. Iliuta, F. Larachi, B.P.A. Grandjean, I. Iliuta, A. Sayari, *Ind. Eng. Chem. Res.*, 2002, 41, 2371–2378.
31. F. Larachi, H. Oudghiri-Hassani, M.C. Iliuta,, B.P.A. Grandjean, P.H. McBreen, *Catal. Lett.*, 2002, 84, 183–192.
32. M.C. Iliuta, B.P.A. Grandjean, F. Larachi, *Ind. Eng. Chem. Res.*, 2003, 42, 323–340.
33. L. Mleczko, T. Ostrowski, T. Wurzel, *Chem. Eng. Sci.*, 1996, 51, 3187–3192.
34. W. Wang, S. Wang, X. Ma, J. Gong, *Chem. Soc. Rev.*, 2011, 40, 3703–3727.
35. T. Nakayama, N. Ichikuni, S. Sato, F. Nozaki, *Appl. Catal. A Gen.*, 1997, 158, 185–199.
36. D. Li, N. Ichikuni, S. Shimazu, T. Uematsu, *Appl. Catal. A Gen.*, 1998, 172, 351–358.

Index

For Product Safety Concerns and Information please contact our EU
representative GPSR@taylorandfrancis.com Taylor & Francis Verlag GmbH,
Kaufingerstraße 24, 80331 München, Germany

Printed and bound by CPI Group (UK) Ltd, Croydon, CR0 4YY
01/05/2025
01858585-0001